^{13}C NMR Data for Organometallic Compounds

ORGANOMETALLIC CHEMISTRY

A Series of Monographs

BRIAN G. RAMSEY: Electronic Transitions in Organometalloids, 1969.
R. C. POLLER: The Chemistry of Organotin Compounds, 1970.
RUSSELL N. GRIMES: Carboranes, 1970.
PETER M. MAITLIS: The Organic Chemistry of Palladium, Volumes I and II, 1971.
DONALD S. MATTESON: Organometallic Reaction Mechanisms of the Nontransition Elements, 1974.
RICHARD F. HECK: Organotransition Metal Chemistry: A Mechanistic Approach, 1974.
P. W. JOLLY AND G. WILKE: The Organic Chemistry of Nickel, Volume I, Organonickel Complexes, 1974. Volume II, Organic Synthesis, 1975.
P. C. WAILES, R. S. P. COUTTS, AND H. WEIGOLD: Organometallic Chemistry of Titanium, Zirconium, and Hafnium, 1974.
U. BELLUCO: Organometallic and Coordination Chemistry of Platinum, 1974.
P. S. BRATERMAN: Metal Carbonyl Spectra, 1974.
L. MALATESTA AND S. CENINI: Zerovalent Compounds of Metals, 1974.
THOMAS ONAK: Organoborane Chemistry, 1975.
R. P. A. SNEEDEN: Organochromium Compounds, 1975.
A. G. SHARPE: The Chemistry of Cyano Complexes of the Transition Metals, 1976.
ERNST A. KOERNER VON GUSTORF, FRIEDRICH-WILHELM GREVELS, AND INGRID FISCHLER: The Organic Chemistry of Iron, Vol. 1, 1978.
G. DEGANELLO: Transition Metal Complexes of Cyclic Polyolefins, 1978.
BRIAN E. MANN AND BRIAN F. TAYLOR: ^{13}C NMR Data for Organometallic Compounds, 1981.

^{13}C NMR Data for Organometallic Compounds

BRIAN E. MANN
BRIAN F. TAYLOR

Department of Chemistry
University of Sheffield, UK

1981

Academic Press

A Subsidiary of Harcourt Brace Jovanovich, Publishers
London New York Toronto Sydney San Francisco

ACADEMIC PRESS INC. (LONDON) LTD.
24-28 Oval Road, London NW1

United States Edition published by
ACADEMIC PRESS INC.
111 Fifth Avenue, New York, New York 10003

British Library Cataloging in Publication Data

Mann, B.E.
^{13}C nmr data for organometallic compounds.
1. Spectroscopy 2. Chemistry, Physical organic
3. Nuclear magnetic resonance spectroscopy
4. Carbon – Isotopes
I. Title II. Taylor, B.F.
547'.3'085 QD476

ISBN 0-12-469150-1
LCCCN 81-66696

Printed in Great Britain by Whitstable Litho Ltd., Whitstable, Kent

PREFACE

During the 1970s, ^{13}C n.m.r. spectroscopy has changed for organometallic chemists from a rare technique in the hands of a few specialists to a routine analytical technique. Consequently, the growth of ^{13}C n.m.r. data in the literature has increased rapidly. Despite this vast data base, no general satisfactory theoretical or empirical relationships have been established for organometallic compounds. Thus at present, the only satisfactory method of estimating the ^{13}C chemical shifts of organometallic compounds is by analogy with known compounds. It is the object of this monograph to provide a convenient source of data. It is hoped that this compilation will stimulate the derivation of empirical relationships by which the ^{13}C chemical shifts of organometallic compounds can be predicted and encourage theoretical treatments

There are numerous books on the ^{13}C n.m.r. spectra of organic compounds, and in many respects, the ^{13}C n.m.r. spectra of organometallic compounds are similar, apart from the chemical shifts and coupling constants of carbon atoms directly attached to metals. Consequently the emphasis of the book is on these carbon atoms rather than the carbon atoms more distant from the metal. For this monograph, a metal is taken as being any element apart from hydrogen, carbon, nitrogen, oxygen, sulphur, the halogens, and the noble gases, but due to the large quantity of data available for silicon and phosphorus containing compounds, these groups of compounds have been treated superficially. Also the ligands, cyanide and organo-isocyanide, have been omitted. Organic anions which may be associated with the metal, *e.g.*, sodium allyl, provide a marginal case and are included where the interaction between the metal and anion appears significant. The literature coverage is up to the end of 1979.

The book has been divided into two chapters. The first chapter is a guide to ^{13}C n.m.r. data for organometallic compounds. It is assumed that the reader is familiar with organic and experimental aspects of ^{13}C n.m.r. The intention is to provide the reader with a review of the extra information necessary to go from organic to organometallic ^{13}C n.m.r. spectroscopy. The bulk of the book is in the form of tables in Chapter 2. These tables are divided by the nature of the ligand and then by the metal. In the cases where even this division produces large subsections, further division is made by ligand or metal oxidation state. As a result the reader will have only small sections of the tables to search when trying to find an example of a particular type of compound.

I (B.E.M.) wish to express my gratitude to the University of Sheffield for granting sabbatical leave and the University of Southern California, Riverside for an invitation to spend the sabbatical leave there. We also wish to thank our wives for considerable assistance during the early stages of manuscript preparation, and last, but not least, we owe a debt of gratitude to Mrs. J. Taylor and especially Mrs. E. Fisher for the monumental task of typing the final camera-ready copy of the manuscript.

June 1981

B.E. Mann and B.F. Taylor
University of Sheffield

CONTENTS

Preface v

Chapter 1

A Guide to the Use of ^{13}C N.m.r. Spectroscopy to Investigate Organometallic Compounds

I Introduction 1

II Experimental 2

III ^{13}C Chemical Shifts of Diamagnetic Organometallic Compounds 6

IV Spin-Spin Coupling Constants 17

V Relaxation Measurements 24

VI Applications 28

Chapter 2

Tables of ^{13}C N.m.r. Data for Organometallic Compounds

Table 2.1 ^{13}C Chemical Shift Data for sp^3 Carbon Atoms Attached to Metals 38

Table 2.2 ^{13}C Chemical Shift Data for Vinyl Groups Attached to Metals 86

Table 2.3 ^{13}C Chemical Shift Data for Acetylenic Groups Attached to Metals 103

Table 2.4 ^{13}C N.m.r. Data for Some Phenyl Derivatives 107

Table 2.5 ^{13}C N.m.r. Data for Some Carbene Derivatives 133

Table 2.6 ^{13}C N.m.r. Data for Some Carbyne Derivatives 144

Table 2.7 ^{13}C N.m.r. Data for Some Acyl, Thioacyl and Selenoacyl Derivatives 147

Table 2.8 ^{13}C Chemical Shift Data for Some Carbonyl Derivatives 151
Table 2.9 ^{13}C Chemical Shift Data for Some CS and CSe Derivatives 183
Table 2.10 Chemical Shift Data for Some Olefins and Aceylenes π-Bonded to Metals 184
Table 2.11 ^{13}C Chemical Shift Data for Some Allyl Groups π-Bonded to Metals 200
Table 2.12 ^{13}C Chemical Shift Data for Cyclobutadienes π-Bonded to Metals 210
Table 2.13 ^{13}C N.m.r. Chemical Shift Data for Some Dienes and Trimethylenemethane π-Bonded to Metals 211
Table 2.14 ^{13}C N.m.r. Chemical Shift Data for Some Cyclopentadienyl Groups π-Bonded to Metals 219
Table 2.15 ^{13}C N.m.r. Chemical Shift Data for Some Dienyl Groups π-Bonded to Metals 244
Table 2.16 ^{13}C N.m.r. Chemical Shift Data for Some Arenes π-Bonded to Metals 247
Table 2.17 ^{13}C N.m.r. Chemical Shift Data for Some Trienes π-Bonded to Metals 256
Table 2.18 ^{13}C Chemical Shift Data for Some Cycloheptatrienyl Groups π-Bonded to Metals 258
Table 2.19 ^{13}C N.m.r. Data for Some Paramagnetic Organometallic Compounds 259

References 260

Chapter 1

A GUIDE TO THE USE OF ^{13}C N.M.R. SPECTROSCOPY TO INVESTIGATE ORGANOMETALLIC COMPOUNDS

I INTRODUCTION

^{13}C N.m.r. spectroscopy is now established as a routine tool in the characterisation and investigation of new compounds and numerous books have appeared describing experimental details and spectral interpretation for organic compounds. To a very large extent the approach for organometallic compounds is identical to that for organic compounds and the reader is referred to the organic compound based texts.[1-6] The differences arise predominantly with the additional types of bonding available for organometallic compounds which do not have analogues in organic chemistry and to the extra information that may be obtained from ^{13}C coupling to n.m.r. active metal nuclei. It is in these areas that the stress is laid in this book and little or no attention is paid to ^{13}C n.m.r. data for carbon atoms which are essentially in an organic framework and have a metal atom at least two bonds away.

During the period 1974-1976 several reviews of the ^{13}C n.m.r. spectra of organometallic compounds appeared.[7-10] At the time of these reviews the data available were very restricted because ^{13}C n.m.r. spectroscopy was accessible to only a few organometallic chemists. Now that most organometallic chemists have ^{13}C on site, the output of data is rapidly increasing. However, in spite of the mass of chemical shift information reported in Chapter 2, there are at present no quantitative theoretical treatments of chemical shifts for carbon atoms attached directly to a metal. There are numerous qualitative theoretical treatments based on only one factor changing, but it is unusual for changes in any single factor to dominate. Consequently the validity of these qualitative treatments have not been tested and are losing their original popularity in the discussion sections of papers. In view of this, emphasis is placed on empirical relationships which at least apply to a restricted group of compounds.

II EXPERIMENTAL

The experimental procedures are generally identical to those applied to organic compounds, and the reader is referred to texts on this subject. A number of areas pose problems not commonly encountered for organic compounds, and merit discussion here.

Chemical Shift Range

For organometallic compounds there is a very large chemical shift range from $\overline{W(C_5H_4CH_2\mathit{C}}H_2)(CO)_3$ at δ -61.7,[11] to *trans*-$Fe_2(\mu\text{-}\mathit{C}C_5H_5)(\mu\text{-}CO)Cp_2(CO)_2$ at δ 448.3.[12] The observation of this full range can present problems. The standard spectral range for organic compounds is *ca* δ -10 to δ 220. Therefore if the n.m.r. spectroscopist is not familiar with organometallic compounds, signals may be lost or folded back into the spectrum. An additional difficulty occurs with the modern high field n.m.r. spectrometers. At 9.4 T (400 MHz ^1H n.m.r. spectrometer) this 500 p.p.m. chemical shift range corresponds to a 50,000 Hz spectral width, which in the quadrature detection mode sets an upper limit of 10 μs for the length of the observing r.f. pulse to avoid intensity distortions.

Coupling to Quadrupolar Nuclei

In general quadrupolar nuclei do not cause problems for

organic ^{13}C n.m.r. spectroscopy. Only coupling to the relaxing ^{14}N nuclei is occasionally observed as broadening of the n.m.r. signal or more rarely as a 1:1:1 triplet. For the organometallic chemist the range of quadrupolar nuclei which can cause problems is far more extensive with line broadening having been noted for carbon atoms attached to ^{10,11}B, ^{27}Al, ^{51}V, ^{55}Mn, ^{59}Co, and ^{93}Nb.[13-18]

In principal, all magnetically active nuclei should split the signal of the attached carbon atom into $2I + 1$ lines. Thus for $[Co(CO)_4]^-$, the ^{13}C n.m.r. spectrum consists of eight lines of equal intensity and equal spacing due to coupling between ^{59}Co and ^{13}C, (^{59}Co has a nuclear spin quantum number, I, of $\frac{7}{2}$ and an isotopic abundance of 100%).

However, when $I > \frac{1}{2}$, the nucleus also possesses an electric quadrupole moment and can interact with an electric field gradient. It is this interaction that normally dominates the spin-lattice relaxation time, Eq. 1.1.

$$T_{1_q}{}^{-1} = \frac{3\pi^2}{10} \frac{(2I + 3)}{I^2(2I - 1)} \chi^2(1 + \frac{1}{3}\eta^2)\,\tau_c \qquad 1.1$$

where T_{1_q} is the quadrupolar spin-lattice relaxation time, τ_c is the molecular correlation time, χ is the nuclear quadrupole coupling constant as defined by Eq. 1.2.

$$\chi^2 = \frac{e^2 q_{zz} Q}{h} \qquad 1.2$$

e is the charge on the electron, q_{zz} is the largest component of the electric field gradient at the nucleus, Q is the nuclear electric quadrupole moment, and η is the asymmetry parameter of the electric field gradient at the nucleus

$$\eta = \frac{q_{yy} - q_{xx}}{q_{zz}} \qquad 1.3$$

For most quadrupolar nuclei, the quadrupolar relaxation mechanism is dominant, causing rapid relaxation of the quadrupolar nucleus. This usually effectively decouples the quadrupolar nucleus and the attached carbon atom gives a sharp singlet. For example, for ^{127}I, $I = \frac{5}{2}$ and it is 100% abundant, but as relaxation is fast, the signal due to an attached ^{13}C nucleus is a sharp singlet.

Problems arise when quadrupolar relaxation is occurring at an intermediate rate i.e. T_{1_q} is between 0.1 s and 10^{-6} s and the coupling constant between ^{13}C and the quadrupolar nucleus is substantial. Under these conditions the ^{13}C resonance can have a line width of up to $2I_Q.J$ (X, ^{13}C) where I_Q is the spin quantum number of the coupled quadrupolar nucleus and J (X, ^{13}C) is the coupling constant between the quadrupolar and ^{13}C nuclei. This linewidth may be substantial, *e.g.* 1 kHz, with a consequential decrease in the signal to noise ratio.

A number of approaches to this problem are possible.

(a) Change the molecular correlation time, τ_c. If τ_c is decreased, it follows from Eq. 1.1, that T_{1_q}(X) will lengthen and hence we are more likely to resolve J(X, ^{13}C). Alternatively, if τ_c is increased, T_{1_q}(X) will decrease and hence the broadened ^{13}C resonance will move towards being a sharp singlet. It is this latter approach that has been successfully applied to sharpening ^{13}C signals for carbonyls attached to ^{55}Mn and ^{59}Co.[18-21]

How can the value of τ_c be changed? For a spherical solute molecule in a medium of viscosity, η, the rotational correlation time is given by

$$\tau_c = \frac{\eta V f}{kT} \qquad 1.4$$

where f is a microviscosity factor which is about 0.16 for pure liquids and V is the volume of the solute molecule. Thus for a given molecule in order to sharpen the ^{13}C resonance τ_c can be increased by using a viscous solvent, *e.g.*, ethylene glycol, and reducing the temperature. In order to try to resolve any coupling to ^{13}C τ_c may be decreased by the reverse changes. Frequently there is some choice in the organic substituents, so that the volume of the molecule may be changed, *e.g.* by changing a PMe_3 ligand to $PBu^n{}_3$, the volume increases, increasing τ_c and decreasing T_{1q}.

(b) Decouple the quadrupolar nucleus. This is the ideal solution, but has not yet been done for organometallic ^{13}C n.m.r. spectra. There are two technical problems associated with this approach. Firstly, to decouple a nucleus with a short T_1 value and a large coupling constant to ^{13}C nuclei, *e.g.*, $^1J(^{59}Co, ^{13}C)$ in $[Co(CO)_4]^-$ is 287 Hz,[22] requires relatively high power. Secondly, the nuclei that the organometallic chemist usually wishes to decouple lie close to the ^{13}C resonance frequency making it essential to use well matched filters to prevent stray beats destroying the ^{13}C n.m.r. spectrum, necessitating the purchase of additional equipment.

(c) In principle, chemical modification of the substituents on the metal will change $\chi^2(1 + \frac{1}{3}\eta^2)$ in equation 1.1 with consequential changes in T_{1q}. However large ligand changes are necessary to have a significant effect; this results in a different group of compounds with different properties being studied.

Although the coupling of ^{13}C nuclei to quadrupolar nuclei can be very inconvenient, it is in principle possible to determine the coupling constant using the Carr-Purcell-Meiboom-Gill pulse sequence. This method has been applied to molecules such as PBr_3 to determine $^1J(^{81}Br, ^{31}P)$,[23] but not as yet to organometallic compounds.

Long Spin-Lattice Relaxation Times

For proton bearing carbon atoms, dipole-dipole relaxation

due to neighbouring protons is the major contribution. This term is dependent on the (carbon-hydrogen distance)$^{-6}$. For metal carbonyls especially, the carbonyl carbon atoms are very distant from hydrogen atoms resulting in long T_1 values, and the nucler Overhauser enhancement is generally small. Consequently metal carbonyl carbon atoms can be very difficult to detect and require small pulse angles and/or long repetition times. The situation can be improved by the addition of a relaxation reagent, *e.g.* $Cr(acac)_3$ or $Fe(acac)_3$.[24] The approach is satisfactory for spectral acquisition, but line broadening at low temperatures can lead to errors in rate measurements for fluxional molecules.[25]

III ^{13}C CHEMICAL SHIFTS OF DIAMAGNETIC ORGANOMETALLIC COMPOUNDS

Theoretical Treatments

For organic compounds, ^{13}C chemical shifts can be calculated theoretically with a moderate degree of success. However when carbon is attached to a heavy atom or an atom with accessible *d*-orbitals, no reliable calculations exist. For organometallic compounds, the few calculations that exist are restricted to carbon attached to the lighter elements, *e.g.*, silicon.

In theoretical calculations, it is usual to use nuclear screening terminology developed by Ramsey instead of chemical shifts. The magnetic field, B, experienced by a nucleus in a molecule differs from the applied field, B_o, as given by

$$B = B_o(1 - \sigma) \quad 1.5$$

Note that the sign convention for the nuclear screening, σ, is the reverse to that adopted for chemical shift, δ. The nuclear screening, σ, is a scalar quantity which is the trace of a second rank tensor, **σ**, *i.e.*,

$$\sigma = \tfrac{1}{3}(\sigma_{xx} + \sigma_{yy} + \sigma_{zz}) \qquad 1.6$$

where σ_{ii}'s are the three principal tensor components. It is σ that is measured in a typical high resolution n.m.r. experiment due to the averaging of the σ_{ii}'s by rapid molecular tumbling in solution.

The nuclear screening may be split into two components, the diamagnetic term, σ^d, and the paramagnetic term, σ^p, *i.e.*,

$$\sigma = \sigma^d + \sigma^p \qquad 1.7$$

where σ^d involves the free rotation of electrons about the nucleus in question and σ^p describes the hindrance to rotation caused by the other electrons and nuclei in the molecule.

Calculations of ^{13}C chemical shifts are frequently based on the independent electron model as developed by Pople. Using this approach, the nuclear screening is expressed as the sum of local, non-local, and interionic contributions

$$\sigma = \sigma^d(\text{loc}) + \sigma^d(\text{non-loc}) + \sigma^d(\text{inter}) + \sigma^p(\text{loc}) + \sigma^p(\text{non-loc}) + \sigma^p(\text{inter}) \qquad 1.8$$

The local terms, $\sigma^d(\text{loc})$ and $\sigma^p(\text{loc})$, arise from electronic currents localised on the atom containing the nucleus of interset, $\sigma^d(\text{non-loc})$ and $\sigma^p(\text{non-loc})$ from currents on neighbouring atoms, and $\sigma^d(\text{inter})$ and $\sigma^p(\text{inter})$ from non-localised currents, *e.g.*, aromatic ring currents. For carbon attached to hydrogen or a first row element, $\sigma^d(\text{loc})$ and $\sigma^p(\text{loc})$ are dominant with a small (typically 5 p.p.m.) contribution from $\sigma^p(\text{non-loc})$. As $\sigma^d(\text{loc})$ is approximately constant (generally 257 and 261 p.p.m.) for carbon, emphasis is placed on the calculation of $\sigma^p(\text{loc})$. This approach may be justified for organic molecules, but for organometallic compounds the presence of large metal atoms with available *d*-orbitals and, possibly, *f*-orbitals can make other terms significant. There is a proven heavy atom effect, *e.g.*, CI_4 at δ -293.[26] This very large low frequency shift is believed to arise from the σ^p term.[27] For transition metal hydrides, a similar low frequency shift in the ^1H spectrum of up to 50 p.p.m.[28] is observed which has been attributed to the presence of partially filled *d*-orbitals.[29] Although this treatment of metal hydrides works qualitatively, the quantitative application is still awaited. The presence of such an effect in ^{13}C n.m.r. spectra of related transition metal organometallic compounds is to be expected with shifts of a similar magnitude. In spite of the expected magnitude of this effect, due to the lack of understanding of the ^{13}C shifts of carbon attached to transition metal, there have been no cases where it has proven possible to clearly demonstrate a large low frequency shift due to the anisotropy of the attached transition metal.

There are numerous levels of sophistication in the calculation of ^{13}C chemical shifts. For organometallic compounds, the equations most commonly adopted are based on the average energy approximation. In the derivation of the equations, two major approximations are made:- overlap is neglected and each energy difference between filled and empty orbitals is averaged, then

$$\sigma_A p(\text{loc}) = -\frac{\mu_o e^2 \hbar^2}{8m^2} \frac{\langle r^{-3}\rangle_{np}}{\Delta E} \sum_B Q_{AB} \qquad 1.9$$

where the bond-order, charge density terms, Q_{AB}, are given by

$$Q_{AB} = \frac{4}{3}\,\delta_{AB}(P_{x_A x_B} + P_{y_A y_B} + P_{z_A z_B})$$

$$-\frac{2}{3}\,(P_{x_A x_B}P_{y_A y_B} + P_{x_A x_B}P_{z_A z_B} + P_{y_A y_B}P_{z_A z_B})$$

$$+\frac{2}{3}\,(P_{x_A y_B}P_{x_B y_A} + P_{x_A z_B}P_{x_B z_A} + P_{y_A y_B}P_{y_B z_A}) \qquad 1.10$$

and $$P_{\mu\nu} = 2\sum_{j}^{occ} C_{\mu j}C_{\nu j} \qquad 1.11$$

$C_{\alpha\beta}$'s are the LCAO coefficients. The qualitative use of Eq. 1.9 has proven to be very popular in the discussion of ^{13}C chemical shifts of organometallic compounds. For a series of closely related compounds, it is generally possible to attribute chemical shift changes to variations in one of the parameters in Eq. 1.9 with any anomalies being attributed to the changes in a second parameter. Such treatments may look impressive in a paper, but in view of the inadequacies shown in the average excitation energy approximation for small molecules where more sophisticated methods are possible, any relationships between carbon chemical shifts and experimental parameters for organometallic compounds using Eq. 1.9 are best considered as being empirical.

Empirical Relationships

Although there has been little success with theoretical treatments of ^{13}C chemical shifts in organometallic compounds, numerous empirical relationships exist between ^{13}C chemical shifts and various molecular parameters. In view of the dependence of ^{13}C chemical shifts on a wide variety of parameters, it is of no surprise that any one empirical relationship applies only to a narrow range of compounds. Nevertheless, these relationships are invaluable in assigning signals and estimating chemical shifts. Typical applications of these relationships are

(a) The assignment of a structure to a compound, or to restrict the range of possible structures.

(b) To estimate the chemical shift, *e.g.*, for signals which may be exchange averaged.

(c) To estimate the molecular parameters which is related to the particular ^{13}C chemical shift.

Caution must be exercised in the trust to be placed in conclusions drawn using these methods. These relationships are too numerous to catalogue every one, but the principles can be outlined and the reader is left to apply them to the compounds of his own interest.

Additive relationships. For many organic compounds, the ^{13}C chemical shift of a given carbon atom can be estimated with considerable accuracy by Eq. 1.12

$$\delta_C(k) = B + \sum_l A_l\, n_{kl} \qquad 1.12$$

where $\delta_C(k)$ is the chemical shift of the k th carbon atom, B is a constant, n_{kl} is the number of like substituents in the l th position relative to the k th carbon atom and A_l is the additive chemical shift parameters assigned to the l th substituent. Due to the great variety of organometallic compounds, this type of analysis has been rarely carried out for the lack of data on a large number of closely related compounds. However, this approach has proved successful for a number of tin compounds.[30,31] The most sophisticated approach uses 11 parameters for a wide range of organotin derivatives but errors of up to 9 p.p.m. were found.[31] It would appear this approach is too ambitious to apply to the wide range of organometallic compounds currently available.

A simpler and more readily applicable approach is to use

substituent effects. For example, the method has been applied to some $Sn(CH_3)_3$ derivatives.[32] The substituent parameter is simply the chemical shift difference for the required carbon atom between R—H and R—$Sn(CH_3)_3$. For saturated carbon atoms α to the tin, the value varies between -7.4 p.p.m. for $Sn(CH_3)_4$ to +2.4 p.p.m. for $Sn(CH_3)_3$-7-anti-norbornenyl, but β parameters are much more constant being in the range +2.0 to +5.6 p.p.m. As significant medium effects are found, an accuracy of 2 p.p.m. in a predicted chemical shift is acceptable. The attraction of the approach is that the parameters are easily derived but great care must be exercised in deciding how reliable it is. For example, when this method is applied to the carbon atom in $Mo(\mathit{C}H_3)Cp(CO)_3$ and $Mo(\mathit{C}H_2CH_3)Cp(CO)_3$ values of -20.0 and -9.8 p.p.m.,[33] respectively are obtained for the α-parameters. A corollory of this is that great caution is necessary when applying organic compound derived substituent constants to organometallic compounds. For example replacing a hydrogen with a methyl group on an sp^3 carbon atom in an organic compound normally produces about an 8 p.p.m. shift but for $Mo(\mathit{C}H_3)Cp(CO)_3$ and $Mo(\mathit{C}H_2CH_3)Cp(CO)_3$ the observed shift is 18.2 p.p.m. However, this simple approach appears to be adequate for most sp^3 carbon atoms with the exception of methyl derivatives.

A refinement on this approach comes from the observation that there is a linear relationship as described by Eq. 1.13

$$\delta_\alpha[Sn(CH_3)_3\text{-}R] = 1.13\ \delta_\alpha[H\text{-}R] - 5.0 \qquad 1.13$$

where $\delta_\alpha[Sn(CH_3)_3\text{-}R]$ is the ^{13}C chemical shift for the carbon atom in R directly attached to tin and $\delta_\alpha[H\text{-}R]$ is the shift for the corresponding carbon atom in the hydrocarbon.[32] From Eq. 1.13 it is clear why the simple treatment described in the previous paragraph does not work as the gradient is non-unity. A similar relationship has been noted for Grignard reagents.[34] These relationships are far from ideal and for most purposes it is probably just as useful to omit the case of R = Me and use the simple additive relationship described in the previous paragraph.

Correlations with other experimental parameters. For a tightly defined series of compounds it is possible to obtain linear relationships between ^{13}C chemical shifts and almost any other experimental quantity. For example, if $\delta(^{13}CO)$ is plotted against the stretching force constant, k(CO), for a series of metal carbonyls, a linear relationship is obtained provided that the metal is kept constant and the changes in ligand are somewhat restricted. These relationships will be discussed further in each appropriate sub-section.

Shifts of Alkyl Carbon Atoms Attached to Metals

For organic compounds, it is customary to pay considerable attention to the substituent atom of a methyl group in estimating the ^{13}C chemical shift and then a secondary correction is made for further substituents. For organic compounds the correction for a β-substituent typically lies between 0.7 and 11.2 p.p.m.[35] In contrast, Hg(CH$_3$)X compounds cover the range δ −3.0 to δ 40.1.[36,37] If we allow the coordination number of the metal to change, then this range is considerably extended, e.g., W(CH$_3$)$_6$ at δ 83.6[38] and W(CH$_3$)Cp(CO)$_3$ at *ca* δ −35;[39] a chemical shift range of 118 p.p.m. for a methyl group attached to one element. It is therefore not surprising that at present few correlations exist. To a very large extent, chemical shifts become more predictable when the attached metal atom is light and a non-transition metal. When these restrictions are coupled with the lack of wide ranges of data on relevant compounds, it is found that correlations have been reported mainly for tin compounds and have already been discussed, pp 10 and 11.

Shifts of Phenyl Carbon Atoms Attached to Metals

It is established for organic compounds that the chemical shifts of the *ortho*-, *meta*-, and *para*-carbon atoms of a phenyl ring are proportional to the CNDO/2 calculated charge densities.[40] Similarly it has been shown that the chemical shift of the *para*-carbon atom in monosubstituted benzenes is linearly related to the total π-electron density at the *para*-position, while the *meta*-carbon atom is only slightly perturbed by the substituent.[41,42] It is therefore often possible to assign the signals of the phenyl group. The *meta*-carbon atom signal is normally between δ 127.5 and δ 129.5 while the *ortho*-carbon atom signal can lie well outside this range and the unique *para*-carbon atom gives a signal approximately half the intensity of those of the *ortho*- and *meta*-carbon atoms. The *ipso*-carbon atom can readily be identified by use of off-resonance decoupling and often by the intensity being reduced by the longer spin-lattice relaxation time compared to the proton bearing *ortho*-, *meta*-, and *para*-carbon atoms and frequently a non-maximum nuclear Overhauser enhancement.

As a consequence of the facile assignment of phenyl carbon signals and the use of the *para*-carbon atom shift to provide information on the interaction between the metal and the phenyl group, the shifts of phenyl groups attached to metal atoms has attracted considerable interest. However the major problems is the interpretation of the results. The chemical shift of the *ipso*-carbon atom is not explained. The chemical shift of the remaining carbon atoms are dominated by the charge on them, but the problem is in deciding how the metal atom affects the charge. The π-mechanism is readily understood and is therefore frequently invoked. The metal atom can act as either a π-donor, I and II, or as a π-acceptor, III and IV, affecting the π-electron density at the *ortho*-, I and III, and *para*-, II and

IV, carbon atoms.

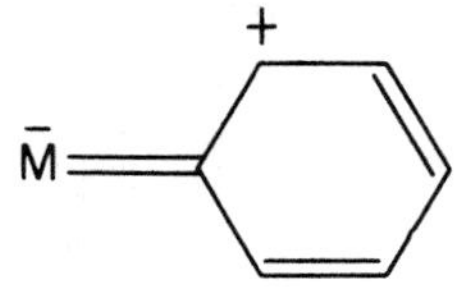

I 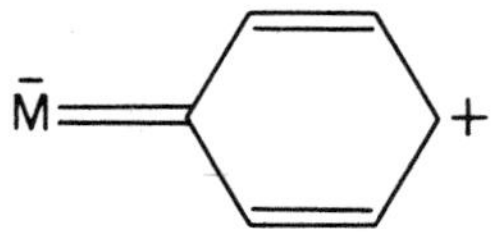 II

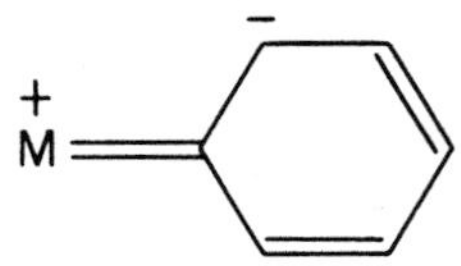

III 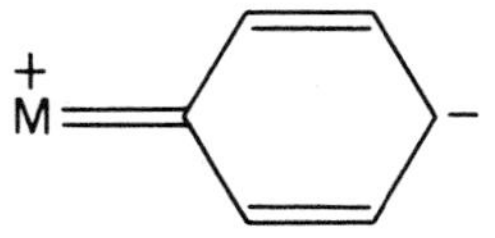 IV

The complications are caused by the σ-mechanism. The metal changes the electron density on the *ipso*-carbon atom. This effect is transmitted by both the σ- and π-framework of the phenyl ring. The π-electron density is perturbed by the change in σ-electron density on the *ipso*-carbon atom and to a lesser extent on the other carbon atoms.

Shifts of Carbonyls Attached to Metals

Although the chemical shift of the metal carbonyls covers a range in excess of 130 p.p.m., the chemical shifts are predictable to a reasonable accuracy. The shifts are dependent on the metal as illustrated for $Cr(CO)_6$, *ca* δ 212.5, $Mo(CO)_6$, *ca* δ 202, and $W(CO)_6$, *ca* δ 192.5.[43] This movement to lower frequency on descending a triad is typical, and can be used to estimate the chemical shifts between the same compound of different elements within a triad. For related compounds of a given element there is a good correlation between the C—O stretching force constant and the carbonyl chemical shift, see Figure 1.1. but the compound $WCp(CO)_3Me$ lies off the plot

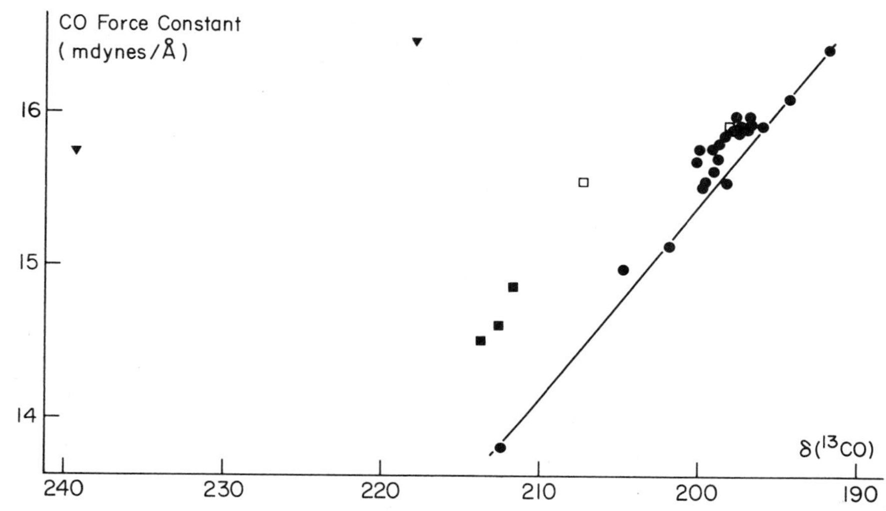

Figure 1.1. A plot of the ^{13}C chemical shift, $\delta(^{13}CO)$, against the force constant for the carbonyl group in some tungsten carbonyl compounds. ●, $W(CO)_{6-n}L_n$; □, $W(CO)_5CMe(SMe)$; ■, (arene)$W(CO)_3$ or $(C_7H_8)W(CO)_3$; and ▼, $C_5H_5W(CO)_3Me$. (Reproduced with permission from *J. Chem. Soc., Dalton Trans.*, 1973, 2012.)

established for $W(CO)_{6-n}L_n$ type compounds. However it is generally possible to use this relationship to make a very good estimate of the ^{13}C chemical shift of carbonyls.

Bridging and terminal carbonyls can frequently be distinguished on the basis of chemical shift since bridging carbonyls are markedly shifted to high frequency.

Shifts of Carbenes and Carbynes Attached to Metals

Metal carbenes and carbynes, are more analogous in shift to carbonium ions, coming between δ 200 and δ 400. As for organic carbonium ions, the presence of π-electron donating substituents, *e.g.*, OR or NR_2, produces a marked low frequency shift of the signal, *e.g.*, for $Cr(CO)_5\mathit{C}Ph_2$ at δ 399.4,[44]

$Cr(CO)_5\mathit{C}Ph(NMe_2)$ at δ 277.5,[45] and $Cr(CO)_5CPh(OMe)$ at δ 350.9.[45] Dual substitution can produce further low frequency shifts, *e.g.*, $Cr(\overline{\mathit{C}NMeCH_2CH_2N}Me)(CO)_5$ at δ 219.6[46] and $Cr\{\mathit{C}(OEt)_2\}(CO)_5$ at δ 266.4.[47] The same behaviour appears to be true for carbyne complexes with $Cr(\mathit{C}Ph)(CO)_4Br$ at δ 318.2[48] and $Cr(\mathit{C}NEt_2)(CO)_4Br$ at δ 264.1.[49]

Shifts of π-Bonded Carbon Ligands Attached to Metals

When a carbon ligand is π-bonded to a metal very marked shifts may occur. Numerous factors have been identified which contribute to the shift. Normally for ligands which exist in the absence of a metal, the free ligand is taken as the reference and the shift induced by the metal is discussed. There are two *major* contributions to the shifts.

(a) Coordination of the ligand to the metal causes the bonding to move from sp^2 towards sp^3 in character, *e.g.*, using the valence bond formalism for ethylene

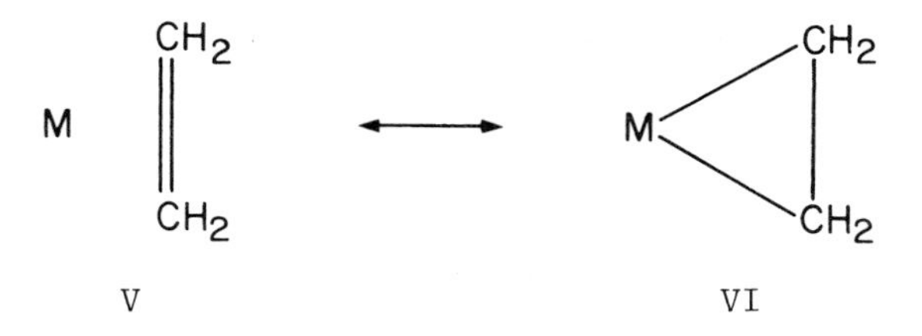

V will have a ^{13}C chemical shift close to free ethylene, δ 123.3, while VI will have a ^{13}C chemical shift close to cyclopropane, δ −2.6. The chemical shift depends on the relative importance of the two canonical forms.

(b) The metal atom will influence the charge on the carbon ligand. It has been shown that for sp^2 carbon atoms, the ^{13}C chemical shift is proportional to charge on the carbon atom and the resonance moves to low frequency with increasing negative charge.[50,51] Thus the charge on the complex influences the ^{13}C chemical shift and in complexes with a high positive charge, it is known for the carbon ligand to move to high frequency from the position of the free ligand, *e.g.*, arene-Hg^{2+} complexes.[52]

In addition these two major factors, minor contributions to the ^{13}C chemical shifts can be expected from changes to ring current contributions, and metal anistropy.

IV ^{13}C SPIN-SPIN COUPLING IN ORGANOMETALLIC COMPOUNDS

In contrast to the previous section on chemical shifts, coupling constants are more amenable to prediction, although once again *ab initio* theoretical calculations have rarely been attempted for organometallic compounds. Unlike chemical shifts, coupling constants are magnetic field independent making theoretical calculations considerably easier. However, like chemical shifts, they are the trace of a tensor. Thus for $HgMe_2$, $^1J_{//} - {^1J_{\perp}} \approx 326 \pm 24$ Hz compared with $^1J_{HgC} = 687$ $^1J_{HgC} = 687 \pm 1$ Hz.[53]

For the purposes of comparing coupling constants for different nuclei, it is more convenient to use the reduced coupling constant K_{AB} as defined by

$$J_{AB} = h \frac{\gamma_A}{2\pi} \frac{\gamma_B}{2\pi} K_{AB} \qquad 1.14$$

where γ_A is the gyromagnetic ration of nucleus A. K_{AB} depends only on the electronic environment, and is therefore independent of the isotope for a given pair of elements to the present accuracy of our measurements. Thus for a given coupling constant, the coupling to two different isotopes of a given element may be related by

$$\gamma_C J_{AB} = \gamma_B J_{AC} \qquad 1.15$$

e.g., $$\gamma_{^{117}\text{Sn}} J(^{119}\text{Sn}, {^{13}\text{C}}) = \gamma_{^{119}\text{Sn}} J(^{117}\text{Sn}, {^{13}\text{C}}) \qquad 1.16$$

or $$J(^{119}\text{Sn},\ ^{13}\text{C}) = 1.046\ J(^{117}\text{Sn},\ ^{13}\text{C}) \qquad 1.17$$

As γ may be positive or negative depending on the nucleus concerned, J_{AB} and K_{AB} may be of opposite sign.

Ramsey[54] has described the interaction between two nuclei as the sum of three terms, the orbital term, $K_{AB}{}^{1}$, the dipolar term, $K_{AB}{}^{2}$, and the contact term, $K_{AB}{}^{3}$.

$$K_{AB} = K_{AB}{}^{1} + K_{AB}{}^{2} + K_{AB}{}^{3} \qquad 1.18$$

As for chemical shifts, the values obtained from calculations are frequently in poor agreement with calculations. However, the indications are that for light elements without lone pair electrons $K^{3}{}_{AB}$ is dominant for one bond coupling constants. When nuclei with lone pairs, *e.g.*, ^{19}F, are involved then all the terms are significant. If attention is concentrated on the coupling between nuclei without lone pairs then we find a tolerable rationalization of one bond coupling constants using the very simplified equation

$$K^{3}{}_{AB} = \frac{16}{9}\pi\ \mu_{o}\mu_{B}{}^{2}(^{3}\Delta E)^{-1}\ S^{2}{}_{A}(0)S^{2}{}_{B}(0)P^{2}{}_{S_A S_B} \qquad 1.19$$

where μ_{o} is the permeability in a vacuum, μ_{B} is the Bohr magneton, $^{3}\Delta E$ is the mean triplet excitation energy, $S^{2}{}_{A}(0)$ is the electron density at the nucleus and $P_{S_A S_B}$ is the *s*-electron bond order between A and B. This equation has been derived using the average energy approximation and linear combination of atomic orbitals with all the consequent errors. However, unlike the same treatment of chemical shifts, a very wide range of coupling constants may be rationalized using Eq. 1.19, or more frequently

$$K^{3}{}_{AB} \propto (^{3}\Delta E)^{-1}S_{A}{}^{2}(0)S^{2}{}_{B}(0)\alpha^{2}{}_{A}\alpha^{2}{}_{B} \qquad 1.20$$

This type of equation can be used to estimate the coupling constant between carbon and any metal without a lone pair using Eq. 1.21[55] which is based on Eq. 1.19.

$$^{1}J_{MC} = \frac{\alpha_{M}^{2}}{\alpha_{C_A}^{2}} \cdot \frac{\alpha_{C}^{2}}{\alpha_{C_B}^{2}} \frac{\Delta E_{C_A}}{\Delta E_{C_M}} \left(\frac{Z^{*}_{M}}{Z^{*}_{C_A}}\right)^{3} \frac{n_{C_A}^{3}}{n_{C_M}^{3}} \frac{\gamma_M}{\gamma_C} \, {}^{1}J_{C_AC_B} \qquad 1.21$$

where Z^{*} is the effective nuclear charge of the s-oribital and n is the principal quantum number. Z^{*} may be estimated from $^{2}J(M-C-{}^{1}H)$, atomic beam experiments, optical hyperfine splittings, a formula due to Goldsmit, or Hartree-Fock calculations, with the value obtained from $^{2}J(M-C-{}^{1}H)$ being preferred.[56,57] The use of this treatment is illustrated in Table 1.1. It will be observed that the agreement between calculation and experiment is good enough to estimate where to search in the n.m.r. experiment.

On a simpler scale, the equation is frequently re-written in the form

$$^{1}J_{AB} \propto \alpha^{2}{}_{A}\alpha^{2}{}_{B} \qquad 1.22$$

This equation is a useful guide to estimating the value of coupling constants, *e.g.*, for the series methane, ethene, and ethyne, $^{1}J_{CH}$ increases with values 125, 156, and 248 Hz respectively in line with the nominal $\alpha^{2}{}_{C}$ values of 0.25, 0.33, and 0.5 respectively. This relationship is poor when the effective nuclear charge on the carbon is changed, *e.g.*, $CHCl_3$ has $^{1}J_{CH}$ = 209 Hz, which would require an unreasonable nominal $\alpha^{2}{}_{C}$ value of 0.42. Eq. 1.22 is a useful memory aid rather than a quantitive way of determining $\alpha^{2}{}_{C}$. For example on going from WMe_6 to $W(CO)_6$, $^{1}J_{WC}$ increases from 43.2 Hz[38] to 126 Hz,[43] in line with the doubling of the notional $\alpha^{2}{}_{C}$ and an extra increase due to the electrnegativity of the oxygen substituent on carbon.

TABLE 1.1

A Comparison of Coupling Constants Calculated using Eq. 1.21 and Experimental Values

COMPOUNDS	n	γ (rad sec^{-1} T^{-1} x 10^{-8})	α_M	α_{MC}	Z^*	K	$^1J(M-^{13}C)$	
							OBSERVED	CALCULATED
$[^{13}CH_4]$	1	2.67519	1	0.25	1.00	1.00	125	111
$[^{11}BPh_4]^-$	2	0.85830	0.25	0.33	3.4	1.00	49.5	57.5
$[^{13}CMe_4]$	2	0.67266	0.25	0.256	3.44	1.00	36.2	(36.2)
$[^{27}AlMe_4]^-$	3	0.69707	0.25	0.25	6.0	1.02	71.2	58.7
$[^{29}SiMe_4]$	3	0.53143	0.25	0.291	6.63	1.02	50.2	70.3
$[^{31}PMe_4]^+$	3	1.0829	0.25	0.205	6.72	1.03	55.5	99.1
$[^{51}V(CO)_6]^-$	4	0.7031	0.167	0.50	8.2	1.06	116	88.6
$[^{55}Mn(CO)_6]^+$	4	0.6598	0.167	0.50	9.0	1.07	-	110
$[^{57}Fe(CO)_5]$	4	0.08645	0.20	0.50	9.3	1.07	23.2	19.2
$[^{57}Fe(CO)_4]^{2-}$	4	0.08645	0.25	0.50	9.3	1.07	-	24.0
$[^{59}Co(CO)_4]^-$	4	0.63173	0.25	0.50	9.6	1.075	287	194
$[^{73}GeMe_4]$	4	0.093320	0.25	0.256	11.6	1.11	18.7	26.8
$[^{95}Mo(CO)_6]$	5	0.1743	0.167	0.50	12.0	1.20	68	40
$[^{111}CdMe_2]$	5	0.5673	0.50	0.240	15.4	1.28	512	420
$[^{119}SnMe_4]$	5	0.9971	0.25	0.137	16.5	1.30	330	262
$[^{183}W(CO)_6]$	6	0.11132	0.167	0.50	15.6	1.98	126	137
$[^{187}Os(CO)_5]$	6	0.06161	0.20	0.50	19.0	2.06	-	91.6
$[^{199}HgMe_2]$	6	0.47691	0.50	0.222	20.7	2.25	689	808
$[^{203}TlMe_3]$	6	1.5289	0.333	0.25	21.4	2.32	1930	2410
$[^{207}PbMe_4]$	6	0.55970	0.25	0.196	20.5	2.39	251	432

The qualitative validity of Eq. 1.22 for organometallic compounds has been tested for some platinum(II) compounds.

$$^{1}J_{PtP} = 37200\ \alpha^{2}_{Pt}\alpha^{2}_{P} \qquad 1.23$$

based on $[Pt(PMe_2Ph)_4]^{2+}$ whereas $^{1}J_{PtP}$ = 2325 Hz, α^2_{Pt} = 0.25 (based on *dsp*^{2} hybridisation), and α^2_P = 0.25, while

$$^{1}J_{PtC} = 7805\ \alpha^{2}_{Pt}\alpha^{2}_{C} \qquad 1.24$$

based on *cis*-$PtMe_2(PMe_2Ph)_2$ where $^{1}J_{PtP}$ = 1819 Hz, $^{1}J_{PtC}$ = 594 Hz, α^2_P = 0.25, $\alpha^2_{Pt(P)}$ = 0.196, whence α^2_{Pt-C} = 0.304, and α^2_C = 0.25. When Eq. 1.23 and Eq. 1.24 are applied to a series of platinum(II) compounds, then α^2_{Pt} for all bonds involving platinum(II) is close to 1.0, see Table 1.2.[56] The equation

TABLE 1.2

Values of $^{1}J(^{195}Pt-^{13}C)$, $^{1}J(^{195}Pt-^{31}P)$, and α_{Pt}^2 for some Complexes of Platinum

COMPOUND	$^{1}J(^{195}Pt-^{13}C)$ (Hz)	$^{1}J(^{195}Pt-^{31}P)$ (Hz)	α_{Pt}^2
$[Pt(PMe_2Ph)_4]^{2+}$	-	2325	1.0
$[Pt(PEt_3)_4]$	-	3740	1.61
$[Pt(PEt_3)_3]$	-	4220	1.36
cis-$[PtMe_2(PMe_2Ph)_2]$	594	1819	1.00
cis-$[PtPh_2(PEt_3)_2]$	817	1771	1.01
trans-$[PtPh_2(PEt_3)_2]$	594	2831	1.07
$[Pt(CN)_4]^{2-}$ †	1036	-	1.06

†H.H. Rupp, quoted by H.H. Keller, *Ber. Bunsenges. Phys. Chem.*, 1972, **76**, 1080.

breaks down completely when applied to platinum(0) with a changing coordination number, *e.g.*, for $Pt(PEt_3)_4$ $^{1}J_{PtP}$ = 3740 Hz, for $Pt(PEt_3)_3$ $^{1}J_{PtP}$ = 4220 Hz, for $Pt\{P(C_6H_{11})_3\}_3$, $^{1}J_{PtP}$ = 4228 Hz, and for $Pt\{P(C_6H_{11})_3\}_2$, $^{1}J_{PtP}$ = 4120 Hz.[57]

The value of these relationships is not as a way to estimate $\alpha_A{}^2$ (which is very questionable) but as a method to estimate a coupling constant. Some authors have concluded when Eq. 1.22 does not work that the orbital and dipolar terms are significant. In view of the gross approximations involved in deriving Eq. 1.22 from the contact term it is surprising that the equation ever works, and deviations may equally well reflect the inadequacies of Eq. 1.22 rather than the lack of dominance by the contact term.

For a suitably defined series of compounds empirical relationships exist between $^1J_{AB}$ and most other relevant parameters, *e.g.*, ν_{AB}, the A—B stretching frequency, or better k_{AB}, the AB force constant, the A—B bond length, and $^2J_{MCH}$. When the carbon is in a methyl group, the linear relationship between $^1J_{MC}$ and $^2J_{MCH}$ is particularly valuable, see Figure 1.2.

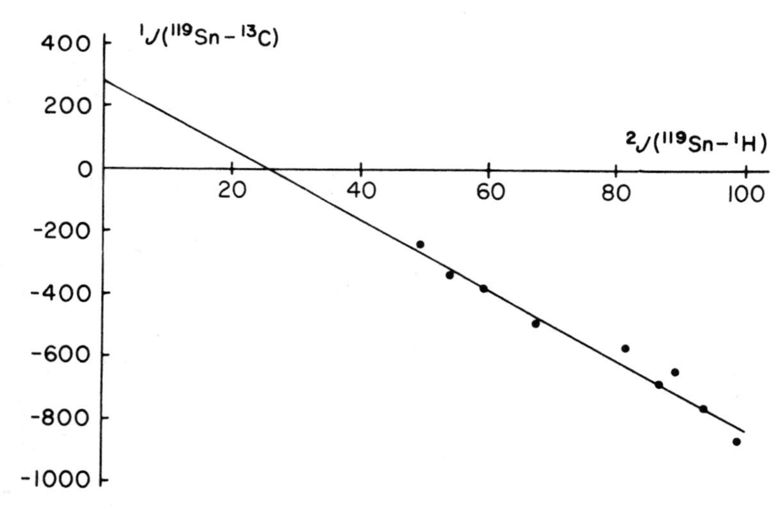

Figure 1.2. A plot of $^1J(^{119}Sn, ^{13}C)$ against $^2J(^{119}Sn, ^1H)$ for some methyl tin complexes. (Reproduced with permission from *J. Chem. Soc., A*, 1967, 528.)

It should be noted that it is possible to show from Eq. 1.20 that there should be a zero intercept. This is often not the case. Although these treatments are very useful for metal-carbon σ bonds, their applicability to metal-carbon π bonds has yet to be established.

Two Bond Coupling Constants

Considerable stereochemical information is often available from $^{2}J_{CMX}$ coupling constants where M is a transition metal. For the later heavier transition metals there is a marked difference in values of $^{2}J_{CMX}$ depending on whether the relationship is *cis* or *trans*, *e.g.*, in *cis*-$PtMe_2(PMe_2Ph)_2$, $^{2}J(^{31}P, {}^{13}C)$(*trans*) is 104 Hz while $^{2}J(^{31}P, {}^{13}C)$(*cis*) is 9 Hz.[58] This relationship works well for complexes of Ru, Os, Rh, Ir, Pd, and Pt, but becomes more difficult to apply outside this region of the Periodic Table.

Dihedral Angle Dependence

In saturated M—C—C—C fragments, the value of $^{3}J_{MC}$ depends critically upon the M—C—C—C dihedral angle, θ, and is described by the Karplus equation, Eq. 1.25.

$$^{3}J_{MC} = A\cos2\theta + B\cos\theta + C \qquad 1.25$$

where A, B, and C are empirical constants.[59-61] For $SnMe_3$ attached to a bicyclo[2.2.1]heptane and related rigid compounds, Eq. 1.26 has been derived[62]

$$^{3}J_{SnC} = 25.2\cos2\theta - 7.6\cos\theta + 30.4 \qquad 1.26$$

It is probable that similar equations apply to other $^{3}J_{MC}$ but as yet no such equations have been derived.

V RELAXATION MEASUREMENTS

Few measurements of ^{13}C relaxation times for organometallic compounds have so far been made, even though they can yield extra information. There are two relaxation times, the spin-lattice, T_1, and the spin-spin relaxation time, T_2. For relatively non-viscous samples, *i.e.*, for most organometallic systems, T_1 and T_2 are equal. However if there is a chemical exchange process going on then there will be an extra contribution to T_2, see section 1.

There are several mechanisms for spin-lattice relaxation to occur.

Dipole-Dipole Interactions

For ^{13}C this mechanism is frequently dominant. It arises from the interaction between two nuclear spins, *e.g.*, ^{13}C and ^{1}H. When the extreme narrowing condition $\omega_C^2\tau_c^2 << 1$ is met, the dipole-dipole relaxation time, T_{1dd} is given by Eq. 1.27

$$T_{1dd}^{-1} = \frac{\mu_o^2\gamma_C^2\gamma_S^2\hbar^2 S(S+1)\tau_c}{12\pi^2\sum_i r^6_{CS_i}} \qquad 1.27$$

where ω_C is the observation frequency in rad s^{-1} for carbon, τ_C is the correlation time, and r_{CS_i} is the distance between the carbon and the i th spin. The contribution of this term may be determined from the relationship in Eq. 1.28

$$T_{1dd} = \frac{\gamma_S}{2\gamma_C} \cdot \frac{T_1}{\eta} \qquad 1.28$$

where T_1 is the experimental spin-lattice relaxation time and η the nuclear Overhauser enhancement for the particular carbon atom. It is therefore easy to determine the contribution to T_{1dd} due to ^{13}C$-$^{1}H interactions. Owing to the r^{-6} dependence of Eq. 1.27 the hydrogen bearing carbons are most efficiently relaxed by the dipole-dipole mechanism, and the contribution is inversely proportional to the number of attached hydrogen atoms.

Although the principal contribution to T_{1dd} in ^{13}C n.m.r. spectra is due to protons, any paramagnetic compounds, *e.g.*, O_2, and also other n.m.r. active nuclei can make a significant contributions.

Spin-Rotation Interactions

For small molecules rotation is fast enough for direct transfer of energy from the spin system to molecular rotation to occur directly. As a consequence, the nuclear relaxation rate $(T_{1sr})^{-1}$ is directly proportional to temperature, and if molecular rotation is isotropic, Eq. 1.29 applies

$$T_{1sr}^{-1} = \frac{2IkTC^2\tau_{sr}}{3\hbar^2} \qquad 1.29$$

where I is the molecular moment of inertia, τ_{sr} is the spin-rotation interaction correlation time and C^2 is the average of the square of the spin-rotation tensor. This term is important for small molecules, *e.g.*, $Si(CH_3)_4$ and for freely rotating groups such as CH_3 and CF_3. This term can be identified by the temperature dependence of Eq. 1.29.

Nuclear Screening Anisotropy

An anisotropic screening tensor may act as a source of nuclear spin relaxation, T_{1sa} which is given by Eq. 1.30

$$T_{1sa}^{-1} = \frac{\mu_o}{30}\gamma_I^2 B_o^2 \Delta\sigma^2 \tau_c \qquad 1.30$$

in the motional narrowing limit. The dependence of T_{1sa} on the applied magnetic field, B_o, should be noted. This term may therefore be derived from a field dependent study of T_1.

At present there are few reports of this term contributing to the relaxation of organometallic compounds. This lack merely reflects the absence of measurements rather than the absence of the effect. The effect is expected to be most significant for non-proton bearing carbon atoms where the T_{1dd} mechanism is inefficient, *e.g.*, metal carbonyls. For example for solid carbon monoxide, $\Delta\sigma$ has been reported as 406 p.p.m.[63] Thus in $Ni(CO)_4$ and $Fe(CO)_5$, this effect has been detected, and used to determine $\Delta\sigma$ of 440 p.p.m. and 408 p.p.m. respectively.

Scalar Coupling

If the spin-spin coupling constant for the interaction between two nuclei becomes time dependent as a result of chemical exchange or internal rotation, nuclear relaxation can be induced. This is scalar coupling of the first kind and is discussed further on pp 3, 4.

If the relaxation rate of the nucleus S is fast compared to $2\pi J_{CS}$ then scalar relaxation of the second kind may occur for carbon. This situation may arise if the nucleus S is quadrupolar, *i.e.*, $I > \frac{1}{2}$. The scalar contribution to T_1 is given by Eq. 1.31

$$T_{1sc}^{-1} = \frac{8}{3}\pi^2 J^2 S(S+1) T_1^{S}[1 + (\omega_I - \omega_S)^2 (T_1^{S})^2]^{-1} \qquad 1.31$$

At present there is no evidence for this process occurring in organometallic compounds. However, once again it is the lack of measurements, rather than the lack of occurrence of this relaxation process. When it is remembered that $^1J_{MC}$ is generally large for organometallic compounds and that T_1 for many quadrupolar metals is short, then the scalar contribution to T_1 is expected to be often significant.

The scalar coupling contribution to T_2 is important when $(T_1^{S})^{-1}$ is comparable to $2\pi J_{MC}$. This is the cause of the broadening frequently observed for carbon atoms attached to manganese, vanadium, niobium, or cobalt.

Uses of Relaxation Measurements

While line shape analysis is useful for giving information on dynamic processes with lifetimes of 1 to 10^{-6}s, relaxation measurements provide information on the faster processes, with the fastest process being the most important. Thus for most molecules, it is the tumbling at that position in the molecule that dominates the T_1 measurement. It is therefore possible to determine the rate of molecular motion at each carbon atom in the molecule.[64] It is generally easiest to use T_{1dd} and Eq. 1.27 to determine τ_c. This approach is convenient to investigate segmental motion, *e.g.*, in alkyl lithium compounds,[6] $[Et_3Bhexyl]^-$,[66] and some phosphorus heterocycles.[67] In the case of $Hg(CH_3)(O_2CCH_3)$ it was possible to show that the motion parallel to the mercury-carbon bond is 121 times faster than perpendicular to it.[68] This behaviour results in nonexponential T_1 relaxation, and is due to a competition between spin-rotational and dipolar relaxation.

For the organometallic chemist, a more interesting application of T_1 measurements is to investigate the rotation of groups such as methyl, phenyl, and cyclopentadienyl where the rate is comparable with or faster than molecular reorientation. Thorough treatments have been performed on $Si(CH_3)_4$, $Sn(CH_3)_4$ and $Pb(CH_3)_4$,[69] and activation energies for methyl group rotation were determined. A similar study has been performed on phenyl group rotation in Ph_nSiH_{4-n}, n = 1, 2, 3.[70] In each case the studies were reinforced by additional investigations of appropriate nuclei from ^1H, ^2H, ^{29}Si, ^{119}Sn, and ^{207}Pb. An alternative approach has been to derive T_{1sr} for methyl groups in organometallic compounds and then to use an empirical equation, which had been derived for organic compounds, to determine the activation energy.

Only one π-system has so far been investigated in this way. By determining T_{1dd} for the substituted and unsubstituted cyclopentadienyl groups in $Fe(C_5H_5)(C_5H_4Bu^n)$ it was possible to show that the unsubstituted C_5H_5 ring is rotating about 7 times faster.[71]

In addition to these measurements, relaxation measurements have been reported for Ph_2SiH_2,[72] 5-deoxyadenosylcobalamin,[73] Me_3SiSMe,[74] and $(Me_3Si)_2CHCH(SiMe_3)_2$.[75]

VI APPLICATIONS

To a very large extent the applications of ^{13}C n.m.r. spectroscopy to organometallic compounds is the same as for organic compounds and require only superficial treatment. The major area that ^{13}C n.m.r. spectroscopy has opened up for organometallic chemists is the investigation of metal carbonyls, and especially dynamic processes in metal carbonyls.

Structure determination. As is organic chemistry, ^{13}C n.m.r. spectroscopy readily gives six pieces of valuable structural information

(a) The purity of the sample can be readily determined. It is generally quite easy to detect impurities at the 1% level.

(b) The number of symmetry unrelated carbon position in the molecule can be seen. It must be remembered that fluxionality of the molecule may generate additional equivalence of carbon environments. Also as a result of long T_1 values and low nuclear Overhauser enhancements non proton bearing carbon atoms may be difficult to observe. Carbon atoms which are attached to quadrupolar nuclei may be difficult to detect due to excessive linewidths.

(c) Intensity can be misleading, but coupled with molecular weight determination and a knowledge of starting materials it is frequently possible to ascertain the number of different carbon atoms giving rise to each signal for carbon atoms bearing the same number of protons. Thus for skeletal CH groups intensities may be safely used, but comparison with rapidly rotating CH containing groups, *e.g.*, η-C_5H_5 is dangerous. The rapid rotation will lead to longer T_1 values frequently resulting in the intensity of the rapidly rotating group to be smaller than expected. It is also dangerous to compare intensities of CH, CH_2, and CH_3 groups, as T_{1dd} values are inversely proportional to the number of protons attached to the carbon atoms. For methyl groups, the problem is complicated by rotation affecting the n.O.e. value.

(d) Chemical shifts are a very valuable source of information on the nature of the carbon atom. The shift of a carbon atom directly attached to a transition metal is very dependent on the metal and on the ligand, and it is necessary to estimate chemical shifts by application of the empirical rules outline earlier and by comparision with previously assigned ^{13}C chemical shifts (Chapter 2).

(e) ^{13}C, ^{1}H coupling constants are rarely determined, but off-resonance decoupling provides valuable evidence of the number of hydrogen atoms attached to each carbon atom. There are also numerous techniques to determine the ^{1}H chemical shift(s) of proton(s) attached to each carbon atom. The measurement of ^{1}H coupled spectra can lead to valuable structure assignments. In many cases $^{2}J_{CH}$ is small·(<2 Hz) while $^{3}J_{CH}$ is *ca* 6 to 8 Hz permitting extra assignments to be made, and structures differentiated.

(f) Relaxation data can provide information on the mobility of the carbon atom and hence assist assignments.

Fluxional molecules. It is in the area of fluxional molecules that ^{13}C n.m.r. spectroscopy has made the greatest contribution to advancing the understanding of organometallic compounds.

Before the advent of ^{13}C n.m.r. spectroscopy, fluxional processes among carbonyl ligands could only be suspected with little or no experimental evidence. The first fluxional carbonyl to be examined, $Fe_2Cp_2(CO)_4$, proved to be inexplicable to the investigators,[76] see Figure 1.3. $Fe_2Cp_2(CO)_4$ consists

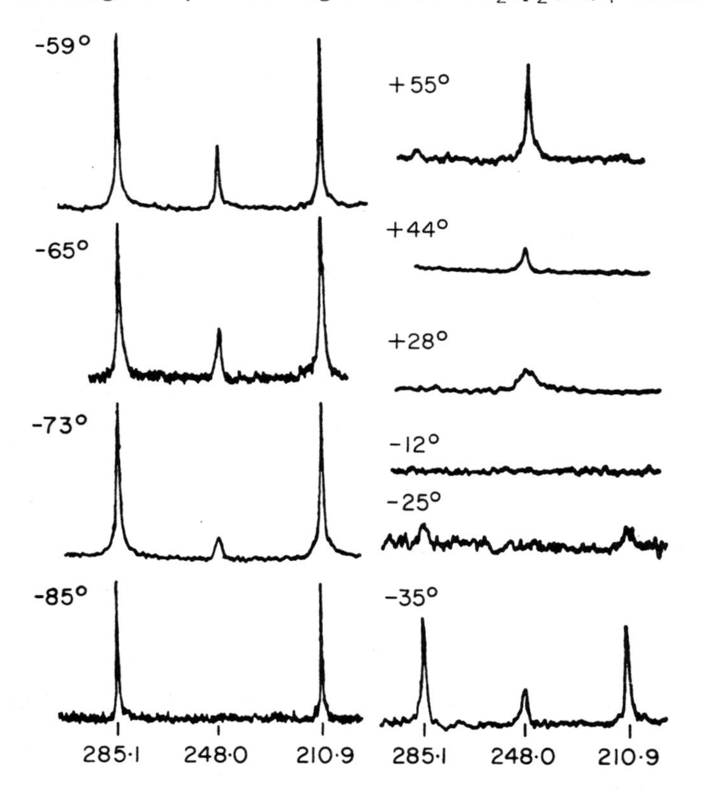

Figure 1.3. The variable temperature ^{13}C n.m.r. spectrum of the carbonyl groups of $[Fe(CO)_2(C_5H_5)]_2$. (Reprinted from *J. Am. Chem. Soc.*, 1972, **94**, 2551. Copyright 1972 by the American Chemical Society. Reprinted by permission of the copyright owners.)

of two isomers, VII and VIII

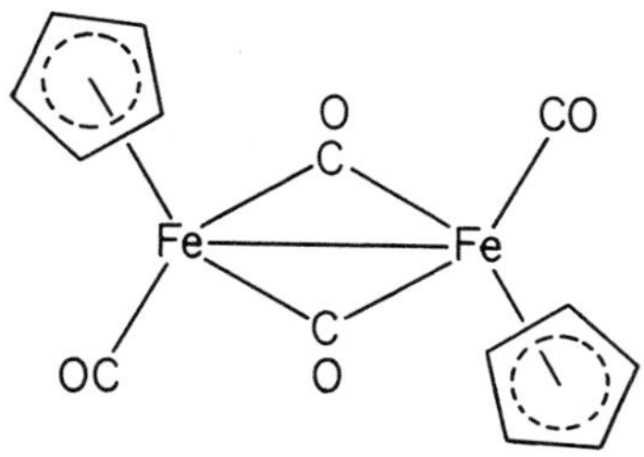

trans

VII

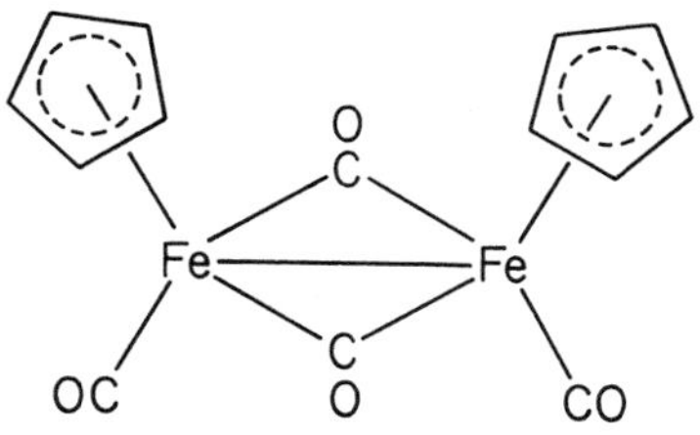

cis

VIII

The ^{13}C n.m.r. spectrum at 55°C consists of one ^{13}CO resonance at δ 248.0, see Figure 1.3. On cooling to -59°C, two additional signals at δ 210.9 and δ 285.1 appear, while on subsequent cooling the signal at δ 248.0 broadens and has vanished by -85°C. Subsequently, this behaviour was interpreted by Cotton.[77] Two postulates were necessary to explain this behaviour.

(a) The carbonyl bridge opens in a *trans* manner.

(b) Rotation about the iron-iron bond is a relatively high energy process.

When the carbonyl bridge opens in the *trans* isomer all the carbonyl groups become equivalent, as in the Newman projection down the iron-iron bond, IX, the carbonyls which were originally bridging are starred. This process is very fast at -59°C causing the two ^{13}CO signals to average to one at δ 248.0 and cooling below this temperature slows the process down further with the broadened signal lost in the noise at -85°C.

When the carbonyl bridge opens in the *cis* isomers, the bridging carbonyl groups do not equilibriate with the terminal carbonyl groups, X, and consequently two separate ^{13}CO signals

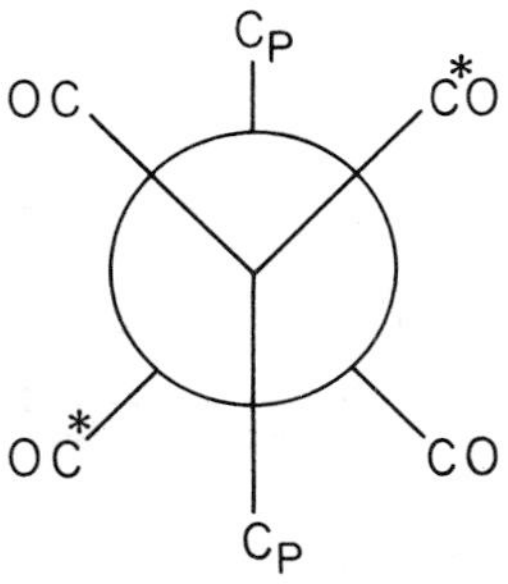

IX 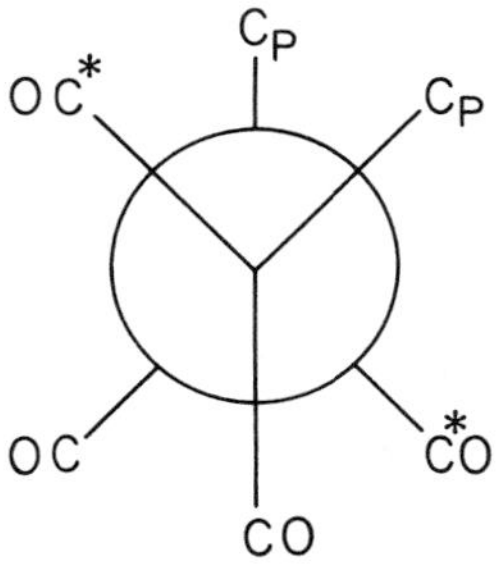 X

are observed for the *cis* isomer at δ 210.9 and δ 285.1 at -59°C. Above -59°C, rotation about the iron-iron bond in the bridge opened intermediate becomes significant, interconverting the *cis*- and *trans*-isomers and causing all the CO groups to exchange.

Subsequently numerous carbonyl scrambling processes have been observed and two comprehensive reviews have appeared.[78,79] In addition to the above carbonyl scrambling mechanism, two other mechanisms have been identified

(a) Local scrambling. For example, in $Mo(\eta^6\text{-}C_7H_8)(CO)_3$, two ^{13}CO signals are observed at low temperatures and one at high temperatures.[80]

(b) The "merry-go-round" mechanism. For example, for $Rh_4(CO)_{12}$, the ^{13}C n.m.r. signal is a quintet at +60°C, but at -80°C it consists of four multiplets of equal intensity, three doublets $^1J(^{103}Rh, ^{13}C)$ due to the three types of terminal carbonyls and one triplet due to the bridging carbonyls.[81] The low temperature spectrum is consistent with the structure XI. It is believed that the mechanism involves the opening of the bridging carbonyl groups to give the tetrahedral XII structure which is the ground state for $Ir_4(CO)_{12}$

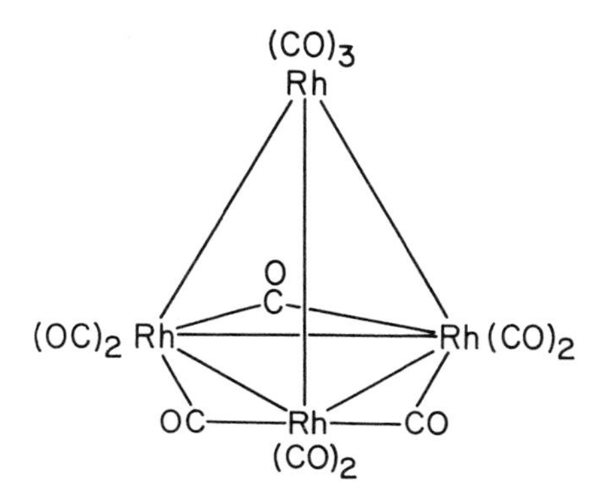

XI

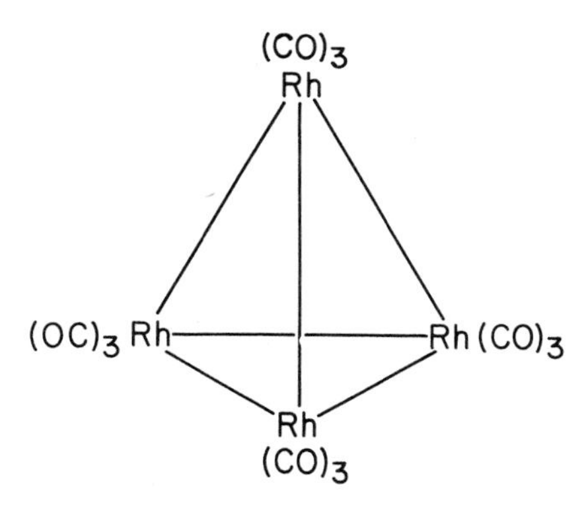

XII

REFERENCES

1. L.F. Johnson and W.C. Jankowski, "^{13}C N.M.R. Spectra", Wiley, New York, 1972.
2. G.C. Levy and G.L. Nelson, "^{13}C N.M.R. for Organic Chemists", Wiley-Interscience, New York, 1972.
3. J.B. Stothers, "^{13}C N.M.R. Spectroscopy", Academic Press, New York, 1972.
4. E. Breitmaier and W. Voelter, "^{13}C N.M.R. Spectroscopy", Verlag Chemie, Weinheim, 1974.
5. F.W. Wehrli and T. Wirthlin, "Interpretation of ^{13}C N.M.R. Spectra", Heyden, London, 1976.
6. R.J. Abraham and P. Loftus, "Proton and ^{13}C N.M.R. Spectroscopy: An Integrated Approach", Heyden, London, 1978.
7. B.E. Mann, *Adv. Organomet. Chem.*, 1974, **12**, 135.
8. M.H. Chisholm and S. Godleski, *Prog. Inorg. Chem.*, 1976, **20**, 299.
9. T.J. Todd and J.R. Wilkinson, *J. Organomet. Chem.*, 1974, **77**, 1.
10. O.A. Gansow and W.D. Vernon, *Top. Carbon-13 NMR Spectrosc.*, 1976, **2**, 270.
11. S. Braun, P. Dahler and P. Eilbracht, *J. Organomet. Chem.*, 1978, **146**, 135.
12. M. Nitay, W. Priester and M. Rosenblum, *J. Am. Chem. Soc.*, 1978, **100**, 3620.
13. L.J. Guggenberger, P. Meakin and F.N. Tebbe, *J. Am. Chem. Soc.*, 1974, **96**, 5420.
14. R. Bramley, B.N. Figgis and R.S. Nyholm, *Trans. Farad. Soc.*, 1962, **58**, 1893.
15. D. Doddrell and A. Allerhand, *Proc. Nat. Acad. Sci., U.S.*, 1971, **68**, 1083.
16. L.F. Farnell, E.W. Randall and E. Rosenberg, *J. Chem. Soc., Chem. Commun.*, 1971, 1078.
17. O.A. Gansow, A.R. Burke and G.N. La Mar, *J. Chem. Soc., Chem. Commun.*, 1972, 456.
18. S. Aime, L. Milone and M. Valle, *Inorg. Chim. Acta*, 1976, **18**, 9.
19. L.J. Todd and J.R. Wilkinson, *J. Organomet. Chem.*, 1974, **80**, C31.
20. S. Aime, G. Gervasio, L. Milone and E. Rosenberg, *Transition Met. Chem.*, 1976, **1**, 177.
21. B.R. Gragg, W.J. Layton and K. Niedenzu, *J. Organomet. Chem.*, 1977, **132**, 29.
22. E.A.C. Lucken, K. Noack and D.F. Williams, *J. Chem. Soc., A*, 1967, 148.
23. A.D. Jordan, R.G. Cavell and R.B. Jordan, *J. Chem. Phys.*, 1972, **56**, 1573.
24. O.A. Gansow, A.R. Burke and W.D. Vernon, *J. Am. Chem. Soc.*, 1972, **94**, 2550.
25. F.A. Cotton, P.L. Hunter and A.J. White, *Inorg. Chem.*, 1975, **14**, 703.

26. O.A. Howarth and R.J. Lynch, *Mol. Phys.*, 1968, **15**, 431.
27. W.D. Litchman and D.M. Grant, *J. Am. Chem. Soc.*, 1968, **90**, 1400.
28. C. Masters, B.L. Shaw and R.E. Stainbank, *J. Chem. Soc., Chem. Commun.*, 1971, 209.
29. A.D. Buckingham and P.J. Stephens, *J. Chem. Soc.*, 1964, 2747.
30. T.N. Mitchell, *Org. Magn. Reson.*, 1976, **8**, 34.
31. D.E. Axelson, S.A. Kandil and C.E. Holloway, *Canad. J. Chem.*, 1974, **52**, 2968.
32. H.G. Kuivila, J.L. Considine, R.H. Sarma and R.L. Mynott, *J. Organomet. Chem.*, 1976, **111**, 179.
33. S. Braun, P. Dahler and P. Eilbracht, *J. Organomet. Chem.*, 1978, **146**, 135.
34. D. Liebfritz, B.O. Wagner and J.D. Roberts, *Justus Liebigs Ann. Chem.*, 1972, **24**, 227.
35. See for example, reference 6, p.25.
36. A.J. Brown, O.W. Haworth and P. Moore, *J. Chem. Soc., Dalton Trans.*, 1976, 1589.
37. T.N. Mitchell and H.C. Marsmann, *J. Organomet. Chem.*, 1978, **150**, 171.
38. A.L. Galyer and G. Wilkinson, *J. Chem. Soc., Dalton Trans.*, 1976, 2235.
39. L.J. Todd, J.R. Wilkinson, J.P. Hickey, D.L. Beach and K.W. Barnett, *J. Organomet. Chem.*, 1978, **154**, 151.
40. G.L. Nelson, G.C. Levy and J.D. Cargioli, *J. Am. Chem. Soc.*, 1972, **94**, 3089.
41. D.T. Clark, *J. Chem. Soc., Chem. Commun.*, 1966, 390.
42. G.E. Maciel and J.J. Natterstad, *J. Chem. Phys.*, 1965, **42**, 2427.
43. B.E. Mann, *J. Chem. Soc., Dalton Trans.*, 1973, 2012.
44. E.O. Fischer, W. Held, F.R. Kreissl, A. Frank and G. Huttner, *Chem. Ber.*, 1977, **110**, 656.
45. J.A. Connor, E.M. Jones, E.W. Randall and E. Rosenberg, *J. Chem. Soc., Dalton Trans.*, 1972, 2419.
46. J.R. Bartels-Keith, M.T. Burgess and J.M. Stevenson, *J. Org. Chem.*, 1977, **42**, 3725.
47. E.O. Fischer, K. Scherzer and F.R. Kreissl, *J. Organomet. Chem.*, 1976, **118**, C33.
48. E.O. Fischer and G. Kreis, *Chem. Ber.*, 1976, **109**, 1673.
49. E.O. Fischer, G. Huttner, W. Kleine and A. Frank, *Angew. Chem. Internat. Edn. Engl.*, 1975, **14**, 760.
50. E.A. LaLancette and R.E. Benson, *J. Am. Chem. Soc.*, 1965, **87**, 1941.
51. H. Spiesecke and W.G. Schneider, *Tetrahedron Letts.*, 1961, 468.
52. P.A.W. Dean, D.G. Ibbott and J.B. Stothers, *Can. J. Chem.*, 1976, **54**, 166.
53. C. Schumann, D. Dreeskamp and K. Hildenbrand, *J. Magn. Reson.*, 1975, **18**, 97.
54. N.F. Ramsey, *Phys. Rev.*, 1953, **91**, 303.

55. F.J. Weigert, M. Winokur and J.D. Roberts, *J. Am. Chem. Soc.*, 1968, **90**, 1566.
56. B.E. Mann in "Nuclear Magnetic Resonance Spectroscopy of Nuclei other than Protons", ed. T. Axenrod and G.A. Webb, Wiley, New York and London, 1974.
57. B.E. Mann and A. Musco, *J. Chem. Soc., Dalton Trans.*, 1980, 776.
58. A.J. Cheney, B.E. Mann and B.L. Shaw, *J. Chem. Soc., Chem. Commun.*, 1971, 431.
59. G.W. Buchanan and C. Benezra, *Canad. J. Chem.*, 1976, **54**, 231.
60. I. Morishima, T. Inubushi, S. Uemura and H. Miyoshi, *J. Am. Chem. Soc.*, 1978, **100**, 354.
61. Y. Kashman and A. Rudi, *Tetrahedron Letts.*, 1976, 2819.
62. D. Doddrell, I. Burfitt, W. Kitching, M. Bullpitt, C.-H. Lee, R.J. Mynott, J.L. Considine, H.G. Kuivila and R.H. Sarma, *J. Am. Chem. Soc.*, 1974, **96**, 1640.
63. A.A.V. Gibson, T.A. Scott and E. Fukushima, *J. Magn. Res.*, 1977, **27**, 29.
64. F. Heatley, *J. Chem. Soc., Faraday Trans. II*, 1974, **70**, 148.
65. F.W. Wehrli, *Org. Magn. Res.*, 1978, **11**, 106.
66. W.T. Ford, *J. Am. Chem. Soc.*, 1976, **98**, 2727.
67. G.A. Gray and S.E. Cremer, *J. Magn. Res.*, 1973, **12**, 5.
68. A.J. Brown, O.W. Howarth, P. Moore and A.D. Bain, *J. Magn. Res.*, 1977, **28**, 317.
69. C.R. Lassigne and E.J. Wells, *J. Magn. Reson.*, 1977, **26**, 55.
70. R.K. Harris and B.J. Kimber, *Adv. Mol. Relaxation Processes*, 1976, **8**, 23.
71. G.C. Levy, *Tetrahedron Letts.*, 1972, 3709.
72. G.C. Levy, *J. Am. Chem. Soc.*, 1972, **94**, 4793.
73. D. Doddrell and A. Allerhand, *Proc. Nat. Acad. Sci., U.S.*, 1971, **68**, 1083.
74. R.K. Harris and B. Lemarié, *J. Magn. Res.*, 1976, **23**, 371.
75. S. Brownstein, J. Dunogues, D. Lindsay and K.U. Ingold, *J. Am. Chem. Soc.*, 1977, **99**, 2073.
76. O.A. Gansow, A.R. Burke and W.D. Vernon, *J. Am. Chem. Soc.*, 1972, **94**, 2550.
77. R.D. Adams and F.A. Cotton, *J. Am. Chem. Soc.*, 1973, **95**, 6589.
78. S. Aime and L. Milone, *Prog. N.M.R. Spectroscopy*, 1977, **11**, 183.
79. E. Band and E.L. Muetterties, *Chem. Rev.*, 1978, **78**, 639.
80. C.G. Kreiter and M. Land, *J. Organomet. Chem.*, 1973, **55**, C27.
81. J. Evans, B.F.G. Johnson, J. Lewis, J.R. Norton and F.A. Cotton, *J. Chem. Soc., Chem. Commun.*, 1973, 807.

Chapter 2

^{13}C N.M.R. DATA OF ORGANOMETALLIC COMPOUNDS

In the following tables, ^{13}C n.m.r. data of organometallic compounds, published in the literature up to 1979 are collected. The tables are ordered by the nature of the organic ligand, and by the metal. Each table is subdivided as much as possible to make the location of information on each group of compounds as easy as possible. It is not possible to correct errors in the literature, some of which are apparent by comparison with data on related compounds. The carbon atoms on which information is given are identified whenever possible by italicising or by numbering. Coupling constants are given in parentheses, *i.e.*,

[ ] coupling to the metal nucleus specified at the beginning of the subsection;

{ } coupling to ^{1}H;

( ) coupling to ^{31}P.

TABLE 2.1

^{13}C Chemical Shift Data for sp^3 Carbon Atoms Attached to Metals

COMPOUND	$\delta(^{13}C)$/p.p.m., J/Hz	REFERENCE
Lithium, [^7Li]		
$[Li\mathit{C}H_3]_4$	-9.2 to 16.2^*, -15.5	109, 110
	-15.5 [+15] {+96.2}	362
	{96 to 116^*}	90, 109, 110
$Li\mathit{C}H_2Pr^n$	1.9, 11.8, 11 [14] {98 to 100^*}	108, 551, 1329
$Li\mathit{C}H_2Ph$	18.3 to 29.8^* {116 to 135^*}	159, 160
$Li\mathit{C}H_2CH{=}CHCH_2Bu^t$	20.0, $20.7^{\S,\dagger}$ $30.7^{\dagger,\#}$	1720
$Li\mathit{C}H_2CMe{=}CHCH_2Bu^t$	24.2, $27.8^{\S,\dagger}$ 33.3, $36.7^{\#,\dagger}$	1720
$Li\mathit{C}HMe_2$	10.2	551
$Li\mathit{C}HMeEt$	17.0	551
$Li\mathit{C}HPh_2$	79 {142}	159
$Li\mathit{C}Me_3$	10.5, 11 [11]	108, 551
$Li\mathit{C}Ph_3$	91	159
$Li\mathit{C}_5H_5$	103.6 {159.1, 5.7, 8.2}	934
$Li\mathit{C}_5H_4Me$	113.1, 103.1, 101.7	1722
$Li\overline{\mathit{C}PhS(CH_2)_2CHRS}$	39.0, 43.4	1326
$Li\mathit{C}MePhCH_2Bu^t$	36.4	1721
$Li_4\mathit{C}_4$	57.2	1331
Sodium		
$Na\mathit{C}H_2CH{=}CHCH_2Bu^t$	34.5, $35.7^{\S}$	1720
$Na\mathit{C}H_2CMe{=}CHCH_2Bu^t$	36.7	1720
$Na\mathit{C}_5H_5$	103.4 {156.5, 7.0}	934
Potassium		
$K\mathit{C}H_2CH{=}CHCH_2Bu^t$	45.0	1720
$K\mathit{C}H_2CMe{=}CHCH_2Bu^t$	45.8	1720
$K\mathit{C}_5H_5$	104.8 {155.9, 6.9}	934
$K\mathit{C}_5H_4Me$	113.1, 103.1, 101.7	1722
$K\mathit{C}_5H_4CHMePh$	125.9, 102.2, 103.7	1723
Rubidium		
$Rb\mathit{C}H_2CH{=}CHCH_2Bu^t$	47.4	1720
$Rb\mathit{C}H_2CMe{=}CHCH_2Bu^t$	49.0	1720

*solvent dependent, §*cis* and *trans* isomers, †in benzene, $^{\#}$in Et_2O

COMPOUND	$\delta(^{13}C)$/p.p.m., *J*/Hz	REFERENCE
Caesium		
$CsCH_2CH{=}CHCH_2Bu^t$	51.4	1720
$CsCH_2CMe{=}CHCH_2Bu^t$	53.7	1720
Beryllium, [^9Be]		
$Be(CH_3)_2$	{105.5}	90
$Be(C_5H_5)Ph$	104.7 [1.1] {177.4, 6.7}	934
$Be(C_5H_5)Br$	105.5 {179.1, 6.3}	934
Magnesium		
$Mg(CH_3)I$	-14.5 {107.7}	1697, 1698
$Mg(C^1H_2C^2H_3)Br$	C^1 -2.9, C^2 12.2	1697
$Mg(CH_2Et)Br$	11.3	1697
$Mg(CH_2Pr^n)Br$	5.9	1697
$Mg(C_5H_{11})Br$	7.4	1697
$Mg(CH_2Ph)Cl$	22.2, 22.9	1332, 1697
$Mg(CH_2Ph)_2$	21.9	1697
$Mg(CH_2CH_2Ph)Br$	10.8	1697
$Mg(CH_2CH_2CH_2Ph)Br$	7.9	1697
$Mg\{C^1H_2C^2(C^3H_3)_3\}_2$	C^1 33.0, C^2 30.8, C^3 36.6	1163
$Mg(CHMe_2)Br$	8.9	1697
Mg(cyclohexyl)Br	23.4	1697
$Mg(4\text{-}Bu^tC_6H_{10})Br$	22.9	1697
$Mg(C_5H_5)_2$	108.0 {167.5, 6.9}	934
Barium		
$Ba(CH_2Ph)_2$	57.3	1332
Yttrium, [^{89}Y]		
$[Y(CH_2SiMe_3)_4]^-$	34.6 [28]	1334
$[Y(C_5H_4Me)_2CH_2Pr^n]_2$	38.7 [22.8]	1335
$[Y(C_5H_4Me)_2(CH_2C_7H_{15})]_2$	39.3 [23.0]	1335
$[Y(\sigma\text{-}\pi C_5H_4Me)(\eta^5\text{-}C_5H_4Me)]_2$	29.5, 30.5	1335
Titanium		
$Ti(CH_3)_2(C_5Me_5)_2$	48.8	316, 328
$Ti(CH_3)Cp(OPr^i)_2$	35.0 {124}	955
$Ti(CH_3)(NEt_2)_3$	30.0	521, 955
$Ti(CH_3)(OPr^i)_3$	42.9 {123}	955
$Ti_2\{(CH_2)_2PMe_2\}_2(OMe)_4(\mu\text{-}OMe)_2$	18.7 (39)	1337
$Cp_2\overline{TiCH_2SiMe_2N}SiMe_3$	71.9	386
$Ti(CH_2Ph)_4$	98.8 {132.0}	166
Zirconium		
$Zr(CH_2Ph)_4$	74.5 {136.0}	166
Niobium		
$Nb(CH_3)(S_2)Cp_2$	7.7	1342

COMPOUND	$\delta(^{13}C)$/p.p.m., J/Hz	REFERENCE
$Nb(C^1H_2C^2H_3)Cp(C_2H_4)$	C^1 11.2 {122}, C^2 20.3 {130.5}	313, 1389
$Nb(CH_2Bu^t)_3CHBu^t$	96.9	1345
$Nb_2(CH_2SiMe_3)_4(CSiMe_3)_2$	64.2	897
Tantalum		
$Ta(CH_3)CH_2Cp_2$	-4, -5	591, 1344
$Ta(CH_3)Cp_2(C_2H_4)$	-5.4	1344
$Ta(CH_3)Cp_2(CHSiMe_2)$	3.8 or -0.5	1346
$Ta(CH_2)_4CpCl_2$ (cyclic)	C_1 89.7 {123}, C_2 33.5 {126}	1102
$Ta(CH_2CHMeCHMeCH_2)CpCl_2$ (cyclic)	95.6	1343
$Ta(CH_2Bu^t)_3CHBu^t$	113.7 {107}, 114	309, 1345
E-$Ta(CH_2Bu^t)_3Cl(OCMe{=}CHBu^t)$	102.0	1345
Z-$Ta(CH_2Bu^t)_3Cl(OCMe{=}CHBu^t)$	105.2	1345
$[Ta(CH_2Bu^t)Cp_2PMe_3]^+$	32.8 (3)	1346
$Ta(CH_2Ph)(CHPh)Cp_2$	28.7	1346
$Ta(CHPh_2)(CHBu^t)Cp_2$	34.7	1346
$Ta_2(CH_2SiMe_3)_4(CSiMe_3)_2$	79.3	897
Chromium		
$[Cr_2(CH_3)_8]^{4-}$	9.0	1114
$Cr\{C(OMe)MePh\}(CO)_5$	86.3	1036
$Cr\{C(OMe)Ph_2\}(CO)_5$	95.0	1036
Molybdenum		
$[Mo_2(CH_3)_8]^{4-}$	4.7	1114
$Mo_2(CH_3)_4(PMe_3)_2$	2.7 (6.0)	1373
$Mo(CH_3)_2(C_6H_6)(PMe_3)_2$	-8.6 (24.0)	1388
$Mo(CH_3)_2(C_6H_4Me)(PMe_3)_2$	-8.2 (23.3)	1388
$Mo(CH_3)_2(o$-$C_6H_4Me_2)(PMe_3)_2$	-7.5 (22.8)	1388
$Mo(CH_3)_2(C_6H_6)(PMe_2Ph)_2$	-7.0 (22.5)	1388
$Mo(CH_3)_2(C_6H_5Me)(PMe_2Ph)_2$	-6.3 (22.2)	1388
$Mo(CH_3)_2(o$-$C_6H_4Me_2)(PMe_2Ph)_2$	-5.7 (21.8)	1388
$Mo(CH_3)_2(p$-$C_6H_4Me_2)(PMe_2Ph)_2$	-4.6 (22.0)	1388
$Mo(CH_3)Cp(CO)_3$	-22.4, -22.1	1394, 1396
trans-$Mo(CH_3)Cp(CO)_2(PPh_3)$	-19.2 (9.6)	1394
$Mo(CH_3)(C_5H_4Me)(CO)_3$	-18.6	1396
$Mo(CH_2Me)Cp(CO)_3$	-3.9	1396
$Mo(C^1H_2C^2H{=}C^3H_2)Cp(NO)$-$(S_2CNMe_2)$	C^1 32.8 or 38.3 or 38.8, C^2 144.9, C^3 106.1	376
$Mo_2(C_8H_8)Cp_2(CO)_2$*	119.1	739
$Mo_2(C_8H_8)Cp_2(CO)_2$*	98.8, 123.5	739
$Mo(\eta^5$-$C_5H_4CH_2CH_2)(CO)_3$ (cyclic)	-49.6	1396
$Mo(\eta^5$-$C_5H_4CH_2CH_2CH_2)(CO)_3$ (cyclic)	4.8	1396
$Mo(CH_2Bu^t)_3(CBu^t)$	88.2	1402
$Mo_2(CH_2Bu^t)_2(OAc)_2(PMe_3)_2$	21.8 (30.0, 5.0)	1373
$Mo_2(CH_2SiMe_3)_6$	65.0	897

*Isomers

COMPOUND	$\delta(^{13}C)$/p.p.m., J/Hz	REFERENCE
$Mo_2(\mathit{C}H_2SiMe_3)_2(OAc)_2(PMe_3)_2$	9.0 (29.2)	1373
$Mo_2(C^1H_2SiMe_3)_2\{(C^2H_2)_2SiMe_2\}$-$(PMe_3)_3$	C^1 43.8, C^2 73.6	1373
$Mo(\sigma$-$\mathit{C}_5H_5)(NO)(S_2CNBu^n{}_2)Cp$	53.9	1387
$Mo(\mathit{C}HPhNMeCPhNMe)Cp(CO)_2$	69.2	1377
Tungsten, $[^{183}W]$		
$W(\mathit{C}H_3)_6$	83.8 [400]*	177
	83.6 [43.2] {124.8}	895
$[W(\mathit{C}H_3)_8]^{2-}$	66.3	895
$W_2(\mathit{C}H_3)_8$	2.2	1113
$W(\mathit{C}H_3)Cp(CO)_3$	-28.9	60, 1060
	-34.6	1396
	-35.1	1394
$W(\mathit{C}H_3)(C_5H_4Me)(CO)_3$	-30.7	1396
$[W(\mathit{C}H_3)(CO)_5]^-$	-31.8	1401
$W(\mathit{C}H_3)Cp_2(C_2H_4)$	-18.5	343
$W_2(\mathit{C}H_3)_2(NEt_2)_4$	27.7, 30.0§	676
$W(C_5H_4CH_2\mathit{C}H_2)(CO)_3$	-61.7	1396
$W(C_5H_4CH_2CH_2\mathit{C}H_2)(CO)_3$	-7.4	1396
$[W(\mathit{C}H_2Ph)(CO)_5]^-$	4.7	1401
$W(\mathit{C}H_2CH{=}CH_2)Cp(CO)_3$	-6.6 [295]	449
$W(\mathit{C}H_2C{\equiv}CH)Cp(CO)_3$	-33.2 [295]	449
$W(\mathit{C}H_2Bu^t)_3CBu^t$	103.7 [89]	1402
$W(\mathit{C}H_2Bu^t)(CHBu^t)(CBu^t)(PMe_3)_2$	53.5 [80] (7)	1402
$W(\mathit{C}H_2Bu^t)(CHBu^t)(CBu^t)$-$(Me_2PCH_2CH_2PMe_2)$	54.7 (4)	1402
$W_2(\mathit{C}H_2SiMe_3)_6$	79.6 [79]	897
$W_2(\mathit{C}H_2SiMe_3)_4(CSiMe_3)_2$	69.0 [74]	897
$W(\mathit{C}H_2PPh_3)_3$	16.9 (79.4)	950
$W(\mathit{C}Ph_2PMe_3)(CO)_5$	40.6	450
$W\{\mathit{C}MePh(OMe)\}(CO)_5$	81.8	1036
$W(\mathit{C}Ph_2OMe)(CO)_5$	94.8	1036
$W(\mathit{C}Ph_2PMe_3)(CO)_5$	40.6 (17.1)	1062
$W(\mathit{C}R^1R^2PMe_3)(CO)_5$ (R^1, R^2 = Ph, 2-furanyl, 2-thiophenyl)	24.3 to 33.2 (22.0 to 29.3)	1062
$WRe(\mu$-$\mathit{C}PhPMe_3)(CO)_8(\mu$-$CO)$	80.4 (9.8)	1029
Manganese		
$Mn(\mathit{C}H_3)(CO)_5$	-9.4	67
$Mn(\mathit{C}H_2CH{=}CH_2)(CO)_5$	9.0	1414
$Mn(\mathit{C}H_2Ph)(CO)_5$	8.7	336
$Mn_2(\mu$-$\mathit{C}H_2)(CO)_4Cp_2$	153.1	533
Rhenium		
$[Re_2(\mathit{C}H_3)_8]^{2-}$	16.1	1114

*subsequently corrected ref. 895, §two isomers

COMPOUND	δ(^{13}C)/p.p.m., J/Hz	REFERENCE
Re(*C*H$_3$)(CO)$_5$	-38.0	514
Re$_2$(*C*H$_2$Ph)$_4$(OAc)$_2$	40.4	1421
Re$_2$(*C*H$_2$CMe$_2$Ph)$_4$(OAc)$_2$	59.2	1421
Re$_2$(*C*H$_2$Bu^t)$_4$(OAc)$_2$	60.5	1421
Re$_2$(*C*H$_2$SiMe$_3$)$_4$(OAc)$_2$	33.7	1421
Re{*C*Ph(PMe$_3$)$_2$}Cp(CO)$_2$	6.2 (22)	1019
ReW(μ-*C*PhPMe$_3$)(CO)$_8$(μ-CO)	80.4 (9.8)	1029
Iron		
[Fe(*C*H$_3$)(CO)$_4$]$^-$	-16.4	1144
Fe(*C*H$_3$)Cp(CO)$_2$	-23.4	60
	-23.5	763, 1394
Fe(*C*H$_3$)Cp(CO)(PPh$_3$)	-22.3 (22.2)	766, 1394
Fe(*C*H$_2$Me)Cp(CO)(PPh$_3$)	-2.7	776
Fe(*C*H$_2$Et)Cp(CO)(PPh$_3$)	7.6	776
Fe(*C*H$_2$Pr^n)Cp(CO)(PPh$_3$)	4.1 (18.6)	776
Fe(*C*H$_2$Pr^i)Cp(CO)(PPh$_3$)	14.1 (18.5)	776
Fe(*C*H$_2$CHMeCH$_2$CH$_3$)Cp(CO)(PPh$_3$)	11.8	1117
Fe(*C*H$_2$CH$_2$Pr^i)Cp(CO)(PPh$_3$)	1.7 (17.9)	1117
{FeCp(CO)$_2$*C*H$_2$}$_2$CH$_2$	7.8	763
{FeCp(CO)$_2$*C*H$_2$CH$_2$}$_2$	3.8	763
Fe(*C*H$_2$CH$_2$CH$_2$SiMe$_2$)(CO)$_4$ (ring)	22.5 or 26.1 or 31.3	1436
Fe{*C*H$_2$CHMeCH(CN)CO$_2$Et}Cp(CO)$_2$	7.1	1232
Fe(*C*H$_2$CH$_2$C$_6$H$_9$O)Cp(CO)$_2$	1.4	1232
Fe(*C*H$_2$CHMeC$_6$H$_9$O)Cp(CO)$_2$	7.5, 9.9*	1232
Fe(*C*H$_2$CHMeCMe$_2$CHO)Cp(CO)$_2$	4.4	1232
Fe(*C*H$_2$CHPhC$_6$H$_9$O)Cp(CO)$_2$	5.3, 9.9*	1232
Fe(C^{1}H$_2$C^{2}H=C^{3}H$_2$)Cl(CNBu^t)$_4$	C^1 14.4, C^2 151.4, C^3 103.0	1182
Fe(C^{1}H$_2$C^{2}H=C^{3}H$_2$)Cp(CO)$_2$	C^1 32.7, C^2 172.2, C^3 134.9	34
[Fe(*C*H$_2$PPh$_3$)Cp(CO)$_2$]$^+$	-23.3 (36.5)	101
[Fe(*C*H$_2$PPh$_3$)Cp(CO)(PPh$_3$)]$^+$	9.6 (43.0, 102.0)	101
[Fe(*C*H$_2$CH$_2$CH$_2$PPh$_3$)Cp(CO)(PPh$_3$)]$^+$	4.2 (19.1, 8.2)	101
Fe{*C*H$_2$CH$_2$PO(OMe)$_2$}Cp(CO)(PPh$_3$)	-7.8	101
Fe(*C*H$_2$C$_3$H$_4$)(CO)$_3$H	-1.9 {$^2J_{CH}$ = 73.7}	679, 785
Fe(*C*HMeEt)Cp(CO)$_2$	3.5 (18.0)	776
Fe(*C*HMePh)Cp(CO)$_2$	28.3	356
Fe{*C*HMeCH$_2$CH(CN)CO$_2$Et}Cp(CO)$_2$	13.9, 15.0*	1232
Fe(*C*HMeCH$_2$C$_6$H$_5$O)Cp(CO)$_2$	16.7, 20.7*	1232
Fe(*C*HMeCH$_2$Ph)Cp(CO)$_2$	22.0	1232
Fe{C^{1}HC^{2}H=C^{3}HC^{3}H=C^{2}H)Cp(CO)$_2$ (ring)	C^1 28.5, C^2 145.3, C^3 121.6	34, 78
Fe(*C*HPhSiMe$_3$)Cp(CO){P(OPh)$_3$}	2.9 (18), 1.6 (18)*	609, 1109
Fe(*C*H[CH{CH(CO$_2$Me)$_2$}CH$_2$]$_2$)Cp(CO)$_2$ (ring)	26.5	1232
Fe(*C*HCH$_2${C(CN)(CF$_3$)}$_2$CH$_2$)Cp(CO)$_2$ (ring)	12.5	1312
Fe(*C*HCMe$_2${C(CN)(CF$_3$)}$_2$CH$_2$)Cp(CO)$_2$ (ring)	29.2	1312

*Enantiomers

COMPOUND	$\delta(^{13}C)$/p.p.m., J/Hz	REFERENCE
$[(\sigma\text{-}1\text{-}\eta^3\text{-}2\text{-}4\text{-}C_4H_4)Fe(CO)_3H]^+$	9.0 {191.4, $^2J_{CH}$ 81.12}	785
$[(\sigma\text{-}1\text{-}\eta^3\text{-}2\text{-}4\text{-}C_6H_8)Fe(CO)_3H]^+$	19.0 or 36.4	679
$(\eta^4\text{-}1,3\text{-}5\text{-}C_7H_{10})Fe(CO)_3$	14.0	816
$(\eta^4\text{-}1,4\text{-}6\text{-}C_8H_{12})Fe(CO)_3$	51.8	1198
$[(\sigma\text{-}2\text{-}\eta^3\text{-}3,5,6\text{-norbornadiene})\text{-}Fe(CO)_3H]^+$	{$^2J_{CH}$ = 37.8 averaged	785
$(C_7H_8NCO_2Et)Fe(CO)_3$	-11.1 to 17.4	1440
$(\eta^4\text{-}2\text{-}4,7\text{-}C_8H_9O)Fe(CO)_3$	56.6	1473
$Fe(CO)_3(C_9H_8Ph_2O)$	22.1	1460
$[Fe\{\mathit{C}H(CO_2Et)Et\}(CO)_4]^-$	32.0	1452
$Fe\{C_9H_6(CN)_4O\}(CO)_3$	5.2, 17.1*	1149
$Fe(\mathit{C}MeCH_2\{C(CN)(CF_3)\}_2CH_2)Cp\text{-}(CO)_2$	30.7	1312
$Fe(\mathit{C}MeCH_2\{C(CN)_2\}_2CH_2)Cp(CO)_2$	33.7	1312
$Fe(\mathit{C}F_2C_2F_5)(C_5H_4I)(CO)_2$	149.3	1447
$Fe(\mathit{C}F_2C_2F_5)(C_5Cl_4I)(CO)_2$	145.0	1447
Ruthenium		
cis-$Ru(\mathit{C}H_3)_2(PMe_3)_4$	-3.4 (12.6, 63.8)	1373
$Ru\{(\mathit{C}H_2)_2SiMe_2\}(PMe_3)_4$	-30.0 (46.0, 7.7)	1373
$Ru_2(CO)_6C_{10}H_{12}$	21.6	1215
Osmium		
$HOs_3(\mathit{C}H_3)(CO)_{10}$	-14.9 {121}	1097
$Os_3(C\mathit{C}H_2)H_2(CO)_9$	65.5 {160}	1093
$Os_3(\mathit{C}H_2)H_2(CO)_{10}$	25.8 {143, 140, 3}	1097
$Os_3(\mathit{C}H)H_3(CO)_{10}$	68.2 {171, 3}	1097
Cobalt		
$Co(C^1H_3)_2(CH^2{}_2PMe_2CH_2)(PMe_3)_2$	C^1 7.0 (16), C^2 -21.8	366
$Co(\mathit{C}H_3)(dmgH)_2NH_2Ph$	9.0	572
$Co(\mathit{C}H_3)(dmgH)_2py$	11.2	572
$(H_3\mathit{C})$cobalamin	6.7	1111
$(H_3\mathit{C})$cobalaminH^+	-0.7	1111
$Co(\mathit{C}H_2Me)_2(acac)(PMe_2Ph)_2$	-4.3	838
$Co(\mathit{C}H_2Me)(dmgH)_2NH_2Ph$	23.4	572
$Co(\mathit{C}H_2Me)(dmgH)_2NH_2Bu^n$	25.6	572
$Co(\mathit{C}H_2Me)(dmgH)_2py$	24.8	572
$Co(\mathit{C}H_2Et)(dmgH)_2py$	34.6	572
$Co(\mathit{C}H_2Bu^n)(dmgH)_2py$	31.7	572
$Co(\mathit{C}H_2C_6H_{13})(dmgH)_2py$	31.7	572
$Co\{\mathit{C}H_2(CH_2)_3CH{=}CH\}(dmgH)_2py$	31.6	572
$Co(\mathit{C}H_2CH_2Pr^i)(dmgH)_2py$	30.0	572
$Co(\mathit{C}H_2CH_2Ph)(dmgH)_2py$	31.6	572
$Co(\mathit{C}H_2CH_2CN)(dmgH)_2py$	19.3	572
$Co(\mathit{C}H_2CH_2CO_2Me)(dmgH)_2py$	21.3	572
$Co(\mathit{C}H_2C_6H_4X\text{-}4)(dmgH)_2py$	28.2 to 35	572
$Co(\mathit{C}H_2Ph)(dmgH)_2NH_2Ph$	29.0	572

*Isomers

COMPOUND	δ(^{13}C)/p.p.m., *J*/Hz	REFERENCE
Co(*C*H$_2$Ph)(dmgH)$_2$NH$_2$Bu^n	30.8	572
Co(*C*H$_2$C$_3$H$_5$)(dmgH)$_2$py	38.1	1489
HO$_2$C*C*H$_2$cobalamin	1.2 (pH 0) 13.1 (pH 5) 21.0 (pH 10) ($^1J_{CC}$ = 55 Hz)	385
EtO$_2$C*C*H$_2$cobalamin	2.0 (pH 0) 13.5 (pH 5)	385
Co(*C*HMe$_2$)(dmgH)$_2$py	37.6	572
Co(*C*HMeEt)(dmgH)$_2$py	45.0	572
Co(C$_6$H$_{11}$)(dmgH)$_2$py	50.0	572
Co$_2$(Me*C*P)(CO)$_6$	144.4 (258)	1497
Co$_2${μ-*C*H(CO$_2$Et)}Cp$_2$(CO)$_4$	102.1	1494
Co$_2${μ-*C*(CO$_2$Et)$_2$}Cp$_2$(CO)$_4$	124.1	1494
Rhodium, [^{103}Rh]		
Rh(*C*H$_3$){C(NHBu^t)NBuC(NHBu^t)}-(CNBu^t)$_2$I	-7.1 [20]	381
fac-Rh(*C*H$_3$)$_3$(PMe$_3$)$_3$	7.6 (100.0)	1373
Rh(*C*H$_3$)(dmgH)$_2$PBu$^n_3$	18.4 [16.9] (94.6)	1503
Rh(*C*H$_3$)(dmgH)$_2$P(OMe)$_3$	16.2 (132.4)	1503
Rh(*C*H$_3$)(dmgH)$_2$PPh$_3$	14.3 [19.1] (75.0)	1503
Rh(*C*H$_3$)(dmgH)$_2$AsPh$_3$	13.3 [22.1]	1503
Rh(*C*H$_3$)(dmgH)$_2$OH$_2$	12.1 [22.5]	1503
[Rh(*C*H$_3$)(dmgH)$_2$CN]$^-$	8.5 [19.1]	1503
[{Rh(*C*H$_3$)(dmgH)}$_2$CN]$^-$	7.1 [22.5]	1503
[Rh(*C*H$_3$)(dmgH)$_2$(S$_2$CNEt$_2$)]$^-$	5.8 [22.1]	1503
[Rh(*C*H$_3$)(dmgH)$_2$Br]$^-$	5.0 [23.5]	1503
[Rh(*C*H$_3$)(dmgH)$_2$OH]$^-$	1.4 [25.0]	1503
[Rh(*C*H$_3$)(dmgH)$_2$Cl]$^-$	1.4 [25.0]	1503
[Rh(*C*H$_3$)(dmgH)$_2$(NCO)]$^-$	0.5 [26.5]	1503
Rh(*C*H$_3$)(dmgH)$_2$(imidazole)	0.3 [22.1]	1503
Rh(*C*H$_3$)(dmgH)$_2$(NC$_5$H$_4$CN-4)	0.2 [22.1]	1503
Rh(*C*H$_3$)(dmgH)$_2$(NH$_2$Ph)	-0.2 [22.1]	1503
Rh(*C*H$_3$)(dmgH)$_2$(NC$_5$H$_5$)	-0.6 [23.5]	1503
[Rh(*C*H$_3$)(dmgH)$_2$(N$_3$)]$^-$	-2.2 [26.5]	1503
[Rh(*C*H$_3$)(dmgH)$_2$(NO$_2$)]$^-$	-3.3 [23.5]	1503
Rh{(*C*H$_2$)$_2$PBu$^t_2$}(cod)	-26.9 [19.5] (41.5)	1512
Rh$_2$(C$_7$H$_8$)Cp$_2$	-20.5 [11]	299
Rh$_2$(C$_8$H$_9$)Cp$_2$	17.9 [19]	299
[Rh$_2$(C$_8$H$_9$)Cp$_2$H]$^+$	25.7 [15]	299
[Rh(η^3-1,2,5-C$_8$H$_{13}$)Cp]$^+$	26.4 [28]	299
Rh$_2$(μ-*C*Ph$_2$)$_2$Cp$_2$	188.2 [42.7]	382
Rh$_2$(μ-*C*Ph$_2$)$_2$Cp$_2$(CO)	156.0 [23.3]	382
Rh$_2$(μ-*C*Ph$_2$)$_2$Cl$_2$py$_2$(CO)	185.2 [18.1]	382
Iridium		
fac-Ir(*C*H$_3$)$_3$(PEt$_3$)$_3$	-8.9	330
Ir(*C*H$_3$)ClI(CO)(AsMe$_2$Ph)$_2$	-12.3	120
Ir(*C*H$_3$)ClI(CO)(AsEt$_2$Ph)$_2$	-13.8	120

COMPOUND	$\delta(^{13}C)$/p.p.m., J/Hz	REFERENCE
$Ir\{\mathit{C}H_2CMe_2PBu^tC_6H_3OMeO\}$-$(PBu^tC_6H_3OMeO)$	-7.0 (23)	1540
$Ir(\mathit{C}H_2C_6H_3R)H\{P(OMe)_3\}_3$	1.5 to 3.0 (15 to 16; 88 to 94)	1521
$Ir(C^1H_2C^2Me{=}C^3H_2)Cl_2(CO)(AsMe_2Ph)_2$	C^1 10.6, C^2 145.2, C^3 107.9	118
Nickel		
$Ni(\mathit{C}H_2Me)(bipy)$	10.6 {132}	791
$Ni\mathit{C}H_2CH(CH_2)_4CH\mathit{C}H_2(PPh_3)_2$	4.0, 4.9*	1526
$Ni(\mathit{C}H_2CHMeCHPr^iN{=}CHCH{=}NCHPr^i{}_2)Br$	22.2	812
$Ni(\mathit{C}H_2PMe_3)(CO)_3$	1.5 (32.5) {123}	416
$Ni_2\{(\mathit{C}H_2)_2PMe_2\}_4$	-17.7 (46.4)	367
$Ni_2\{(C^1H_2)_2PMe_2\}_2\{\mu\text{-}(C^2H_2)_2PMe_2\}_2$	C^1 3.3 (33.0), C^2 -17.7 (45.3)	367
$Ni\{(\mathit{C}H_2PMe_2)_2N\}_2$	12.6 (34.2)	1039
$Ni(\mathit{C}H_2Me)(acac)(PPh_3)$	6.8 {125.4}	317
$Ni(C_6H_4Me_4)bipy$	33.4	978
Palladium		
trans-$Pd(\mathit{C}H_2Ph)Cl(PEt_3)_2$	15.0	1532
trans-$Pd(\mathit{C}H_2Ph)Br(PEt_3)_2$	17.6	1532
$[Pd\{(\mathit{C}H_2)_2PMe_2\}(\mathit{C}H_2PMe_3)_2]^+$	-2.1 (30.9), -18.8 (51.5)	1531
$Pd_2\{(\mathit{C}H_2)_2P(CH_2)_4\}_4$	-13.0 (41.2), 6.7 (36.8)	1531
$[Pd(\mathit{C}H_2C_6H_6MeR)Cl]_2$ R = H	52.5	1086
R = CO_2Me	50.8	1086
$[Pd(\mathit{C}H_2CHPhC_5Me_5)Cl]_2$	46.4	1177
$Pd(\mathit{C}H_2CHPhC_5Me_5)(acac)$	41.1	1177
$Pd(\eta^3$-2,3,5-norbornenyl)(hfac)	25.9	199
$[Pd(\eta^3$-2,3,5-6-phenylnorbornenyl)-$Cl]_2$	31.6	1013
$[Pd(\eta^3$-1,2,5-cyclooctenyl)$Cl]_2$	57.1 {140.8}	643
$Pd\{\mathit{C}H(CO_2Me)CHR^1CH{=}CHR^2\}L_2X$	25.3 to 30.5	694
$[Pd\{\mathit{C}H(CO_2Me)CR^1R^2CR^3{=}CR^4H\}X]_2$	-9.2 to 8.1	862
$[Pd\{7\text{-}\mathit{C}H(CO_2Me)\text{-}7\text{-}X\text{-norbornene}\}Cl]_2$	49.0, 56.0	300
Pd(nortricyclin)$Clpy_2$	39.0 or 43.9 or 50.2	1013
$Pd(\eta^3$-1,2,5-4-MeO_2C-cyclooctenyl)-(hfac)	49.0	199
$Pd(C_7H_7C_2H_3)$(hfac)	64.3	199
$Pd\{\mathit{C}H(COMe)PMe_2Ph\}_2Cl_2$	30.4 (53.0)	1667
$Pd\{\mathit{C}H(COMe)PMePh_2\}_2Cl_2$	28.0 (51.0)	1667
$Pd\{\mathit{C}H(COPh)PMe_2Ph\}_2Cl_2$	27.3 (56.0)	1667
$Pd\{\mathit{C}H(COMe)AsPh_3\}_2Cl_2$	41.1, 41.6	1665

COMPOUND	$\delta(^{13}C)$/p.p.m., J/Hz	REFERENCE
Platinum, [^{195}Pt]		
A. Platinum(IV)-Methyl		
$[Pt(CH_3)_6]^{2-}$	-10.6 [434]	1105
cis-$Pt(CH_3)_4(AsMe_2Ph)_2$	4.4 [613], -13.4 [413]	30
$[Pt(CH_3)_3I]_4$	13.2 [686] {138}	351, 1105
fac-$Pt(CH_3)_3Cl(AsMe_2Ph)_2$	8.1 [573]*, -9.4 [652]§	30
fac-$Pt(CH_3)_3Br(AsMe_2Ph)_2$	6.9 [567]*, -4.8 [646]§	30
fac-$Pt(CH_3)_3I(AsMe_2Ph)_2$	4.5 [562]*, 2.2 [625]§	30
fac-$Pt(CH_3)_3I(AsMe_3)_2$	2.0 [556]*, -0.7 [623]§	400
fac-$[Pt(CH_3)_3(NCC_6H_4OMe\text{-}4)\text{-}(AsMe_3)_2]^+$	3.5 [545]*, -13.0 [650]§	400
fac-$[Pt(CH_3)_3(NC_5H_4Me\text{-}4)\text{-}(AsMe_3)_2]^+$	8.3 [570]*, -14.4 [595]§	400
fac-$[Pt(CH_3)_3(AsPh_3)(AsMe_3)_2]^+$	2.4 [543]*, 4.9 [523]§	400
fac-$[Pt(CH_3)_3(CNEt)(AsMe_3)_2]^+$	-2.0 [515]*, -9.3 [515]§	400
fac-$[Pt(CH_3)_3(PMe_2Ph)(AsMe_3)_2]^+$	-0.9 [530]*, +0.3 [478] (103)§	400
fac-$[Pt(CH_3)_3(PMe_2Ph)_3]^+$	-7.4 [484] (5, 105)	400
fac-$Pt(CH_3)_3I(PMe_2Ph)_2$	2.9 [501]*, 5.8 [625]§	30
fac-$[Pt(CH_3)_3(py)(PMe_2Ph)_2]^+$	12.1 [522] (4, 108)*, -8.7 [98] (4)§	400
fac-$[Pt(CH_3)_3(CNEt)(PMe_2Ph)_2]^+$	0.5 [462] (6, 108)*, -4.3 [514] (7)§	400
fac-$Pt(CH_3)_3(sal{=}NMe)py$	-14.6 [756], -4.3 [688], -10.7 [706]	349
cis,trans-$Pt(CH_3)_2Br_2(NMe{=}CHC_6H_4O\text{-}2)$	-18.0 [559], -8.9 [506]	349
cis,trans-$Pt(CH_3)_2Br_2(NPh{=}CHC_6H_4O\text{-}2)$	-15.1 [559], -8.1 [521]	349
cis,trans-$Pt(CH_3)_2Br_2(O_2C_6H_4\text{-}2)$	-16.0 [559], -14.5 [588]	349
$Pt(CH_3)_2Br_2py_2$	-9.7 [512]	349
cis,cis,trans-$Pt(CH_3)_2(AsMe_3)_2\text{-}Cl(COMe)$	3.5 [583]	400
cis,cis,cis-$[Pt(CH_3)_2(PMe_2Ph)_2\text{-}(NCC_6H_4OMe)_2]^{2+}$	-2.3 [560] (5)	400
cis,cis,cis-$[Pt(CH_3)_2(PMe_2Ph)_2\text{-}(bipy)]^{2+}$	-4.3 [528] (4)	400
cis,cis-cis-$[Pt(CH_3)_2(PMe_2Ph)_2\text{-}(CNEt)_2]^{2+}$	-5.1 [428] (4)	400
B. Platinum(IV)-Alkyl		
$Pt(C^1H_2C^2H_3)_2I_2(4\text{-}Mepy)$	C^1 -2.4 [507], C^2 31.5 [<8]	351
$Pt(C^1H_2C^2H_2CH_2)Cl_2py_2$	C^1 -15.2 [335] {148}, C^2 30.0 [105] {135}	351
$Pt(C^1H_2C^2H_2CH_2)Cl_2(4\text{-}Mepy)$	C^1 3.8 [338], C^2 29.9 [105]	351

*intensity 2, §intensity 1

COMPOUND	$\delta(^{13}C)$/p.p.m., J/Hz	REFERENCE
$\overline{Pt(C^1H_2C^2H_2CH_2})Cl_2en$	C^1 -19.9 [332], C^2 29.6 [108]	351
$\overline{Pt(C^1H_2C^2H_2CH_2})Br_2py_2$	C^1 -17.9 [325], C^2 30.4 [110]	351
$\overline{Pt(C^1H_2C^2H_2CH_2})Br_2$(4-Mepy)	C^1 -18.5 [323], C^2 30.5 [103]	351
$\overline{Pt(C^1H_2C^2H_2CH_2})Br_2en$	C^1 -21.4 [317], C^2 30.4 [105]	351
$\overline{Pt(C^1H_2C^2HPhCH_2})Cl_2py_2$	C^1 -4.9 [369], C^2 48.1 [100.5]	767
$\overline{Pt(C^1H_2C^2HMeCH_2})Cl_2py_2$	C^1 1.0 [344], C^2 42.6 [98]	1548
$\overline{Pt(C^1H_2C^2HBuCH_2})Cl_2py_2$	C^1 -5.1 [344], C^2 43.4 [95]	1548
$\overline{Pt(C^1HMeC^2H_2C^3H_2})Cl_2py_2$	C^1 5.6, C^2 45.2, C^3 -8.0	1548
$\overline{Pt(C^1HPhC^2H_2C^3H_2})Cl_2py_2$	C^1 5.6 [326], C^2 35.1 [112], C^3 -11.3 [354]	767
$\overline{Pt(C^1HMeC^2HPhC^3H_2})Cl_2py_2$	C^1 10.6 [360], C^2 57.5 [100], C^3 -2.9 [370]	1548
$\overline{Pt(C^1HPhC^2HMeC^3H_2})Cl_2py_2$	C^1 16.9 [333], C^2 41.2 [103], C^3 -0.2 [364]	1548
$\overline{Pt(C^1HPhC^2HPhC^3H_2})Cl_2(4\text{-}Bu^tpy)$	C^1 15.1 [343], C^2 50.6 [105], C^3 -1.8 [377]	1548
$\overline{Pt(C^1Hp\text{-}tolC^2H_2CHp\text{-}tol})Cl_2$-(4-$Bu^tpy)_2$	C^1 6.3 [338], C^2 14.6 [125]	1548
$\overline{Pt(C^1Hp\text{-}tolC^2Hp\text{-}tolC^3H_2})Cl_2$-(4-$Bu^tpy)_2$	C^1 15.7 [341], C^2 50.5 [105], C^3 -1.6 [383]	1548
C. Platinum(II)-Methyl		
cis-Pt($CH_3)_2$(1,5-cod)	4.7 [773]	32, 351, 403
	4.8 [777]	573
cis-Pt(CH_3)Et(1,5-cod)	7.7 [828]	1538
cis-Pt($CH_3)_2$(nbd)	5.9 [816]	403
cis-Pt($CH_3)_2(PMe_2Ph)_2$	-3.3 [594]	29, 403
	3.3 [594]	31
cis-Pt($CH_3)_2(AsMe_3)_2$	-4.1 [689]	29
	-4.6 [668]	403
cis-Pt($CH_3)_2(AsMe_2Ph)_2$	-3.0 [685]	29
	3.0 [685]	403
trans-Pt($CH_3)_2(PEt_3)$-$\overline{CNMeCH_2CH_2NMe}$	-16.6 (9.8)	202
cis-Pt($CH_3)_2(PEt_3)\overline{CNMeCH_2CH_2NMe}$	11.6 *trans* carbene	202
	-18.9 (12.2) *trans* phosphine	202

COMPOUND	$\delta(^{13}C)$/p.p.m., *J*/Hz	REFERENCE
$Pt(CH_3)_2(diars)$	-8.8 [690]	402
$Pt_2(CH_3)_4(SEt_2)_2$	-6.5 [784]	777
$Pt_2(CH_3)_4(O_2CCF_3)_2(SEt_2)_2$	-10.3 [768, 19]	777
$Pt_2(CH_3)_4(OAc)_2(Mepy)$	-12.1 [784, 27]	777
$Pt_2(CH_3)_4(O_2CPr^i)_2(SEt_2)_2$	-12.6 [756, 20.6]	777
cis-$Pt(CH_3)_2(CNptol)_2$	-5.7 [590]	403
$Pt_2(CH_3)_4(O_2CCF_3)_2(SEt_2)$	-9.5 [735, 28]	1243
$Pt_2(CH_3)_4(O_2CCF_3)_2(picoline)$	-6.7 [680, 17], -9.6 [821, 15]	1243
$[Pt(CH_3)_2(OAc)(AsPh_3)]_2$	-12.2	1243
$[Pt(CH_3)_2(OAc)(SbPh_3)]_2$	-14.7	1243
$Pt(CH_3)(\sigma\text{-}Cp)(cod)$	6.2 [724]	790
$Pt(CH_3)(\eta^5\text{-}Cp)(CO)$	-44.1 [691]	790
$Pt(CH_3)(\eta^5\text{-}Cp)P(OMe)_3$	-47.8 [795]	790
trans-$Pt(CH_3)Ph(PEt_3)_2$	4.3	857
trans-$Pt(CH_3)Cl(PMe_2Ph)_2$†	-18.7 [673] (7)	203
trans-$Pt(CH_3)I(PMe_2Ph)_2$†	-7.3 [664] (6)	203
trans-$Pt(CH_3)(NCS)(PMe_2Ph)_2$†	-27.8 [632] (6)	203
trans-$Pt(CH_3)(NO_2)(PMe_2Ph)_2$†	-26.1 [564] (6)	203
trans-$Pt(CH_3)(CN)(PMe_2Ph)_2$†	-11.1 [495] (8)	203
trans-$[Pt(CH_3)(CO)(PMe_2Ph)_2]^+$	0.0 [509] (6)	203
trans-$[Pt(CH_3)(CNMe)(PMe_2Ph)_2]^+$	-7.1 [510] (7)	203
trans-$[Pt(CH_3)(\overline{CCH_2CH_2CH_2O})(PMe_2Ph)_2]^+$	-10.1 [395] (9)	203, 569
trans-$[Pt(CH_3)(C_2H_4)(PMe_2Ph)_2]^+$	5.6 [615] (6)	32, 203
trans-$[Pt(CH_3)(MeC_2Me)(PMe_2Ph)_2]^+$	-4.2 [632] (6)	32, 203
trans-$[Pt(CH_3)(NC_6H_4X\text{-}4)(PMe_2Ph)_2]^+$	-22.7 to -23.6 [589 to 600] (6)	203, 1313
trans-$[Pt(CH_3)(NCC_6H_4OMe\text{-}4)(PMe_2Ph)_2]^+$	-19.1 [652] (6)	203
$[Pt(CH_3)(PMe_2Ph)_3]^+$	0.7 [457] (75)	203
trans-$[Pt(CH_3)(AsPh_3)(PMe_2Ph)_2]^+$	0.0 [548] (6)	203
trans-$[Pt(CH_3)(OCHNMe_2)(PMe_2Ph)_2]^+$	-27.8 [698] (5)	203
trans-$PtMeCl(AsMe_3)_2$	-28.4 [643]	31, 203
trans-$[Pt(CH_3)(CO)(AsMe_3)_2]^+$	-6.8 [470]	31, 203
trans-$[Pt(CH_3)(CNMe)(AsMe_3)_2]^+$*	-14.0 [475]	31, 203
trans-$[Pt(CH_3)(CMeNH_2)(AsMe_3)_2]^+$	-18.6 [380]	569
trans-$[Pt(CH_3)(CMeNHMe)(AsMe_3)_2]^+$§	-17.6 [381]	569
trans-$[Pt(CH_3)(CMeNMe_2)(AsMe_3)_2]^+$	-19.8 [385]	569
trans-$[Pt(CH_3)(CMeNHC_6H_4X\text{-}4)(AsMe_3)_2]^+$	-18.9 to -19.3 [358 to 381]	580
trans-$[Pt(CH_3)(CMeOMe)(AsMe_3)_2]^+$	-15.9 [360]	31, 569
trans-$[Pt(CH_3)(CMeOEt)(AsMe_3)_2]^+$	-16.5 [354]	569
trans-$[Pt(CH_3)(\overline{CCH_2CH_2CH_2O})(AsMe_3)_2]^+$	-15.9 [366]	569
trans-$[Pt(CH_3)(NC_5H_4Me\text{-}4)(AsMe_3)_2]^+$	-30.8 [566]	203
trans-$[Pt(CH_3)(NC_5H_4OMe\text{-}4)(AsMe_3)_2]^+$	-26.9 [613]	203

*also CNEt, §also CMeNHEt,
† $^1J_{CH}$ also reported for *trans*-$PtMeX(PPh_3)_2$, {129.5 to 133.0}[40]

COMPOUND	$\delta(^{13}C)$/p.p.m., *J*/Hz	REFERENCE
trans-$[Pt(\mathit{C}H_3)(CNC_6F_5)(AsMe_3)_2]^+$	-25.5 [616]	31, 203
trans-$[Pt(\mathit{C}H_3)(AsPh_3)(AsMe_3)_2]^+$	-6.8 [512]	203
$Pt(\mathit{C}H_3)Cl(cod)$	4.5 [620]	573
$[Pt(\mathit{C}H_3)(CNEt)(cod)]^+$	-2.7 [538]	573
$[Pt(\mathit{C}H_3)(NCC_6H_4OMe\text{-}4)(cod)]^+$	1.8 [575]	573
$[Pt(\mathit{C}H_3)(NC_5H_4Me\text{-}4)(cod)]^+$	5.7 [640]	573
$[Pt(\mathit{C}H_3)(PPh_3)(cod)]^+$	5.5 [592] (5)	573
$[Pt(\mathit{C}H_3)(AsPh_3)(cod)]^+$	2.4 [560]	573
$Pt(\mathit{C}H_3)(HBpz_3)(C_2H_4)$	-16.9 [628]	866
$Pt(\mathit{C}H_3)(HBpz_3)(C_2H_3CN)$*	-9.2 [596], -12.2 [576]	866
$Pt(\mathit{C}H_3)(HBpz_3)(NCCH{=}CHCN)$	-4.6 [550]	866
$Pt(\mathit{C}H_3)(HBpz_3)(C_2H_3CO_2Me)$	-13.1 [607]	866
$Pt(\mathit{C}H_3)(HBpz_3)(EtO_2CCH{=}CHCO_2Et)$	-8.6 [572]	866
$Pt(\mathit{C}H_3)(HBpz_3)$(maleic anhydride)	-6.8 [571]	866
$Pt(\mathit{C}H_3)(HBpz_3)(C_2F_4)$	-5.7 [615]	866
$Pt(\mathit{C}H_3)(HBpz_3)(CNBu^t)$	-22.2 [606]	667
$Pt(\mathit{C}H_3)(HBpz_3)P(OMe)_3$	-20.4 [610] (12)	667
$Pt(\mathit{C}H_3)(diars)Cl$	-5.4 [555]	402
$[Pt(\mathit{C}H_3)(diars)CO]^+$	-14.1 [453]	402
$[Pt(\mathit{C}H_3)(diars)(CNEt)]^+$	-15.6 [500]	402
$[Pt(\mathit{C}H_3)(diars)(NCC_6H_4OMe\text{-}4)]^+$	-8.5 [523]	402
$[Pt(\mathit{C}H_3)(diars)(NC_5H_4Me\text{-}4)]^+$	-4.0 [575]	402
$[Pt(\mathit{C}H_3)(diars)(PPh_3)]^+$	-8.0 [533]	402
$[Pt(\mathit{C}H_3)(diars)(AsPh_3)]^+$	-10.9 [508]	402
$[Pt(\mathit{C}H_3)Cl_2(CO)]^-$	-15.5 [568]	402
cis-$Pt_2(\mathit{C}H_3)_2(\mu\text{-}Cl)_2(PMe_2Ph)_2$	-13.7 [708] (30)	1166
D. Platinum(II)-Alkyl		
$Pt(C^1H_2C^2H_3)_2(cod)$	C^1 19.4 [843], C^2 15.7 [32]	573
$Pt(C^1H_2C^2H_3)(Me)(cod)$	C^1 19.2 [833], C^2 15.7 [35]	351
$[Pt(C^1H_2C^2H_3)Cl_2(CO)]^-$	C^1 3.0 [568], C^2 18.8 [14]	1562
$[Pt(C^1H_2C^2H_2C^3H_3)Cl_2(CO)]^-$	C^1 13.3 [577], C^2 26.9 [11], C^3 17.8 [79]	1562
$[Pt(C^1H_2C^2H_2C^3HMe)Cl_2(CO)]^-$	C^1 10.7 [576], C^2 36.1 [12], C^3 26.4 [79]	1562
$Pt(C^1H_2C^2H{=}C^3H_2)Br(PEt_3)_2$	C^1 4.8, C^2 144.2, C^3 108.3	1279
$Pt(\mathit{C}H_2Ph)_2(cod)$	33.3 [740]	573
$Pt(\mathit{C}H_2Ph)Cl(cod)$	28.7 [577]	573
$\overline{Pt\{C^1H_2C^2(C^3H_3){=}C(CH_3)CH_2}\}(cod)$	C^1 40.7 [787.4], C^2 136.4 [0], C^3 20.1 [111.4]	720
$\overline{Pt\{C^1H_2C^2(C^3H_3){=}C(CH_3)CH_2}\}$-$(CNBu^t)_2$	C^1 30.9 [608.8], C^2 136.2 [46.2], C^3 21.4 [96.1]	720

*two isomers

COMPOUND	$\delta(^{13}C)$/p.p.m., *J*/Hz	REFERENCE
$Pt\{C^1H_2C^2H{=}C^3H(CH_2)_2C_3H_4\}PMe_3$	C^1 4.4 [640.4] (4.9), C^2 134.2 [78.1] (2.0), C^3 112.6 [59.6]	720
$\overline{Pt(CH_2C_9H_6N})(PPh_3)Br$	20.5 [741] (5.5)	564
$[Pt\{C^1H(C^2H_3)_2\}Cl_2(CO)]^-$	C^1 20.7 [592], C^2 28.8 [4]	1562
$Pt(\eta^3\text{-}1,2,5\text{-}C_8H_{13})(hfac)$	27.6 [711]	1200
$Pt(\eta^3\text{-}1,2,5\text{-}6\text{-}AcO\text{-}C_8H_{12})(hfac)$	25.9 [730]	199
$[Pt(\eta^3\text{-}1,2,5\text{-}C_8H_{13})(cod)]^+$	46.1 [543]	1200
$Pt(\sigma\text{-}C_5H_5)Me(cod)$	115.2 [87.8] fluxional	790
$Pt(\sigma\text{-}C_5H_5)_2(cod)$	116.5 [96.8] fluxional	790
$Pt(\eta^3\text{-}2,3,5\text{-norbornenyl})(hfac)$	0.3 [470]	199
Pt(2-allyl-3-norbornyl)(hfac)	37.4 [758]	199
$Pt(AcOC_{10}H_{12})(hfac)$	32.8 [730]	199
$\overline{Pt\{CH(CH{=}CH_2)CH_2CH_2CH}(CH{=}CH_2)\}$-(cod)	56.6 [76.3]	1543
$\overline{Pt(CHMeC_6H_4P}Bu^t_2)(acac)$	20.6 [785]	240
$Pt\{OC(CF_3)_2C_8H_{12}\}(cod)$	48.4 [709.5]	1174
$Pt\{C(CN)_2C(CF_3)_2C_8H_{12}\}(cod)$	42.7 [677.5]	1174
$\overline{Pt\{C(CN)_2CMe_2O}_2\}PPh_3$	119.0 [70] (10)	581
Silver, [^{109}*Ag*]		
$Ag(CH_2PPh_3)_2Cl$	4.4 [96.5]	520
Gold		
$Au\{CH(SiMe_3)_2\}PEt_3$	22.9 (65)	1592
$Au\{CH(SiMe_3)_2\}PPh_3$	22.8 (65)	1592
$Au\{CH(SiMe_3)_2\}AsPh_3$	21.1	1592
$Au\{C(SiMe_3)_3\}PEt_3$	7.3 (51)	1592
$Au\{C(SiMe_3)_3\}PPh_3$	7.8 (52)	1592
$Au\{C(SiMe_3)_3\}AsPh_3$	22.2	1592
Zinc		
$Zn(CH_3)_2$	-4.2	163
$Zn(C^1H_2C^2H_3)_2$	C^1 6.8, C^2 10.4	1239
$Zn(C^1H_2C^2H_3)_2$.dme	C^1 4.6, C^2 11.4	1239
$Zn(CH_2Pr^n)_2$	16.1	1557
$Zn(CH_2Pr^n)_2$bipy	12.4	1557
$Zn(CH_2Pr^n)_2$.tmed	11.1	1557
$Zn(C^1H_2C^2H_2CH{=}CH_2)_2$	C^1 14.9, C^2 30.6	1155
$Zn(C^1H_2C^2H_2CH_2CH{=}CH_2)_2$	C^1 13.9, C^2 27.1	1155
$Zn(C^1H_2C^2H_2CH_2CH{=}CH_2)_2$.bipy	C^1 12.3, C^2 30.3	1155
$Zn(C^1H_2C^2H_2CH_2CH_2CH{=}CH_2)_2$	16.1, 26.3	1155
$Zn\{CH_2(CH_2)_2OMe\}_2$	10.5	1557
$Zn\{CH_2(CH_2)_2OMe\}_2$.bipy	5.4	1557
$Zn\{CH_2(CH_2)_2OMe\}_2$.tmed	5.6	1557
$Zn\{CH_2(CH_2)_3OMe\}_2$	14.0	1557
$Zn\{CH_2(CH_2)_3OMe\}_2$.bipy	12.4	1557
$Zn\{CH_2(CH_2)_3OMe\}_2$.tmed	11.1	1557
$Zn\{CH_2(CH_2)_2SMe\}_2$	13.5	1557

COMPOUND	$\delta(^{13}C)$/p.p.m., J/Hz	REFERENCE
$Zn\{CH_2(CH_2)_2SMe\}_2$.bipy	7.8	1557
$Zn\{CH_2(CH_2)_2SMe\}_2$.tmed	10.3	1557
$Zn\{CH_2(CH_2)_2NMe_2\}_2$	11.9	1557
$Zn\{CH_2(CH_2)_2NMe_2\}_2$.bipy	4.0	1557
$Zn\{CH_2(CH_2)_2NMe_2\}_2$.tmed	4.1	1557
$Zn(CH_2SiMe_3)_2$	3.6	1555
$Zn\{(CH_2PMe_2)_2CH\}_2$	5.1 (N_{PC} = 48.8 Hz)	1041
$Zn\{C^1(C^2H_3)_3\}_2$	C^1 28.4, C^2 31.0	1239
Cadmium, [^{113}Cd]		
$Cd(CH_3)_2$	1.0 [537.5], {126.6}	163
$Cd(C^1H_2C^2H_3)_2$	C^1 12.5 [519.3], C^2 12.5 [17.5]	1239
$Cd(CH_2SiMe_3)_2$	7.5 [420]	1555
$Cd\{(CH_2PMe_2)_2BH_2\}_2$	1.2 [281]	1556
$Cd\{(CH_2PMe_2)_2CH\}_2$	2.6 (N_{PC} = 48.8 Hz)	757, 1041
$Cd\{(CH_2PMe_2)_2N\}_2$	7.1 [271] (N_{PC} = 48.8 Hz)	1039
Mercury, [^{199}Hg]		
$Hg(CH_3)_2$	22.1 to 23.7 [687 to 692] {129.6, -1.8} (J_{CC} = +22.4)	46, 49, 112, 125, 162, 163, 246, 861, 1562, 1565, 1719
$Hg(CH_3)(C_5H_5)$	15.6	80
$Hg(CH_3)CN$	4.7, 6.0 [1235] {132}	861, 916
$Hg(CH_3)C(N_2)CO_2Et$	9.5 [1146]	1709
$\{Hg(CH_3)\}_2CN_2$	12.4 [1061]	1709
$Hg(CH_3)C(SiMe_3)_3$	20.5 [799]	1270
$Hg(CH_3)SiMe_3$	40.1	1565
$Hg(CH_3)GeMe_3$	36.7	1565
$[Hg(CH_3)L]^+$ L = pyridine	4.7	687
various pyridines	1.7 to 7.5	687
glycine	-1.3 [1588]	861
L-2-phenyl-alanine	-2.7 [1583]	861
tyrosine	-3.0 [1591]	861
methionine	5.8 [1601.0]	861
selenomethio-nine	8.6 [1510.2]	861
cysteine	9.8 [1257]	861
thioglycollic-acid	9.6 [1270]	861
penicillamine	8.4 [1336]	861
dmso	1.4 [1937]	861
OH_2	0.3 [1764]	861
PPh_3	1.2 [1710]	861
tetrahydro-thiophene	0.7 [1690]	861

COMPOUND		$\delta(^{13}C)$/p.p.m., J/Hz	REFERENCE
$Hg(\mathit{C}H_3)X$	X = NO_3*	5.9 to -3.7 [1654 to 1954] {139.0 to 146.4}	687, 916
	OAc	0.8 [1695]	861
	OHgMe	-0.8 [1309]	861
	methionato	0.9	1167
	salicylaldehydato	-2.1 [1559]	1212
	X = quinolinato	-3.9 [1690]	1212
	SCN	0.0 [1710]	861
	Cl	8.4 [1674], 6.0 [1665]	861, 943
	Br	11.7 [1631], 1.1 [1530] {135.6}§, 9.5 [1625]	861, 916, 943
	I	17.1 [1540], 14.9 [1543]	861, 943

COMPOUND	C^1	C^2	C^3	REFERENCE
$Hg(C^1H_2C^2H_3)_2$	36.0 [642]	13.0 [25]		162, 246
	35.6	13.4 [24]		1562, 1565
	36.1 [647]			
	35.6 [648]	12.9 [25]		1719
$Hg(C^1H_2Me)SiMe_3$	17.2			1565
$Hg(C^1H_2Me)GeMe_3$	50.5			1565
$Hg(C^1H_2C^2H_3)Cl$	22.6 [1689]	13.7 [95.5]		943
	25.1 [1474]	13.9 [92]		1562
$Hg(C^1H_2C^2H_3)Br$	28.8 [1414]	13.9 [89]		1562
$Hg(C^1H_2C^2H_3)I$	34.8	14.2		1562
$Hg(C^1H_2C^2H_2C^3H_3)_2$	47.3			162, 571
	47.2 [660]	22.3 [26]	19.9 [102]	1562, 1565
$Hg(C^1H_2Et)SiMe_3$	63.8			1565
$Hg(C^1H_2Et)GeMe_3$	60.6			1565
$Hg(C^1H_2C^2H_2C^3H_3)Cl$	35.8 [1469]	22.2 [87]	19.1 [187]	1562
$Hg(C^1H_2C^2H_2C^3H_3)Br$	39.6 [1407]	22.4 [85]	19.1 [182]	1562
$Hg(C^1H_2C^2H_2C^3H_2Me)_2$	44.0			1565
	44.2 [659]	31.1 [27]	28.4 [100]	571, 1562
	[+656]	[-26.3]	[100]	161, 162
$Hg(C^1H_2Pr^n)SiMe_3$	60.7			1565
$Hg(C^1H_2Pr^n)GeMe_3$	57.7			1565
$Hg(C^1H_2C^2H_2C^3H_2Me)Cl$	33.2 [1444]	30.6 [86]	27.9 [177]	346, 1562
$Hg(C^1H_2C^2H_2C^3H_2Me)Br$	36.9 [1401]	30.9 [82]	27.8 [175]	1562

*solvent dependent, §in pyridine

COMPOUND	$\delta(^{13}C)$/p.p.m., J/Hz			REFERENCE
	C^1	C^2	C^3	
$Hg(C^1H_2C^2H_2C^3H_2Me)I$	42.3	31.3	27.6	1562
$Hg\{C^1H_2C^2H(C^3H_3)_2\}_2$	56.4	29.1	28.1	571
	[671]	[29]	[82]	
$Hg(C^1H_2C^2H_2C^3H_2Et)_2$	44.3	28.4	37.8	571
	[656]	[26]	[95]	
$Hg(C^1H_2C^2H_2C^3H_2Pr^n)_2$	44.3	28.7	35.1	571
	[659]	[19]	[95]	
$Hg(C^1H_2C^2H_2C^3H_2Bu^n)_2$	44.3	28.7	35.4	571
	[654]	[30]	[94]	
$Hg(C^1H_2C^2H_2C^3H{=}CH_2)_2$	42.8	33.1	143.1	1155
	[688]	[27]	[81]	
$Hg(C^1H_2C^2H_2C^3H_2CH{=}CH_2)_2$	43.0	28.9	39.3	1155
	[680]	[30]	[89]	
$Hg(C^1H_2C^2H_2C^3H_2C_3H_5)_2$	44.7	28.9	35.4	1155
	[654]	[26]	[100]	
$Hg(C^1H_2C^2H_2OMe)Cl$	[1607]	[91]		425
$Hg(C^1H_2C^2H_2OH)(O_2CCF_3)$	23.7	74.8		1263
$[Hg(O_2CCF_3)(C^1H_2C^2H_2O-)]_2$	23.9	72.8		1263
	[1710]			
$[HgCl(C^1H_2C^2H_2O-)]_2$	25.2	73.6		1263
		[43]		
$Hg(C^1H_2C^2HMeOH)(O_2CCF_3)$	31.3	80.1		1263
$[Hg(O_2CCF_3)(C^1H_2C^2HMeO-)]_2$	31.3*	78.8		1263
	[1699]	[137]		
	31.6*	78.7		
	[1699]			
$Hg(C^1H_2C^2HPhOH)(OAc)$	29.2	85.9		1263
	[1780]	[90]		
$Hg(C^1H_2C^2HPhOH)(O_2CCF_3)$	30.7	85.6		1263
$Hg(C^1H_2C^2HPhOH)Cl$	34.4	86.1		1263
$[Hg(O_2CCF_3)(C^1H_2CHPhO-)]_2$	30.7	84.2		1263
$Hg\{C^1H_2C^2H(C^3H_3)OH\}Cl$	[1630]	[108]	[132]	425
$Hg\{C^1H_2C^2H(C^3H_2Me)OH\}Cl$	[1635]	[106]	[118]	425
$Hg\{C^1H_2C^2H(CH_3)OMe\}Cl$	[1606]	[109]		1236
	{136}	{134}		
$(C_3H_4O_2)(CH_2HgNO_3)_2$	26.5, 26.7§			1559
$(C_3H_4O_2)(CH_2HgCl)_2$	31.6 [1783]			1559
$(C_4H_6O_2)(CH_2HgNO_3)_2$	25.9, 28.1§			1559
$(C_4H_6O_2)(CH_2HgCl)_2$	28.8, 30.8 [1757]			1559
$Hg(C^1H_2C^2H_2C^3H_2Ph)_2$	41.8	40.3	30.6	571
	[690]	[27]	[155]	
1,7-dimercuracyclododecane	44.8	28.6	42.0	571
	[672]	[31]	[79]	
$Hg(C^1H_2C^2H_2Ph)_2$	44.9	34.7		571
	[709]	[26]		
$Hg(CH_2CH{=}CH_2)_2$	45.5 [626]			1726
$Hg(C^1H_2C^2H{=}C^3H_2)(OAc)$	28.6	135.1	115.1	1566
	[1461]	[192]	[222]	
$Hg(CH_2CH{=}CHCH_2Bu^t\text{-}cis)_2$	39.6			1720

*enantiomers, §*cis*, *trans* isomers

COMPOUND	$\delta(^{13}C)$/p.p.m., J/Hz			REFERENCE
	C^1	C^2	C^3	
$Hg(\mathit{C}H_2CH{=}CHCH_2Bu^t\text{-}\mathit{trans})_2$	43.7			1720
$Hg(\mathit{C}H_2CMe{=}CHCH_2Bu^t\text{-}\mathit{cis})_2$	45.1			1720
$Hg(CH_2CMe{=}CHCH_2Bu^t\text{-}\mathit{trans})_2$	52.8			1720
$Hg(\mathit{C}H_2Ph)_2$	46.0, 46.6 [632], 47.0			166, 246, 405, 571, 1332
	{135.0} [620 to 698]*			1073
$Hg(\mathit{C}H_2Ph)Cl$	34.9			405, 459
$Hg\{C^1H_2C^2(C^3H_3)_3\}_2$	63.0 [690] {126.0}	33.9 [29]	35.7 [72]	150, 458
$Hg\{C^1H_2C^2(C^3H_3)_3\}Me$	62.6 [690] {126.5}	33.4 [29.5]	35.3 [71.0]	150
$Hg\{C^1H_2C^2(C^3H_3)_3\}(CH{=}CH_2)$	55.8 [820] {127.5}	33.8 [31.0]	35.7 [75.0]	150
$Hg\{C^1H_2C^2(C^3H_3)_3\}CN$	47.7§ [1374] [1404] {134}	33.3 [51.0]	34.8 [120.0]	150, 458
$Hg\{C^1H_2C^2(C^3H_3)_3\}Cl$	52.5§ [1472] [1514] {137}	35.5 [70.0]	34.3 [149.0]	150, 458
$Hg\{C^1H_2C^2(C^3H_3)_3\}Br$	56.6§ {137}	33.9 [69.5]	34.4 [148.0]	150
$Hg\{C^1H_2C^2(C^3H_3)_3\}OAc$	45.3§ [1546]	33.1 [71.0]	34.0 [150.0]	150, 458
$Hg\{C^1H_2C^2(C^3H_3)_3\}NO_3$	47.0§ {138}	32.8 [85.0]	33.3 [170.0]	150
$Hg(\mathit{C}H_2CF_3)_2$	41.1 [896]	39.4 [1212]		246, 1719
$Hg(\mathit{C}H_2CO_2Me)_2$	88.1 [753]			246
$Hg(C^1H_2C^2Me_2OH)(O_2CCF_3)$	37.6	83.5 [127]		1263
$[Hg(O_2CCF_3)(C^1H_2C^2Me_2O\text{-})]_2$	38.2	82.5		1263
$Hg(O_2CCF_3)(C^1H_2C^2MePhOH)$	38.0	86.4 [121]		1263
$[Hg(O_2CCF_3)(C^1H_2C^2MePhO\text{-})]_2$	39.1† 39.4†	85.4		1263
$Hg(O_2CCF_3)(C^1H_2C^2Me_2O$ $Hg(O_2CCF_3)(C^4H_2C^3HPhO$	38.5 C^4 31.0	83.6 C^3 84.4		1263
$Hg\{C^1H_2C^2(C^3H_3)_2OMe\}Cl$	47.3 [1642] {137}	75.3 [103]	28.0 [142]	425, 426
$Hg\{C^1H_2C^2(C^3H_3)_2OMe\}Br$	51.5 [1579] {136}	76.0 [103]	28.9 [142]	426

*solvent dependent, §concentration dependent, †enantiomers

COMPOUND	$\delta(^{13}C)$/p.p.m., J/Hz			REFERENCE
	C^1	C^2	C^3	
$Hg\{C^1H_2C^2(C^3H_3)_2OMe\}I$	55.7 [1504] {136}	76.1 [102]	28.7 [134]	426
$Hg\{C^1H_2C^2(C^3H_3)_2OMe\}CN$	44.7 [1452] {134}	77.3 [63]	29.3 [107]	426
$Hg\{C^1H_2C^2(C^3H_3)_2OMe\}OAc$	40.3 [1705] {136}	75.7 [106]	28.2 [142]	426
$Hg\{C^1H_2C^2(C^3H_3)_2OMe\}SCN$	50.8 [1524] {136}	76.0 [103]	28.0 [138]	426
$[Hg(C^1H_2C^2Ph_2)(O_2CCF_3)]^+$	141.9	226.6		424
$Hg(\mathit{C}H_2CHO)_2$	50.3 [768]			1726
$Hg(CH_2CEtO)_2$	48.4 [742]			1726
$Hg(CH_2CPhO)_2$	45.2 [753]			1726
$Hg(\mathit{C}H_2SiMe_3)_2$	28.3 [545.6]			1270, 1555
$Hg(CH_2SiMe_3)Cl$	22.3 [1173.6]			1270
$Hg(CH_2SiMe_3)Br$	22.1 [1108.1]			1270
$Hg(CH_2GeMe_3)_2$	30.1 [600.1]			1270
$Hg(CH_2GeMe_3)Cl$	18.0 [1240.0]			1270
$Hg(CH_2Cl)SiMe_3$	83.8			1565
$Hg(CH_2Cl)GeMe_3$	79.5			1565
$Hg\{C^1H(C^2H_3)_2\}_2$	49.7 49.2 [636]	23.4 [27]		162, 1562, 1565
	32.2 [635]			1719
$Hg\{C^1H(C^2H_3)_2\}Cl$	43.1 [1555]	24.0 [62]		1562
$Hg\{C^1H(C^2H_3)_2\}Br$	46.6 [1528]	24.2 [64]		1562
$Hg\{\mathit{C}H(CH_3)_2\}SiMe_3$	66.4			1565
$Hg\{\mathit{C}H(CH_3)_2\}GeMe_3$	63.7			1565
$Hg\{C^1H(C^4H_3)C^2H_2C^3H_2Me\}_2$	55.9 [649]	40.8 [29] C^4 20.6 [30]	24.4 [70]	571
$Hg(C_5H_5)_2$	{160} (averaged)			128
$Hg(C_5H_5)Cl$	117.7 (averaged)			80
Hg(sugar)Cl HgCH<	49.2 to 56.2 [1825 to 1879.7]*			1323
Hg(*anti*-2-norbornyl)Cl	58.8 [1502]			943
Hg(*syn*-2-norbornyl)Cl	60.5 [1710]			943
Hg(norbornylderivative)OAc	48.7 to 56.3 [1660 to 1798]			1162, 1564
Hg(norbornylderivative)Cl	56.6, 61.7 [1864, 1938]			1162
Hg(cyclohexyl)XL	44.0 to 62.0 [1387 to 1719]			270, 1240

*$^2J_{HgC}$ and $^3J_{HgC}$ are useful tools in structure determination

COMPOUND	$\delta(^{13}C)$/p.p.m., J/Hz			REFERENCE
	C^1	C^2	C^3	
$Hg(C_6H_9O)_2$	62.0 [697]			1726
$Hg(C_6H_9O)Cl$	58.3 [1613]			1726
$Hg\{$7-norbornadiene-2,3-$(CO_2Me)_2\}$	109.5			638
$Hg\{C^1H(C^4H_3)C^2H(C^3H_3)OMe\}Cl$				
erythro	[1705]	[98]	[107]	425
		C^4 [76]		
threo	[1705]	[117]	[98]	425
		C^4 [68]		
$Hg\{C^1H(C^{2'}H_3)C^2(C^3H_3)_2OMe\}Cl$	[1724]	[113,53]	[98]	425
$Hg\{\mathit{C}H(CO_2Et)_2\}_2$	53.5 [917]			975
$Hg\{\mathit{C}H(CF_3)_2\}_2$	56.4 [1212]			1719
$Hg\{\mathit{C}H(SiMe_3)_2\}_2$	35.5 [423.0]			1270
$Hg\{\mathit{C}H(SiMe_3)_2\}Cl$	30.0 [867.7]			1270
$Hg\{\mathit{C}H(SiMe_3)_2\}OAc$	24.8 [905.6]			1270
$Hg\{C^1(C^2H_3)_3\}_2$	60.3	31.5		1239, 1565
	59.5	[29.2]		
	59.4	31.3		1562
	[637]	[28]		
$Hg\{C^1(C^2H_3)_3\}Cl$	57.9	32.4		1562
	[1619]	[17]		
$Hg(\mathit{C}Me_3)SiMe_3$	75.5			1565
$Hg(\mathit{C}Me_3)GeMe_3$	73.8			1565
$Hg\{C^1(C^{2'}H_3)_2C^2(C^3H_3)_2OMe\}Cl$	[1814]	[97,76]	[143]	425
$Hg(C_7H_7Me_2O)_2$	81.5 [1068]			1726
$Hg(C_7H_7Me_2O)Cl$	73.7 [2507]			1726
$Hg\{\mathit{C}(SiMe_3)_3\}_2$	38.7 [334.0]			1270
$Hg\{\mathit{C}(SiMe_3)_3\}Me$	34.0 [264.4]			1270
$Hg\{\mathit{C}(SiMe_3)_3\}Cl$	39.4 [786.7]			1270
$(ClHg)_2\mathit{C}(CO_2Et)CN$	31.4 [852]			975
$Hg(\mathit{C}F_3)_2$	160.5 [2098]			1719
$Hg(\mathit{C}F_2CF_3)_2$	145.7 [2149]			1719
$Hg(\mathit{C}ClFCF_3)_2$	127.1 [1997]			1719
$Hg\{\mathit{C}F(CF_3)_2\}_2$	113.4 [1684]			1719
Boron, $[^{11}B]$				
$[B(\mathit{C}H_3)_4]^-$	6.2, 7 [22] {110}			262, 566
$[B(\mathit{C}H_3)_3hexyl]^-$	17.1			278
$[B(\mathit{C}H_3)Bu^nC_8H_{14}]^-$	5.6			1707
$B(\mathit{C}H_3)_3$	11 [45] {114}, 14.8 [52], [46.7], 18.4 [+47]			262, 566, 549, 1046
$B(\mathit{C}H_3)_3.NMe_3$	8 [35] {111}, 7.2 [35], 7.2 [+36]			262, 566, 1046
$B(\mathit{C}H_3)_3.NH_2Pr^i$	12.2 [36]			566
1,1-$B_2(\mathit{C}H_3)_2H_4$	[61.3]			549
$B(\mathit{C}H_3)_2Et$	11.3 [<58]			1002
$B(\mathit{C}H_3)_2C_8H_{13}$	12			1228
$B(\mathit{C}H_3)_2\overline{C{=}CBu^tSnMe_2CBu^tC}Me$	18.4			1625
$B(\mathit{C}H_3)_2NMe_2$	4.0 [54], 3.0			566, 669

COMPOUND	δ(^{13}C)/p.p.m., *J*/Hz	REFERENCE
$B(CH_3)_2NC(CF_3)_2$	3.2	774
$MeB\{PMe_2B(CH_3)_2\}_2PMe_2$	4.0 [42]	670
$B(CH_3)_2OMe$	6.3 [64]	566
$B(CH_3)_2SMe$	9.9 [50]	566
$1\text{-}(CH_3)B_5H_9$	12.2 [72.6], [72.7], 11.6 [74]	53, 161, 548, 549
$2\text{-}(CH_3)B_5H_9$	20.5 [64]	548
$1,2\text{-}(CH_3)_2B_5H_8$	10.9 [73], 18.8 [63]	548
$B(CH_3)Et_2$	7.7	1002
$B(CH_3)EtC_8H_{13}$	19	1228
$B(CH_3)PrC_8H_{13}$	5	1228
$B(CH_3)Bu^tC_8H_{13}$	9	1228
$B(CH_3)Bu^tCBu^t{=}CHSnMe_3$	14.5	1730
$B(CH_3)Bu^nC_8H_{13}$	8.5	1228
$B(CH_3)C_8H_{14}$	13.6	1707
$Tl[C_5H_5BCH_3]$	4.1	1576
$\{B(CH_3)\}_6(CH)_6$	11.6	213
$B(CH_3)(NMe_2)CH_2CO_2Me$	2.5	1038
$B(CH_3)(NMe_2)OC({=}CH_2)OMe$	2.5	1038
$B(CH_3)(NMe_2)_2$	-1.0 [+59]	547, 566
$\overline{B(CH_3)NMeCH_2CH_2NMe}$	-2.0 [+62]	1046
$B(CH_3)(OMe)NMe_2$	-3.0 [68]	566
$B(CH_3)(SMe)NMe_2$	2.7 [60]	566
$B(CH_3)(OMe)_2$	-2.0 [76]	566
$B(CH_3)(SMe)_2$	-7.2 [64]	566
$B(CH_3)(SeMe)_2$	9.2 [54]	566
$B(C^1H_2C^2H_3)_3$	C¹ 19.7, 20.8 [<52]; C² 7.9	278, 999, 549, 567
$[B(C^1H_2C^2H_3)_3hexyl]^-$	C¹ 18.5; 20.2, C² 11.7, 13.2	278, 724
$B(C^1H_2C^2H_3)_2Me$	C¹ 19.7 [<47], C² 6.1	1002
$B(C^1H_2C^2H_3)_2(C_4H_2Pr^iSnEt_2)$	C¹ 18.9, C² 9.7	999
$B(C^1H_2C^2H_3)_2\{C_4(SiMe_3)_2\text{-}EtSnMe_2\}$	C¹ 23.4	1625
$B(C^1H_2CH_3)_2\{C_4MeBu^tEtSnMe_2)\}$	C¹ 23.5	1625
$B(CH_2Me)_2CMe(SnMe_3)CEt{=}C{=}C\text{-}(SnMe_3)_2$	15.4	1631
$B(CH_2Me)_2(N_2C_3H_3)$	13.0	1569
$[B(CH_2Me)_2(N_2C_3H_3)]_2$	17.3	1718
$[B(CH_2Me)_2(N_2C_3HMe_2)]_2$	9.1	1718
$B(CH_2Me)_2(N_2C_3HMe_2)_2BH_2$	14.5	1718
$B(CH_2Me)_2NH\text{-}2\text{-}pyridyl$	8.0	1579
$B(C^1H_2C^2H_3)_2N(NO_2)CO_2Et$	C¹ 9.9, C² 6.4	1700
$B(C^1H_2C^2H_3)_2N(NO_2)CO_2Et.py$	C² 8.4	1700
$BEt_2N(BEt_2NMe_3)_2$	10.3, 11.9	1571
$1\text{-}EtB_5H_8$	[72.1]	549
$BMe(CH_2C^2H_3)C_8H_{13}$	C² 8.7	1228
$B(C^1H_2C^2H_3)Me_2$	C¹ 21.6 [<64], C² 6.1	1002
$B(CH_2Me)(CEt{=}CMe)_2SnEt_2$	16.0	1574
$\overline{B(CH_2Me)O(CH_2)_3O}$	8.6	567
$B(CH_2Et)_3$	31.7, 31.1	162, 278

COMPOUND	$\delta(^{13}C)$/p.p.m., J/Hz	REFERENCE
$B(\mathit{C}H_2Et)_3$hexyl	33.6	278
$[B(\mathit{C}H_2Pr^n)_4]^-$	29.9	1707
$B(\mathit{C}H_2Pr^n)_3$	27.8, 29.8, 31.7	162, 278, 567, 1707
$[B(\mathit{C}H_2Pr^n)_3$hexyl$]^-$	29.4	278
$[B(\mathit{C}H_2Pr^n)_2C_8H_{14}]^-$	29.2	1707
$B(\mathit{C}H_2Pr^n)_2C_8H_{13}$	24.8	1228
$B(\mathit{C}H_2Pr^n)MeC_8H_{13}$	27.2	1228
$[B(\mathit{C}H_2Pr^n)Bu^t{}_3]^-$	30.8	1707
$B(\mathit{C}H_2Pr^n)C_8H_{14}$	28.9	1707
$[B(\mathit{C}H_2Pr^n)(C_8H_{14})Me]^-$	28.4	1707
$B(\mathit{C}H_2Pr^n)(OH)_2$	14.5	1076
$B(\mathit{C}H_2Pr^n)O(CH_2)_3O$ (ring)	16.4	567
$[B(\mathit{C}H_2C_5H_{11})Me_3]^-$	36.3 [39.7 to 40.8]	278
$[B(\mathit{C}H_2C_5H_{11})Et_3]^-$	27.9, 29.2	278, 724
$[B(\mathit{C}H_2C_5H_{11})Pr^n{}_3]^-$	28.3	278
$[B(\mathit{C}H_2C_5H_{11})Bu^n{}_3]^-$	29.8	278
$B(\mathit{C}H_2C_5H_{11})_2Cl$	30.4	567
$B(\mathit{C}H_2C_5H_{11})_2OH$	21.3	567
$B(\mathit{C}H_2C_5H_{11})(1,2\text{-}O_2C_6H_4)$	12.2	567
$B\{\mathit{C}H_2(CH_2)_3C{=}CHSnMe_3\}Ph$ (ring)	25.2, 25.7	1730
$B(C^1H_2C^2H{=}C^3H_2)_3$	C^1 35 {114}, C^2 122 {149}, C^3 115 {156}	263
$B(C^1H_2C^2Me{=}C^3H_2)_3$	C^1 37 {114}, C^2 109 {154}	263
$BPh(NMe_2)\mathit{C}H_2CO_2Me$	29.8	1038
$BMe(NMe_2)\mathit{C}H_2CO_2Me$	29.2	1038
$BPr^i(NMe_2)\mathit{C}H_2CO_2Me$	25.0	1038
$B(\mathit{C}H_2Ph)_3$	37.5 {116}	166
$B(\mathit{C}HMe_2)_3$	20.5	162, 999
$Me_2\mathit{C}HB_5H_8$	[75.0]	549
$BMe\{C^1H(C^2H_3)_2\}C_8H_{13}$	C^2 17.9	
$B\{C^1H(C^2H_3)_2\}_2C_4H_2Pr^iSnMe_2$	C^1 25.2, C^2 19.6	999
$Me_2\mathit{C}HB(CPr^i{=}CMe)SnMe_2$	21.5	1574
$B\{C^1H(C^2H_3)_2\}(CH_2CO_2Me)NMe_2$	C^1 14.5, C^2 19.0	1038
$B\{C^1H(C^2H_3)_2\}(NMe_2)_2$	C^1 16.1, C^2 19.2	1038
$B\{C^1H(C^2H_3)_2\}(NMe_2)OC(OMe){=}CH_2$	C^1 14.5, C^2 19.0	1038
$B\{C^1H(C^2H_3)_2\}(NMe_2)Br$	C^1 12.6, C^2 19.8	1038
$[B(\mathit{C}HMeEt)_3Bu^n]^-$	30.8	1707
$B(\mathit{C}HMeEt)_3$	31.0	1707
$B[\mathit{C}H\{(CH_2)_3\}_2\mathit{C}H]Me$ (ring)	34.0	1707
$B[\mathit{C}H\{(CH_2)_3\}_2\mathit{C}H]Bu^n$ (ring)	33.8	1707
$[B[\mathit{C}H\{(CH_2)_3\}_2\mathit{C}H]MeBu^n]^-$ (ring)	28.4	1707
$[B[\mathit{C}H\{(CH_2)_3\}_2\mathit{C}H]Bu^n{}_2]^-$ (ring)	29.2	1707
$B[\mathit{C}H\{(CH_2)_3\}_2\mathit{C}H]R$ (ring)	30 to 31.7	1229
$B[\mathit{C}H\{(CH_2)_3\}_2\mathit{C}H]HNH_3$ (ring)	21.5	1229
$[B[\mathit{C}H\{(CH_2)_3\}_2\mathit{C}H]NH_2]_2$ (ring)	25.3	1229

COMPOUND	$\delta(^{13}C)$/p.p.m., J/Hz	REFERENCE
$[B[CH\{(CH_2)_3\}_2CH]HNH_2]_2$	30.5	1229
$B\{CH(CHMe)_2CH_2\}C_8H_{14}$	52.0	1229
$B\{CH(CHMeCH_2)_2\}C_8H_{14}$	55.1	1229
$B\{CH(CHMeCH_2)_2CH_2\}C_8H_{14}$	54.0	1229
$B\{CH(CHMe)_2C_6H_4\}C_8H_{14}$	56.5	1229
$B(CHMePr^i)_2CH{=}CHBu^n$	37.6	567
$(BMe)_6(CH)_6$	82.2	213
$B\{C^1(C^2H_3)_3\}_3$	C^1 30.3, C^2 31.5	1080
$B\{C^1(C^2H_3)_3\}_3NH_3$	C^1 25.6, C^2 33.1	1080
$B\{C^1(C^2H_3)_3\}(NMeCH_2)_2$	C^1 17.7	1567
$B(CMe_3)_2CMe{=}CHSnMe_3$	28.0	1730
$B(CMe_3)MeCBu^t{=}CHSnMe_3$	28.0	1730
$BEt_2CMe(SnMe_3)CEt{=}C{=}C$-$(SnMe_3)_2$	56.0	1631

<u>*Aluminium*, $[^{27}Al]$</u>

COMPOUND	$\delta(^{13}C)$/p.p.m., J/Hz	REFERENCE
$Al_2(CH_3)_6$	-8.2 [110]*, -5.6 [19]§	575, 1717
average	-4.6, -7.0 {113}	125, 669, 1715
$Al(CH_3)_2(\mu\text{-}CH_3)_2ScCp_2$	-6.3*, 20.7§	911
$Al(CH_3)_2(\mu\text{-}CH_3)_2YCp_2$	-7.9*, 7.9§	911
$Al(CH_3)_3$(pyridine)	-5.8	669
$Al(CH_3)_3(NCCMe{=}CH_2)$	-7.3	1715
$[Al(CH_3)_3]_2(OEt_2)$	[91]	575
$Al(CH_3)_2(C{\equiv}CMe)$	-7.8	1586
$Al(CH_3)_2(C{\equiv}CMe)(OMe_2)$	-9.9	1586
$B_6(NMe_2)_{12}Al_6(CH_3)_{12}$	-11.6	669
$[Al(CH_3)_2NMe_2]_2$	-11.3	669
$Al(CH_3)_2N{=}CMeCMe{=}CH_2$	-8.2	1715
$\{Al(CH_3)_2\}_2\{C_2O_2(NMe)_2\}$	-12.2	477
$[Al(CH_3)_2Cl]$	[105]	575
$[Al(CH_3)_2Br]$	[105]	575
$Al_2(C^1H_2C^2H_3)_6$ bridge	C^1 0.5; 1.1; -0.1, C^2 8.3, 7.8	348, 887, 1239, 1717
terminal	C^1 0.0; -0.6, C^2 9.2; 9.5; 9.1	
$[Al(CH_2CH_3)_4]$	$[^2J_{AlC} = 1.0]$	214
$[Al(C^1H_2C^2H_3)_2Cl]_2$	C^1 2.8 [105], C^2 8.5	575, 887
$Al(C^1H_2C^2H_3)_2C_5H_5$	C^1 2.4 {113}, C^2 9.6 {124}	934
$\{Al(CH_2Me)_2\}_2\{C_2O_2(NMe)_2\}$	2.3	477
$\{Al(CH_2Me)_2.thf\}_2PhC_4Ph$	1.1	1588
$\{Al(CH_2Me)_2.OEt_2\}_2PhC_2Ph$	2.3	1588
$[Al(C^1H_2C^2H_3)Cl_2]_2$	C^1 2.5, C^2 7.7	887
$Al_2(C^1H_2C^2H_2Me)_6$ bridge	C^1 14.0, C^2 18.8	348
terminal	C^1 12.5, C^2 19.6	
$[Al(CH_2Pr^n)_4]^-$	[71.6]	214
$Al(C^1H_2C^2H_2CH{=}CH_2)_3$	C^1 9.7, C^2 29.9	1155
$Al(C^1H_2C^2H_2CH_2CH{=}CH_2)_3$	C^1 10.3, C^2 26.4	1155
$Al(C^1H_2C^2H_2CH_2CH{=}CH_2)_3$.-bipy	C^1 8.9, C^2 27.0	1155
$Al(CH_2Ph)_3$	21.0	166

*terminal, §bridge

COMPOUND	$\delta(^{13}C)$/p.p.m., J/Hz	REFERENCE
$[Al(\mathit{C}HMe_2)_4]^-$	26.4	1584
$[Al_2(\mathit{C}HMe_2)_6]^{2-}$	27.2	1584
$Al(\mathit{C}HMe_2)_3$	25.3	1584
$Al_2(C_3H_5)_6$	C^1 -11.0*, -15.8§, $C^{2,3}$ 1.4*, 12.1§	1717
$AlMe_2(\mathit{C}_5H_5)$	111.6 {167.6, 6.6}	934
$AlEt_2(\mathit{C}_5H_5)$	111.7 {168.3, 6.7}	934
$Al\{C^1(C^2H_3)_3\}_3$	C^1 26.4, C^2 30.6	1239
Gallium		
$Ga(\mathit{C}H_3)_3$	{122}	125
$Ga(\mathit{C}H_3)_3CH_2PMe_3$	-3.0	478
$Ga(\mathit{C}H_3)_2C{\equiv}CMe$	-3.3	1586
$\{Ga(\mathit{C}H_3)_2\}_2\{C_2O_2(NMe)_2\}$	-6.2, -7.5, -9.4†	477
$Ga(C^1H_2C^2H_3)_2(C_5H_5)$	C^1 6.8 {120.0}, C^2 10.6 {123.0}	934
$\{Ga(CH_2Me_3)_2\}\{C_2O_2(NMe)_2\}$	2.9	477
$Ga(\mathit{C}H_2PMe_3)Me_2$	5.1	478
$GaMe_2(\mathit{C}_5H_5)$	112.7 {165.1, 6.8}	934
$GaEt_2(C_5H_5)$	112.0 {164.7, 6.8}	934
Indium		
$In(\mathit{C}H_3)_3$	{126}	125
$In(\mathit{C}H_3)_3CH_2PMe_3$	-8.1	478
$In(\mathit{C}H_3)_2C{\equiv}CMe$	-5.4	1586
$In(\mathit{C}^1H_2C^2H_3)_2Cp$	C^1 9.7, C^2 13.0 {122.0}	934
$In(\mathit{C}H_2PMe_3)Me_2$	-0.8	478
$InEt_2(\mathit{C}_5H_5)$	111.5 {163.3, 7.0}	934
Thallium, $[^{205}Tl]$		
$Tl(\mathit{C}H_3)_3$	[1930]	1014
$Tl(\mathit{C}H_3)_3CH_2PMe_3$	1.6	478
$Tl(\mathit{C}H_3)_2NO_3/D_2O$	24.8 [2503.2]	1221
$Tl(\mathit{C}H_3)_2NO_3$ solvent dependence	22.2 to 26.8 [2478 to 3080]	836, 1221
$Tl(\mathit{C}H_3)_2OPh$ solvent dependence	22.1 to 23.1 [2475 to 2971]	836, 977, 1014
$Tl(\mathit{C}H_3)_2OAc/H_2O$	26.7 [2513]	1014
$Tl_2(\mathit{C}H_3)_4(\mu\text{-}OC_6H_4Cl\text{-}2)_2$	[2516]	977
$Tl(\mathit{C}H_3)_2I/C_5H_5N$	25.5 [3012]	836
$Tl(\mathit{C}H_3)_2I$/dmso	27.5 [2934]	836
$Tl(\mathit{C}H_3)_2\{(NC_4Me_2CO_2Et)_2CH\}$	15.3 [2808]	800
$Tl(\mathit{C}H_3)(CN)(OAc)$	15.6 [5914]	1014
$Tl(CH_3)(OAc)_2$/MeOH	17.7 [5976]	1014
$Tl(CH_3)(OAc)_2/CHCl_3$	18.7 [5631]	1014
2-$\{(AcO)_2Tl\}$-3-AcO-norbornane	C^2 76.6 [5754], 76.4 [5750]	1162, 1564
2-$\{(AcO)_2Tl\}$-3-AcO-norborn-5-ene	C^2 72.0 [5764], 72.9 [5471]	1162, 1564

*terminal, §bridge, †conformational isomers

COMPOUND	$\delta(^{13}C)$/p.p.m., J/Hz	REFERENCE
2-{(AcO)$_2$Tl}-3-Aco-5,6-benzo-norbornene	C^2 73.9 [5645]	1564
Tl(CH$_2$PMe$_3$)Me$_2$	2.2	478
5-{(AcO)$_2$Tl}-2,COO-6-norbornane	C^5 67.7 [6655]	1162
Tl(C$_5$H$_5$)	107.5	934

Silicon, [^{29}*Si*]

A. Selected Methyl Silicon Compounds

COMPOUND	$\delta(^{13}C)$/p.p.m., J/Hz	REFERENCE
Si(CH$_3$)$_4$	[40.5, -50 to -52] {118.1 to 120} 0.0*	45, 48, 52, 83, 92, 111, 125, 152, 154, 163, 462, 466, 600
Si(CH$_3$)$_3$H	-2.6 [50.8] {119.3, 7.2}	52, 83, 144, 462
Si(CH$_3$)$_3$NMe$_2$	-1.6	1308
Si(CH$_3$)$_3$OMe	-1.4 {118.0}	83, 1308
	1.9	56
Si(CH$_3$)$_3$F	0.2, -0.3 [60.5] {118.9, 120.0}	52, 83, 462, 1308
Si(CH$_3$)$_3$Cl	3.3, 4.1 [-57.4], 3.4 [57.7], 14 {120.5, 120.8}	52, 83, 462, 600, 1308
Si(CH$_3$)$_3$Br	4.2, 4.6 [56.0] {121.2}	51, 462, 1308
Si(CH$_3$)$_3$I	4.8, 6.5 [54.0] {122}	126, 462, 1308
Si(CH$_3$)$_3$Ph	-1.1, -0.6, -1.0, -0.6 [52.2] {119.0}	83, 462, 565, 880, 914, 1308
Si(CH$_3$)$_3$SMe	-1.0, 0.7 [53.7] {120.1}	820, 1308
M(CO)$_4${Si(CH$_3$)$_3$}$_2$ (M = Fe, Ru, Os)	6.7 to 8.0	727, 1130
Al{Si(CH$_3$)$_3$}$_3$	2.8	1239
P{Si(CH$_3$)$_3$}$_3$	4.2 (11.1)	1239
Hg{Si(CH$_3$)$_3$}$_2$	6.7 [J_{HgC} = 92.3], 0.3	1239, 1545
Sn{Si(CH$_3$)$_3$}Me$_3$	1.1, 0.5 [J_{SnC} = 60.1]	457, 547
Re{Si(CH$_3$)$_3$}(CO)$_5$	6.6	515
Ru{η^6-C$_8$H$_7$Si(CH$_3$)$_3$}(CO)$_3$SiMe_3	6.3	517
Si$_2$(CH$_3$)$_6$	-2.1 [43.6]	462
Hg{Si(CH$_3$)$_3$}R	0.1, 0.2	1545
(CH$_3$)$_3$SiFe(CO)$_3$(η^3-CH$_2$CHSiMe$_2$)	-1.0	708
M-CH$_2$Si(CH$_3$)$_3$	1.8 to 6.9	609, 897, 1109, 1270, 1334, 1346, 1373, 1421, 1555
M-CH$_{3-n}${Si(CH$_3$)$_3$}$_n$	1.8 to 7.8	1592, 1614
Ru{C$_8$H$_8$Si(CH$_3$)$_3$}(CO)$_2$Si(CH$_3$)$_3$	4.3	1478
{Si(CH$_3$)$_3$}$_3$AuAsPh$_3$	7.3	1592
N{Si(CH$_3$)$_3$}$_3$	6.0 [56.8] {118.4}	820

*by definition

COMPOUND	$\delta(^{13}C)$/p.p.m., J/Hz	REFERENCE
$(\mathit{C}H_3)_3$Si-imidazole	16.6	668
$Si(\mathit{C}H_3)Ph_3$	-2.9, -1.5	545, 1308
$Si(\mathit{C}H_3)_2PhCl$	2.4	545
$Si(\mathit{C}H_3)PhCl_2$	5.1	545
$Si(\mathit{C}H_3)_2Ph(OEt)$	-1.8	545
$Si(\mathit{C}H_3)Ph(OEt)_2$	-4.9	545
$Si(\mathit{C}H_3)_2PhF$	-1.5	545
$Si(\mathit{C}H_3)PhF_2$	-5.8	545
$Si(\mathit{C}H_3)_2Cl_2$	8.0 [-68.3], 6.7	83, 600, 1308
$Si(\mathit{C}H_3)Cl_3$	11.0 [-86.6], 9.8	83, 600, 1308
$Mo_2(CH_2SiMe_3)_2(PMe_3)_3\{(CH_2)_2$-$Si(\mathit{C}H_3)_2\}$	8.9	1373
$Mo_2(CH_2SiMe_3)_2\{P(OMe)_3\}_3\{(CH_2)_2$-$Si(\mathit{C}H_3)_2\}$	8.4	1373
$Ru\{(CH_2)_2Si(\mathit{C}H_3)_2\}(PMe_3)_4$	10.4 (2.7)	1373
$Rh\{CH_2Si(\mathit{C}H_3)_3\}(PMe_3)_3\{(CH_2)_2Si$-$(CH_3)_2\}$	10.8, 8.4	1373
$\overline{Fe(CO)_4Si(\mathit{C}H_3)_2(CH_2)_3}$	4.2	1436
$Si(\mathit{C}H_3)H\{(C_5H_5)Fe(CO)_2\}_2$	12.0	1455
$Si(\mathit{C}H_3)Cl_3$	-9.4	1589
$Si(\mathit{C}H_3)H_3$	13.3	1589
$Si(\mathit{C}H_3)_3OEt$	-0.5 [-59.0]	466
$Si(\mathit{C}H_3)_2(OEt)_2$	-3.1 [-73.0], -3.7	466, 1308
$Si(\mathit{C}H_3)(OEt)_3$	-6.9 [-96.2], -7.6	466, 1308
$(\mathit{C}H_3)_2\overline{SiCMe_2C}Me_2$	-9.7	463
$Si(\mathit{C}H_3)_2I_2$	13.0	1308
$Si(\mathit{C}H_3)_2(OMe)_2$	-4.7, -5.1	56, 1308
$Si(\mathit{C}H_3)_2(SMe)_2$	0.3	1308
$Si(\mathit{C}H_3)_2(NMe_2)_2$	-3.8	1308
$Si(\mathit{C}H_3)_2Ph_2$	-2.4	1308
$Si(\mathit{C}H_3)I_3$	21.1	1308
$Si(\mathit{C}H_3)(OMe)_3$	-9.4, 10.3	56, 1308
$Si(\mathit{C}H_3)(SMe)_3$	1.5	1308
$Si(\mathit{C}H_3)(NMe_2)_3$	-6.4	1308
$Si(\mathit{C}H_3)_3SPh$	2.0	285
$(OC)_4\overline{FeSi(\mathit{C}H_3)_2CH_2CH_2Si}(\mathit{C}H_3)_2$	8.4	393
$(OC)_4\overline{RuSi(\mathit{C}H_3)_2CH_2CH_2Si}(\mathit{C}H_3)_2$	4.7	393
$(OC)_4\overline{OsSi(CH_3)_2CH_2CH_2Si}(CH_3)_2$	3.9	393
$Me\overline{P=C\{Si(CH_3)_2CH_2Si(\mathit{C}H_3)_2C}H_2)_2\}_2$	16.1, 16.4, 17.1	759
$M(CO)_4\{Si(\mathit{C}H_3)_2Cl\}_2$ (M = Fe, Ru, Os)	11.4 to 13.4	727
$M(CO)_4\{Si(\mathit{C}H_3)Cl_2\}_2$	16.5 to 19.3	727
$Si(\mathit{C}H_3)_3PPh(COBu^t)$	0.0	1024

B. Other Compounds

COMPOUND	$\delta(^{13}C)$/p.p.m., J/Hz	REFERENCE
Me_3Si-carbon	-6.2 to 5.0, 62.2, 58.4	78-80, 126, 219, 264, 282 332, 350, 360

COMPOUND	$\delta(^{13}C)$/p.p.m., J/Hz	REFERENCE
		407, 419, 421, 451, 459, 465, 467, 468, 514, 521, 565, 711, 835, 880, 884, 922, 944, 967, 1027, 1063, 1100, 1108, 1157, 1161, 1193, 1253, 1308, 1322, 1324, 1429, 1478, 1594, 1596, 1611, 1614, 1616, 1623, 1625, 1709, 1711, 1724, 1731, 1725, 1735, 1737, 1738, 1739
Me_3Si-nitrogen	-7.0 to 11.6	52, 142, 308, 411, 462, 469, 750, 923, 944, 967, 989, 1015, 1022, 1030, 1044, 1054, 1090, 1122, 1193, 1204, 1255, 1256, 1287, 1600, 1603, 1604, 1612, 1613, 1621, 1732
Me_3Si-oxygen	-1.9 to 3.0	52, 56, 83, 273, 460, 462, 466, 770, 771, 788, 794, 803, 831, 900, 902, 1024, 1045, 1052, 1054, 1074, 1075, 1122, 1247, 1250, 1308, 1322, 1359, 1603, 1608, 1617
Miscellaneous MeSi compounds	-7.7 to 11.6	17, 27, 52, 56, 78, 80, 83, 91, 102, 127, 144, 166, 252, 282, 324, 332, 386, 421, 427, 460, 465, 565, 708, 711, 719, 759, 772, 788, 794, 796, 822, 835, 854, 900, 902, 914, 989, 1005, 1010, 1022, 1024, 1146, 1158, 1246, 1247, 1249, 1250, 1286, 1287, 1322, 1527, 1590, 1593, 1596, 1601, 1605, 1607, 1619-1621, 1733, 1734, 1736, 1768
$Si\mathit{C}H_2R$ compounds		80, 154, 166, 219, 225, 252, 264, 282, 324, 350, 357, 386, 393, 459, 460, 465, 469, 634, 638, 759, 835, 854, 897, 1005, 1146, 1157, 1158, 1161, 1193, 1239, 1241, 1249, 1259, 1270, 1290, 1322, 1373, 1421, 1555, 1562, 1589, 1596, 1605, 1607, 1611, 1614, 1619, 1620, 1711, 1720, 1725, 1735, 1737-1739
$Si\mathit{C}HR_2$ compounds		78 to 80, 324, 332, 360, 419, 467, 517, 638, 1005, 1108, 1157, 1246, 1270, 1429, 1589, 1602, 1623
$Si\mathit{C}R_3$ compounds		78, 252, 421, 463, 464, 467, 753, 822, 992, 1010, 1239, 1246, 1589, 1592, 1606, 1620

COMPOUND		δ(^{13}C)/p.p.m., *J*/Hz	REFERENCE
Germanium, [^{73}*Ge*]			
Ge(CH$_3$)$_4$		-0.8, -1.4, [18.7] {124}	44, 48, 152, 154, 157, 163
Ge(CH$_3$)$_3$H		{126.2}	157
Ge(CH$_3$)$_3$Cy		-2.4	1207
Ge(CH$_3$)$_3$Ph		-1.8	880
Ge(CH$_3$)$_3$CH$_2$Ph		-2.3, -2.6	459, 1161
Ge(CH$_3$)$_3$CN		-1.7	967
Ge(CH$_3$)$_3$C$_6$H$_{11}$		-4.5	922
{Ge(CH$_3$)$_3$CH$_2$}$_2$Hg		2.0 [J_{HgC} = 38.2]	1270
Ge(CH$_3$)$_3$CH$_2$HgCl		1.4 [J_{HgC} = 68.5]	1270
(CH$_3$)HGeCHMe(CH$_2$)$_3$ (ring)	Z	-8.6	1005
	E	-5.4	1005
(CH$_3$)$_3$Ge(Me$_3$M)C=C=O (M = Si, Ge, Sn)		2.2 to 3.0	264
Me$_3$P=CHGe(CH$_3$)$_3$		4.3	1063
Me$_3$P=C{Ge(CH$_3$)$_3$}$_2$		6.4	1063
Ge(CH$_3$)$_3$CH$_2$PMe$_2$=C(GeMe$_3$)$_2$		0.6, 6.4	1063
Ge(CH$_3$)$_3$CF$_3$		-5.2	452
Re(CO)$_5$Ge(CH$_3$)$_3$		-4.9	514
cis-Fe(CO)$_4${Ge(CH$_3$)$_3$}$_2$		7.0	727
cis-Ru(CO)$_4${Ge(CH$_3$)$_3$}$_2$		7.0	727
cis-Os(CO)$_4${Ge(CH$_3$)$_3$}$_2$		5.2	727
trans-Os(CO)$_4${Ge(CH$_3$)$_3$}$_2$		6.2	727
Ru{C$_7$H$_8$Ge(C^1H$_3$)$_3$}(CO)$_2$Ge(C^2H$_3$)$_3$		C^1 -4.2, C^2 4.7	375, 517
Ru(C$_7$H$_9$)(CO)$_2$Ge(CH$_3$)$_3$		4.7	518
Ru{C$_{10}$H$_8$Ge(C^1H$_3$)$_3$}(CO)$_2$Ge(C^2H$_3$)$_3$		C^1 -2.2, C^2 6.8	516
cis-Fe(CO)$_4${Ge(CH$_3$)$_2$H}$_2$		-3.0	1438
Hg{Ge(CH$_3$)$_3$}X		0.2 to 0.4	1565
(CH$_3$)$_3$GeSnMe$_3$		-0.4 [J_{SnC} = 56.4]	457
1,2-B$_{10}$H$_{10}$C$_2$HGe(CH$_3$)$_3$		-0.3	453
1,7-{(CH$_3$)$_3$Ge}$_2$-1,7-C$_2$B$_{10}$H$_{10}$		-1.2	944
Ge(CH$_3$)$_3$R		-4.5 to 1.6	78, 80, 360 407, 421, 451, 521, 854, 880, 996, 1004, 1207 1286, 1429, 1628, 1709, 1724, 1725
(CH$_3$)R^{1}GeCH$_2$CR^{2}CR^{3}CH$_2$ (ring)		-2.0 to -4.5	1158
(CH$_3$)(Et$_2$N)GeCHMe(CH$_2$)$_3$ (ring)	Z	-2.8	1629
	E	-5.7	1629
(OC)$_5$MCNGe(CH$_3$)$_3$ (M = Cr, Mo, W)		1.6 to 2.0	967
(CH$_3$)$_3$(MeO)GeCHMe(CH$_2$)$_3$ (ring)	Z	-3.1	1629
	E	-5.4	1629
Me$_2$NCO$_2$Ge(CH$_3$)$_3$		3.5	788
Me$_2$NCOSGe(CH$_3$)$_3$		2.5	788
Me$_3$NCS$_2$Ge(CH$_3$)$_3$		3.1	788

COMPOUND		$\delta(^{13}C)$/p.p.m., J/Hz	REFERENCE
$(CH_3)(MeS)\overline{GeCHMe(CH_2)_3}$	Z	-0.8	1629
	E	-3.6	1629
$Ge(C^1H_2C^2H_3)_3CHCO$		C^1 8.2, C^2 10.2	264
$Ge(\mathit{C}H_2Me)_3C{\equiv}CH$		5.8	1725
$\{Ge(C^1H_2C^2H_3)_3\}_2CCO$		C^1 8.4, C^2 9.6	264
$GeMe_3(C^1H_2C^2H_2C^3H_2Pr^n)$		C^1 16.8, C^2 25.1, C^3 33.2	1207
$GeMe_3(C^1H_2C^2H_2Ph)$		C^1 18.6, C^2 31.3	1207
$GeMe_3(\mathit{C}H_2Ph)$		26.3, 26.8	459, 1161
$Hg(\mathit{C}H_2GeMe_3)_2$		30.1 [J_{HgC} = 600.1]	1270
$Hg(\mathit{C}H_2GeMe_3)Cl$		18.0 [J_{HgC} = 1240.0]	1270
$Ge(\mathit{C}H_2Sp\text{-}tol)Ph_3$		16.8	341
$Ge\{\mathit{C}H_2PMe_2{=}C(GeMe_3)_2\}Me_3$		23.9	1063
$\overline{Ge(C^1H_2C^2H_2C^3H_2C^2H_2C^1H_2)}Me_2$		C^1 15.4, C^2 25.9, C^3 30.6	854
$\overline{Ge(C^1HMeCH_2CH_2C^2H_2)}MeH$		C^1 12.7, C^2 19,4*; 22.7§	1005
$\overline{Ge(\mathit{C}H_2CH{=}CH\mathit{C}H_2)}Me_2$		19.2	1158
$\overline{Ge(\mathit{C}H_2CH{=}CHCH_2)}Me(C_2H_3)$		18.1	1158
$\overline{Ge(\mathit{C}H_2CH{=}CHCH_2)}Ph_2$		18.0	1158
$\overline{Ge(C^1H_2CMe{=}CHC^2H_2)}Me_2$		C^1 24.1, C^2 20.6	1158
$\overline{Ge(C^1H_2CMe{=}CHC^2H_2)}Ph_2$		C^1 22.5, C^2 19.2	1158
$\overline{Ge(\mathit{C}H_2CMe{=}CMe\mathit{C}H_2)}Me_2$		27.1	1158
$\overline{Ge(\mathit{C}H_2CMe{=}CMe\mathit{C}H_2)}Ph_2$		25.8	1158
$\overline{Ge\{CH_2CH(N_3C_2PhHO_2)CH{=}CH\}}Me_2$		14.3 {130}	1158
$\overline{Ge\{CH_2CH(N_3C_2PhHO_2)CH{=}CH\}}Ph_2$		13.4 {130}	1158
Ge(cyclohexyl)Me_3		27.9	1628
Ge(4-Mecyclohexyl)Me_3		27.0, 27.4†	1628
Ge(4-Bu^tcyclohexyl)Me_3		27.2, 27.5†	1628
Ge(2-HOcyclohexyl)Me_3		36.8	1628
$\overline{Ge(C^1HC^2H{=}C^3HC^3H{=}C^2H)}Me_3$		C^1 52.1; 51.3, C^2 133.9, C^3 129.8	78 to 80
$Ge(\mathit{C}_5H_4Me)Me_3$		52.8, 55.6	80
Ge(1-indenyl)Me_3		46.2	360
$Ge\{\mathit{C}_8H_9Ru(CO)_3GeMe_3\}Me_3$		24.5	517
$Ge\{\mathit{C}H(SCH_2)_2CH_2\}Me_3$		33.6	1429
$Ge\{\overline{\mathit{C}HS[Fe(CO)_4](CH_2)_3S}\}Me_3$		43.9	1429
5,5-$(Me_3Ge)_2\mathit{C}_5H_4$		55.4	421
5,5-$(Me_3Si)(Me_3Ge)\mathit{C}_5H_4$		56.4	421
5,5-$(Me_3Sn)(Me_3Ge)\mathit{C}_5H_4$		58.9	421
$Ge(\mathit{C}F_3)Me_3$		132.6	452
Tin, [^{119}Sn]			
$Sn(\mathit{C}H_3)_4$		-9.6 to -8.6	48, 88, 100, 152, 154, 156, 163

*Z-isomer, §E-isomer, †*cis* and *trans* isomers

COMPOUND	δ(^{13}C)/p.p.m., *J*/Hz	REFERENCE
	[330 to 340, 350] {126 to 128}	227, 456, 875, 882, 921, 1238, 1634
$Sn(\mathit{C}H_3)_3H$	-11.8 [351.8] {128.5}	63, 227, 875
$Sn(\mathit{C}H_3)_3Li$	-1.5 [+155 ±5] {+122 ±1}	373
$Sn(\mathit{C}H_3)_3Et$	-11.2 [320.8], -11.0 [319]	875, 882, 921
$Sn(\mathit{C}H_3)_3CH_2R$	-11.2 to -8.4 [314 to 343]	47, 88, 456, 459, 598, 875, 879, 882, 921, 948, 1161
$Sn(\mathit{C}H_3)_3Pr^i$	-12.1 [306.7], -12.0 [305.2]	875, 879
$Sn(\mathit{C}H_3)_3C_6H_{11}$ axial	-9.2 [295.5]	653, 922, 1628,
equatorial	-11.9 [299.4]	653, 922, 1628,
$Sn(\mathit{C}H_3)_3$(cyclopropyl)	-11.3 [341.6]	875
cis-$Sn(\mathit{C}H_3)_3$-$\overline{CHCH(CO_2Et)CH_2CH_2CH_2}$	-9.1 [332]	948
$Sn(\mathit{C}H_3)_3CHR_2$	-13.0 to -9.1 [290 to 322]	88, 145, 146, 325, 360, 459, 655, 875, 922, 1316, 1628, 1638
$Sn(\mathit{C}H_3)_3Bu^t$	-12.2 [295.7], -12.1 [295.3]	457, 875, 882
$Sn(\mathit{C}H_3)_3CH_2Cl$	-10.0 [356]	456
$Sn(\mathit{C}H_3)_3CHCl_2$	-9.3 [362]	456
$Sn(\mathit{C}H_3)_3CCl_3$	-7.2 [363]	227, 456
$Sn(\mathit{C}H_3)_3CF_3$	-10.8	452
1-$Sn(\mathit{C}H_3)_3$adamantyl	-8.3	231
$Sn(\mathit{C}H_3)_3C_{16}H_{13}$	-8.2 [324]	459
$Sn(\mathit{C}H_3)_3CH{=}CHMe$ *cis*	-8.9 [346.9], -9.9	875, 1634
trans	-9.9 [352.0]	1634
$Sn(\mathit{C}H_3)_3CH{=}CH_2$	1.0 [353]	458
$Sn(\mathit{C}H_3)_3CH(SCH_2)_2CH_2$	-10.4 [350]	1429
$Sn(\mathit{C}H_3)_3CH{=}CBPh(CH_2)_4$	-7.8 [343.6], -8.5 [336.2]*	1660
$Sn(\mathit{C}H_3)_3CH{=}CMeBBu^t{}_2$	-8.1 [334]	1730
$Sn(\mathit{C}H_3)_3CH{=}CBu^tBMeBu^t$	-7.4 [339]	1730
1,7-$\{(CH_3)_3Sn\}_2$-1,7-$B_{10}C_2H_{10}$	-8.0	944
$Sn(CH_3)_3\{CH(SCH_2)_2CH_2\}Fe(CO)_4$	-7.8 [359]	1429
$Sn(\mathit{C}H_3)_3CMe{=}CH_2$	-10.3 [339.9]	1634
$\{(\mathit{C}H_3)_3Sn\}_2C{=}C{=}CEtCMe(BEt_2)$-$SnMe_3$	-7.8 [332], -7.5 [301]	1631
$Sn(\mathit{C}H_3)_3CH{=}C{=}CH_2$	-10.4 [356]	47
$\{(CH_3)_3Sn\}(Me_3M)C{=}C{=}O$ (M = Si, Ge, Sn)	-4.5 to -5.1	264
$(\mathit{C}H_3)_3SnCMe{=}C{=}CHMe$	-9.8 [342.5]	875
$\{(CH_3)_3Sn\}_2C{=}PMe_3$	-2.8 [324.7]	1063

* *cis* and *trans* isomers

COMPOUND	δ(^{13}C)/p.p.m., *J*/Hz	REFERENCE
$(CH_3)_3SnCN_2(CO_2Et)$	-8.3 [385.2]	1709, 1724
$\{(CH_3)_3Sn\}(Me_2As)CN_2$	-8.2	996
$\{(CH_3)_3Sn\}_2CN_2$	-7.9 [365.2]	996, 1709, 1724
$Sn(CH_3)_3Ph$	-9.8, -10.4 [351.4]	47, 455, 880
$Sn(CH_3)_3(2,6\text{-}R_2C_6H_3)$	-3.8 to -10.4	455
$Sn(CH_3)_3Ar$	-7.2 to -11.6 [343 to 357]	47, 229, 451, 455, 880
$Sn(CH_3)_3$-9-anthracyl	-4.6	880
$Sn(CH_3)_3C_5H_5$	-6.6, -7.2	78, 80
$\{Sn(CH_3)_3\}_2C_5H_4$	-1.5, -8.2 [342]	78, 421
$\{Sn(CH_3)_3\}(MMe_3)C_5H_4$ (M = Si, Ge)	-6.9, -7.2	421
$Sn(CH_3)_3C{\equiv}CR$	-7.6 to -8.1 [399.8 to 405.1]	875, 882, 1238, 1643
$(CH_3)_3SnW(CO)_2PR_3(C_5H_5)$	-6	71
cis-$Fe(CO)_4\{Sn(CH_3)_3\}_2$	-3.7 [274]	727, 1128
cis-$Ru(CO)_4\{Sn(CH_3)_3\}_2$	-4.7 [267]	727
cis-$Os(CO)_4\{Sn(CH_3)_3\}_2$	-6.6 [278]	727
trans-$Os(CO)_4\{Sn(CH_3)_3\}_2$	-6.8	727
$(CH_3)_3SnSiMe_3$	-12.3 [245.8]; -11.5	457, 547
$(CH_3)_3SnGeMe_3$	-11.6 [255.1]	457
$(CH_3)_3SnSnMe_3$	-10.2 [244.9; 56.9]	105, 227, 457
$(CH_3)_3SnSnPh_3$	-8.4	547
$Re(CO)_5Sn(CH_3)_3$	-7.5 [252]	514
$Sn(CH_3)_3$(norbornene)	-7.2	88
$Sn(CH_3)_3(NMe_2)$	-8.7 [381.2]	457
$Sn(CH_3)_3(NEt_2)$	-6.7 [378.8], [380]	227, 457
$Sn(CH_3)_3(\overline{NCH_2CH_2})$	-8.2 [347.4]	457
$Sn(CH_3)_3(N{=}C{=}NR)$	-0.2 to 3.7	1035
$Sn(CH_3)_3(NR_2)$	-8.5 to -1.5 [372 to 382]	457, 1127, 1325
$Sn(CH_3)_3(NC_5H_5)Cl$	2.1 [472]	227
$Sn(CH_3)_3(OMe)$	-5.9 [398.0]; -5.7 [398.0]	457, 1127
$Sn(CH_3)_3(OEt)$	-3.1 [416]	227, 457
$\{(CH_3)_3Sn\}_2O$	-1.9	454
$Sn(CH_3)_3(O_2CNMe_2)$	-2.0	788
$Sn(CH_3)_3(OSCNMe_2)$	-3.2	788
$Sn(CH_3)_3(S_2CNMe_2)$	-0.7	788
$Sn(CH_3)_3(O_2SeR)$	2.8 to 2.9	1637
$Sn(CH_3)_3(SMe)$	-6.0 [355.8] -5.4	156, 457
$Sn(CH_3)_3(SEt)$	-5.5 [354.1]	457
$Sn(CH_3)_3(SPh)$	-3.0	285
$\{(CH_3)_3Sn\}_2S$	-0.9 [356]	454
$Sn(CH_3)_3(SeMe)$	-5.2	232
$(CH_3)_3SnSePh$	-3.7	232
$\{(CH_3)_3Sn\}_2Se$	-1.0 [340]	454
$\{(CH_3)_3Sn\}_2Te$	-1.9	454
$Sn(CH_3)_3Cl$	0.0 [386]; 0.7 [379] [494]* {133}	17, 227, 458
$Sn(CH_3)_3Br$	-0.1 [372], 0.8 [365] {131}	100, 227, 458
$Sn(CH_3)_2Bu^n_2$	-11.5 [302.1]	882

*solvent dependent

COMPOUND	δ(^{13}C)/p.p.m., *J*/Hz	REFERENCE
$(CH_3)_2\overline{Sn(CH_2)_n}$ (*n* = 4 to 6)	-10.3 to -11.4 [300.4 to 306.9]	875
$(CH_3)_2\overline{SnCR^1{=}CR^2CR^3{=}CR^4}$	-8.7 to -4.1 [289 to 328]	999, 1286, 1625
$EtB(Pr^iC{=}CMe)_2Sn(CH_3)_2$	-8.5 [315]	1574
$(CH_3)_2Sn(CH{=}CHMe)_2$	-10.4 to -8.4 [357 to 364.0]	1634
$(CH_3)_2Sn(C{\equiv}CH)_2$	-6.5 [502.7]	1643
$(CH_3)_2Sn(C{\equiv}CBu^n)_2$	-6.6 [495.5]	1238
$(CH_3)_2SnRCl$	-2.1 to 1.6 [347 to 350]	921
$(CH_3)_2SnCl(CH_2)_nCOMe$ (*n* = 2,3)	0.0, 0.7 [463.8; 436.6]	598
$(CH_3)_2Sn(NEt_2)_2$	-5.9 [472]	227
$(CH_3)_2Sn(OEt)_2$	0.6 [682]	227
$(CH_3)_2Sn(acac)_2$	7.8 [966]	227
$(CH_3)_2Sn(SMe)_2$	-2.6	156
$(C^1H_3)_2\overline{SnSSn(C^2H_3)_2Sn(C^2H_3)_2S}$	C^1 3.4 [388] {132}, C^2 -1.5 [259, 75] {136}	1217
$Sn(CH_3)_2Cl$(cysteine ethyl ester)	7.1 [581]	1639
$Sn(CH_3)_2Cl$(penicillamine)	5.2 [600]	1623
$Sn(CH_3)_2(SeMe)_2$	-1.3	232
$Sn(CH_3)_2(SePh)_2$	0.1	232
$(C^1H_3)_2\overline{SnSeSn(C^2H_3)_2Sn(C^2H_3)_2Se}$	C^1 3.6 [345] {134}; C^2 -2.5 [238, 69] {134}	1217
$(C^1H_3)_2\overline{SnTeSn(C^2H_3)_2Sn(C^2H_3)_2Te}$	C^1 2.9 [285] {133}, C^2 -3.6 [218, 75] {132}	1217
$Sn(CH_3)_2Cl_2$	[566 to 864]	100
$Sn(CH_3)_2Br_2$	8.6 [440]	227
$Sn(CH_3)Et_3$	-14.4 [287.7]	882
$Sn(CH_3)Pr^n{}_3$	-12.6 [285.3]	882
$Sn(CH_3)Pr^i{}_3$	-16.8 [248.7]	882
$Sn(CH_3)Bu^n{}_3$	-12.7 [286.1]	882
$Sn(CH_3)(CH_2CMe_2Ph)PhFeCp(CO)_2$	-4.8	1729
$Sn(CH_3)Ph_3$	-10.6 [374.6]	882
$Sn(CH_3)(CH{=}CHMe)_3$	-7.7 to -10.9 [377.8]	1634
$Sn(CH_3)(C{\equiv}CBu^n)_3$	-5.0 [613.2]	1238
$Sn(CH_3)RCl_2$	5.8 to 7.1 [406, 415]	921
$Sn(CH_3)(SMe)_3$	1.0	156
$Sn(CH_3)(SeMe)_3$	3.0	232
$Sn(CH_3)(SePh)_3$	8.4	232
$Sn(CH_3)Br_3$	[-640] {141.2}	100
$Sn(C^1H_2C^2H_3)_4$	C^1 0.0 [320], C^2 11.1 [23]	154, 161, 227, 875, 882, 921
$Sn_2(C^1H_2C^2H_3)_6$	C^1 1.8 [247; 41], C^2 12.5 [19]	227
$Sn(C^1H_2C^2H_3)_3H$	C^1 -0.3 [347], C^2 11.8 [25]	227

COMPOUND	$\delta(^{13}C)$/p.p.m., *J*/Hz	REFERENCE
cis-$Fe(CO)_4\{Sn(C^1H_2C^2H_3)_3\}_2$	C^1 6.3 [272], C^2 11.8 [23]	1128
$(C^2H_3C^1H_2)_3Sn(C^3H_2C^4H_2-)_2$	C^1 0.4 [318], C^2 11.2 [24], C^3 8.0 [312], C^4 32.2 [21]	921
$\{(C^2H_3C^1H_2)_3Sn\}_2C=C=O$	C^1 4.9, C^2 11.6	264
$Sn(C^1H_2C^2H_3)_3Bu^t$	C^1 -0.5 [294.4], C^2 11.3 [23.7]	882
$Sn(C^1H_2C^2H_3)_3CH=CHPh$	C^1 1.1 [349.5], C^2 11.2 [24.4]	882
$Sn(\mathit{C}H_2Me)_3C\equiv CH$	2.3	1725
$Sn(C^1H_2C^2H_3)_3C\equiv CPh$	C^1 2.6 [387.6], C^2 11.1 [23.5]	882
$Sn(C^1H_2C^2H_3)_3NEt_2$	C^1 3.6 [370], C^2 10.2 [25]	227
$Sn(C^1H_2C^2H_3)_3OAc$	C^1 8.1 [370], C^2 9.3 [25]	227
$Sn(C^1H_2C^2H_3)_3Cl$	C^1 9.3 [352], C^2 9.4 [26]	227, 921
$Sn(C^1H_2C^2H_3)_3Br$	C^1 9.0 [340], C^2 10.4 [29]	227
$Sn(C^1H_2C^2H_3)_3I$	C^1 7.9 [326], C^2 11.4 [29]	227
$\overline{Sn(C^1H_2C^2H_3)_2CH=C(BEt_2)CEt=CH}$ (ring)	C^1 4.1 [342.6], C^2 12.3	999
$Sn(C^1H_2C^2H_3)_2(CMe=CEt)_2BEt$	C^1 1.6 [336], C^2 11.2 [24]	1574
$[SnCl(C^1H_2C^2H_3)_2(C^3H_2C^4H_2CH_2)]_2$	C^1 9.8 [353], C^2 9.9 [28], C^3 17.2, C^4 25.6 [24]	921
$Sn(C^1H_2C^2H_3)_2Br_2$	C^1 19.6 [400], C^2 10.1 [40]	227
$Sn(C^1H_2C^2H_3)Me_3$	C^1 2.8 [373.0], C^2 10.7 [24.0]	875, 882, 921
$Sn(C^1H_2C^2H_2C^3H_3)_4$	C^1 12.3 [316], C^2 20.9 [30], C^3 19.2 [51.0]	227, 875, 882, 921
cis-$Fe(CO)_4\{Sn(C^1H_2C^2H_2C^3H_3)_3\}_2$	C^1 17.3 [263], C^2 21.4 [21], C^3 19.6 [67]	1128
$Sn(C^1H_2C^2H_2C^3H_3)_3Bu^t$	C^1 11.3 [288.5], C^2 20.9 [20.5], C^3 19.4	882
$Sn(C^1H_2C^2H_2C^3H_3)_3Br$	C^1 20.2 [331], C^2 19.8 [32], C^3 18.3	227, 921
$Sn(C^1H_2C^2H_2C^3H_3)_2Br_2$	C^1 29.9 [390], C^2 19.5 [32], C^3 17.4	227, 921
$Sn(C^1H_2C^2H_2C^3H_3)Me_3$	C^1 13.9 [369], C^2 20.3 [21], C^3 18.6 [53]	875, 882

COMPOUND	$\delta(^{13}C)$/p.p.m., *J*/Hz			REFERENC
	C^1	C^2	C^3	
$Sn(C^1H_2C^2H_2C^3H_2Me)_4$	9.1 [314]	29.6 [20]	27.6 [52]	227, 875 882, 921 1633, 17
$Sn(C^1H_2C^2H_2C^3H_2Me)_3H$	8.3 [355]	30.2 [22]	27.3 [53]	227
$Sn_2(C^1H_2C^2H_2C^3H_2Me)_6$	10.3 [241]	31.0 [16]	27.8 [53]	227
cis-$Fe(CO)_4\{Sn(C^1H_2C^2H_2C^3H_2Me)_3\}_2$	14.5 [263]	29.9 [21]	28.2 [66]	1128
$Sn(C^1H_2C^2H_2C^3H_2Me)_3CH_2$-1-naphthyl	9.9	28.9	27.2	879
$Sn(C^1H_2C^2H_2C^3H_2Me)_3CH_2CONMe_2$	10.4	$C^{2,3}$ 29.0,	27.3	651
$Sn(C^1H_2C^2H_2C^3H_2Me)_3C_5H_7$	9.2	$C^{2,3}$ 29.6,	27.8	655
$Sn(C^1H_2C^2H_2C^3H_2Me)_3Bu^s$	8.7 [297]	29.8 [19]	27.9	1316
$Sn(C^1H_2C^2H_2C^3H_2Me)_3C_6H_9Br_2$	9.3 [320]	$C^{2,3}$ 28.6,	30.5	346
$Sn(C^1H_2C^2H_2C^3H_2Me)_3Bu^t$	8.1 [291]	29.7	27.9	882
$Sn(C^1H_2C^2H_2C^3H_2Me)_2C{\equiv}CH$	11.2 [388]	29.2 [24.2]	27.1 [58.8]	1633
$\{Sn(C^1H_2C^2H_2C^3H_2Me)_3\}_2C_2$	11.0	29.0	27.2	1633
$Sn(C^1H_2C^2H_2C^3H_2Me)_3N_3C_2R_2$	15.6 to 16.6	27.8 to 28.0	26.9 to 27.3	1325
$Sn(C^1H_2C^2H_2C^3H_2Me)_3NHCN$	18.3 [502]	28.6 [27]	27.2 [74]	1727
$Sn(C^1H_2C^2H_2C^3H_2Me)_3OMe$	14.1 [370]	28.4 [20]	27.4 [62]	227
$\{Sn(C^1H_2C^2H_2C^3H_2Me)_3\}_2O$	16.6 [370]	28.5 [20]	27.4 [60]	227
$Sn(C^1H_2C^2H_2C^3H_2Me)_3OR$	14.2 to 16.8 [360 to 376]	28.2 to 28.4 [22, 45]	27.3 to 27.5 [62, 105]	346, 1727
$Sn(C^1H_2C^2H_2C^3H_2Me)_3O_3SC_6H_4COOH$	20.3 [406]	27.8	26.9 [74]	1727
$Sn(C^1H_2C^2H_2C^3H_2Me)_3SR$	13.0 to 20.7 [326 to 350]	28.6 to 29.3 [22 to 40]	26.0 to 27.3 [52 to 96]	1727
$Sn(C^1H_2C^2H_2C^3H_2Me)_3Cl$	17.2 [341]	27.5 [46]	26.4 [67]	1727
$Sn(C^1H_2C^2H_2C^3H_2Me)_3Br$	17.2 [330]	28.5 [23]	26.9 [60]	227, 921
$Sn(C^1H_2C^2H_2C^3H_2Me)_2H_2$	6.9 [375]	30.6 [24]	26.9 [57]	227
$Sn(C^1H_2C^2H_2C^3H_2Me)_2HCl$	17.6 [397]	28.1 [27]	26.6 [70]	227
$Sn(C^1H_2C^2H_2C^3H_2Me)_2Me_2$	10.4 [348.7]	29.4 [20.6]	27.4 [54.1]	882

COMPOUND	$\delta(^{13}C)$/p.p.m., J/Hz			REFERENCE
	C^1	C^2	C^3	
$Sn(C^1H_2C^2H_2C^3H_2Me)_2Bu^tCl$	15.5	28.3	27.2	833
$Sn(C^1H_2C^2H_2C^3H_2Me)_2(OMe)_2$	19.5	27.8	27.2	227
	[642]	[30]	[85]	
$Sn(C^1H_2C^2H_2C^3H_2Me)_2(OPr^i)_2$	19.2	27.5	27.0	227
	[520]	[38]	[88]	
$Sn(C^1H_2C^2H_2C^3H_2Me)_2(OCy)_2$	19.4 [515]			227
$Sn(C^1H_2C^2H_2C^3H_2Me)_2(OBu^t)_2$	C^1 21.4 [496]			227
$Sn(C^1H_2C^2H_2C^3H_2Me)_2(OAc)_2$	25.0	27.0	26.5	227, 1728
	[601, 630]	[35]	[92]	
$Sn(C^1H_2C^2H_2C^3H_2Me)(acac)_2$	27.7	27.4	26.5	227
	[914]	[41]	[130]	
	27.3	27.7	28.4	1728
	[910]	[42]	[122]	
$Sn(C^1H_2C^2H_2C^3H_2Me)_2(bzac)_2$	27.9	27.3	28.4	1728
	[901]			
$Sn(C^1H_2C^2H_2C^3H_2Me)_2(OCH_2CH_2O)_2$	22.7	27.7	27.1	1728
	[653]			
$Sn(C^1H_2C^2H_2C^3H_2Me)_2(O_2CC_6H_4OEt)_2$	25.5	26.8	26.4	1728
	[586]	[20]	[98]	
$Sn(C^1H_2C^2H_2C^3H_2Me)_2(OAc)(OMe)$	24.9	27.4	27.0	227
	[666]	[37]	[110]	
$Sn(C^1H_2C^2H_2C^3H_2Me)_2(acac)(OMe)$	27.7	27.8	27.0	227
$Sn(C^1H_2C^2H_2C^3H_2Me)_2(SBu^n)_2$	17.6	28.5	28.9	1728
	[386]	[23]		
$Sn(C^1H_2C^2H_2C^3H_2Me)_2(S\text{-}p\text{-}tol)_2$	18.4	27.9	26.4	1728
	[365]	[26]		
$Sn(C^1H_2C^2H_2C^3H_2Me)_2(SCH_2CHMeS)$	20.7	28.2	20.5	1728
		[27]	[70]	
$Sn(C^1H_2C^2H_2C^3H_2Me)_2(SC_5H_3NNO_2)_2$	26.4	28.4	26.8	1728
	[523]	[34]		
$Sn(C^1H_2C^2H_2C^3H_2Me)_2(SCNSC_6H_4)_2$	28.9	28.2	26.3	1728
	[505]			
$Sn(C^1H_2C^2H_2C^3H_2Me)_2(OMe)Cl$	26.3	27.2	26.7	227
	[622]	[36]	[94]	
$Sn(C^1H_2C^2H_2C^3H_2Me)_2(OAc)Cl$	26.1	27.0	26.5	227
	[516]	[34]	[90]	
$Sn(C^1H_2C^2H_2C^3H_2Me)_2$(cysteine-ethylester)Cl	26.0	28.1	26.5	1639
	[539]		[68.4]	
$Sn(C^1H_2C^2H_2C^3H_2Me)_2Cl_2$	27.0	27.0	26.3	227, 921
	[424]	[36]	[85]	
	27.1	27.1	26.3	1728
	[402]			
$Sn(C^1H_2C^2H_2C^3H_2Me)_2Cl_2.py_2$	38.4	28.4	26.0	227
	[860]	[45]	[150]	
$Sn(C^1H_2C^2H_2C^3H_2Me)Me_3$	11.1	29.2	27.3	875, 882, 921
	[368]	[21]	[53]	
$Sn(C^1H_2C^2H_2C^3H_2Me)Me_2Cl$	18.7	27.7	26.5	921
	[409]	[25]	[68]	

COMPOUND	$\delta(^{13}C)$/p.p.m., J/Hz			REFERENCE
	C^1	C^2	C^3	
$Sn(C^1H_2C^2H_2C^3H_2Me)MeCl_2$	26.0 [490]	26.8 [37]	26.1 [82]	921
$Sn(C^1H_2C^2H_2C^3H_2Me)Cl_3$	33.7 [645]	26.7 [60]	25.4 [120]	227, 921
$Sn(C^1H_2C^2H_2C^3H_2R)Me_3$	11.2 [368]	27.0 [21]	34.3 to 36.5 [53]	921
$Sn(C^1H_2C^2H_2C^3H_2R)Me_2Cl$	18.9 [409]	25.2 to 25.7 [25, 26]	33.3 to 35.7 [64 to 68]	921
$Sn(C^1H_2C^2H_2C^3H_2R)MeCl_2$	26.1 to 27.1 [490]	24.5 to 24.8 [37]	32.7 to 35.1 [84 to 88]	921
$Sn(C^1H_2C^2H_2C^3H_2R)Cl_3$	33.4 [650]	24.8 [60]	32.0 to 34.3 [107 to 110]	227, 921
$Sn(CH_2Ph)_4$	19.0 [257.9] {132, 133}			158, 166, 882
$Sn(CH_2Ph)Me_3$	20.1 [296], 20.4 [298], 20.3 [285.4]			47, 405, 459, 879, 882, 1161
$Sn(C^1H_2Ph)PhBu^t(CH_2{}^2CMe_2Ph)$	18.9	29.4		879
$Sn(CH_2Ar)Me_3$	17.2 to 20.8 [234 to 287.0]			47, 166, 879, 1161
$Sn(C^1H_2C^2H_2Ph)_4$	11.1 [309.8]	32.8 [18.3]		882
$Sn(C^1H_2C^2H_2Ph)Me_3$	12.6 [358.6]	32.7 [18.2]		822
$Sn\{C^1H_2C^2H(C^3H_3)_2\}_3Bu^s$	21.3 [28.9]	26.9 [16]	26.8 [32]	1316
$Sn\{C^1H_2C^2H(C^3H_3)_2\}_2Cl_2$	38.5 [405]	25.9 [39]	25.7 [78]	227
$Sn\{C^1H_2C^2H(C^3H_3)_2\}Me_3$	23.2 [368.1]	27.2 [19.8]	26.6 [42.1]	875, 882
$Sn\{C^1H_2C^2H(C^3H_3)_2\}Cl_3$	44.1 [627]	26.2 [50]	24.9 [106]	227
$Sn\{C^1H_2C^2(C^3H_3)_3\}_4$	33.1 [302]	32.5 [15]	33.8 [32]	458
$Sn_2\{C^1H_2C^2(C^3H_3)_3\}_6$	33.3 [222]	32.5	33.9 [29]	458
$Sn\{C^1H_2C^2(C^3H_3)_3\}_3Bu^s$	31.6 [281]	32.2	34.1 [29]	1316
$Sn\{C^1H_2C^2(C^3H_3)_3\}_3Cl$	40.5 [327]	32.3 [24]	33.2 [42]	458
$Sn\{C^1H_2C^2(C^3H_3)_3\}Me_3$	31.5 [369]	30.8 [20.7]	33.3 [35.5]	875

COMPOUND	$\delta(^{13}C)$/p.p.m., J/Hz			REFERENCE
	C^1	C^2	C^3	
$Sn(C^1H_2C^2H{=}C^3H_2)_4$	16.2 [264.9] [250]	136.3 [48.3]	111.2 [51.3]	80, 875
$Sn(C^1H_2C^2H{=}C^3H_2)Me_3$	18	137.1	110.4	47
$\overline{Sn(C^1H_2C^2H_2C^2H_2C^1H_2)}Me_2$	10.7 [334.8]	29.8 [17.0]		875
$Sn\{(C^1H_2C^2H_2)_2C^3H_2\}Me_2$	10.8 [322.1]	28.6 [29.6]	32.8 [46.4]	875
$Sn(C^1H_2C^2H_2C^3H_2)_2Me_2$	11.3 [339.7]	31.4 [11.9]	25.7 [23.0]	875
$Sn(\mathit{C}H_2CONMe_2)Bu^n{}_3$	16.6			651
$Sn(C^1H_2C^2H_2COMe)Me_3$	3.5 [369.6]	39.0 [28.8]		598
$Sn(C^1H_2C^2H_2COMe)Me_2Cl$	10.2 [490.8]	39.7 [29.0]		598
$Sn(C^1H_2C^2H_2C^3H_2COMe)Me_3$	10.6 [359.8]	21.3 [17.8]	47.4 [57.0]	598
$Sn(C^1H_2C^2H_2C^3H_2COMe)Me_2Cl$	20.0 [490.5]	20.7 [23.8]	44.8 [19.0]	598
$Sn(\mathit{C}H_2C_5H_8OH)Me_3$ *cis*	10.9 [371]			948
trans	15.9 [371]			948
$Sn(\mathit{C}H_2CMe_2Ph)MePhFeCp(CO)_2$	36.0			1729
$(Bu^n{}_3Sn\mathit{C}H_2)_2NMe$	51.0			357
$Ph_3Sn\mathit{C}H_2Sp$-tol	12.0			341
$Sn(\mathit{C}H_2Cl)Me_3$	24.9 [324]			456
$Sn\{C^1H(C^2H_3)_2\}_4$	14.3 [301.0]	22.6 [15.0]		882
$Sn\{C^1H(C^2H_3)_2\}_3Bu^s$	14.5 [305.0]	22.6 [15]		1316
$Sn\{C^1H(C^2H_3)_2\}_3Bu^t$	14.7 [291]	22.6		882
$Sn\{C^1H(C^2H_3)_2\}Me_3$	13.7 [409.0]	22.2 [12.9]		882
	15.2 [410.2]	21.1 [13.4]		875
$Sn\{C^1H(C^4H_3)C^2H_2C^3H_3\}_4$	23.3 [299]	29.7 [12]	14.6 [40]	1316
		C^4 18.6 [15]		
$Sn\{C^1H(C^4H_3)(C^2H_2C^3H_3\}Me_3$	22.9 [408.1]	29.3 [13.4]	14.4 [35.6]	875, 1316
		C^4 18.2 [13.2]		
$Sn\{C^1H(C^4H_3)C^2H_2C^3H_3\}Bu^n{}_3$	22.8	29.6	14.6 [32]	1316
		C^4 18.7 [15]		
$Sn\{C^1H(C^4H_3)C^2H_2C^3H_3\}Pr^i{}_3$	22.6	29.6	14.5	1316
		C^4 18.8		

COMPOUND	δ(^{13}C)/p.p.m., *J*/Hz C^1	C^2	C^3	REFERENCE
$Sn\{C^1H(C^4H_3)C^2H_2C^3H_3\}Bu^i{}_3$	22.6	28.8 C^4 17.9 [12]	14.0	1316
$Sn\{C^1H(C^4H_3)C^2H_2C^3H_3\}Ph_3$	25.4 [434]	28.7 [14] C^4 18.3 [18]	14.3 [40]	1316
$Sn\{C^1H(C^4H_3)C^2H_2C^3H_3\}$-$(CH_2Bu^t)_3$	23.5	28.5 [13] C^4 17.3 [12]	14.3	1316
$Sn\{C^1H(C^4H_3)C^2H_2C^3H_3\}$-$\{C^5H(C^6H_2C^7H_3)_2\}_3$	24.4	29.5 [12] C^4 18.3 [16]	14.5	1316
	C^5 33.2 [289]	C^6 26.0 [12]	C^7 15.2 [35]	
$[Sn(C^1HPhC^2H_2CO_2Et)Me_3]^-$	36.8 [727]	30.0		1636
$SnMe_3CHCl_2$	58.9 [296]			456
$Sn(C^1HC^2H_2C^2H_2)Me_3$	-7.3 [502.8]	1.3 [18.8]		875
$Sn(C^1HC^2H_2C^3H_2C^2H_2)Me_3$	21.2 [389.7]	27.8 [23.3]	25.0 [57.6]	875
$Sn\{C^1H(C^2H_2C^3H_2)_2\}Me_3$	24.0 [405.6]	30.8 [<6]	26.4 [51.2]	875
$Sn(C^1HC^2H_2CH{=}CHC^2H_2)Me_3$	20.2 [400.0]	37.0 [6.8]		875
$Sn(C^1H_2C^2H_2C^3H_2CH{=}CH)Me_3$	C$^{1-3}$ 27.8, 32.6, 32.8			655
$Sn(C^1H_2C^2H_2C^3H_2CH{=}CH)Bu^n{}_3$	C$^{1-3}$ 28.8, 32.2, 32.9			655
Sn(1-norbornane)Me_3	33.8 [459]			875
Sn(2-*exo*-norbornane)Me_3	27.6 [416]			325
Sn(2-*endo*-norbornane)Me_3	28.7 [432]			325
Sn(7-norbornane)Me_3	31.8 [411.8]			325
	38.8 [405.6]			875
	37.9			88
Sn(7-norbornene)Me_3	51.2 [373.6]*			875
	50.4*			88
Sn(nortricyclin)Me_3	31.8 [411.8]			875
$Sn\{$7-norbornadiene-2-$SnMe_3$-5,6-$(CF_3)_2\}Me_3$	82.1 [385]			638
$Sn(5\text{-}C_5H_5)Me_3$	114.3 [19]§			80
$[Sn(o\text{-}C_5H_5)_2(C_4Me_4)Br]^-$	114.3 [30.2]§, 114.8 [26.2]§			475
Sn(1-indenyl)Me_3	45.1	134.7	126.3	145, 360
Sn(1-indane)Me_3	33.4 [333]			405, 459
Sn(1-indane-5-fluoro)Me_3	32.2 [340]			405, 459

*two isomers, §averaged

COMPOUND		$\delta(^{13}C)$/p.p.m., J/Hz	REFERENCE
$Sn\{C^1HCH(CO_2Et)(CH_2)_3\}Me_3$		28.7 [404]	1091
$Sn(C_6H_{11})Me_3$	equatorial	C^1 24.7 [403.8], C^2 30.9, C^3 29.0 [65.0], C^4 26.9	653
	average	C^1 25.9 [407.4], C^2 31.4 [14.4], C^3 29.0 [57.5], C^4 27.3	875
trans-$Sn(4\text{-}MeC_6H_{10})Me_3$		24.8 [390]	1628
cis-$Sn(4\text{-}MeC_6H_{10})Me_3$		26.7 [403]	1628
trans-$Sn(4\text{-}Bu^tC_6H_{10})Me_3$		25.3	1628
cis-$Sn(4\text{-}Bu^tC_6H_{10})Me_3$		27.8	1628
trans-$Sn(2\text{-}HOC_6H_{10})Me_3$		35.4 [300]	1628
$Sn(3\text{-}C_6H_9O)Me_3$		25.3 [376]	1636
Sn(1-tetrahydronaphthalene)Me_3		31.1 [340]	405, 459
Sn(1-tetrahydronaphthalene-5-F)Me_3		30.1 [340]	405, 459
Sn(1-adamantyl)Me_3		28.3 [451.7]	325, 875
$Sn(C_6H_9Br_2)Bu^n{}_3$		18.6	346
$Sn(C_7H_7)Ph_3$		34.6 [375]	1636
$Sn(C_7H_9)Ph_3$		37.7 [346]	1636
$Sn\{C^1(C^2H_3)_3\}_2(OCH_2CH_2)_2NR$		C^1 37.5 to 39.0, C^2 30.3 to 30.6	1091
$Sn\{C^1(C^2H_3)_3\}Me_3$		C^1 21.1 [437.0], C^2 29.9 [0.0]	457, 875, 882
$Sn\{C^1(C^2H_3)_3\}Et_3$		C^1 24.1 [378.0], C^2 31.0 [0.0]	882
$Sn\{C^1(C^2H_3)_3\}Pr^n{}_3$		C^1 23.4 [378.0], C^2 30.9 [0.0]	882
$Sn\{C^1(C^2H_3)_3\}Bu^n{}_3$		C^1 23.5, C^2 30.9 [0.0]	882
$Sn\{C^1(C^2H_3)_3\}Pr^i{}_3$		C^1 27.5 [469.0], C^2 31.8 [0.0]	882
$Sn\{C^1(C^2H_3)_3\}Vi_3$		C^1 26.0 [469.0], C^2 30.4 [0.0]	882
$Sn\{C^1(C^2H_3)_3\}Ph(CH_2Fh)CH_2CMe_2Ph$		C^1 26.6, C^2 30.2	
$Me_3Sn\mathit{C}Me(BEt_2)CEtCC(SnMe_3)_2$		56.0	1631
$Sn\{C^1(C^2H_3)_3\}Bu^n{}_2Cl$		C^1 31.5, C^2 29.4	833
$(Me_3Sn)_2C_5H_4$		52.6 [194]	421
$(Me_3Sn)(Me_3Si)C_5H_4$		59.0	421
$(Me_3Sn)(Me_3Ge)C_5H_4$		58.9	421
Sn(1-adamantyl)Me_3		40.0 [406.6]	325, 875
$Sn(C_{10}H_9)Me_3$		43.0	405, 598
$Sn(\mathit{C}F_3)Me_3$		133.8	452
$Sn(\mathit{C}Cl_3)Me_3$		90.9 [264]	456
Lead, [^{207}Pb]			
$Pb(\mathit{C}H_3)_4$		-2.6 [250]; -3.4	48, 63, 104, 152, 154, 163
		-3.2 [251] {134}	870
$Pb(\mathit{C}H_3)_3$cyclohexyl	axial	-3.2 [147.1]	653, 922
	equatorial	-5.3 [148.3]	653, 922
$Pb(\mathit{C}H_3)_3CH_2Bu^t$		-3.0 [194]	458

COMPOUND	δ(^{13}C)/p.p.m., *J*/Hz	REFERENCE
$Pb(CH_3)_3CH_2Ph$	-2.3	1161
$Pb(CH_3)_3\{CH(SCH_2)_2CH_2\}$	-1.1 [258]	1429
$Pb(CH_3)_3\{CH(SCH_2)_2CH_2$-$Fe(CO)_4\}$	2.9 [274]	1429
$Pb(CH_3)_3CH$=CH_2	-3.0 [279]	458
$Pb(CH_3)_3Ph$	-2.5 [269.7]; -2.2 [274]	870, 880
$Pb(CH_3)_3$(1-naphthyl)	-1.2 [271.1]	880
$Pb(CH_3)_3$(2-naphthyl)	-2.1 [271.6]	880
$\{Pb(CH_3)_3\}_2C$=PMe_3	3.4 [214.8]	1063
$\{Pb(CH_3)_3\}_2CN_2$	1.0 [278.5, 299.7]	996, 1709, 1724
$\{Pb(CH_3)_3\}(AsMe_2)CN_2$	1.7 [291.6]	966
$\{Pb(CH_3)_3\}C(CO_2Et)N_2$	3.2 [316.3]	1709, 1724
$Re(CO)_5Pb(CH_3)_3$	-5.0 [88]	514
$Pb_2(CH_3)_6$	6.0 [+28, +92] {+134.4}	37, 458, 844
$Pb(CH_3)_3Br$	15.2 [246]	458
$Pb(CH_3)_3SOCNMe_2$	10.2	788
$Pb(CH_3)_3S_2CNMe_2$	13.6	788
cis-$Ru(CO)_4\{Pb(CH_3)_3\}_2$	-0.9 [108, 12]	727
cis-$Os(CO)_4\{Pb(CH_3)_3\}_2$	-3.2 [138, 13]	727
$Pb(CH_3)_3SPh$	10.3	285
$(CH_3)_2\overline{PbCPh=CPhCPh=C}Ph$	4.6 [222]	1646
$Pb(CH_3)_2Ph_2$	-1.0 [295]	870
$Pb(CH_3)_2(C_6H_4X$-4$)_2$	0.0 to 0.8 [321 to 357]	870
$Pb(CH_3)_2$2-$C_6H_4C_6H_4$-2 (ring)	1.6 [300]	1646
$Pb(C^1H_2C^2H_3)_4$	C^1 9.6 [200] 10.6, C^2 13.8 [30] 15.7	154, 818
$Pb(C^1H_2C^2H_3)_3OAc$	C^1 32.0 [235], C^2 12.6 [42]	818
$Pb(C^1H_2C^2H_3)_3Cl$	C^1 32.3 to 36.2 [205 to 285]*	1645
$Pb(C^1H_2C^2H_3)_2(OAc)_2$	C^1 52.7 [490], C^2 11.7 [84]	818
$Pb(CH_2Pr^n)_3$-2-pyridyl	21.9	451
$Pb(CH_2Pr^n)_3$-3-pyridyl	21.3	451
$Pb(CH_2Pr^n)_3$-4-pyridyl	21.5	451
$Pb\{C^1H_2C^2(C^3H_3)_3\}_4$	C^1 43.7 [168], C^2 33.7 [33], C^3 33.7 [48]	458
$Pb_2\{C^1H_2C^2(C^3H_3)_3\}_6$	C^1 45.6 [58, 19], C^2 34.0 [24], C^3 34.0 [40]	458
$Pb\{C^1H_2C^2(C^3H_3)_3\}_3Me$	C^1 41.8 [222], C^2 33.4 [34], C^3 33.4 [51]	458
$Pb\{C^1H_2C^2(C^3H_3)_3\}_3C_2H_3$	C^1 44.4 [241], C^2 33.1, C^3 33.5 [54]	458
$Pb\{C^1H_2C^2(C^3H_3)_3\}_3Br$	C^1 61.3 [133], C^2 34.9 [34], C^3 33.9 [66]	458
$Pb\{C^1H_2C^2(C^3H_3)_3\}Me_3$	C^1 39.6 [366], C^2 33.1 [37], C^3 33.1 [61]	458
$Pb(CH_2Ph)Me_3$	25.2 [219]	405, 459, 1161
$Pb(CH_2$-1-naphthyl$)Me_3$	23.0 [194]	1161
$Pb(CH_2$-2-naphthyl$)Me_3$	25.6 [196]	1161

*solvent dependent

COMPOUND		$\delta(^{13}C)$/p.p.m., J/Hz	REFERENCE
$Pb(C_6H_{11})Me_3$	equatorial	C^1 35.0 [415.6], C^2 33.7, C^3 30.1 [121.8], C^4 26.8	653
	axial	C^1 38.7, C^2 32.2, C^3 25.9, C^4 26.8	653
$Pb\{\mathit{C}H(SCH_2)_2CH_2\}Me_3$		28.8 [102]	1429
$Pb\{\overline{\mathit{C}HS[Fe(CO)_4](CH_2)_3S}\}Me_3$		39.2 [27]	1429
Phosphorus, [^{31}P]			
A. Tetramethyl Derivatives			
$[P(\mathit{C}H_3)_4]^+$		11.3 (56) {132.5}	106, 1769
$P(\mathit{C}H_3)_4OMe$		19.4 (116) {127.7}*, 34.3 (7.3) {120.7}§	310
B. Trimethyl Derivatives			
$P(\mathit{C}H_3)_3$		14.3 (-13.6), 16.4 (-13.6), {126.9}, 17.1 (13.5)	106, 116, 162
$M\text{-}P(\mathit{C}H_3)_3$		12.6 to 25.5 (9.3 to 45.8)	71, 366, 440, 490, 720, 851, 890, 904, 1016, 1034, 1037, 1202, 1346, 1347, 1363, 1373, 1388, 1402, 1403, 1404, 1408, 1529, 1539
$P(\mathit{C}H_3)_3(CF_3)_2$		8 (102); 8.6 (102)	526, 1231
$P(\mathit{C}H_3)_3CH_2$		18.6 (56.6), 19.7 (56.0), 14.2 (+56)	143, 523, 1769
		18.5 (57) {128}, 14.2 (+56.0), 18.9 (56.0)	143, 416, 1063
$P(\mathit{C}H_3)_3{=}CR^1R^2$		11.3 to 23.9 (52 to 59)	435, 527, 759, 851, 1063, 1359
$[P(\mathit{C}H_3)_3R]^+$		7.5 to 18.8 (51.3 to 62.3)	416, 450, 478, 906, 928, 1019, 1029, 1042, 1053, 1062, 1363, 1531, 1650
$\overline{P(\mathit{C}H_3)_3CH_2CH_2CH_2O}$		21.1 (113.5)*, 32.9§	808
$P(\mathit{C}H_3)_3{=}NR$		14.9 to 22.2 (64.1 to 69.2)	411, 758, 1040, 1122, 1287, 1604
$P(\mathit{C}H_3)_3O$		17.4 (58.3), 18.6 (68.3)	143, 1770
$P(\mathit{C}H_3)_3S$		22.8 (+56.1) {129.3}	106
$P(\mathit{C}H_3)_3Se$		23.2 (+48.5) {130.5}	106

*equatorial, §axial

COMPOUND	$\delta(^{13}C)$/p.p.m., J/Hz	REFERENCE
C. Dimethyl Derivatives		
$P(\mathit{C}H_3)_2H$	6.9 (11.6)	162
$M\text{-}P(\mathit{C}H_3)_2R$	8.7 to 22.5 (14 to 41.2)	71, 76, 162, 203, 236, 400, 406, 443, 445, 483, 490, 569, 570, 582, 585, 586, 838, 864, 1166, 1171, 1341, 1388, 1480, 1529, 1539, 1547, 1774
$W\{P(\mathit{C}H_3)_2CH_2CH_2P(\mathit{C}H_3)_2\}\text{-}(CBu^t)(CHBu^t)(CH_2Bu^t)$	33.0	1402
$P(\mathit{C}H_3)_2(BXY)_2$	4.2 to 14.4 (14 to 35)	670
$P(\mathit{C}H_3)_2(BH_2)R$	20.1	1556
$P(\mathit{C}H_3)_2(CF_3)_3$	7, 8.8 (88)	526, 1231
	15.8 (80)	1659
$P(\mathit{C}H_3)_2R{=}CR^1{}_2$	10.6 to 24.7 (53.7 to 65.9)	706, 1057, 1063, 1289, 1605, 1665
$[P(\mathit{C}H_3)_2R_2]^+$	6.5 to 23.8 (26.8 to 65.9)	76, 77, 106, 366, 367, 757, 764, 804, 1041, 1053, 1337, 1531, 1605, 1651, 1665
$P(\mathit{C}H_3)_2R$	8.4 to 17.1 (8 to 17)	106, 404, 443, 483, 928, 1233, 1650
$P(\mathit{C}H_3)_2({=}CH_2)N{=}PMe_3$	24.2 (72.8, 5.1)	758, 1040
$P(\mathit{C}H_3)_2R{=}NR^1$	14.9 (69), 22.2 (64.7, +69.1)	411, 1039, 1193, 1604
$\overline{P(\mathit{C}H_3)_2(OMe)(CH_2)_5}$	14.3, 18.3 (107.4)	1775
$P(\mathit{C}H_3)_2PhO$	18.0 (70.8)	1770
$P(\mathit{C}H_3)_2RS$	18.5 to 21.8 (52 to 54.3)	404, 928, 1650
$P(\mathit{C}H_3)_2(CF_3)_2Cl$	22.4 (92)	1231
$P(\mathit{C}H_3)_2N(SiMe_3)_2$	19.3 (22.0)	1122
$P(\mathit{C}H_3)_2N(SiMe_3)N(SiMe_3)_2$	31.4 (75.7)	1122
$P_4(\mathit{C}H_3)_8N_4$	20.6 (+100) {127.5}	106
$P(\mathit{C}H_3)_2N(SiMe_3)OSiMe_3$	20.8 (95.2)	1122
$P(\mathit{C}H_3)_2(N_3)O$	17.2 (85)	525
$P(\mathit{C}H_3)_2(N_3)S$	24.2 (68)	525
$P(\mathit{C}H_3)_2N(SiMe_3)_2S$	28.8 (67.1)	1122
$P_2(\mathit{C}H_3)_4$	10.3 (−22, +14), 11.2	162, 289, 863
$P(C^1H_3)_2(O)P(C^2H_3)_2(S)$	C^1 12.5 (50.4, 11.0), C^2 16.4 (59.9, 16.5)	1045

COMPOUND	$\delta(^{13}C)$/p.p.m., *J*/Hz	REFERENCE
$[P(\mathit{C}H_3)_2S]_2$	16.9 (48, 12)	863
$P(\mathit{C}H_3)_2O_2SiMe_3$	16.8 (97.1)	1045
$P(C^1H_3)_2(O)OP(C^2H_3)_2(S)$	C^1 19.0 (93.3), C^2 26.5 (72.5)	1657
$P(\mathit{C}H_3)_2(S)OP(\mathit{C}H_3)_2(S)$	26.9 (72.2, 1.7)	1657
$P(\mathit{C}H_3)_2(S)SP(\mathit{C}H_3)_2(S)$	32.9 (56.0, 0.3)	1657
$P(\mathit{C}H_3)_2(O)Cl$	24.3 (80)	525
$P(\mathit{C}H_3)_2(S)Cl$	30.3 (61)	525
<u>D. Monomethyl Derivatives</u>		
$P(\mathit{C}H_3)H_2$	-4.6 (9.3)	162
$[P(\mathit{C}H_3)H(CH_2)_5]^+$	5.3	854
$P(\mathit{C}H_3)HCH_2CH_2P(NMe_2)_2$	13.9 (14)	1233
$M\text{-}P(\mathit{C}H_3)R_2$	5.7 to 21.8 (16.9 to 39)	193, 198, 279, 443, 445, 490, 570, 585, 586, 864, 1176, 1539
$P(\mathit{C}H_3)\{(CH_2)_4\}_2$	19.6 (65.9)	1018
$P(\mathit{C}H_3)(CF_3)_3CN$	15.8 (80)	1231
$P(\mathit{C}H_3)R_2(=CR^1{}_2)$	8.3 to 37.9 (48.4 to 69.1)	759, 1067, 1289, 1665, 1685, 1686
$[P(\mathit{C}H_3)R_3]^+$	4.2 to 12.7 (51.5 to 61.0)	76, 162, 234, 472, 623, 858, 1649, 1665, 1769, 1771, 1776
$P(\mathit{C}H_3)R_2$	0.5 to 15.0 (10 to 24.4)	61, 76, 116, 133, 237, 277, 404, 443, 627, 804, 854, 1233, 1265, 1651, 1664, 1771
$P(\mathit{C}H_3)=CBu^t(OSiMe_3)$	9.0 (38.0)	1250
$P(\mathit{C}H_3)(SiMe_3)(COBu^t)$	2.0 (13.0)	1250
$P(\mathit{C}H_3)(CF_3)_3NMe_2$	11.5 (87), 11.7	1231, 1659
$P(\mathit{C}H_3)(CH_2)_5{=}NP(CH_2)(CH_2)_5$	5.0 (63.5)	1057
$P(\mathit{C}H_3)Bu^t{}_2NSiMe_3$	10.6 (+51.5)	411
$P(\mathit{C}H_3)Ph_2NCN$	13.9 (72)	1604
$\{P_2(\mathit{C}H_3)R\}_2$	2.6 to 6.9 (9.3 to 33.7)	863
$\{P_2(\mathit{C}H_3)RS\}_2$	15.1 to 17.9 (41 to 52.5, 11.5 to 17)	863
$P(\mathit{C}H_3)(CF_3)_3OMe$	11.4 (71)	1231, 1659
$P(\mathit{C}H_3)R_2(O)$	11.5 to 20.4 (60 to 73.5)	76, 801, 804, 854, 1138, 1651, 1770, 1771

COMPOUND	$\delta(^{13}C)$/p.p.m., J/Hz	REFERENCE
$P(CH_3)(OMe)_2O$	9.0 (+142.2) {127.0}	106
$P(CH_3)\{OCH(CF_3)_2\}\{CH_2C(CF_3)_2O\}$-{$OCH(CF_3)_2R$}	22.9; 24.2 (128.8, 133.3)	1179
$P(CH_3)\{OCH(CF_3)_2\}\{CH_2C(CF_3)_2O\}F$	21.0 (134.9)	1179
$P(CH_3)(CF_3)_3(SMe)$	12.5 (72)	1231, 1659
$P(CH_3)R_2(S)$	15.9 to 26.1 (50 to 56)	277, 404, 854, 1651, 1679, 1771
$\{P(CH_3)Ph(Se)\}_2CH_2$	21.2 (52.5), 22.7 (51.9)	1664
$P(CH_3)(CF_3)_3F$	27.5 (100)	1231, 1659
$P(CH_3)(CF_3)_3Cl$	21.1 (72)	1231, 1659
$P(CH_3)(OMe)_2$	19.3 (23)	404
$P(CH_3)(OEt)_2O$	11.6 (143.2)	73
$P(CH_3)(OCH_2CH_2)_2O$	22.6 (12.5)	1658
$SP(CH_3)(OCH_2CH_2)_2O$	21.4 (125.1)	1658
$P(CH_3)OF_2$	21.5 (147)	123
$P(CH_3)F_2OC(CF_3)_2C(CF_3)_2O$	16.8 (185) {122}	699
$P(CH_3)OCl_2$	31.5 (+104) {+130}	123, 898
$P(CH_3)(S)S(CH_2)_3S$	27.0 (51.5)	1071
$P(CH_3)SCl_2$	(+75) {+135}; 87.5 (41.0)	123, 898
$P(CH_3)Cl_2$	15.0 (45)	123

E. Ethyl Derivatives

61, 73, 106, 116, 117, 193, 202, 234, 235, 277, 404, 406, 411, 416, 443, 523, 525, 570, 584, 616, 681, 724, 857, 860, 863, 1171, 1287, 1310, 1328, 1530, 1532, 1545, 1666, 1770, 1771, 1774

F. Other Alkyl Derivatives

15, 16, 61, 73, 74, 76, 77, 95, 107, 116, 123,

COMPOUND	$\delta(^{13}C)$/p.p.m., J/Hz	REFERENCE
		162, 188, 193, 202, 234, 240, 241, 277, 315, 378, 380, 404, 416, 421, 478, 483, 520, 526, 570, 584, 599, 616, 626, 627, 634, 672, 681, 698, 724, 757, 759, 786, 801, 804, 806, 808, 824, 847, 854, 858, 860, 863, 864, 898, 928, 930, 1001, 1018, 1039, 1041, 1057, 1062, 1106, 1133, 1136, 1138, 1179, 1193, 1199, 1231, 1233, 1239, 1256, 1260, 1265, 1277-9, 1287, 1289, 1297, 1298, 1307, 1308, 1310, 1337, 1356, 1371, 1420, 1446, 1512, 1531, 1539, 1540, 1554, 1556, 1604, 1605, 1613, 1648, 1654, 1657, 1659, 1662, 1664, 1665, 1667, 1672, 1673, 1675-1679, 1681, 1683, 1687, 1720, 1731, 1737, 1738, 1752, 1769, 1771-1781
Arsenic		
$As(CH_3)_5$	17.0*, 26.0§	473
$As(CH_3)_3$	11.4	617
$M\text{-}As(CH_3)_3$	6.6 to 11.8	31, 203, 400, 403, 409, 569, 580, 617
$Cr\{As(CH_3)_3\}(C_7H_8)(CO)_2$	15.8	1765
$\overline{As(CH_3)_3CH_2CH_2CH_2O}$	18.8 {134.1}*, 23.8 {130.0}§	808
$M\text{-}As(CH_3)_2Ph$	5.7 to 17.0	29, 30, 120, 305, 582
$M\text{-}o\text{-}\{As(CH_3)_2\}_2C_6H_4$	10.0 to 15.6	402, 1407, 1752
$[As(CH_3)_2(CH_2)_4CH_2]^+$	6.4	854
$As(CH_3)_2C_5H_5$	22.4	706
$As(CH_3)_2C_6H_4NCH$	10.8	1784

*equatorial, §axial

COMPOUND	$\delta(^{13}C)$/p.p.m., J/Hz	REFERENCE
$As(\mathit{C}H_3)_2CRN_2$	9.2 to 10.4	474, 641, 996, 1709
$\{As(\mathit{C}H_3)_2F_2\}_2CH_2$	23.0 (J_{FC} = 18)	1782
$As(\mathit{C}H_3)_2NMe_2$	10.1	471
$As(\mathit{C}H_3)_2(OMe)_3$	17.6	1783
$Mo[o\text{-}\{As(\mathit{C}H_3)Ph\}_2C_6H_4](CO)_4$	15.6, 15.8	396
$[As(\mathit{C}H_3)Ph_3]^+$	7.6 to 9.9	472
$As(\mathit{C}H_3)(CH_2)_4CH_2$	5.1	845
$As(\mathit{C}H_3)\{C(N_2)CO_2Et\}_2$	6.8	471, 1709
$As(\mathit{C}H_3)\{C(N_2)CO_2Et\}NMe_2$	9.7	471, 1709
$As(\mathit{C}H_3)\{(CH_2)_4CH_2\}O$	14.6	854
$As(\mathit{C}H_3)\{(CH_2)_4CH_2\}S$	17.5	854
$As(\mathit{C}H_3)\{(CH_2)_4CH_2\}Se$	18.6	854
$As(\mathit{C}H_3)Ph_2F_2$	22.7 (J_{FC} = 21.5)	1782
$As(\mathit{C}H_3)\{(CH_2)_4CH_2\}Cl_2$	22.4	854
$As(\mathit{C}H_3)\{(CH_2)_4CH_2\}Br_2$	22.2	854
$As(\mathit{C}H_3)(NMe_2)_2$	12.4	471
$[As(C^1H_2C^2H_3)_4]^+$	C^1 14.7, C^2 7.8	617
$As(C^1H_2C^2H_3)_3$	C^1 16.5, C^2 10.8	617
$M\text{-}As(C^1H_2C^2H_3)_3$	C^1 12.2 to 18.1, C^2 8.8 to 11.3	202, 406, 617, 681
$\{As(\mathit{C}H_2Me)_2F_2\}_2CH_2$	32.1 (J_{FC} = 15.5)	1782
$IrClIMeCO\{As(C^1H_2C^2H_3)_2Ph\}_2$	C^1 14.1, 17.1, C^2 9.3, 9.8	120
$[As(C^1H_2C^2H_2Me)_4]^+$	C^1 23.4, C^2 17.2	617
$As(C^1H_2C^2H_2Me)_3$	C^1 28.1, C^2 20.6	617
$M\text{-}As(C^1H_2C^2H_2Me)_3$	C^1 21.6 to 26.1, C^2 18.1 to 19.1	617
$[As(C^1H_2C^2H_2Et)_4]^+$	C^1 21.8, C^2 25.4	617
$As(C^1H_2C^2H_2Et)_3$	C^1 25.1, C^2 29.4	617
$M\text{-}As(C^1H_2C^2H_2Et)_3$	C^1 19.2 to 23.7, C^2 26.4 to 27.0	617
$[As\{(C^1H_2C^2H_2)_2CH_2\}Me_2]^+$	C^1 22.8, C^2 22.8	854
$As\{(C^1H_2C^2H_2)_2CH_2\}Me$	C^1 22.4, C^2 23.4	854
$As\{(C^1H_2C^2H_2)_2CH_2\}Ph$	C^1 22.8, C^2 24.7	854
$As\{(C^1H_2C^2H_2)_2CH_2\}MeO$	C^1 33.0, C^2 25.5	854
$As\{(C^1H_2C^2H_2)_2CH_2\}MeS$	C^1 34.0, C^2 24.9	854
$As\{(C^1H_2C^2H_2)_2CH_2\}MeSe$	C^1 32.1, C^2 23.1	854
$As\{(C^1H_2C^2H_2)_2CH_2\}MeCl_2$	C^1 51.1, C^2 23.9	854
$As\{(C^1H_2C^2H_2C^3H_2)_2CH_2\}MeBr_2$	C^1 47.2, C^2 24.2	854
$As(C^1H_2C^2H_2C^3H_2O)Me_3$	C^1 24.9 {137.9}, C^2 26.6 {129.4}, C^3 60.6 {136.7}	808
$(AsF_2Ph_2\mathit{C}H_2)_2CH_2$	C^1 40.3 (J_{FC} = 18.0)	1782
$Ph_2As\mathit{C}H_2CH_2PPh_2$	29.4 (16.2)	1752
$W(Ph_2As\mathit{C}H_2CH_2PPh_2)(C_3H_4R)\text{-}(CO)_3$	29.2 to 29.7 (19.1 to 20.6)	1752
$F_2Ph_2As\mathit{C}H_2\mathit{C}H_2AsPh_2F_2$	34.8 (J_{FC} = 4.8, 20.7)	1782
$[As(\mathit{C}H_2COMe)Ph_3]^+$	44.2	1665

COMPOUND	$\delta(^{13}C)$/p.p.m., J/Hz	REFERENCE
$[As(\mathit{C}H_2COPh)Ph_3]^+$	42.6	1665
$(AsMe_2F_2)_2\mathit{C}H_2$	45.5 (J_{FC} = 22)	1782
$(AsEt_2F_2)_2\mathit{C}H_2$	40.7 (J_{FC} = 22.5)	1782
$(AsPr^i_2F_2)_2\mathit{C}H_2$	36.9 (J_{FC} = 23.4)	1782
$(AsPh_2F_2)_2\mathit{C}H_2$	48.9 (J_{FC} = 25.6)	1782
$[As\{C^1H(C^2H_3)_2\}_3Pr^n]^+$	C^1 26.5, C^2 18.7	617
$As\{C^1H(C^2H_3)_2\}_3$	C^1 22.6, C^2 21.5	617
$M\text{-}As\{C^1H(C^2H_3)_2\}_3$	C^1 24.1 to 28.6, C^2 20.2	617
$\{As(\mathit{C}HMe_2)_2F_2\}_2CH_2$	42.9 (J_{CF} = 15.0)	1782
$\{As(\mathit{C}HCOMe)Ph_3\}_2PdCl_2$	41.1, 41.6	1665
Antimony		
$Sb(\mathit{C}H_3)_3(CH_2Ph)_2$	13.1	470
$Sb(\mathit{C}H_3)_3(C_6H_4Cl\text{-}4)_2$	13.3	470
$\{Sb(\mathit{C}H_3)_2\}_2CN_2$	0.7	474, 1709
$Sb(\mathit{C}H_3)_2C(N_2)CO_2Et$	0.5	474, 1709
$Sb(\mathit{C}H_3)_2\{As(CH_3)_2\}CN_2$	-0.6	996
$Sb(\mathit{C}H_3)(CH_2SiMe_3)_2{=}C(SiMe_3)_2$	5.2	1614
$Sb(\mathit{C}H_3)Ph_2F$	11.0 (J_{CF} = 16.7)	1782
$Sb(\mathit{C}H_2Me)_3(C_6H_4Cl\text{-}4)_2$	25.0	470
cis-$\{(Pr^n\mathit{C}H_2)_3Sb\}_2PtCl_2$	15.1	617
trans-$\{(Pr^n\mathit{C}H_2)_3Sb\}_2PtCl_2$	12.5	617
$Sb(\mathit{C}H_2Pr^n)_3Br_2$	46.2	470
$Sb(\mathit{C}H_2SiMe_3)_2MeC(SiMe_3)_2$	14.1	1614
Bismuth		
$Bi(\mathit{C}H_3)_2C(N_2)CO_2Et$	11.9	474, 1709
Selenium, $[^{77}Se]$		
$[Se(\mathit{C}H_3)_3]^+$	-50 {145.8}	103, 243, 1291
$Se(\mathit{C}H_3)_2$	-62 {139.0, 140.3}	50, 103, 243, 1291
$\overline{Se(\mathit{C}H_3)_2(CH_2)_4CH_2}$	15.7	854
$Se(\mathit{C}H_3)_2O$	-60	1291
$Se(\mathit{C}H_3)_2Br_2$	-68	1291
$Se(\mathit{C}H_3)H$	-46, -48 {142.7}	50, 243, 1291
$\{Se(\mathit{C}H_3)\}_2BCH_3$	3.6	566
$Se(\mathit{C}H_3)CMeW(CO)_5$	21.8	178
$Se(\mathit{C}H_3)CN_4Ph$	8.6, -56.5	1226, 1713
$Se(\mathit{C}H_3)C{\equiv}CCH{=}CH_2$	9.6	1785
$Se(\mathit{C}H_3)Ph$	6.8, 7.1, -63.4	1291, 1710, 1786
$Se(\mathit{C}H_3)PhBr_2$	-53.8	1291
$Se(\mathit{C}H_3)C_6H_4C_4H_2SCO_2Me$	4.9	807
$\{Se(\mathit{C}H_3)\}_4Sn$	3.5	232
$\{Se(\mathit{C}H_3)\}_3SnMe$	1.7	232
$\{Se(\mathit{C}H_3)\}_2SnMe_2$	-0.7	232
$\{Se(\mathit{C}H_3)\}SnMe_3$	-3.2	232
$\{Se(\mathit{C}H_3)\}SnPh_3$	-1.7	232
$Se(\mathit{C}H_3)O_2D$	-89.8	1291

COMPOUND	δ(^{13}C)/p.p.m., *J*/Hz	REFERENCE
$[Se(CH_3)O_3]^-$	-13.0	1291
$Se_2(CH_3)_2$	-73.7, -75 {141.7}	103, 232, 1291
$[Se(CH_3)]^-$	-58.2	1291
$[Se(CH_2Me)_3]^+$	-56.4	1291
$Se(C^1H_2C^2H_3)_2$	C^1 -61.2, C^2 8.5	1291
$Ru\{Se(C^1H_2C^2H_3)_2\}_2(NO)Br_3$	C^1 27.9, C^2 14.0	799
$Se(CH_2Me)_2Br_2$	-54.4	1291
$Se(CH_2Me)C{\equiv}CCH{=}CH_2$	21.9	1785
$Se(C^1H_2C^2H_3)Ph$	C^1 20.9, C^2 15.4	1710
$Ru\{Se(C^1H_2C^2H_3)Ph\}_2(NO)Cl_3$	C^1 29.2; 29.5, C^2 12.8	799
$Ru\{Se(C^1H_2C^2H_3)Ph\}_2(NO)Br_3$	C^1 31.0; 31.6, C^2 13.3	799
$Se(CH_2Me)O_2SnMe_3$	5.9	1632
$[Se(C^1H_2C^2H_3)O_3]^-$	C^1 -18.5, C^2 7.1	1291
$Se(CH_2Et)C{\equiv}CCH{=}CH_2$	31.6	1785
$Se(C^1H_2C^2H_2Me)Ph$	C^1 23.4, C^2 14.4	1710
$[Se\{(C^1H_2C^2H_2)_2CH_2\}H]^+$	C^1 41.8, C^2 23.8	854
$[Se\{(C^1H_2C^2H_2)_2CH_2\}Me]^+$	C^1 34.1, C^2 20.5	854
$Se(C^1H_2C^2H_2)_2CH_2$	C^1 20.2, C^2 29.1	854
$Se\{(C^1H_2C^2H_2)_2CH_2\}O$*	C^1 42.1, C^2 18.6	854
$Se\{(C^1H_2C^2H_2)_2CH_2\}O$§	C^1 39.4, C^2 16.8	854
$Se\{(C^1H_2C^2H_2)_2CH_2\}O_2$	C^1 57.5, C^2 25.1	854
$Se\{(C^1H_2C^2H_2)_2CH_2\}Br_2$	C^1 51.2, C^2 20.9	854
$Se\{(C^1H_2C^2H_2)_2CH_2\}I_2$	C^1 29.7, C^2 25.7 or 26.0	854
$\overline{Se(CH_2CH_2})(p\text{-tol})Cl$	38.1 [18.5] {155.2, 4.4} 65.2 [59.3] {152.0, 3.2} (J_{CC} = 37.3)	286, 610
$[Se(C^1H_2C^2H_2NH_3)_2]^{2+}$	C^1 20.5 [-65.9], C^2 40.7	1226
$Se(C^1H_2C^2H_2Cl)Ph$	C^1 29.0 [65.8], C^2 43.1 [7.7]	1787
$Se(C^1H_2C^2H_2Cl)(p\text{-tol})$	C^1 29.3 [66.3] {145.6, 4.8, 7.8}, C^2 66.3 (J_{CC} = 28.8)	286
$Se(C^1H_2C^2H_2Cl)(p\text{-tol})Cl_2$	C^1 65.0 [59.4], C^2 38.1 [18.1]	610
$Se(CH_2CHClMe)Ph$	37.3 [63.2]	1787
$Se(CH_2CClMe_2)Ph$	44.4 [66.6]	1787
$Se(CH_2CClMeEt)Ph$	43.6	1787
$Se_2(CH_2Ph)_2$	32.6	1227
$Se\{C^1H(C^2H_3)_2\}Ph$	C^1 33.1, C^2 24.1	1710
$\overline{SeCHMeCH_2C_6H_4\text{-}2}$	45.5	743
Z-$Se(CHMeCHClMe)Ph$	45.9 [20]	1787
E-$Se(CHMeCHClMe)Ph$	46.4 [63.9]	1787
$Se(\overline{CHClCH_2CCl_2})C_6H_4X$	26.7 to 28.0	1788
$Se\{C^1(C^2H_3)_3\}H$	C^1 -47.6, C^2 9.7	1291
$Se\{C^1(C^2H_3)_3\}Ph$	C^1 41.8, C^2 32.0	1710
$Se_2\{C^1(C^2H_3)_3\}_2$	C^1 -70, C^2 6.5	1291
$[SeCMe_3]^-$	-60.6	1291

*average of axial and equatorial oxygen, §axial oxygen

COMPOUND	$\delta(^{13}C)$/p.p.m., J/Hz	REFERENCE
$SeCMe_2CH_2C_6H_4$-2	45.9	743
$SeCCH_2CH_2C(O)(CH_2)_3C{=}NNR$	103.8 to 109.5	1789
$MeNC_6H_4SeCCMeCHC_6H_2(NO_2)(OMe)O$	113.2	1070
Tellurium, [^{125}Te]		
$[Te(CH_3)_3]^+$	{146.0}	103
$Te(CH_3)_2$	[+158.5, 162] {139.7, 140.7}	50, 103
$[Te(CH_3)(CH_2)_4CH_2]^+$	-0.6	854
$Te(CH_3)C{\equiv}CCH{=}CH_2$	-14.5	1785
$Te(CH_3)Ph$	-16.8, -16.4	1786, 1790
$Te(CH_2Me)C{\equiv}CCH{=}CH_2$	17.2	1785
$Te(C^1H_2C^2H_3)Ph$	C^1 -0.4, C^2 16.3	1790
$[Te\{(C^1H_2C^2H_2)_2CH_2\}H]^+$	C^1 24.0, C^2 25.0	854
$[Te\{(C^1H_2C^2H_2)_2CH_2\}Me]^+$	C^1 17.7, C^2 19.9	854
$Te(C^1H_2C^2H_2)_2CH_2$	C^1 -2.1, C^2 29.9	854
$Te\{(C^1H_2C^2H_2)_2CH_2\}Br_2$	C^1 36.9, C^2 20.3	854
$Te\{(C^1H_2C^2H_2)_2CH_2\}I_2$	C^1 33.2, C^2 21.4	854
$Te(CH_2CO)_2CH_2$	15.2 [139.7]	1791
$Te(CH_2CO)_2CMe_2$	10.7 [134.7], 6.3 [134.7]	817, 1791
$Te(CH_2CO)_2CHEt$	15.4 [139.7]	1791
$TeC^1H_2COCHMeCOC^2HMe$	C^1 16.2 [145.9], C^2 26.4 [118.4]	1791
$Te_2(CH_2Ph)_2$	6.6	1227
$Te\{C^1H(C^2H_3)_2\}Ph$	C^1 14.9, C^2 26.6	1790
$Te(CHMeCO)_2CH_2$	27.6 [126.0]	1791
$Te(CHMeCO)_2CHMe$	27.7	1791

TABLE 2.2

^{13}C Chemical Shift Data for Vinyl Groups Attached to Metals

COMPOUND	$\delta(^{13}C)$/p.p.m., *J*/Hz	REFERENCE
Titanium		
$Ti(C_5H_4FeCp)(NMe_2)_3$	109.6	184
$Ti(C_5H_4FeCp)(NEt_2)_3$	109.8	184
$Fe\{C_5H_4Ti(NEt_2)_3\}_2$	C^1 109.8, $C^{2,5}$ 75.0, $C^{3,4}$ 69.5	1251
$Ru\{C_5H_4Ti(NEt_2)_3\}_2$	C^1 112.6, $C^{2,5}$ 76.8, $C^{3,4}$ 71.5	1251
Chromium		
$Cr_2(C_8H_8)Cp_2$	178.6	1350
$Cr_2(\mathit{C}CPhCH\mathit{C}Ph)Cp_2CO$	194, 207	1349
Molybdenum		
$Mo(C^1H{=}C^2HBu^t\text{-}\mathit{trans})Cp\{P(OMe)_3\}_3$	C^1 151.3 (4.5), C^2 128.1	1096
$Mo\{C^1H{=}C^2(CN)_2\}Cp(CO)_3$	C^1 207.3, C^2 99.2	280
$\overline{Mo(C^1H{=}C^2HCMe{=}O})Cp(CO)_2$	C^1 247.1, C^2 131.7	749
$Mo\{\mathit{C}Cl{=}\mathit{C}(CN)_2\}Cp(CO)_3$	217.3, 100.7	441
$\overline{Mo\{C^1{=}C^2PhC(CF_3)_2OCH_2}\}Cp(CO)_3$	C^1 148.7, C^2 135.7	1312
Tungsten		
$W\{C^1H{=}C^2(CN)_2\}Cp(CO)_3$	C^1 206.7, C^2 98.9	280
$\overline{W(C^1H{=}C^2HCMeO})Cp(CO)_2$	C^1 237.7 [73], C^2 131.7	749, 1050
$W\{C^1(CN){=}C^2(CN)_2\}Cp(CO)_3$	C^1 168.4, C^2 107.0	280
$W\{C^1Cl{=}C^2(CN)_2\}Cp(CO)_3$	C^1 195.7, C^2 101.6	441
trans-$W\{\mathit{C}(p\text{-}tol){=}C{=}O\}Cp(CO)_2PMe_3$	-15.3 (7.3)	1404
$W(\mathit{C}Ph{=}C{=}O)Cp(CO)(PMe_3)_2$	5.5 (12.2)	1037
$W\{\mathit{C}(p\text{-}tol){=}C{=}O\}Cp(CO)(PMe_3)_2$	5.0 (14.7)	1037
$W\{\mathit{C}(C_6H_4OMe\text{-}4){=}C{=}O\}Cp(CO)(PMe_3)_2$	3.6 (13.4)	1037
$W(\mathit{C}Fc{=}C{=}O)Cp(CO)(PMe_3)_2$	-6.6 (13.4)	1037
$W\{\mathit{C}(NCy){=}CMeOMe\}(CO)_5$	227.6	1368
Manganese		
$Mn\{C^1H{=}C^2(CN)_2\}(CO)_5$	C^1 208.1, C^2 101.3	280
$Mn\{C^1(CN){=}C^2(CN)_2\}(CO)_5$	C^2 104.8	280
$Mn_2(\mu\text{-}C^1C^2PhH)Cp_2(CO)_4$	C^1 284.2, C^2 125.2	970

COMPOUND	$\delta(^{13}C)$/p.p.m., J/Hz	REFERENCE
$Mn\{C^1{=}C^2PhC(CN)_2C(CN)_2CH_2\}(CO)_5$	C^1 164.2, C^2 133.6	1312
$Mn\{C^1{=}C^2PhC(CN)(CF_3)C(CN)(CF_3)CH_2\}$-$(CO)_5$	C^1 166.4, C^2 134.3	1312
$Mn(C^1PhC^2O)Cp(CO)_2$	C^1 -4.1, C^2 237.5	754
$Mn(C^1PhC^2O)(C_5H_4Me)(CO)_2$	C^1 -5.1, C^2 236.4	754
Iron		
$Fe\{C^1H{=}C^2(CN)_2\}Cp(CO)_2$	C^1 212.8, C^2 99.6	280
$Fe(C^1H{=}C^2HC^2H{=}C^1H)Fe(CO)_6$	C^1 153.0, C^2 108.7	949
$Fe(C^1Me{=}C^2HC^3H{=}C^4H)Fe(CO)_6$	C^1 179.1, C^2 108.7, C^4 151.1, C^3 108.7	949
$Fe(C^1H{=}C^2MeC^3H{=}C^4H)Fe(CO)_6$	C^1 149.8, C^2 129.3, C^4 152.7, C^3 110.4	949
$Fe\{C^1(C_2Me){=}C^2MeC^3(C_2Me){=}C^4Me\}$-$Fe(CO)_6$	C^1 131.5, C^2 140.0, C^4 176.4, C^3 84.9	847
$Fe\{\mathit{C}{=}CMeCOC(CN)Bu^tCH_2\}Cp(CO)_2$	193.9	1213
$Fe\{\mathit{C}{=}CPhCOCPh_2CH_2\}Cp(CO)_2$	199.9	1213
$Fe\{C^1{=}C^2PhC(CN)(CF_3)C(CN)(CF_3)CH_2\}$-$Cp(CO)_2$	C^1 160.2, C^2 135.5	1312
$Fe\{C^1{=}C^2PhC(CN)_2C(CN)_2CH_2\}Cp(CO)_2$	C^1 164.4, C^2 135.5	1312
$Fe\{C^1{=}C^2MeC(CN)_2C(CN)_2CH_2\}Cp(CO)_2$	C^1 156.2, C^2 128.0	1312
$Fe\{C^1{=}C^2PhC(CF_3)_2OCH_2\}Cp(CO)_2$	C^1 153.7, C^2 133.5	1312
$Fe\{C^1{=}C^2PhCON(SO_2Cl)CH_2\}Cp(CO)_2$	C^1 186.6, C^2 146.6	1312
$Fe\{\mathit{C}(CO_2Me{=}C(CO_2Me)SCS\}(CO)_2(PMe_3)_2$	140.2 (5)	1433
$Fe_2\{\mathit{C}(CO_2Me)C(CO_2Me)NR\}(CO)_6$	112.0 or 125.8	1431
$Fe\{C^1(CF_3){=}C^2(CF_3)CH_2CH{=}CHMe\}Cp(CO)$	C^1 156.3 (J_{CF} = 30, 3) C^2 138.1 (J_{CF} = 30, 6)	1116
$Fe\{\mathit{C}{=}NBu^tCMe{=}NBu^t\}I(CNBu^t)_3$	194.5	1182
$Fe_2(\mathit{C}Ph{=}CPhS)(CO)_6$	154.7	1457
$Fe_2(\mathit{C}Ph{=}CPhS)(CO)_5PPh_3$	151.5	1457
Cobalt		
$Co_3\mathit{C}_2(NEt_2)_2Cp_3$	321	684
$Co_3(CO)_9\mathit{C}Y$		
Y = H	263	664
Me	296	664
C_5H_{11}	305.6	537
C_9H_{19}	306.0	537
$C_{10}H_{21}$	306.4	537
Ph	286	664
CF_3	255	664

COMPOUND	$\delta(^{13}C)$/p.p.m., J/Hz	REFERENCE
Y = CO_2Me	268	664
CH_2^+	286.2	537
$CHMe^+$	273.5	537
$CHPh^+$	267.0	537
$CH(C_9H_{19})^+$	273.3	537
CMe_2^+	257.8	537
$CH=CHSiMe_3$	283.0	537
COEt	258.4	537
COC_6H_{13}	276.3	537
COC_9H_{19}	276.2	537
$COC_{10}H_{21}$	257.6	537
F	309 (J_{CF} = 498)	664
Cl	276	664
Br	269	664
I	234	664
Rhodium, [^{103}Rh]		
Rh(*C*H=CH=2-pyridyl)$Cl_2(PBu^n{}_3)_2$	186.0 [27] (8)	208
$Rh_3\{C_2(NEt_2)_2\}Cp_3$	280.6 [40]	684
$[Rh_6{}^{13}C(CO)_{15}]^{2-}$	264.7	306
Iridium		
Ir($C^1H=C^2$MeN=NMe)HCl$(PPh_3)_2$	C^1 161.9, C^2 155.4	604
Ir($C^1H=C^2$HCH=NPr)HCl$(PCy_3)_2$	C^1 181.9, C^2 168.3	636
Ir(*C*H=CClCH=NPr)HCl$(PCy_3)_2$	167.1	636
Ir(*C*H=CBrCH=NPr)HBr$(PCy_3)_2$	173.6	636
Ir($C^1H=C^2HCH_2PBu^t{}_2$)H(acac)py	C^1 131.6, C^2 130.0 (9.0)	988
Ir($C^1H=C^2MeCH_2PBu^t{}_2$)H(acac)py	C^1 122.4 (1.0), C^2 136.3 (10.2)	988
Ir($C^1H=C^2MeCH_2PCy_2$)H(acac)py	C^1 122.2, C^2 135.6 (9.5)	988
Ir($C^1H=C^2HCH_2PBu^t{}_2$)H(acac)CO	C^1 157.2, C^2 133.3 (8.1)	988
Palladium		
Pd(*C*Bu^t=CMeCMe=CBu^tCl)(acac)	from 84.0, 124.9, 127.3, 141.3	1194
Pd{*C*(*p*-tol)C(*p*-tol)C(*p*-tol)-C(*p*-tol)Ph}(S_2CNMe_2)	147.4	1529
Pd{*C*(*p*-tol)C(*p*-tol)C(*p*-tol)-C(*p*-tol)Ph}(S_2CNEt_2)	147.7, 149.9	1529
Pd{*C*(*p*-tol)C(*p*-tol)C(*p*-tol)-C(*p*-tol)Ph}$(S_2CNPr^i{}_2)$	149.9, 150.1	1529
Pd{*C*(*p*-tol)C(*p*-tol)C(*p*-tol)-C(*p*-tol)Ph}$(S_2CNMe_2)\{P(OMe)_3\}$	160.1 (5)	1529
Pd{*C*(*p*-tol)C(*p*-tol)C(*p*-tol)-C(*p*-tol)Ph}$(S_2CNMe_2)PMe_2Ph$	161.4 (6)	1529
Pd{*C*(*p*-tol)C(*p*-tol)C(*p*-tol)-C(*p*-tol)Ph}$(S_2CNPr^i{}_2)PMe_2Ph$	161.5	1529
Pd{*C*(*p*-tol)C(*p*-tol)C(*p*-tol)-C(*p*-tol)Ph}(acac)PMe_2Ph	152.9 (5)	1529
Pd{*C*(NCy)C(NCy)$C_6H_4N_2Ph$}-(CNCy)Cl	134.8 or 158.2	636

COMPOUND	$\delta(^{13}C)$/p.p.m., J/Hz	REFERENCE
$[Pd(CPhC_6H_8NOH)Cl]_2$	168.9 or 168.3	994
$Pd\{C{=}CHCOC({=}CHCO_2CH_2OH)N\}Cl(PEt_3)_2$	181.4	1530
$[Pd\{C^1C^2(NPr^i_2)C^2(NPr^i_2)\}Cl]_2$	C^1 106.5, C^2 148.3	1740
$Pd\{C^1C^2(NPr^i_2)C^2(NPr^i_2)\}(PBu^n_3)Cl_2$	C^1 128.4 (4.9), C^2 147.4	1740
$[Pd\{C^1C^2(NPr^i_2)C^2(NPr^i_2)\}$-$(PBu^n_3)_2Cl]^+$	C^1 125.6 (11.3), C^2 147.8	1740
Platinum, $[^{195}Pt]$		
$Pt(C^1H{=}C^2H_2)Cl_3(PEt_2Ph)_2$	C^1 169 (11), C^2 116.6	1171
$Pt(C^1H{=}C^2H_2)Br_3(PEt_2Ph)_2$	C^2 119.4	1171
$Pt(C^1H{=}C^2H_2)I_2Me(PEt_2Ph)_2$	C^1 172, C^2 116.0	1171
$Pt(C^1H{=}C^2H_2)Cl(PMe_2Ph)_2$	C^1 132 [880], C^2 119.5 [279] (4.5)	1171
$Pt(C^1H{=}C^2H_2)Cl(PEt_2Ph)_2$	C^1 131 [870], C^2 118.7 [216] (4.0)	1171
$Pt(C^1H{=}C^2H_2)I(PEt_2Ph)_2$	C^1 138.5 [872] (10), C^2 118.4	1171
$Pt(C^1H{=}C^2H_2)Ph(PEt_3)_2$	C^1 140.2, C^2 121.9	857
$[Pt(C^1H{=}C^2H_2)\{C(NHBu^n)NHC_6H_4OMe\text{-}p\}$-$(PEt_2Ph)_2]^+$	C^1 156, C^2 120.7	1171
$Pt(CH{=}CPhCOCO)(AsPh_3)_2$	183.3 [1056]	1176
$Pt(CH{=}CPhCOCO)(PPh_3)_2$	191.1 [920] (90, 8)	1176
$Pt(CH{=}CPhCOCO)(PEt_2Ph)_2$	185.7 [882] (90, 9)	1176
$Pt(CH{=}CPhCOCO)(PEt_3)_2$	188.8 [871] (93, 10)	1176
$Pt(CH{=}CPhCOCO)(PMePh_2)_2$	190.9 [892] (93, 8)	1176
$Pt\{C^1(CO_2Me){=}C^2(CO_2Me)OMe\}$-$CO_2Me(PPh_3)_2$	C^1 156.7 (12.0), C^2 144.3	1064
$Pt\{C^1(CO_2Me){=}C^2(CO_2Me)OH\}$-$CO_2Me(PPh_3)_2$	C^1 120.8 (12.8), C^2 155.9	1064
Pt(2-thiophenyl)Cl(cod)	C^2 137.2 [1060], C^3 127.3 [61]	1541
Pt(2-furanyl)Cl(cod)	C^3 115.5	1541
$Pt(2\text{-thiophenyl})_2(cod)$	C^2 145.7 [1180], C^3 126.9 [82]	1541
$Pt(2\text{-furanyl})_2(cod)$	C^2 164.1, C^3 117.2 [162]	1541
$Pt(2\text{-benzofuranyl})_2(cod)$	C^2 168.3, C^3 113.9	1541
$[Pt\{C^1C^2(NPr^i_2)C^2(NPr^i_2)\}$-$(PBu^n_3)_2Cl]^+$	C^2 144.1	1740
Gold		
$Fe(C_5H_4AuPPh_3)Cp$	72.4	245

COMPOUND	$\delta(^{13}C)$/p.p.m., J/Hz	REFERENCE
Mercury, [^{199}*Hg*]		
$Hg(\mathit{C}^1H{=}C^2H_2)_2$	C^1 168.2 [1133], C^2 134.3 [38]	246
$Hg(C^1H{=}C^2H_2)(CH_2Bu^t)$	C^1 178.8 [934], C^2 133.4 [38]	150
$Hg(C^1H{=}C^2HCl\text{-}\mathit{trans})_2$	C^1 156.6 [1330], C^2 130.7 [191]	246
$Hg(C^1H{=}C^2HCl\text{-}\mathit{cis})_2$	C^1 156.3 [1256], C^2 132.0 [53]	246
$Hg(C^1H{=}C^2HMe)OAc$	C^1 128.3 [2196], C^2 142.6 [18]	1566
$Hg(2\text{-furanyl})_2$	C^2 189.5 [1880], C^3 121.2 [242]	399
$Hg(2\text{-thiophenyl})_2$	C^2 164.7, C^3 134.8	399
$Hg\{C^1{=}C^2Bu^tP(O_2H)OCHBu^t\}Cl$	160.5 (20), 146.5 (137)	1654
$Hg\{\mathit{C}(N_2)CO_2Et\}_2$	72.6 [2033]	1709
$Hg\{\mathit{C}(N_2)CO_2Et\}Me$	38.1	1709
Boron		
$B(\mathit{C}H{=}CH_2)_2(OBu^n)_2$	127.8	567
$B(\mathit{C}H{=}CHBu^n\text{-}\mathit{trans})(CHMePr^i)_2$	134.6	567
$B(\mathit{C}H{=}CHBu^n\text{-}\mathit{trans})_2OMe$	129.1	567
$B(\mathit{C}H{=}CHBu^n\text{-}\mathit{trans})O_2C_6H_4$	114.3	567
$B(\mathit{C}H{=}CHBu^n\text{-}\mathit{trans})_2Cl$	133.0	567
$B(C^1Me{=}C^2HSnMe_3)Bu^t_2$	C^1 149.0, C^2 119.4	1730
$B(C^1Bu^t{=}C^2HSnMe_3)MeBu^t$	C^1 168.0, C^2 121.8	1730
$B(C^1{=}C^2HSnEt_2CH{=}CEt)Et_2$	C^1 177.0, C^2 127.6	999
$B(C^1{=}C^2HSnMe_2CH{=}CPr^i)Pr^i_2$	C^1 175.2, C^2 125.7	999
$B(\mathit{C}{=}CBu^tSnMe_2CBu^t{=}CMe)Me_2$	164.2	1625
$B(\mathit{C}{=}CBu^tSnMe_2CMe{=}CEt)Et_2$	161.5	1625
$B\{\mathit{C}{=}C(SiMe_3)SnMe_2C(SiMe_3){=}CEt\}Et_2$	182.1	1625
$EtB(C^1Et{=}C^2Me)_2SnEt_2$	C^1 163.4, C^2 148.8 (J_{SnC} = 438)	1574
$EtB(C^1Pr^i{=}C^2Me)_2SnEt_2$	C^1 169.3, C^2 137.8	1574
$[B\{\mathit{C}H(CH)_3\mathit{C}H\}Me]Tl$	131.9	1576
$[B\{\mathit{C}H(CH)_3\mathit{C}H\}Ph]Tl$	128.6	1576
$\{B(C^1H{=}C^2HCH_2CH_2C^2H{=}C^1H)Ph\}Cr(CO)_4$	C^1 112.8, C^2 91.3	1365
$BPh(CH_2)_4\mathit{C}{=}CHSnMe_3$	168.2, 169.1	1730
$\{B(C^1H{=}C^2HCH_2CH_2C^2H{=}C^1H)Ph\}Mo(CO)_4$	C^1 108.4, C^2 93.3	1365
$\{B(C^1H{=}C^2HCH_2CH_2C^2H{=}C^1H)Ph\}W(CO)_4$	C^1 100.5, C^2 89.2	1365
Carboranes		
$C_2B_3H_5$	103.3	1717

COMPOUND	$\delta(^{13}C)$/p.p.m., J/Hz	REFERENCE
	102.4 {$^1J_{CH}$ = 192; $^2J_{CH}$ = 19.7} 123.8 [$^1J_{CB}$ = 18; $^2J_{CB}$ = 15.0]	155, 304
1,6-$C_2B_4H_6$	77.2, 75.8	155, 1717
4,5-$C_2B_4H_8$	122	1717
3,6-$C_2B_5H_7$	80.0	155, 1717
4,5-$C_2B_7H_9$	69.3	155
1,6-$C_2B_8H_{10}$	32.4, 57.1	155
1,10-$C_2B_8H_{10}$	102.6	1741
2,3-$C_2B_9H_{11}$	86.5	155
$[7,8\text{-}C_2B_9H_{12}]^-$	38.0	38
$[7,9\text{-}C_2B_9H_{12}]^-$	38.0	38
$[CpFe(CO)(CNC_6H_{11})_2][7,8\text{-}C_2B_9H_{12}]$	41.8	487
$[Co(1,2\text{-}C_2B_9H_{11})_2]^-$	51.5	296
$[Co(9\text{-}Br\text{-}1,2\text{-}C_2B_9H_{10})_2]^-$	46.6, 50.0	296
$[Co(2,9\text{-}Br_2\text{-}1,2\text{-}C_2B_9H_9)_2]^-$	45.9	296
$[Co(7,8\text{-}B_9C_2H_{10})_2C_6H_4]^-$	38.6	65
$[CB_{11}H_{12}]^-$	45.3, 54.6	38, 944
1,2-$C_2B_{10}H_{12}$	49.1, 54.7, 55.5, 56.4, 56.6, 56.9	38, 155, 550, 768, 944, 1555, 1717, 1741
1-Me-1,2-$C_2B_{10}H_{11}$	C^1 71.5, 67.9, C^2 62.9, 62.2	944, 1704, 1555
1-Vi-1,2-$C_2B_{10}H_{11}$	C^1 74.6, C^2 61.4	944
1-$EtCO_2CH_2$-1,2-$C_2B_{10}H_{11}$	C^1 73.8, C^2 61.2	1717
1-$EtSCH_2$-1,2-$C_2B_{10}H_{11}$	C^1 76.7, C^2 61.5	1717
1-$ClCH_2$-1,2-$C_2B_{10}H_{11}$	C^1 72.2, C^2 59.5	1555
1-$BrCH_2$-1,2-$C_2B_{10}H_{11}$	C^1 72.8, C^2 62.9	944
1-Ph-1,2-$C_2B_{10}H_{11}$	C^1 77.7, C^2 61.4	944
1-PhCO-1,2-$C_2B_{10}H_{11}$	C^1 77.0, C^2 60.3	944
1-HO_2C-1,2-$C_2B_{10}H_{11}$	C^1 70.7, C^2 59.0	944
1-MeO_2C-1,2-$C_2B_{10}H_{11}$	C^1 69.9, C^2 58.8	944
1-Me_3Si-1,2-$C_2B_{10}H_{11}$	C^1 65.0, C^2 60.9	944
1-H_2N-1,2-$C_2B_{10}H_{11}$	C^1 92.3, C^2 69.5	944
3-Benzoyloxy-1,2-$C_2B_{10}H_{11}$	$C^{1,2}$ 56.2 {148}	775
4-Benzoyloxy-1,2-$C_2B_{10}H_{11}$	$C^{1,2}$ 51.1, 54.0 {192}	775
1-Me-4-benzoyloxy-1,2-$C_2B_{10}H_{11}$	C^1 70.0, C^2 62.2	775
1,2,3-$C_2BrB_{10}H_{11}$	C^1 59.6, C^2 59.6	768
1,2,4-$C_2BrB_{10}H_{11}$	C^1 56.6, C^2 55.7	768
1,2,8-$C_2BrB_{10}H_{11}$	C^1 53.2, C^2 53.2	768
1,2,9-$C_2BrB_{10}H_{11}$	C^1 53.7, C^2 47.2	768
1,2-$CPB_{10}H_{11}$	68.4 {210}	296, 1741
1,2-$CPB_{10}H_9Br_2$	60.9	296
1,2-$CAsB_{10}H_{11}$	70.0, 69.2	944, 1741
1,2-Me_2-1,2-$C_2B_{10}H_{10}$	74.6	944
1,2-Ph_2-1,2-$C_2B_{10}H_{10}$	85.8, 86.5	944, 1701
1,2-C_6H_{12}-1,2-$C_2B_{10}H_{10}$	74.5	944
1,2-C_6H_{10}-1,2-$C_2B_{10}H_{10}$	71.4	944
1,2-$(Me_3Si)_2$-1,2-$C_2B_{10}H_{10}$	76.3	944

COMPOUND	$\delta(^{13}C)$/p.p.m., *J*/Hz	REFERENCE
$1,2-(MeS)_2-1,2-C_2B_{10}H_{10}$	73.9	944
$[1,2-Ph_2-1,2-C_2B_{10}H_{10}]^{2-}$	81.0	1701
$[Ph_2-1,2-C_2B_{10}H_{11}]^-$	90.2	1701
$1,7-C_2B_{10}H_{12}$	49.0, 55.4, 56.8, 56.3	38, 155, 550, 1555, 1704, 1717, 1741
$1-Me-1,7-C_2B_{10}H_{11}$	C^1 71.9; 71.3, C^2 56.1, 57.3	944, 1555, 1704
$1-Ph-1,7-C_2B_{10}H_{11}$	C^1 78.9, C^2 56.6	944
$1-MeCO-1,7-C_2B_{10}H_{11}$	C^1 81.1, C^2 56.6	944
$1-HO_2C-1,7-C_2B_{10}H_{11}$	C^1 73.4, C^2 55.8	944
$1-MeO_2C-1,7-C_2B_{10}H_{11}$	C^1 72.7, C^2 56.1	944
$1-Me_3Si-1,7-C_2B_{10}H_{11}$	C^1 66.3, C^2 59.2	944
$1-AcO-1,7-C_2B_{10}H_{11}$	C^1 97.2, C^2 53.0	944
$1-MeS-1,7-C_2B_{10}H_{11}$	C^1 72.9, C^2 57.3	944
$1-MeSO_2-1,7-C_2B_{10}H_{11}$	C^1 86.9, C^2 57.2	944
$1,7-CPB_{10}H_{11}$	67.0	1741
$1,7-CAsB_{10}H_{11}$	63.0, 62.8	944, 1741
$1,7-Me_2-1,7-C_2B_{10}H_{10}$	72.1	944
$1,7-Ph_2-1,7-C_2B_{10}H_{10}$	78.7, 79.3	944, 1701
$1,7-(MeO_2C)_2-1,7-C_2B_{10}H_{10}$	73.0	944
$1,7-(Me_3Si)_2-1,7-C_2B_{10}H_{10}$	66.8, 67.4, 68.6	155, 550, 944
$1,7-(Me_3Ge)_2-1,7-C_2B_{10}H_{10}$	68.5	944
$1,7-(Me_3Sn)_2-1,7-C_2B_{10}H_{10}$	52.6	944
$1,7-(MeS)_2-1,7-C_2B_{10}H_{10}$	72.7, 73.6	155, 550, 944
$9,10-Br_2-1,7-C_2B_{10}H_{10}$	53.1	155, 550
$[1,7-Ph_2-1,7-C_2B_{10}H_{10}]^{2-}$	81.0	1701
$1,12-C_2B_{10}H_{12}$	62.5, 64.0, 63.5, 64.6	155, 550, 944, 1555, 1704, 1741
$1-Me-1,12-C_2B_{10}H_{11}$	C^1 79.7, C^2 56.8	1555, 1704
$1,12-CPB_{10}H_{11}$	84.6	1741
$1,12-CAsB_{10}H_{11}$	86.0, 84.9	944, 1741
$1,12-Ph_2-1,12-C_2B_{10}H_{10}$	83.2	1701
$[1,12-Ph_2-1,12-C_2B_{10}H_{10}]^{2-}$	85.6	1701
Aluminium		
$Et_2Al(OEt_2)$*C*Ph=*C*PhAlEt$_2$-(OEt$_2$)	175.8	1588
$Et_2Al(thf)$*C*Ph=C=C=*C*PhAlEt$_2$-(OEt$_2$)	147.0	1588
Silicon, [^{29}Si]		
$Si(C^1H=C^2H_2)_4$	134.3 C^1 135.7 [70]; 134.1 C^2 136.3; 135.3	97, 113, 600, 1241
$\overline{Si(C^1H=C^2H_2)_2CH_2CH=CHCH_2}$	135.2, 134.2	1158

COMPOUND	$\delta(^{13}C)$/p.p.m., J/Hz	REFERENCE
$Si(C^1H=C^2H_2)Me_3$	C^1 139.0, C^2 129.9; 130.3, 139.5, 130.6 139.7 [64]	97, 1241, 1735 1322, 1324
$Si(C^1H=C^2H_2)Me_2Et$	C^1 138.8, C^2 131.4	1322
$Si(C^1H=C^2H_2)Me_2Ph$	C^1 137.9, C^2 132.8	1241
$Si(C^1H=C^2H_2)MeEt_2$	C^1 137.0, C^2 131.5	1322
$\overline{Si(C^1H=C^2H_2)Me(CH_2CH=CH_2CH_2})$	132.5, 137.4	1158
$Si(C^1H=C^2H_2)MePh_2$	C^1 135.8, C^2 134.9	1241
$Si(C^1H=C^2H_2)Et_3$	C^1 135.2, C^2 131.3	1322
$Si(C^1H=C^2H_2)MeHPh$	C^1 134.8, C^2 134.7	1241
$Si(C^1H=C^2H_2)Me_2Cl$	C^1 136.0; 135.6, C^2 133.9; 133.1	1241, 1322
$Si(C^1H=C^2H_2)Me_2OEt$	C^1 137.5; 137.1, C^2 132.7; 131.4	1241, 1322
$Si(C^1H=C^2H_2)Me_2OBu^t$	C^1 139.9, C^2 130.2	1322
$Si(C^1H=C^2H_2)Me_2OSiMe_3$	C^1 139.4, C^2 131.0	1322
$Si(C^1H=C^2H_2)MePhOEt$	C^1 135.5, C^2 134.6	1241
$Si(C^1H=C^2H_2)Ph_2OEt$	C^1 133.7, C^2 136.6	1241
$Si(C^1H=C^2H_2)MePhCl$	C^1 134.2, C^2 135.8	1241
$Si(C^1H=C^2H_2)Me(NMe_2)_2$	C^1 137.2, C^2 136.7	1241
$Si(C^1H=C^2H_2)Me(OEt)_2$	C^1 133.8, C^2 134.8	1241
$Si(C^1H=C^2H_2)Me(OBu^t)_2$	C^1 138.3, C^2 130.6	1322
$Si(C^1H=C^2H_2)Me(OSiMe_3)_2$	C^1 137.7, C^2 132.0	1322
$Si(C^1H=C^2H_2)Me(OR)_2$	C^1 136.0 to 136.3, C^2 132.4 to 133.0	772
$Si(C^1H=C^2H_2)Ph(OEt)_2$	C^1 132.1, C^2 136.8	1241
$Si(C^1H=C^2H_2)MeF_2$	C^1 128.8, C^2 137.6 ($^2J_{CF}$ = 20)	1322
$Si(C^1H=C^2H_2)MeCl_2$	C^1 132.2, C^2 136.8; 134.0, 137.3; 138.0 135.2 [92]	600, 1241, 1322
$Si(C^1H=C^2H_2)(OEt)_3$	C^1 129.3; 129.5; 130.2 C^2 136.6; 134.7; 136.1	1241, 1322, 1734
$\overline{Si(C^1H=C^2H_2)(OCH_2CH_2)_3N}$	C^1 140.6, C^2 129.6	1734
$Si(C^1H=C^2H_2)(OBu^t)_3$	C^1 134.0, C^2 134.0	1322
$Si(C^1H=C^2H_2)(OSiMe_3)_3$	C^1 132.7, C^2 131.7	1322
$Si(C^1H=C^2H_2)Cl_3$	C^1 130.9; 131.6; 131.4, C^2 138.2; 138.5; 138.7 131.2; 133.1 [113], 139.2; 140.1	113, 600, 1735 1241, 1322
$Si(\mathit{C}H=CHCO_2R)Me_3$	150.9	1626
trans-$Me_3Si\mathit{C}H=\mathit{C}HSiMe_3$	150.0	1324
cis-$Me_3SiCH=CHSiMe_3$	150.5	1324
$Si(C^1H=C^2HCl)Me_3$	132.2, 138.3	1324
$Si(C^1H=C^2Me_2)Me_3$	C^1 124.9, C^2 151.4	80
$Si(C^1H=C^2Me_2)MeCl_2$	C^1 120.5, C^2 153.3	80
$Si\{\mathit{C}H=C(NMe_2)OLi\}Me_3$	56.0	1611
$Si\{C^1H=C^2(SiMe_2Cl)_2\}Me_2Cl$	C^1 164.1 {146}, C^2 162.7	1246
$[Si(\mathit{C}HPh)Me_3]^-$	55.7 {130}	350
$Si(C^1H=C^2=O)Me_3$	C^1 -0.1, C^2 179.2	264

COMPOUND	$\delta(^{13}C)$/p.p.m., J/Hz	REFERENCE
$Si(C^1H{=}C^2{=}O)Et_3$	C^1 -4.9, C^2 179.2	264
$Si(\mathit{C}HN_2)Me_3$	19.1	1709, 1724
$Si(\mathit{C}H{=}PMe_3)Me_3$	0.7 (88.2) {134.6}	1063
$Me_2\overline{Si\mathit{C}H{=}PMe_2CH_2SiMe_2CH_2}$	3.1 (90.3)	1605
$\overline{Si(\mathit{C}H{=}CHCH_2})Me_2$	142.0	1607
$\overline{Si(\mathit{C}H{=}CMeCH_2})Me_2$	132.1	1607
$\overline{Si(\mathit{C}H{=}CHCH_2)CH_2CH_2CH_2CH_2}$	139.3	1607
$\overline{Si(C^1H{=}C^2HCH_2CH_2})MeH$	C^1 131.3 {142}, C^2 153.9 {154}	1158
$\overline{Si(C^1H{=}C^2HCH_2CH_2})PhH$	C^1 127.2 {155}, C^2 156.0 {155}	1158
$\overline{Si(C^1H{=}C^2HCH_2CH_2})(C_2N_3O_2HPh)Ph$	C^1 133.4 {160}, C^2 155.1 {162}	1158
$\overline{Si(C^1H{=}C^2HCH_2CH_2})MeBr$	C^1 133.2 {141}, C^2 153.8 {165}	1158
$\overline{Si(C^1H{=}C^2HCH_2CH_2})PhBr$	C^1 130.2, C^2 156.7	1158
$Si(C^1Et{=}C^2H_2)Me_3$	C^1 153.5, C^2 123.5	80
Si(1-cyclohexene)Me_3	C^1 139.9, C^2 136.2	80
2-Me_3Si-5,6-F_2-5,6-$(CF_3)_2$-7-Me_3Si-norbornene	C^2 143.6, C^3 149.8	638
$Si(1\text{-}\mathit{C}_5H_5)Me_3$	C^1 147.9, C^2 142.8	80
$Si(1\text{-}\mathit{C}_5H_5)Me_2Cl$	C^1 144.3, C^2 145.7	80
$Si(1\text{-}\mathit{C}_5H_5)MeCl_2$	C^1 141.0, C^2 148.3	80
Si(2-indenyl)Me_3	C^2 148.6, C^3 141.2	360, 419
Si(3-indenyl)Me_3	C^3 148.2, C^2 144.0	360, 419
$[Si(C_7H_7)Me_3]^+$	C^1 181.3	1616
$Si(C^1Et{=}C^2HCy)Me_3$ - *E*	C^1 140.1, C^2 145.9	1256
- *Z*	C^1 138.8, C^2 147.8	
$Si\{C^1(C_5H_{11}){=}C^2HPr^i)\}Me_3$ - *E*	C^1 137.9, C^2 147.7	1256
- *Z*	C^1 136.3, C^2 150.4	
$Si\{C^1(C_5H_{11}){=}C^2H(C_{10}H_{21})Me_3$ - *E*	C^1 140.9, C^2 140.5	1256
- *Z*	C^1 139.2, C^2 143.1	
$Si\{C^1H(C_7H_{15}){=}C^2H(C_{10}H_{21})\}Me_3$ - *E*	C^1 140.9, C^2 140.4	1256
- *Z*	C^1 139.3, C^2 143.2	
$(Me_3Si)_2C^1{=}C^2{=}O$	C^1 1.7, C^2 166.8	264
$(Me_3Si)(Me_3Ge)C^1{=}C^2{=}O$	C^1 0.5, C^2 166.2	264
$(Me_3Si)(Me_3Sn)C^1{=}C^2{=}O$	C^1 -5.8, C^2 164.4	264
$(Me_3Si)RC^1{=}C^2{=}S$	C^1 67.8 to 73.4, C^2 236.7 to 237.9	1253
$Me_2\overline{Sn\{Me_3Si\mathit{C}{=}C(BEt_2)CEt{=}\mathit{C}}SiMe_3\}$	145.6, 139.7	1625
$Me_2\overline{Si\mathit{C}(SiMe_3){=}\mathit{C}}(SiMe_3)$	189.3	711
$Me_2\overline{Si\mathit{C}Me{=}\mathit{C}}Me$	152	719
$Me_2\overline{SiMe_2Si\mathit{C}(SiMe_3){=}C}(SiMe_3)$	200.8	1010
$Me_2Si(\mathit{C}Me{=}\mathit{C}Me)_2SiMe_2$	148.3	711

COMPOUND		$\delta(^{13}C)$/p.p.m., J/Hz	REFERENCE
$Si\{\mathit{C}(N_2)CO_2Et\}Me_3$		42.9	1709, 1724
$(Me_3Si)_2\mathit{C}{=}PMe_3$		0.3 (63.3)	1063
$Me_2SiCH_2Me_2Si\mathit{C}{=}PMe_3$		13.2 (95), 12.9 (95.2)	759, 1605
$MeP{=}\mathit{C}(SiMe_2CH_2SiMe_2CH_2)_2$		3.0 (65.9)	1605
$(Me_3Si)_2\mathit{C}{=}SbMe(CH_2SiMe_3)_2$		9.8	1614
Germanium			
$Ge(\mathit{C}H{=}\mathit{C}H_2)MeCH_2CH{=}CHCH_2$		130.6, 138.2	1158
$Ge(C^1H{=}C^2HCH_2CH_2)Me(C_2HN_3O_2Ph)$		C^1 137.4 {158}, C^2 146.7 {161}	1158
$Ge(C^1H{=}C^2HCH_2CH_2CH_2)Ph(C_2HN_3O_2Ph)$		C^1 133.2 {165}, C^2 149.7 {165}	1158
Ge(2-indenyl)Me_3		C^2 150.3, C^3 139.3	360
Ge(3-indenyl)Me_3		C^3 148.3, C^2 141.3	360
$Ge(C^1H{=}C^2{=}O)Et_3$		C^1 -4.8, C^2 179.4	264
$Ge(\mathit{C}HN_2)Me_3$		19.3, 19.8	1004, 1709 1724
$Ge(\mathit{C}H{=}PMe_3)Me_3$		1.3 (87.9) {144.0}	1063
$Ge(1,2\text{-}C_2B_{10}H_{11})Me_3$		31.9	453
$Ge\{C_{10}H_9Ru(CO)_2GeMe_3\}Me_3$		78.6	516
$(Me_3Ge)_2C^1{=}C^2{=}O$		C^1 0.0, C^2 167.0	264
$(Et_3Ge)_2C^1{=}C^2{=}O$		C^1 -8.5, C^2 165.7	264
$(Me_3Si)(Me_3Ge)C^1{=}C^2{=}O$		C^1 0.5, C^2 166.2	264
$(Me_3Ge)(Me_3Sn)C^1{=}C^2{=}O$		C^1 -6.3, C^2 164.6	264
$(Me_3Ge)\mathit{C}(N_2)CO_2Et$		43.7	1709, 1724
$(Me_3Ge)_2\mathit{C}N_2$		17.3	1709, 1724
$(Me_3Ge)(Me_2As)\mathit{C}N_2$		20.6	996
$(Me_3Ge)_2\mathit{C}{=}PMe_3$		1.9 (65.9)	1063
$(Me_3Ge)_2\mathit{C}{=}PMe_2CH_2GeMe_3$		2.9 (65.9)	1063
Tin, [^{119}*Sn*]			
$Sn(C^1H{=}C^2H_2)_4$		C^1 135.3 [519.3] 136.0, C^2 135.8 [<6] 136.4	113, 875
$Sn(C^1H{=}C^2H_2)_3Bu^t$		C^1 138 [435.5], C^2 138.0	882
$Sn(C^1H{=}C^2H_2)_2Cl_2$		C^1 135.0, C^2 140.4	113
$Sn(C^1H{=}C^2HMe)_4$	*cccc**	C^1 128.5, C^2 143.8	1634
	ccct	C^1 126.9 to 128.0, C^2 144.0, 144.8	1634
	cctt	C^2 144.2, 145.2	1634
	cttt	C^1 126.8, C^2 145.7	1634
$Sn(C^1H{=}C^2HMe)_3Me$	*ccc*	C^1 129.3, C^2 143.3	1634
	cct	C^1 128.7; 128.4, C^2 143.6; 144.2	1634
	ctt	C^1 128.4, 128.0, C^2 143.8, 144.6	1634
	ttt	C^1 128.0, C^2 145.0	1634

**c* = *cis*, *t* = *trans* referring to the olefin

COMPOUND		$\delta(^{13}C)$/p.p.m., *J*/Hz	REFERENCE
$Sn(C^1H{=}C^2HMe)_2Me_2$	*cc**	C^1 129.5, C^2 143.1	1634
	ct	C^1 129.2; 129.0, C^2 143.3; 143.9	1634
	tt	C^1 128.8, C^2 144.2	1634
$Sn(C^1H{=}C^2HMe)Me_3$	*cis*	C^1 129.9 [464.6], 129.7 C^2 142.9, 143.4	875, 1634
	trans	C^1 129.7 [478.4], C^2 143.5	1634
$Sn(C^1H{=}C^2HPh)Et_3$		C^1 127.7 [387.6], C^2 139.2 [64.1]	882
$Sn(\mathit{C}H{=}\overline{CBPhCH_2CH_2CH_2C}H_2)Me_3$		150.6 [430.0] 139.8 [489.3]	1660
$Sn(C^1H{=}C^2MeBBu^t{}_2)Me_3$		C^1 119.4 [506], C^2 149.0	1730
$Sn(C^1H{=}C^2Bu^tBMeBu^t)Me_3$		C^1 121.8 [465], C^2 168.0	1730
$Me_2\overline{Sn\{C^1H{=}C^2(BPr^i{}_2)C^3Pr^i{=}C}^4H\}$		C^1 125.7 [409], C^2 175.2, C^4 119.7 [432], C^3 170.1 [86.6]	999
$Et_2\overline{Sn\{C^1H{=}C^2(BEt_2)C^3Et{=}C}^4H\}$		C^1 127.6 [364.6], C^2 177.0, C^4 120.8 [433.8], C^3 164.5 [74.0]	999
$Et_2Sn(C^1Me{=}C^2Et)_2BEt$		C^1 148.8 [438], C^2 163.4	1574
$Et_2Sn(C^1Me{=}C^2Pr^i)_2BEt$		C^1 137.8, C^2 169.3	1574
$Bu^n{}_2Sn(C^1H{=}C^2H)_2CBu^tOMe$		C^1 130.2	233
$Sn(\mathit{C}H{=}C{=}CH_2)Me_3$		75.2 [382]	47
$Sn(C^1Me{=}C^2H_2)Me_3$		C^1 149.6, C^2 125.3	1634
$Sn(C^1Me{=}C^2{=}C^3HMe)Me_3$		C^1 87.5 [431.4], C^2 204.3 C^3 76.1	875
$Sn(2\text{-furanyl})_4$		C^2 152.4 [808.7]; 153.3 [818], C^3 124.3 [100.7]; 124.8 [100.9] C^5 148.5 [39.7]; 149.4 [41.1] C^4 109.8 [42.7], 110.3 [41.6]	399, 915
$Sn(2\text{-furanyl})_2Me_2$		C^2 158.2 [589], C^3 122.4 [77.7], C^5 148.0 [27.8], C^4 109.8 [31.3]	399
$Sn(2\text{-furanyl})Me_3$		C^2 160.9 [508], C^3 121.4 [72.4], C^5 147.5 [24.6], C^4 109.5 [28.6]	399
$Sn(2\text{-thiophenyl})_4$		C^2 131.0 [1997]; 131.6 [647], C^3 137.6 [45.8], 138.2 [45.7], C^5 128.3 [27.5], 128.6 [65.9]	399, 915

*c = *cis*, t = *trans* referring to the olefin

COMPOUND	δ(^{13}C)/p.p.m., *J*/Hz	REFERENCE
	C^4 132.6 [64.1], 133.1 [28.4]	
$Sn(2\text{-thiophenyl})_2Me_2$	C^2 135.3 [477], C^3 136.7 [36.7], C^5 128.6 [51.0], C^4 132.0 [21.4]	399
$Sn(2\text{-thiophenyl})Me_3$	C^2 137.1 [407], C^3 135.6 [31.7], C^5 128.7 [47.0], C^4 131.2 [17.4]	399
$Sn(3\text{-thiophenyl})_4$	C^2 133.8 [57.8], C^3 138.8 [1016.2], C^5 126.2 [128.0], C^4 132.8 [34.5]	399
$5\text{-}Me_3Sn\text{-}2,3\text{-}(CF_3)_2\text{-}7\text{-}(Me_3Sn)$-norbornadiene	$C^{5,6}$ 154.3, 157.1	638
$(Me_3Sn)_2C^1{=}C^2{=}O$	C^1 -13.9, C^2 161.7	264
$(Et_3Sn)_2C^1{=}C^2{=}O$	C^1 -20.5, C^2 161.3	264
$(Me_3Si)(Me_3Sn)C^1{=}C^2{=}O$	C^1 -5.8, C^2 164.4	264
$(Me_3Ge)(Me_3Sn)C^1{=}C^2{=}O$	C^1 -6.3, C^2 164.6	264
$(Me_3Sn)_2\mathit{C}N_2$	5.8 [217.2]	951, 1709, 1724
$(Me_3Sn)(Me_2As)\mathit{C}N_2$	15.5	996
$Me_3SnC(N_2)CO_2Et$	38.4 [300.0]	1709, 1724
$(Me_3Sn)_2C{=}PMe_3$	-6.7 [51.3]	1063
Lead, [^{209}*Pb*]		
$Pb(C^1H{=}C^2H_2)_4$	C^1 145.2, C^2 134.9	113
$Pb(2\text{-furanyl})_4$	C^2 163.6 [933]; 162.5 [924], C^3 123.6 [160.4]; 122.8 [158.7], C^5 148.9 [59.0]; 148.0 [58.0], C^4 110.6 [67.7]; 110.1 [64.1]	399, 915
$Pb(2\text{-thiophenyl})_4$	C^2 142.0 [656]; 141.0 [953], C^3 137.5 [75.3]; 128.1 [109.9], C^5 128.7 [110.3]; 136.9 [76.3], C^4 133.9 [39.1]; 131.9 [36.6]	399, 915
$Me_2\overline{PbC^1Ph{=}C^2Ph{=}C^2PhC^3Ph{=}C^1}Ph$	C^1 153.6 [290], C^2 152.3 [140]	1646
$(Me_3Pb)_2\mathit{C}N_2$	1.9 [228]	996, 1709, 1724
$(Me_3Pb)(Me_2As)\mathit{C}N_2$	13.8	996
$Me_3Pb\mathit{C}(N_2)CO_2Et$	37.8 [195.2]	1709, 1724
$(Me_3Pb)_2C{=}PMe_3$	14.7 [57.4]	1063

COMPOUND	$\delta(^{13}C)$/p.p.m., J/Hz	REFERENCE
Phosphorus		
$[P(C^1H{=}C^2H_2)Bu^n{}_3]^+$	C^1 119 (72.6), C^2 141.2	1772
$[P(C^1H{=}C^2H_2)MePh_2]^+$	C^1 119.3 (80.6), C^2 143.7	1770
$[P(C^1H{=}C^2H_2)Ph_3]^+$	C^1 119.2 (80.3), C^2 145.2	1770, 1772
$P(C^1H{=}C^2H_2)(CH_2CH_2PPh_2)(NEt_2)$	C^1 141.3 (21), C^2 123.1 (16)	1233
$P(C^1H{=}C^2H_2)Ph_2O$	C^1 131.3 (97.8), C^2 134.4	1770
	C^1 131.5 (96.4), C^2 134.4 (11.8)	828
$P\{C^1H{=}C^2H_2Fe(CO)_4\}Ph_2O$	C^1 120.8 (86.9), C^2 117.0	828
$P(CH{=}CR^1R^2)$ derivatives		76, 149, 234, 241, 1275, 1652, 1654, 1679, 1772, 1773, 1792
$P(CR^1{=}CR^2R^3)$ derivatives		241, 806, 1136, 1247, 1652, 1654, 1737
Methylphosphole	$C^{2,5}$ 134.8 (7), $C^{3,4}$ 135.8 (8)	237
Other phospholes		133, 237, 1412, 1419, 1467, 1474
PC_5H_5	$C^{2,6}$ 154.1 (53) {157}, $C^{3,5}$ 133.6 (14) {156}, C^4 128.8 (22) {161}	714
Derivatives of PC_5H_5		84, 153, 271, 620, 706, 714, 1066, 1141, 1258, 1746
Arsenic		
$As(C^1H{=}C^2HCl)Cl_2$	C^1 138.5 {174.6}, C^2 132.5 {197.2}	1068
$As(\mathit{C}HN_2)Me_2$	24.1	641, 951, 1709
$Me_2As\mathit{C}(N_2)CO_2Et$	45.7	474, 951, 1709
$MeAs\{\mathit{C}(N_2)CO_2Et\}_2$	27.2, 27.3	471, 1709
$Me_2As\mathit{C}(N_2)GeMe_3$	20.6	996
$Me_2As\mathit{C}(N_2)SnMe_3$	15.5	996
$Me_2As\mathit{C}(N_2)PbMe_3$	13.8	996
$(Me_2As)_2\mathit{C}(N_2)$	24.7, 24.8	474, 641, 951, 1709

COMPOUND	δ(^{13}C)/p.p.m., J/Hz	REFERENCE
$Me_2As\mathit{C}(N_2)SbMe_2$	11.7	996
$Me(Me_2N)As\mathit{C}(N_2)CO_2Et$	46.4, 46.5	471, 1709
$(Me_2N)_2As\mathit{C}(N_2)CO_2Et$	38.4, 38.5	471, 1709
$(2,5-Me_2$-arsole$)Mn(CO)_3$	$C^{2,5}$ 128.4, $C^{3,4}$ 93.5	1416
$As\mathit{C}_5H_5$	$C^{2,6}$ 167.7 {159}, $C^{3,5}$ 133.2 {157}, C^4 128.2 {161}	714
$(As\mathit{C}_5H_5)Mo(CO)_5$	$C^{2,6}$ 164.5, $C^{3,5}$ 136.1, C^4 126.4	1141
$(As\mathit{C}_5H_5)Mo(CO)_3$	$C^{2,6}$ 110.8, $C^{3,5}$ 94.2, C^4 87.5	1141
$[1-MeAs\mathit{C}_5H_5]^-$	$C^{2,6}$ 72.9, $C^{3,5}$ 132.1, C^4 92.6	706, 1258
$1,1-Me_2As\mathit{C}_5H_5$	$C^{2,6}$ 70.3, $C^{3,5}$ 138.6, C^4 91.8	706
$2-MeAs\mathit{C}_5H_5$	C^2 182.9, $C^{3,5}$ 131.0, 134.0, C^4 129.2, C^6 168.3	714
Antimony		
$Me_2Sb\mathit{C}(N_2)CO_2Et$	34.3	474, 1709
$Me_2Sb\mathit{C}(N_2)AsMe_2$	11.7	1709
$(Me_2Sb)_2\mathit{C}N_2$	-1.2	474
$Sb\mathit{C}_5H_5$	$C^{2,6}$ 178.3, $C^{3,5}$ 134.4, C^4 127.4	714
$(Sb\mathit{C}_5H_5)Mo(CO)_3$	$C^{2,6}$ 114.6, $C^{3,5}$ 95.5, C^4 88.0	1141
$[1-MeSb\mathit{C}_5H_5]^-$	$C^{2,6}$ 69.8, $C^{3,5}$ 134.2, C^4 95.3	1258
Bismuth		
$Me_2Bi\mathit{C}(N_2)CO_2Et$	26.3	474, 1709
Selenium, [^{77}Se]		
$Se(C^1H{=}C^2H_2)(C_6H_4X-4)$	C^1 124.7 to 129.9, C^2 115.6 to 124.6	766, 1220
$MeSe\mathit{C}MeW(CO)_5$	355.5	178
$SeC^1{=}C^2(N{=}N)C_6H_4(CH_2)_2$ (cyclic)	C^1 155.6, C^2 157.8	813
$SeC^1{=}C^2(N{=}N)C_6H_4(CH_2)_3$ (cyclic)	C^1 159.4, C^2 159.8	813
$SeC^1{=}C^2(N{=}N)C_6H_4(CH_2)_4$ (cyclic)	C^1 159.5, C^2 161,9	813
$SeC^1{=}C^2(N{=}N)C_6H_4CO$ (cyclic)	C^1 141.9, C^2 162.9	813
$SeC^1{=}C^2(N{=}N)CH_2C_6H_4$ (cyclic)	C^1 160.0, C^2 167.2	813
$SeC^1{=}C^2(N{=}N)(CH_2)_2C_6H_4$ (cyclic)	C^1 154.9, C^2 157.4	813
$HSe\mathit{C}{=}NN{=}C(NHMe)S$ (cyclic)	168.5	1226

COMPOUND	δ(^{13}C)/p.p.m., *J*/Hz	REFERENCE
HSe*C*=NN=C(NH-*p*-tol)S	170.5	1226
HSeC^1=NN=C^2(NHMe)Se	C^1 171.7, C^2 166.1	1226
HSeC^1=NN=C^2(NH-*p*-tol)Se	C^1 175.2, C^2 161.1	1226
HSC=NN=*C*(NH-*p*-tol)Se	157.2	1226
[SeC^1=NN=C^2(NHMe)Se]$_2$	C^1 140.0, C^2 173.1	1226
HSe*C*=NNC(2-thienyl)S	177.8	1226
Se*C*H=NC$_6$H$_4$	160.6 [-136.7]	1226, 1713
MeSe*C*=NN=NNPh	148.0	1226, 1713
Se*C*=NN=NNPh	165.8	1226, 1713
Se*C*=*C*(N=N)(CH$_2$)$_3$CHOH(CH$_2$)$_2$	159.6, 160.5	1789
Se*C*=*C*(N=N)(CH$_2$)$_3$CHOAc(CH$_2$)$_2$	158.3, 160.1	1789
Se{*C*=*C*(OH)COCMe$_2$CH$_2$CMe$_2$}$_2$	132.5, 143.5	1794
Se*C*=*C*SCSCH$_2$CHMeCH$_2$	142.1, 144.1	1793
Se*C*=CSeCSeCH$_2$CHMeCH$_2$	150.7	1793

Selenophenes

R^2	R^3	R^4	R^5	C^2	C^3	C^4	C^5	
H	H	H	H	131.0	128.8	128.8	131.0	398, 1226
				{189}	{166}	{166}	{189}	
H	H	H	H	130.5	129.2	129.2	130.5	1713
H	H	H	H	131.0	129.8	129.8	131.0	1796
H	H	H	H	131.4	129.5	129.5	131.4	619
H	H	H	H	187.2	164.6	164.6	187.2	134
Me	H	H	H	146.4	128.1	129.6	129.0	1796
Me	H	Br	H	146.9	129.3	108.2	124	619
				[122]			[116]	
CN	H	H	H	115.6	142.3	130.2	141.7	619
				111.4	141.3	130.5	141.0	1796
H	CN	H	H	145.5		130.1	135.2	619
CN	CN	H	H	122.7	120.7	130.4	194.5	619
CN	H	CN	H	114.5	141.3	114.5	153.3	619
CN	H	H	CN	120.2	140.6	140.6	120.2	619
H	CN	CN	H	148.1	113.7	113.7	148.1	619
CN	H	CHO	H		137.7	143.3	153.2	619
CHO	H	CN	H	150.4	139.1	112.6	152.5	619
CH$_2$OH	H	H	H	154.0	126.1	129.3	130.3	619
CHMe(OAc)	H	H	H	152.9	127.7	129.6	131.3	783
CHO	H	H	H	151.1	140.3	131.6	141.5	1796

COMPOUND				$\delta(^{13}C)$/p.p.m., J/Hz				REFERENCE
R^2	R^3	R^4	R^5	C^2	C^3	C^4	C^5	
H	CHO	H	H	147.4	145.9	126.7	134.2	619
CHO	CHO	H	H	155.6	146.1	130.7	141.3	619
CHO	H	CHO	H	150	136.1	144.9	152.9	619
CHO	H	H	CHO	154.6	139.6	139.6	154.6	619
H	CHO	CHO	H	147.3 [124]	141.5	141.5	147.5 [124]	619
COMe	H	H	H	151.9	135.7	131.4	140.4	1796
$CONMe_2$	H	H	H	146.1	131.6	130.6	135.6	783
COOH	H	H	H	140.1	136.7	131.0	139.8	1796
				140.7	136.0	130.8	139.8	619
H	COOH	H	H	141.0	137.0	130.1	132.7	619
COOH	COOH	H	H		138.5	131.8	137.0	619
COOH	H	COOH	H	140.5	134.4	136.5	146.1	619
COOH	H	H	COOH	145.7	135.0	135.0	145.7	619
H	COOH	COOH	H	138.8	135.7	135.7	138.8	619
CO_2Me	H	H	H	140.7	136.4	130.9	139.6	1796
NO_2	H	H	H	161.3	132.7	130.1	142.4	1796
OMe	H	H	H	172.9	106.4	127.8	117.8	1796
OBu^t	H	H	H	165.6	122.2	122.2	126.1	619
H	OBu^t	H	H	112.8	154.3	126.8	128.1	619
OBu^t	H	Br	H	165.6	114.3	106.4	117.4	619
OCOMe	H	H	H	154.9	113.8	124.9	124.8	1796
SMe	H	H	H	144.2	131.9	130.3	133.2	1796
F	H	H	H	171.4	110.9	126.0	121.5	1796
Cl	H	H	H		169.2	167.2	190.0	134
Cl	H	H	H	133.7	130.1	129.5	131.7	1796
Cl	H	H	H	132.8 [140]	128.6	128.0	130.3 [115]	619
Cl	H	H	Cl		171.6	171.6		134
Br	H	H	H	115.3	133.8	130.6	134.4	1796
Br	H	H	H		169.7	167.3	189.6	134
Br	H	H	H	115.2	134.0	130.1	130.3	619
H	Br	H	H	127.9 [127]	109.6	133.7	132.0	619
Br	Br	H	H	112.5 [138]	114.4	133.3	131.2 [117]	619
Br	H	Br	H	115.6 [136]	133.8	108.4	129.3 [120]	619
Br	H	H	Br	115.0 [139]	132.5	132.5	115.0 [139]	619
Br	H	H	Br		171.4	171.4		134
H	Br	Br	H	127.9 [119.5]	112.8	112.8	127.9 [119.5]	619
I	H	H	H	75.9	140.8	132.0	138.4	1796

COMPOUND	$\delta(^{13}C)$/p.p.m., J/Hz	REFERENCE

Tellurium

Tellurophene

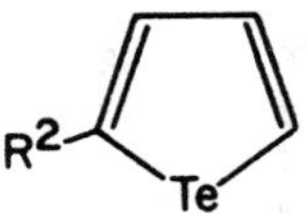

R^2	C^2	C^3	C^4	C^5	
H	126.2			126.2	1795
H	127.3 {183}	138.0 {159}	138.0 {159}	127.3 {183}	398
Me	144.6	137.5	136.8	124.9	783
CH_2OH	155.3	132.2 {158, 5.8, 10.8}	137.4 {160, 3.0, 4.9}	124.9 {183, 5.0, 10.6}	398
CHMe(OAc)	152.7	134.8	137.4	127.1	783
CHO	151.5	148.2 {161, 5.6, 11.0}	139.4 {163, 2.5, 4.5}	138.7 {184, 4.9, 11.2}	398
CMeO	153.5	143.4 {161, 5.3, 10.7}	139.8 {163, 2.5, 4.7}	137.8 {182, 4.6, 10.8}	398
$CONMe_2$	146.6	138.4	138.4	132.6	783
CO_2H	137.6	144.7 {159, 5.6, 11.0}	138.6 {161, 2.4, 4.8}	138.1 {183, 5.0, 10.8}	398
CO_2H	139.0	144.5 {164, 5.6, 10.7	138.6 {163, 3.0, 5.0}	137.4 {163, 4.8, 11.2}	398
SMe	142.1	136.3 {161, 6.0, 10.5}	137.8 {161, 2.7, 5.0}	125.6 {185, 4.0, 11.0}	398
Cl	136.4	139.1	136.0	128.7	783
Br	110.0	142.6	137.4	131.5	783
I	68.9	149.2	139.4	135.0	783

TABLE 2.3

^{13}C Chemical Shift Data for Acetylenic Groups Attached to Metals

COMPOUND	δ(^{13}C)/p.p.m., J/Hz		REFERENCE
Lithium			
LiC≡CLi	75.0		1331
Tungsten, [^{183}W]			
W(*C*≡Ctol)Cp(CO)$_2$PMe$_3$	96.3 [97.1] (41.2)		1403
Iron			
Fe(*C*≡CH)Cp(dppe)	105.7, 68.3		1446
Fe(*C*≡CMe)Cp(dppe)	112.6, 97.5		1446
Ruthenium			
Ru(*C*≡CR)(PPh$_3$)$_2$Cp	230		1483
Platinum, [^{195}Pt]			
	C^1	C^2	
trans-Pt(C^1≡C^2Me)$_2$-(PMe$_2$Ph)$_2$	91.2 [946]	102.1 [264]	1313
trans-[Pt(C^1≡C^2Me)py-(PMe$_2$Ph)$_2$]$^+$	58.9	100.8	1313
trans-[Pt(C^1≡C^2Me)-(NC$_5$H$_4$NMe$_2$-4)(PMe$_2$Ph)$_2$]$^+$	60.9 [1225]	100.6 [355.4]	1313
trans-[Pt(C^1≡C^2Me)-(NC$_5$H$_4$OBu^n-4)(PMe$_2$Ph)$_2$]$^+$	59.7 [1238]	100.5 [359.1]	1313
trans-[Pt(C^1≡C^2Me)-(NC$_5$H$_4$COMe-4)(PMe$_2$Ph)$_2$]$^+$	58.4 [1251]	100.6 [362.2]	1313
trans-Pt(C^1≡C^2Me)Cl-(PMe$_2$Ph)$_2$	64.9 [1084]	95.2	1313
trans-Pt(C^1≡C^2Me)Br-(PMe$_2$Ph)$_2$	68.2 [1397]	95.1 [362]	1313
trans-Pt(C^1≡C^2CMe$_2$OH)-(PMe$_2$Ph)$_2$	97.2 [892] (14.9)	119.3 [255]	582
trans-Pt(C^1≡C^2CMe$_2$OH)-(AsMe$_2$Ph)$_2$	90.9 [915]	116.2 [252]	582
Mercury, [^{199}Hg]			
Hg(C^1≡C^2Ph)$_2$	121.5 [2584]	108.7	246
Boron, [^{11}B]			
B(C^1≡C^2H)$_2$NEt$_2$	84.4 [132] {51.0}	85.0 {240.0}	1059
B(C^1≡C^2H)(OBu)$_2$		91.4 {240.0}	1059
B(C^1≡C^2Bu^n)(OBu)$_2$		104.9	1059

COMPOUND	$\delta(^{13}C)$/p.p.m., *J*/Hz		REFERENCE
	C^1	C^2	
$B(C^1{\equiv}C^2NEt_2)NMeCH_2CH_2NMe$	63.0 [120]	108.9	1581
Aluminium			
$Al(C^1{\equiv}C^2Me)Me_2$	90.2	132.9	1586
$Al(C^1{\equiv}C^2Me)Me_2.OMe_2$	92.5	103.0	1586
Gallium			
$Ga(C^1{\equiv}C^2Me)Me_2$	89.9	122.4	1586
Indium			
$In(C^1{\equiv}C^2Me)Me_2$	90.9	122.4	1586
Silicon, [^{29}*Si*]			
$Si(C^1{\equiv}C^2H)_2Me_2$	86.4 {44.1}	95.4 {239.7}	1711
$Si(C^1{\equiv}C^2H)_2Ph_2$	83.5 {44.1}	98.0 {241.2}	1711
$Si(C^1{\equiv}C^2H)_2p\text{-}tol_2$	84.0 {44.1}	97.7 {241.2}	1711
$Si(C^1{\equiv}C^2H)_2(C_6H_4OMe\text{-}4)_2$	84.2 {42.7}	97.6 {239.7}	1711
$Si(\mathit{C}{\equiv}CH)Me_3$	89.3		1725
$Si(C^1{\equiv}C^2H)Et_3$	86.9 [75]	96.1 [18.6]	1633
	86.8		1725
$Si(C^1{\equiv}C^2H)Ph_3$	85.5 {41.2}	98.5 {241.2}	1711
	84.0		218
$1,4\text{-}(HC^2{\equiv}C^1SiMe_2)_2C_6H_4$	88.3 {38.2}	95.9 {238.2}	1711
$Me_3Si\mathit{C}{\equiv}\mathit{C}SiMe_3$	113.0		1711
$Et_3Si\mathit{C}{\equiv}\mathit{C}SiEt_3$	112.5 [748, 11.5]		1633
$HCl_2CMe_2SiC{\equiv}CSiMe_2CHCl_2$	111.5		1246
$Me_3SiC^1{\equiv}C^2CH{=}CHMe$-Z	100.0	103.0	797, 821
-E	92.2	104.0	797, 821
$Me_3SiC^1{\equiv}C^2CH{=}CHEt$-Z	98.4	101.9	821
-E	92.4	104.2	821
$Me_3SiC^1{\equiv}C^2CH{=}CHPr$-Z	98.3	102.2	821
-E	92.4	104.2	821
$Me_3SiC^1{\equiv}C^2C{\equiv}CMe$	82.7	89.2	797
$Me_3SiC^1{\equiv}C^2C^2{\equiv}C^1SiMe_3$	86.0	88.7	797
$Si(C^1{\equiv}C^2Ph)Me_3$	92.5 [83.6]	85.7 [16.1]	97
$[\text{-}Ph_2SiC_2\text{-}]_n$	83.7 {42.6}	97.6 {241.2}	1211
	83.3	98.3	
	111.4		
Germanium			
$Ge(C^1{\equiv}C^2H)_2Ph_2$	82.5 {44.1}	95.9 {241.2}	1711
$Ge(C^1{\equiv}C^2H)_2(C_6F_5)_2$	78.1 {47.1}	97.1	1711
$Ge(C^1{\equiv}C^2H)Me_3$	89.1	92.3	1725
$Ge(C^1{\equiv}C^2H)Et_3$	86.4	93.6	1725
$Ge(C^1{\equiv}C^2H)Bu^n{}_3$	91.5	84.8	138, 139
$Ge(C^1{\equiv}C^2H)Ph_3$	84.9 {41.2}	96.6 {238.2}	218, 1711
$[\text{-}Ph_2GeC_2\text{-}]_n$	108.3		1711

COMPOUND	$\delta(^{13}C)$/p.p.m., J/Hz		REFERENCE
Tin, [^{119}Sn]			
	C^1	C^2	
$Sn(C^1\equiv C^2H)_2Me_2$	85.4 [585]	97.6 [122.0]	1643
$Sn(C\equiv CH)Et_3$	86.8		1725
$Sn(C^1\equiv C^2H)Bu^n{}_3$	87.5 [327.6]	97.4 [61.5]	1633
$Sn(C^1\equiv C^2H)Ph_3$	84.6	97.5	218
$Sn(C^1\equiv C^2Me)_4$	75.2 [1168]	107.3 [241.3]	1643
$Sn(C^1\equiv C^2Pr)Me_3$	81.7	104.6	882, 1643
$Sn(C^1\equiv C^2Bu^n)_4$	76.7 [1160.4]	110.0 [235.3]	1238
$Sn(C^1\equiv C^2Bu^n)_3Me$	78.2 [877.8]	111.6 [180.8]	1238
$Sn(C^1\equiv C^2Bu^n)_2Me_2$	79.6 [661.5]	110.6 [137.4]	1238
$Sn(C^1\equiv C^2Bu^n)Me_3$	81.4 [506.6]	110.4 [105.2]	882, 1238
	81.9 [513.4]	110.0	875
$Sn(C^1\equiv C^2C_5H_{11})Me_3$	81.4 [505.8]	110.4 [105.3]	1238
$Sn(C^1\equiv C^2Bu^t)Me_3$	77.5 [460]	119.6 [124]	1643
	78.8 [510.4]	118.7 [100.7]	1238
$Sn(C^1\equiv C^2Ph)Me_3$	93.0 [444.1]	109.2 [92.7]	1238
	83.3	97.8	882
$Me_3SnC\equiv CSnMe_3$	115.4 [388.5, 45.2]; 116.1		882, 1643
$Bu^n{}_3SnC\equiv CSnBu^n{}_3$	115.4 [319.6, 28.2]		1633
Lead			
$Pb(C^1\equiv C^2H)Ph_3$	98.6	98.2	218
Phosphorus, [^{31}P]			
$P(C^1\equiv C^2H)_3$	74.8	96.8	138
$P(C^1\equiv C^2H)_3O$	75.8	100.8	138
$Bu^s{}_2PC^1\equiv C^2H$	82.0	91.5	138
$Ph_2PC\equiv C^2H$	82.0 (+13.0)	96.5 (+1.9)	1791
$Et_2P(O)C^1\equiv C^2H$	79.8	92.2	138
$Ph_2P(O)C^1\equiv C^2H$	77.9 (+164.0)	95.3 (+29.0)	374, 1770, 1797
$(Me_2N)P(C^1\equiv C^2H)_2$	86.1 (-9.8)	96.2 (+9.2)	1797
$(Me_2N)_2PC^1\equiv C^2H)$	83.7 (3.5)	92.5 (7.6)	1797
$(Me_2N)_2P(O)C^1\equiv C^2H$	76.4 (+224.0)	91.3 (+37.0)	374, 1797
$Ph_2NPC^1\equiv C^2H$	80.1 (-9.8)	96.2 (+9.2)	374
$(EtO)_2PC^1\equiv C^2H$	85.0 (-50.0)	91.8 (+0.4)	374, 1797
$(EtO)_2P(O)C^1\equiv C^2H$	73.9 (+294)	92.2 (+51)	374, 1797
		(+50) {252}	898
$[Ph_3PC^1\equiv C^2]^+$	60.4 (191.7)	121.8 (33.0)	1772
$Ph_2PC^1\equiv C^2Me$	75.5 (3.0)	105.7 (4.2)	1797
$Ph_2P(O)C^1\equiv C^2Me$	74.3 (+174.4)	105.2 (+31.4)	1797
$P(C^1\equiv C^2Me)_3$	71.4 (+8.8)	102.9 (+10.9)	1797
$P(O)C^1\equiv C^2Me)_3$	75.4 (+239.8)	101.4 (+47.4)	1797
$(Me_2N)_2PC^1\equiv C^2Me$	76.8 (5.8)	101.4 (10.4)	1797
$(Me_2N)_2P(O)C^1\equiv C^2Me$	72.8 (+233.3)	98.4 (+40.2)	1797
$(EtO)_2PC^1\equiv C^2Me$	80.8 (-39.1)	101.6 (+38)	1797
$(EtO)_2P(O)C^1\equiv C^2Me$	71.2 (+299.8)	98.1 (+53.4)	1797
	71.5	100.5	123

COMPOUND	δ(^{13}C)/p.p.m., *J*/Hz		REFERENCE
	C^1	C^2	
$Cl_2PC^1{\equiv}C^2Me$	75.2 (+300.1)	105.9 (+58.8)	1797
$Bu^t{}_2PC^1{\equiv}C^2Ph$	106.6	88.9 (24)	1199
Arsenic			
$Et_2AsC{\equiv}\mathit{C}H$	{132.8}		147
Selenium			
$MeSe\mathit{C}{\equiv}CCH{=}CH_2$	72.6		1785
$EtSe\mathit{C}{\equiv}CCH{=}CH_2$	71.2		1785
$Pr^nSe\mathit{C}{\equiv}CCH{=}CH_2$	71.7		1785
Tellurium			
$MeTe\mathit{C}{\equiv}CCH{=}CH_2$	49.5		1785
$EtTe\mathit{C}{\equiv}CCH{=}CH_2$	46.1		1785

TABLE 2.4

^{13}C N.m.r. Data for Some Phenyl Derivatives

COMPOUND	$\delta(^{13}C)$/p.p.m., *J*/Hz				REFERENCE
	C^1	$C^{2,6}$	$C^{3,5}$	C^4	
Lithium, [^{7}*Li*]					
LiPh	171.7	141.4	126.1	125.4	87
Li-2-$Me_2NCH_2C_6H_4$	122.6 or 187.2				1328
$Ag_2Li_2(2\text{-}Me_2NCH_2C_6H_4)_4$	[7.2]				333
$Au_2Li_2(2\text{-}Me_2NCH_2C_6H_4)_4$	174.4				333
Li-2-$Me_2NCHMeC_6H_4$	122.6 or 185.5				1328
Beryllium					
$BePh(C_5H_5)$		140.0	127.2	127.4	934
Magnesium					
MgPhBr	164.4	139.9	125.8	124.6	87
Titanium					
$TiPh_2Cp_2$	192.9	136.0	127.3	124.3	60
	193.4				184
	190.9	134.7	126.2	123.4	932
Ti(*p*-tol)$_2$Cp$_2$	187.5	134.8	127.0	132.7	339
Ti(*m*-tol)$_2$Cp$_2$	191.2	124.2, 125.9, 131.7, 135.2			339

COMPOUND	$\delta(^{13}C)$/p.p.m., J/Hz				REFERENCE
	C^1	$C^{2,6}$	$C^{3,5}$	C^4	
Tantalum					
$[Ta(p\text{-tol})_6]^-$	217.1	136.1	135.1	129.1	1248
Tungsten					
$[WPh(CO)_5]^-$	163.3	150.6*	128.6*	123.5	1401
Rhenium					
$RePh(CO)_5$	134.4	145.5*	128.7*	124.3	514
Iron					
$FePhCp(CO)_2$	145.4	145.0	127.5	122.8	856
$FePhCp(CO)P(OPh)_3$	150.0	146.5	126.4	121.7	856
$Fe\{P(OPh)_2OC_6H_4\text{-}2\}Cp(CO)$	141.6 (28)	161.7 (22) 144.2	110.9 (15) 122.1*	123.9*	856
$Fe\{P(OPh)_2OC_6H_4\text{-}2\}Cp\{P(OPh)_3\}$	143.0 (11)	163.2 (22) 146.0	110.2 (15) 122.7*	120.7*	856
$Fe\{P(OCH_2CH_2O)OC_6H_4\text{-}2\}Cp(CO)$	141.9 (27)	161.5 (21) 143.8	110.3 (13) 123.6*	121.5*	856
$Fe_2(C_6H_4\text{-}2\text{-}CH_2NMe)(CO)_6$	152.7, 148.2				792, 1457
$Fe_2(C_6H_4\text{-}2\text{-}CH_2NPh)(CO)_6$	150.4				1457
$Fe_2(C_6H_4\text{-}2\text{-}CHPhS)(CO)_6$	155.2				792
$Fe_2(C_6H_4\text{-}2\text{-}CHPhS\text{-}4\text{-}p\text{-}tol)(CO)_6$	153.5				792
$Fe_2(C_6H_4\text{-}2\text{-}CHPhS\text{-}4\text{-}OMe)(CO)_6$	156.2				792
$Fe_2(C_6H_4\text{-}2\text{-}CHPhS\text{-}4\text{-}NMe_2)(CO)_6$	161.4				792

*Relative assignments uncertain

COMPOUND	$\delta(^{13}C)$/p.p.m., *J*/Hz	REFERENCE
Ruthenium		
$Ru(CNp\text{-}tolCH_2CH_2NC_6H_3\text{-}4\text{-}Me\text{-}2)\{C(Np\text{-}tolCH_2)_2\}ClPPh_3$	C^1 158.2 (14.4)	1196
$Ru(CNp\text{-}tolCH_2CH_2NC_6H_3\text{-}4\text{-}Me\text{-}2)(PEt_3)_2Cl$	C^1 157.3 (12.2)	1196
$Ru(CNp\text{-}tolCH_2CH_2NC_6H_3\text{-}4\text{-}Me\text{-}2)(CO)(PEt_3)Cl$	C^1 161.2 (15.2)	1196
$Ru(CNp\text{-}tolCH_2CH_2NC_6H_3\text{-}4\text{-}Me\text{-}2)(CO)(PPh_3)Cl$	C^1 158.2 (15.2)	1196
Osmium		
$Os_3(OC_6H_4\text{-}2)H_2(CO)_9$	C^1 175.0	1485
Rhodium, [^{103}*Rh*]		
$[Rh(C_6H_4CONNC_5H_5)_2(OH_2)_2]^+$	C^1 159.5 [32.5]	563
$[Rh(C_6H_3MeCONNC_5H_5)_2(OH_2)_2]^+$	C^1 160.5 [33.0]	563
$[Rh(C_6H_4NNPh)_2Cl_2]$	C^1 166.0 [37]	564
Iridium		
$Ir(C_6H_4N_2Ph)HCl(PPh_3)_2$	C^1 155.8	604
$Ir(5\text{-}MeC_6H_3N_2Ph)HCl(PPh_3)_2$	C^1 156.1	604
$Ir(5\text{-}MeC_6H_3N_2p\text{-}tol)HCl(PPh_3)_2$	C^1 155.8	604
$Ir(4,6\text{-}Me_2C_6H_2N_2m\text{-}xylyl)HCl(PPh_3)_2$	C^1 155.2	604
$Ir(5\text{-}MeC_6H_3CHNMe)HCl(PPh_3)_2$	C^1 152.2	604
$Ir(C_6H_4CHNMe)HCl(PCy_3)_2$	C^2 146.3, $C^{3,4}$ 125.0; 126.7, C^5 142.5, C^6 143.1	636
$Ir(CH_2C_6H_4)H\{P(OMe)_3\}_3$	C^1 157.8 (31)	1521
$Ir(CH_2C_6H_3\text{-}3\text{-}Me)H\{P(OMe)_3\}_3$	C^1 158.0 (30)	1521
$Ir(CH_2C_6H_3\text{-}4\text{-}Me)H\{P(OMe)_3\}_3$	C^1 156.1 (30)	1521

COMPOUND	$\delta(^{13}C)$/p.p.m., J/Hz				REFERENCE
$Ir(CH_2C_6H_3\text{-}5\text{-}Me)H\{P(OMe)_3\}_3$	C^1 158.1 (33)				1521
$Ir(CH_2C_6H_3\text{-}6\text{-}Me)H\{P(OMe)_3\}_3$	C^1 155.3 (31)				1521
$Ir(CH_2C_6H_2\text{-}4,6\text{-}Me_2)H\{P(OMe)_3\}_3$	C^1 153.2 (30)				1521
Palladium					
$Pd(C_6H_4N_2Ph)(acac)$	163.8*, 157.1*, 151.2*				564
$[Pd(C_6H_4CH_2NMe_2)Cl]_2$	146.7*, 142.9*				564
$Pd(C_{13}H_8N)(acac)$	150.9*, 132.8*, 128.8*, 126.3*, 122.6*				564
$Pd(C_{13}H_8N)Cp$	158.8*, 135.6*, 134.5*, 127.0*				564
	C^1	$C^{2,6}$	$C^{3,5}$	C^4	
Platinum, [^{195}Pt]					
$PtPh_2Cl_2(PEt_3)_2$	130.5 [815]	138.9	127.8	125.6	122
$[PtPh_2(OAc)(SEt_2)]_2$		135.9	125.3	122.6	1243
$Pt_2Ph_4(OAc)_2PEt_3$		136.4	125.2	122.7	1243
		137.0	126.5	124.8	
trans-$PtPh_2(PEt_3)_2$	164.0 [594]	140.7	127.4 [42]	121.4	122
		140.0 [22]	126.9 [40]	120.8	857
cis-$PtPh_2(PEt_3)_2$	165.0 [817]	136.9 [36]	127.1 [64]	121.0	122
		136.4 [32]	126.8 [68]	120.6	857
$PtPh_2(CNBu^t)_2$	153.2	139.1 [43]	126.8 [66]	122.6	1244
$PtPh_2(SEt_2)_2$	148.5	136.2 [26.3]	126.8 [82]	121.5	1244
$[PtPh_2(SEt_2)]_2$	147.7	134.2 [40]	127.4 [83]	122.8	1244
$PtPh_2(CO)(SEt_2)$	159.3	137.4 [50]	127.8 [80]	123.9	1244
	140.2	136.3 [31]	127.5 [64]	123.6	
$[Pt(p\text{-}tol)_2(OAc)(SEt_2)]_2$		135.8	126.7	129.4	1243
$[Pt(p\text{-}tol)_2(SEt_2)]_2$	144.0	134.0 [32]	128.3 [85]	131.5	1244
$Pt(p\text{-}tol)_2(CNBu^t)_2$	149.4	139.1 [45]	127.7 [67]	131.4	1244
$Pt(p\text{-}tol)_2(CO)(SEt_2)$	155.6	137.2 [53]	128.6 [82]	132.8	1244
	138.5	136.0 [33]	128.3 [67]	132.8	
$Pt(p\text{-}tol)_2(SMe_2)_2$	144.6	135.0 [38]	128.0 [84]	130.6	1244

*Relative assignments uncertain

COMPOUND	$\delta(^{13}C)$/p.p.m., *J*/Hz				REFERENCE
	C^1	$C^{2,6}$	$C^{3,5}$	C^4	
Pt(C_6H_4Br-4)$_2$(cod)	153.3	136.0 [40]	130.5 [81]	117.0	1541
Pt(C_6H_4Cl-2)$_2$(cod)	156.7 [1086]	132.7 [31]	133.5 [97]	123.1 [11]	1541
		133.8 [37]	128.7 [86]		
Pt(C_6H_4OMe-3)$_2$(cod)	141.8 [1077]	161.0 [17]	120.9 [76]	123.3	1541
		135.5 [24]	110.4 [37]		
trans-PtPhMe(PEt_3)$_2$		139.9 [20]	126.9 [44]	120.4	857
trans-PtPhEt(PEt_3)$_2$		140.0 [20]	126.8 [44]	120.6	857
trans-PtPh(CF_2H)(PEt_3)$_2$		136.5 [36]	127.7 [76]	122.0	857
trans-PtPhVi(PEt_3)$_2$		140.2 [20]	126.8 [41]	120.7	857
trans-[PtPh($\overline{CCH_2CH_2CH_2O}$)($AsMe_3$)$_2$]$^+$	146.2 [502]	138.4 [16]	127.6 [34]	123.5	409
trans-[PtPh(CO)($AsMe_3$)$_2$]$^+$	141.7 [642]	135.6 [23]	129.1 [46]	125.4 [8]	409
trans-[PtPh(CNEt)($AsMe_3$)$_2$]$^+$	142.0 [646]	136.4 [21]	128.0 [45]	123.5 [8]	409
trans-PtPhCN(PEt_3)$_2$	151.4	138.1 [24]	127.9 [50]	122.3	857
trans-PtPhF(PEt_3)$_2$		137.2 [28]	127.3 [71]	121.3	857
[PtPhCl_2(CO)]$^-$	130.0 [853]	137.8 [35]	128.0 [60]	124.4	1562
PtPhCl(cod)	143.4	133.4 [12]	127.8 [50]	123.7 [8]	1541
trans-PtPhCl($AsMe_3$)$_2$	131.9 [858]	136.7 [28]	127.3 [64]	121.8 [12]	409
trans-PtPhCl(PEt_3)$_2$	139.9 (8)	137.4 [35]	127.9 [74]	121.8	857
cis-PtPhCl(PEt_3)$_2$	157.9	136.2 [19]	127.2 [45]	122.4	857
trans-PtPhBr(PEt_3)$_2$		137.2 [38]	127.9 [75]	121.9	857
trans-PtPhI(PEt_3)$_2$	144.0 (8)	136.9 [39]	128.0 [76]	122.1	857
trans-PtPh(NCO)(PEt_3)$_2$		137.8	127.8 [73]	121.8	857
trans-PtPh(N_3)(PEt_3)$_2$		137.2 [34]	127.9 [72]	122.0	857
trans-PtPh(OMe)(PEt_3)$_2$		138.2 [30]	127.2 [60]	120.9	857
trans-PtPh(SEt)(PEt_3)$_2$		138.1 [30]	127.6 [60]	121.3	857
trans-PtPh(NO_2)(PEt_3)$_2$		137.7 [28]	127.7 [59]	122.7	857
trans-PtPh($SnCl_3$)(PEt_3)$_2$		136.0 [40]	128.9 [73]	124.0	857
trans-[PtPh(dmf)($AsMe_3$)$_2$]$^+$	118.4	135.8	127.4 [66]	122.6 [12]	409
trans-[PtPh(NCC_6H_4OMe-4)($AsMe_3$)$_2$]$^+$	126.8 [818]	135.9 [26]	128.0 [59]	123.3 [8]	409
trans-[PtPh(py)($AsMe_3$)$_2$]$^+$	126.4 [754]	136.6 [25]	127.4 [56]	122.9 [10]	409

COMPOUND	$\delta(^{13}C)$/p.p.m., J/Hz				REFERENCE
	C^1	$C^{2,6}$	$C^{3,5}$	C^4	
trans-$[PtPh(py\text{-}Me\text{-}4)(AsMe_3)_2]^+$	126.9 [752]	136.6 [24]	127.5 [55]	122.7 [8]	409
trans-$[PtPh(AsPh_3)(AsMe_3)_2]^+$	142.7 [692]	136.7 [26]	128.0 [52]	124.6 [10]	409
Copper					
$[CuPh]_4$	137.2	144.4	127.5	131.3	1552
$[Cu(C_6H_4Me\text{-}2)]_4$	139.2	154.8	127.4	131.2	1552
		144.1	124.5		
$[Cu(C_6H_3Me_2\text{-}2,6)]_4$	140.0	155.1	124.7	131.1	1552
Silver, $[^{109}Ag]$					
AgPh		146.4	127.2	131.2	1011
$2AgPh \cdot 2AgNO_3$	148.5	144.4	128.9	133.1	1011
$Ag_2Li_2(C_6H_4CH_2NMe_2\text{-}2)_4$	[136.0] ($^1J_{C}{}^7{}_{Li}$ = 7.2)				209, 333
Gold					
$AuZnPh_3$	146.7	132.0	129.5	127.3	965
$AuZnPh_2Cl$	146.7	132.1	129.4	127.6	1081
AuZn(*p*-tol)$_3$	147.5	141.8	128.2	130.1	965
AuZn(*p*-tol)$_2$Cl	147.6	138.7	127.5	130.2	1081
$Au_2Li_2(C_6H_4CH_2NMe_2\text{-}2)_4$	174.4				333
Zinc					
$ZnPh_2$	148.6	138.1	128.3	128.8	965
$ZnAuPh_3$	146.7*	137.7*	127.6*	128.8*	965
	146.7§	132.0§	129.5§	127.3§	
$ZnAuPh_2Cl$	146.7	132.1	129.4	127.6	1081
Zn(*p*-tol)$_2$	144.1	141.7	129.7	134.9	965
ZnAu(*p*-tol)$_3$	146.6*	141.5*	128.4*	135.5	965
	147.5§	141.8§	128.2§	130.1	
ZnAu(*p*-tol)$_2$Cl	147.8	138.7	127.5	130.2	1081

*terminal, §bridge

COMPOUND	δ(^{13}C)/p.p.m., *J*/Hz				REFERENCE
	C^1	$C^{2,6}$	$C^{3,5}$	C^4	
Mercury, [^{199}*Hg*]					
$HgPh_2$	[1186]	[88]	[101.6]	[17.8]	161, 162
	172.5 [1275]*	139.7 [85.3]	129.4 [101.5]	128.7 [10]	861
	170.6 [1275]*	137.8 [84.5]	127.6 [104.1]	126.9 [20.5]	943
	170.0 [1189]§	137.3 [85.5]	128.5 [101.0]	128.1 [18.5]	943
	169.0†	138.6 [87.6]	128.8 [102.0]	128.3	943
	170.5 [1176]	137.5 [87]			246
	170.4 [1176]	137.5 [85]	128.5 [101]	128.2 [18]	1562
	[1218]	[86.3]	[100.2]	[18.0]	459
$HgPh(C_6H_{11})$-*axial*	182.5	137.4	128.8#	127.6#	1240
-*equatorial*	179.9	137.4	128.7	127.6	1240
HgPh(OAc)	147.1 [2661.1]*	138.6 [119.6]	129.9 [210.0]	129.7 [36.6]	861
	145.1 [2666]*	136.7 [120.4]	128.0 [210.7]	127.8 [37.2]	943
$HgPh(O_2CPh)$	145.2 [2728]*	136.8 [121.1]	128.0 [214.3]	128.0	943
HgPh(lactate)	145.0 [2714]*	136.8 [120.9]	128.0 [212.3]	127.8 [31.4]	943
$(PhHg)_2$phthalate	145.4 [2702]*	136.8 [122.6]	128.1 [213.8]	128.1	943
HgPhCl	151.0 [2634]*	136.5 [118.6]	127.9 [209.0]	127.6 [37.1]	943
	148.9§	135.7	129.3	129.5	943
	156.8†	137.7	129.4	129.4	943
HgPhBr	150.5 [2530]	136.4 [117]	128.0 [205]	127.9 [35]	1562
	156.4	138.2 [122]	129.9 [206]	129.6 [38]	861
	154.6*	136.5	128.1	127.8	943
Hg(*p*-tol)$_2$	166.9§	137.0 [85.1]	129.2 [106.8]	137.8	943
Hg(*p*-tol)Cl	147.6 [2624]*	136.2 [123.7]	128.6 [220.8]	136.8 [34.8]	943
	139.5	135.3	129.9	135.3	943
$Hg(C_6H_4C_3H_4Ph$-4)Cl-*cis*	140.0	129.1#	135.8#	148.3	546
-*trans*	142.3	125.5#	136.7#	148.3	546
$Hg(C_6H_4CO_2H$-4)Cl	157.5 [2694]*	136.6 [124.4]	128.2 [216.6]	129.8 [40.0]	943
$Hg(C_6H_4CF_3$-4)Cl	154.5*	133.1	124.1	128.3	943

*in dmso, §in $CHCl_2 \cdot CHCl_2$, †in d_6-acetone, #assignment uncertain

COMPOUND	δ(^{13}C)/p.p.m., J/Hz				REFERENCE
	C^1	$C^{2,6}$	$C^{3,5}$	C^4	
$Hg(C_6H_4F\text{-}4)$	165.3	138.6			246
$Hg(C_6H_4C_3H_4Ph\text{-}2)Cl$-*cis*	156.4	142.3	127.8	130.0	546
		136.2	125.5		
-*trans*	154.1	146.2	126.0	128.3	546
		136.6	125.6		
$Hg(C_6H_4OH\text{-}2)Cl$	137.5 [3196]*	159.3 [126]	115.1 [118.6]	129.3	943
		137.2 [72]	119.5 [200.5]		
$Hg(C_6F_5)Cl$	124.1	165.2	135.5	141.3	943
$Hg(C_6Cl_5)Cl$	169.1	126.1	114.7	137.7	943
Boron, [^{11}B]					
$[BPh_4]^-$	165.5	137.3 [1.3]	126.6	122.8	330
	[49.5]		[2.6]		161
	[49.4]	[7.3]	[6.2]	[7.2]	215, 549
	[48.4]	[1.5]	[2.7]	[0.5]	549
	[48.8]	[1.4]	[2.8]	[0.6]	292
	[49.2]	[1.3]	[2.8]	[<0.3]	293
	164.8				1163
	136.9	125.9	128.1	122.7	985
BPh_3	144.9	138.8$^{\phi}$	127.6$^{\phi}$	131.5	1163
$BPh_3NC_5H_4NH_2\text{-}2$	151.6	135.5$^{\phi}$	127.5$^{\phi}$	125.5	1163
$BPh_3.py$		134.8$^{\phi}$	127.2$^{\phi}$	125.3	1163
BPh_2NMe_2	143.1	133.7$^{\phi}$	127.4$^{\phi}$	127.9	1163
$BPh_2NC(CF_3)_2$	132.2	135.3$^{\phi}$	128.3$^{\phi}$	131.4	1163
$BPh_2OCH_2CH_2NHCH_2CH_2OH$	147.0	132.6	127.4	126.6	1572
$(BPh_2)_2O$	134.7	135.7$^{\phi}$	127.8$^{\phi}$	131.1	1163
BPh_2Cl		137.0$^{\phi}$	127.9	133	1163
BPh_2Br	139.2	137.9	128.1	133.4	1163
$\overline{BPh(CH_2)_4C}=CHSnMe_3$	142.0	135.2	127.5	131.1	1730

*in dmso, §in $CHCl_2 \cdot CHCl_2$, †in d_6-acetone, #assignment uncertain, $^{\phi}$literature assignments reversed

COMPOUND	$\delta(^{13}C)$/p.p.m., *J*/Hz				REFERENCE
	C^1	$C^{2,6}$	$C^{3,5}$	C^4	
$BPh(NMe_2)_2$	141.2	134.0	127.4	127.4	1163
$BPh(NMeCH_2)_2CH_2$	140.0, 141.1	132.3*	127.5*	126.8	1163, 1580
$\overline{BPh(NHCH_2CH_2CH_2N}Et)$	141.6				1580
$BPh(NMeCH_2CH_2)_2$	134.3	133.2*	127.7*	128.1	1163
$[BPhNH]_2$	135.4	132.3*	127.9*	129.7	1163
$\overline{BPhNHCH_2CH_2CH_2O}$	136.8	133.4	128.5	130.5	1580
$BP\{\overline{NCH(CH_2)_4}CH_2CH_2O\}$	138.3	134.7	128.3	129.5	1580
$BPh(NPr^n{}_2)Cl$	139.2	132.9	128.4	129.4	1580
$BPh(NPr^i{}_2)Cl$	141.8	132.3	128.6	129.2	1580
$BPh(OH)_2$	132.2	133.7*	127.6*	130.5	1163
		134.7	128.1	130.5	1269
$[BPhO]_3$	134.3	133.6*	127.4*	130.3	1163
$BPh(C_5H_4FeCp)I$	131.0, 134.7				885
$C_6H_4B_{10}H_{10}$	73.5, 129.2, 123.9				994
Aluminium					
Al_2Ph_6	121.3§	128.1	155.1	138.2	1717
	145.5†	127.3	137.0	128.6	
$Al_2(C_6H_4Me\text{-}4)_6$	118.0§	129.6	155.2	150.0	1717
	142.6†	128.3	137.6	137.9	
$Al_2(C_6H_4Me\text{-}3)_6$	121.8§	134.2	132.3	151.8	1717
		138.1	156.9		
	136.6†	127.3	146.1	138.9	
		128.5	155.1		
$Al(C_6H_4Me\text{-}2)_3$	143.9	147.4	129.4	125.0	1717
		135.4	129.1		

*literature assignments reversed, §bridging, †terminal

COMPOUND	$\delta(^{13}C)$/p.p.m., J/Hz				REFERENCE
	C^1	$C^{2,6}$	$C^{3,5}$	C^4	
Gallium					
$GaPh_3$	144.1	137.9	128.2	129.7	476
$Ga(C_6H_4Me-4)_3$	140.8, 144.3	138.0	128.9	139.5	476
$Ga(C_6H_4Et-4)_3$	141.2	138.1	127.7	145.8	476
$Ga(C_6H_4Bu^t-4)_3$	141.0	137.9	125.1	152.5	476
$Ga(C_6H_4SiMe_3-4)_3$	144.7	137.2	133.0		476
$Ga(C_6H_4OMe-4)_3$	137.8	138.9	155.0	159.7	476
$Ga(C_6H_4F-4)_3$	138.7	139.6	115.6	164.3	476
$Ga(C_6H_4Cl-4)_3$	141.2	138.9	128.7	136.8	476
$Ga(C_6H_4Br-4)_3$	143.0	139.1	131.6	123.0	476
$Ga(C_6H_4F-3)_3$	C^1 145.2	C^2 123.7	C^3 162.9	C^4 117.1	476
		C^6 133.3	C^5 130.0		
$Ga(C_6H_4Cl-3)_3$	C^1 145.0	C^2 137.0	C^3 134.5	C^4 130.0	476
		C^6 135.6	C^5 129.7		
$Ga(C_6H_4Br-3)_3$	C^1 145.7	C^2 139.7	C^3 123.3	C^4 132.6	476
		C^6 135.9	C^5 130.0		
$Ga(C_6H_4Me-2)_3$	C^1 147.6	C^2 144.6	C^3 129.2	C^4 129.5	476
		C^6 136.7	C^5 125.2		
$Ga(C_6H_4Cl-2)_3$	C^1 146.7	C^2 141.0	C^3 127.4	C^4 130.6	476
		C^6 137.1	C^5 126.5		
$Ga(C_6H_4Br-2)_3$	C^1 145.7	C^2 123.4	C^3 136.0	C^4 135.9	476
		C^6 139.8	C^5 130.1		
Thallium, [^{205}Tl]					
$TlPh(OAc)_2$	153.4	134.9	130.4	130.9	793
	[10718]	[527.3]	[1047.4]	[202.2]	
		{164.1}	{168.4}		
$TlPh(O_2CCF_3)_2$		134.4	129.8	130.1	291, 408
		[+500]	[+1010]	[-185]	
	150	137.5	126.9	130.7	352
	[9902]	[676]	[878]	[183]	

COMPOUND	$\delta(^{13}C)$/p.p.m., J/Hz				REFERENCE
	C^1	$C^{2,6}$	$C^{3,5}$	C^4	
$Tl(p\text{-tol})(O_2CCF_3)_2$		133.9 [+562]	130.2 [+1080]	139.8 [-209]	291, 408
	143 [9756]	137.4 [756]	127.0 [895]	141.0 [195]	352
$Tl(C_6H_4Et\text{-}4)(O_2CCF_3)_2$		134.1 [+556]	128.9 [+1066]	145.9 [-199]	408
$Tl(C_6H_4Pr^n\text{-}4)(O_2CCF_3)_2$	134.1 [+562]	134.1 [+562]	129.6 [+1075]	144.5 [-204]	408
$Tl(C_6H_4Pr^i\text{-}4)(O_2CCF_3)_2$		134.6 [+563]	127.9 [+1072]	151.2 [-206]	408
$Tl(C_6H_4Bu^t\text{-}4)(O_2CCF_3)_2$	133.9 [+560]	133.9 [+560]	126.5 [+1062]	152.8 [-211]	408
$Tl(C_6H_3Me_2\text{-}2,4)(O_2CCF_3)_2$	146.3 [9329]	140.9 [549]	126.9 [793]		352
		137.1 [646]	123.4 [849]		
		141.5 [+567]	131.4 [+1017]	139.9 [-191]	291, 408
		134.4 [+505]	127.2 [+1048]		
$Tl(C_6H_3Me_2\text{-}2,5)(O_2CCF_3)_2$		138.4 [+520]	130.4 [+1047]	130.7 [-196]	291, 408
		134.7 [+461]	135.8 [+1015]		
$Tl(C_6H_3Me_2\text{-}3,4)(O_2CCF_3)_2$	143.7 [9634]	134.0 [565]	137.1 [800]	140.2 [183]	352
		132.0 [*ca.* 500]	125.5 [820]		
		134.6 [552]	137.7 [+1071]	138.5 [-217]	291, 408
		131.5 [+515]	130.4 [+1101]		
$Tl(C_6H_2Me_3\text{-}2,4,6)(O_2CCF_3)_2$	150.3 [8841]	141.6 [512]	129.3 [968]	141.3 [166]	352
		141.6 [+509]	128.9 [+996]	140.2 [-174]	408
$Tl(C_6H_2Et_3\text{-}2,4,6)(O_2CCF_3)_2$		147.7 [+557]	126.8 [+987]	147.0 [-173]	408
Silicon					
$SiPh_4$	134.3	136.4	127.9	129.6	872
	134.5	136.6	128.1	129.9	565
	135.2	137.3	128.7	130.3	914
$SiPh_3H$	134.2	136.7	129.0	130.6	914
	133.3	135.8	128.0	129.8	872
$SiPh_3Me$	136.1	135.3	127.8	129.4	872
	136.4	135.6	128.0	129.6	565
$SiPh_3CH_2OH$	133.1	135.7	128.0	129.6	565
$SiPh_3C(OEt)Cr(CO)_5$	137.7*, 133.7*, 131.1*, 129.2*				907

*Relative assignments unknown

COMPOUND	$\delta(^{13}C)$/p.p.m., J/Hz				REFERENCE
$SiPh_3C(NMe_2)Cr(CO)_5$	137.0*, 135.7*, 130.8*, 129.0*				907
$SiPh_3CWCp(CO)_2$	135.9*, 134.3*, 130.0*, 128.3*				1026
trans-$SiPh_3CW(CO)_4Br$	135.7*, 131.7*, 130.8*, 128.6*				1026
	C^1	$C^{2,6}$	$C^{3,5}$	C^4	
$SiPh_3C_2H$	133.1	135.8	128.4	130.5	1711
$SiPh_3CH_2C_2H$	133.9	136.0	128.4	130.4	1711
$SiPh_3Cl$	133.1	135.3	128.1	130.7	872
$SiPh_2H_2$	132.2	136.5	128.9	130.6	914
	131.4	135.6	128.1	129.8	872
	131.6	136.0	128.4	130.2	565
$SiPh_2HMe$	135.2	134.8	127.9	129.5	872
	136.0	135.6	128.8	130.3	914
$SiPh_2HBu$	134.6	135.1	127.9	129.4	1157
$SiPh_2Me_2$	138.1	134.1	127.8	129.1	872
$SiPh_2MeVi$	136.1	134.8	127.1	129.3	1157
$SiPh_2R_2$	132.5	134.0	127.3	129.0	1249
$SiPh_2(C_3H_5)_2$	134.8	134.9	127.7	129.4	1157
$SiPh_2MeBu$	137.4	134.4	127.7	128.9	1157
$SiPh_2(C_2H)_2$	131.5	134.9	128.6	131.0	1711
$[SiPh_2C_2]_n$	137.5*, 135.4*, 135.2*, 131.2*, 130.7*, 129.5*				1711
$SiPh_2MeOMe$	135.5	134.1	127.6	129.6	872
$SiPh_2MeOCH_2R$	136.1 to 136.3	134.1, 134.2	127.6, 127.7	129.5, 129.6	872
$SiPh_2MeOCHR^1R^2$	136.6 to 136.9	134.2, 134.3	127.7	129.5	460, 794, 872
$SiPh_2MeOBu^t$	138.6	134.2	127.6	129.5	1157
$SiPh_2ViOEt$	134.7	134.9	127.7	129.8	1157
$SiPh_2BuOEt$	135.5	134.6	127.7	129.6	1157
$SiPh_2(OMe)_2$	132.1	134.6	127.6	130.1	1157
$SiPh_2(OEt)_2$	133.1	134.7	127.6	130.0	1157
$SiPh_2(OPr^n)_2$	133.2	134.8	127.6	130.0	872
$\overline{SiPh_2O(CH_2)_4O}$	135.1	133.9	127.6	129.6	794

*Relative assignments unknown

COMPOUND	$\delta(^{13}C)$/p.p.m., J/Hz				REFERENCE
	C^1	$C^{2,6}$	$C^{3,5}$	C^4	
$SiPh_2(OBu^n)_2$	133.2	134.8	127.6	130.0	872
$SiPh_2(OPh)_2$	131.3	135.0	128.0	130.8	794
$SiPh_2MeCl$	134.4	134.0	128.1	130.5	872
$SiPh_2Cl_2$	132.6	134.8	129.1	132.4	914
	131.9	134.0	128.3	131.7	872
$SiPhH_3$	127.5	135.5	127.7	129.4	565
	128.0	135.8	128.1	129.8	872
	128.6	136.4	128.7	130.4	914
$SiPhH_2Me$	133.3	134.8	128.0	129.5	872
$SiPhHMe_2$	137.3	134.0	127.9	129.2	872
SiPhHMeBu		134.3	127.8	129.1	1157
$SiPhHMe(C_2H_3)$	135.2	134.5	127.9	129.4	1157
$SiPhHMe(C_3H_5)$	135.5	134.3	127.8	129.4	1157
$SiPhHBu^n{}_2$	136.1	134.6	127.8	129.1	1157
SiPhHMeOMe	135.2	133.6	127.8	129.9	872
SiPhHMeOEt	135.8	133.6	127.8	129.8	872
$SiPhHMeOPr^n$	135.9	133.7	127.8	129.9	872
$SiPhHMeOPr^i$	136.4	133.7	127.8	129.8	872
$SiPhHMeOBu^s$	136.5	133.7	127.7	129.8	872
SiPhHMeOCHMePh	135.6	133.7	127.7	129.8	794
$(SiPhHMe)_2O$	136.9	133.3	127.8	129.8	794
$SiPhHMeOBu^t$	138.0	133.6	127.8	129.5	1157
SiPhHMeCl	133.4	133.6	128.2	130.8	872
$SiPhHCl_2$	131.0	133.3	128.5	132.2	872
$SiPhMe_3$	140.3	133.3	127.8	128.8	872
	139.4	132.8	127.3	128.3	565
	140.4	133.6	128.0	129.0	399
	140.2	133.4	127.8	128.8	880
	140.2	133.6	128.1	129.2	884
	140.5	133.8	128.4	129.4	914
$SiPhMe_2(C_3H_5)$		133.6	127.7	129.0	1157

COMPOUND	$\delta(^{13}C)$/p.p.m., J/Hz				REFERENCE
	C^1	$C^{2,6}$	$C^{3,5}$	C^4	
$SiPhMe_2Bu$	139.7	133.6	127.7	128.7	1157
$SiPhMe_2(C_2H_3)$	138.3	133.8	127.8	129.0	1157
$SiPhMeBu_2$	138.9	133.8	127.7	128.7	1157
$SiPhMe_2CH_2CHMePh$	139.6	133.5	127.7	128.8	460
$SiPhR_3$	135.7, 135.8	134.3, 135.6	128.2, 128.3	129.5, 131	1158
$SiPhMe_2NR_2$	137.7	133.7	129.5	127.9	989
$SiPhMe_2OMe$	137.3	133.3	127.7	129.4	872
$SiPhMe_2OEt$	137.8	133.0	127.4	129.0	565
	137.9	133.3	127.7	129.4	872
$SiPhMe_2OPr^i$	138.4	133.3	127.6	129.3	872
$SiPhMe_2OBu^t$	140.5	133.3	127.6	129.0	1157
$SiPhMe_2OCHMePh$	138.1	133.4	127.6	129.4	794
	138.6	133.6	127.8	129.6	460
$SiPhMe(C_3H_5)OEt$		133.6	127.7	129.5	1157
$SiPhMe(C_2H_3)OEt$	136.2	133.8	127.7	129.5	1157
$SiPhBu_2OEt$	137.0	133.9	127.7	129.3	1157
$SiPhMe_2F$	136.0	132.9	127.9	130.1	565
$SiPhMe_2Cl$	136.2	133.0	128.1	130.3	872
	136.2	133.2	128.3	130.3	565
$SiPhMeBuCl$	135.6	133.3	128.0	130.2	1157
$SiPhMe(OMe)_2$	133.4	133.7	127.6	129.9	1157
$SiPhMe(OEt)_2$	133.8	132.9	126.7	128.8	565
$SiPhMe(OPr^n)_2$	134.8	133.8	127.5	129.6	872
$SiPhMe(OPr^i)_2$	135.8	133.8	127.4	129.4	872
$SiPhMe(OPh)_2$	132.8	133.9	128.0	130.6	794
$SiPh(C_2H_3)(OEt)_2$	133.0	134.6	127.8	130.1	1157
$SiPhMeF_2$	129.7	133.0	128.0	131.5	565
$SiPhMeCl_2$	132.2	132.4	127.7	131.0	565
	133.3	133.0	128.3	131.6	872
$SiPh(OMe)_3$	130.0				1742
$SiPh(OEt)_3$	132.1	135.2	128.2	130.6	1734

COMPOUND	$\delta(^{13}C)$/p.p.m., J/Hz				REFERENCE
	C^1	$C^{2,6}$	$C^{3,5}$	C^4	
	131.3	134.2	127.2	129.5	565, 1742
	130.8	134.5	127.6	130.1	1157
$SiPh(OPr^i)_3$	133.3				1742
$SiPh(OBu^t)_3$	138.0				1742
$SiPh(OCH_2CH_2)_3N$	142.2				1742
	143.5	135.0	127.8	128.2	1734
$SiPhF_3$	120.1	133.9	128.1	132.7	565
$SiPhCl_3$	131.4	133.1	128.6	132.8	565
	132.0	133.7	129.1	133.3	272
	131.5	133.1	128.6	132.7	872
Miscellaneous aryl silicon compounds					407, 451, 461, 835, 880, 884, 989, 1027, 1100, 1146, 1286, 1711
Germanium					
$GePh_3C{\equiv}CH$	134.8	134.8	128.9	130.1	1711
$GePh_2(C{\equiv}CH)_2$	133.0	133.8	129.0	136.0	1711
$[GePh_2C{\equiv}C]_n$	133.6	133.9	128.9	130.5	1711
$GePhMe_3$	142.6	133.3	128.4	128.7	399
	142.2	133.0	128.0	128.3	880
$GePhRCH_2CMe{=}CMeCH_2$	137.5	134.3	128.5	129	1158
$GePhRCH_2CH_2CH{=}CH$		135	129.2	130.1	1158
$Ge(C_6H_4Me\text{-}4)Me_3$	138.3	132.8	128.7	137.3	407
$Ge(C_6H_4Bu^t\text{-}4)Me_3$	138.2	132.6	124.7	150.5	407
$Ge(C_6H_4Ph\text{-}4)Me_3$	141.1	127.0	133.3	141.0	880
$Ge(C_6H_4NMe_2\text{-}4)Me_3$	127.2	133.3	112.4	149.1	407
$Ge(C_6H_4OMe\text{-}4)Me_3$	132.5	133.9	113.8	160.1	407
$Ge(C_6H_4Cl\text{-}4)Me_3$	140.3	134.1	128.2	134.7	407
$Ge(1\text{-naphthyl})Me_3$	C^1 140.4	C^2 132.0 C^9 137.2	C^3 125.6 C^{10} 134.2	C^4 129.1	880

COMPOUND	$\delta(^{13}C)$/p.p.m., J/Hz				REFERENCE
Ge(2-naphthyl)Me_3	C^1 133.0	C^2 139.6 C^9 133.5	C^3 129.7 C^{10} 133.5	C^4 127.2	880
Ge(2-pyridyl)Me_3		C^2 170.4 C^6 150.1	C^3 127.9 C^5 122.1	C^4 133.5	451
Ge(3-pyridyl)Me_3		C^2 153.7 C^6 149.7	C^3 136.5 C^5 123.3	C^4 140.1	451
Ge(4-pyridyl)Me_3		C^2 148.7	C^3 128.1	C^4 152.7	451
Ge{$(C_6H_4$-2$)_2$S}Me_2	C^1 140.6	C^2 141.2	C^{3-6} 134.8 to 129.1*		1286
	C^1	$C^{2,6}$	$C^{3,5}$	C^4	
Tin, [^{119}Sn]					
$SnPh_4$	138.8	137.8	129.7	129.2	399
$SnPh_3Me$	139.2 [511.0]	128.5	128.9 [51.0]	136.8	882
cis-$Fe(CO)_4(SnPh_3)_2$	140.7 [416]	137.1 [36]	128.9 [47]	129.4	1128
$SnPh_3$CHEtMe	139.1 [442]	137.2 [34]	128.3 [46]	128.6 [5]	1316
$SnPh_3C_7H_7$	140.4 [459]	136.8 [36]	128.0 [43]	128.4 [11]	1638
$SnPh_3C_7H_9$	138.8 [471]	137.9 [35]	129.2 [47]	129.6	1638
$SnPh_2Me_2$	140.6	135.5	128.5	128.7	399
$SnPhMe_3$	141.6	135.6	128.0	128.0	47
	142.1	136.1	128.5	128.5	399
	141.5 [474]	135.7 [36]	128.1 [47]	128.1 [11]	229, 455
	141.6 [473]	135.6 [35.4]	128.0 [45.4]	128.0 [10]	47
	141.9 [450.0]	135.9 [35.4]	128.3 [45.4]	128.2 [10.0]	880
$SnPhCl_3$	136.8	134.9 [75]	131.4 [123]	134.2	272
$Sn(C_6H_4Me$-4$)Me_3$	137.6 [487]	135.6 [37]	129.0 [49]	137.3 [11]	229, 407
	137.4 [488]	135.7 [42]	129.0 [52]	137.6 [<10]	47
$Sn(C_6H_4Me$-4$)Cl_3$	133.3	134.7 [80]	132.0 [128]	144.9	272
$Sn(C_6H_4Bu^t$-4$)Cl_3$	137.6			150.5	407
$Sn(C_6H_4Ph$-4$)Me_3$	141.0	126.9 [42.6]	136.3 [37.5]	141.5	880
$Sn(C_6H_4CF_3$-4$)Me_3$	147.6	135.9 [37]	124.3 [38.6]	131.0 [<10]	47
1,4-$(Me_3Sn)_2C_6H_4$	141.3 [477, 10]	135.4 [34, 45]			229

*4 Signals, Assignment unknown

COMPOUND	$\delta(^{13}C)$/p.p.m., J/Hz				REFERENCE
	C^1	$C^{2,6}$	$C^{3,5}$	C^4	
$Sn(C_6H_4NMe_2-4)Me_3$	126.0 [520]	136.2 [42]	112.7 [51.8]	150.5 [10]	229, 407
$Sn(C_6H_4OMe-4)Me_3$	131.7 [493]	136.7 [42]	114.3 [51.7]	160.3 [11]	229, 407
	131.6 [488]	136.6 [42]	114.1 [52]	160.4 [<10]	47
$Sn(C_6H_4OEt-4)Me_3$	131.5 [496]	136.7 [42]	114.9 [52]	159.6 [11]	229
$Sn(C_6H_4F-4)Me_3$	136.8 [467]	137.2 [41]	115.3 [50]	163.6 [12]	229
$Sn(C_6H_4Cl-4)Me_3$	139.8 [454]	136.8 [39]	128.2 [48]	134.5 [12]	229
	136.9 [452]	133.8 [38]	125.1 [47]	131.5 [11.5]	47
$Sn(C_6H_4Br-4)Me_3$	140.2 [452]	136.9 [39]	130.9 [47]	122.8 [13]	229
$Sn(C_6H_4Me-3)Me_3$	141.2 [480]	136.3 [36]	136.8 [48]	136.8 [11]	229
		132.7 [36]	129.0 [51]		
	141.2 [482]	136.2 [36]	136.7 [47]	128.8 [<10]	47
		132.6 [37]	127.8 [48]		
$Sn(C_6H_4CF_3-3)Me_3$	144.1	132.3	130.6	124.9	47
		139.4 [35]	128.1 [46]		
$Sn(C_6H_4OMe-3)Me_3$	142.9 [471]	121.5 [41]	159.5 [59]	113.6 [10]	229
		127.9 [35]	129.0 [55]		
	143.5 [474]	121.5 [40]	159.4 [52]	122.3 [<10]	47
		128.1 [37]	129.2 [51]		
$Sn(C_6H_4F-3)Me_3$	144.8 [447]	122.1 [56]	162.9 [62]	115.1 [9]	229
		131.2 [53]	129.9 [33]		
$Sn(C_6H_4Cl-3)Me_3$	144.3 [439]	135.2 [39]	134.5 [60]	128.2 [10]	229
		133.4 [48]	129.1 [33]		
	144.6 [442]	135.4 [38.7]	134.5 [61.0]	128.3 [<10]	47
		135.5 [34.6]	129.2 [47.0]		
$Sn(C_6H_4Me-2)Me_3$	140.9	143.7, 135.7	128.9, 125.1	128.5	455
$Sn(C_6H_4Et-2)Me_3$	140.4	150.2, 135.9	127.4, 125.3	128.7	455
$Sn(C_6H_4CF_3-2)Me_3$	140.7	136.8, 137.2	126.1, 131.0	128.6	455
$1,2-(Me_3Sn)_2C_6H_4$	150.7	150.7, 136.5	136.5, 127.4	127.4	455
$Sn(C_6H_4NMe_2-2)Me_3$	141.6	160.3, 136.4	120.2, 125.3	129.6	455
$Sn(C_6H_4OMe-2)Me_3$	129.7	163.8, 136.2	109.1, 121.2	129.9	455
$Sn(C_6H_4OEt-2)Me_3$	129.8	163.0, 136.2	109.6, 121.1	129.9	455

COMPOUND	δ(^{13}C)/p.p.m., *J*/Hz				REFERENCE
	C^1	$C^{2,6}$	$C^{3,5}$	C^4	
$Sn(C_6H_4F\text{-}2)Me_3$	126.8	167.5, 136.6	114.2, 124.2	130.4	455
$Sn(C_6H_4Cl\text{-}2)Me_3$	142.2	142.5, 136.7	128.1, 125.9	129.8	455
$Sn(C_6H_3Me_2\text{-}2,6)Me_3$	140.8	144.1	126.8	128.4	455
$Sn\{C_6H_3(CF_3)_2\text{-}2,6\}Me_3$	142.0	139.0	129.4	128.8	455
$Sn\{C_6H_3(OMe)_2\text{-}2,6\}Me_3$	116.3	165.2	103.4	131.1	455
$Sn\{C_6H_3(OEt)_2\text{-}2,6\}Me_3$	116.1	164.3	103.5	130.7	455
$Sn(C_6H_3F_2\text{-}2,6)Me_3$	114.1	167.9	110.6	131.5	455
$Sn(C_6H_3Cl_2\text{-}2,6)Me_3$	142.3	130.2	126.9	142.8	455
$Sn(C_6H_3Me_2\text{-}2,3)Me_3$	141.5	142.0, 133.6	135.5, 125.6	130.4	455
$Sn(C_6H_3Me_2\text{-}2,4)Me_3$	137.0	143.6, 135.7	129.9, 126.0	137.6	455
$Sn(C_6H_3Me_2\text{-}3,4)Me_3$	138.0	133.4, 136.9	136.0, 129.5	135.6	455
$Sn\{C_6H_3(CF_3)_2\text{-}3,5\}Me_3$	146.6	136.2	131.8	122.8	455
$Sn(C_6H_2Me_3\text{-}2,4,6)Me_3$	137.1	144.2	127.9	137.5	455
$Sn\{C_6H_2(OMe)_3\text{-}2,4,6\}Me_3$	106.8	165.7	90.4	163.4	455
$Sn\{(C_6H_4\text{-}2)_2S\}Me_2$	C^1 143.6	C^2 146.9	C^{3-6} 128.3* to 137.7		1286
$Sn(1\text{-naphthyl})Me_3$	C^1 142.0	C^2 134.6 [31.6]	C^3 125.3 [37.0]	C^4 128.8 [12.2]	880
		C^9 140.1 [35.6]	C^{10} 133.8 [34.1]		
$Sn(2\text{-naphthyl})Me_3$	C^1 136.1 [34.9]	C^2 139.4	C^3 132.3 [*ca* 46]	C^4 127.1	880
		C^9 133.5 [9.1]	C^{10} 133.5 [50.8]		
$Sn(9\text{-anthryl})Me_3$	C^9 143.1, $C^{1a,8a}$	131.7 [39.0], $C^{4a,5a}$	138.5 [30.0], C^{10}	128.5 [18.0]	880
Lead, [^{207}*Pb*]					
$PbPh_4$	150.1 [481]	137.4 [66]	129.4 [81]	128.4 [20]	1645, 1743
	150.1 [481]	137.7 [68]	129.5 [80]	128.6 [20]	870
$PbPh_3Me$	149.6 [439]	137.1 [67]	129.2 [77]	128.3 [19]	870
Pb_2Ph_6	153.0 [116, 175§]	137.8 [71]	129.5 [66]	128.0	1645
$PhPh_3(OAc)$	158.4	136.7 [90]	130.1 [104]	129.6 [23]	1645
$PbPh_3Cl$	155.7	136.1 [89]	130.3 [105]	129.9	1645
$PbPh_3Br$	154.4 [531]	136.7 [89]	130.2 [104]	129.7	1645

*4 Signals. Assignment uncertain, § $^2J_{PbC}$

COMPOUND	$\delta(^{13}C)$/p.p.m., J/Hz				REFERENCE
	C^1	$C^{2,6}$	$C^{3,5}$	C^4	
$PbPh_2Me_2$	149.0 [395]	136.6 [65]	128.7 [72]	127.8 [19]	870
$PbPh_2(OAc)_2$	167.2	133.1	130.2	130.2	1645
$PbPh_2Br_2$	162.8	133.7 [128]	130.4 [183]	130.4	1645
$PbPhMe_3$	148.9	136.7	128.8	127.7	399
	148.6 [364]	136.4 [63.4]	128.4 [68.8]	127.5 [16.2]	459, 880
	148.5 [348]	136.5 [63]	128.5 [66]	127.5 [16]	870
$Pb(C_6H_4Me-4)_4$	[488.4]	[69.6]	[83.2]	[18.8]	459
	146.4 [492]	137.5 [69]	130.2 [83]	138.0 [19]	870, 1743
	146.5 [491]	137.3 [69]	130.0 [84]	137.9 [18]	1645
$Pb(C_6H_4Me-4)_3Me$	145.8 [459]	137.0 [68]	130.0 [78]	137.6 [18]	870
$Pb(C_6H_4OMe-4)_4$	140.4 [522]	138.4 [79]	115.2 [89]	160.0 [19]	870, 1743
$Pb(C_6H_4OMe-4)_3Me$	140.0 [471]	138.1 [77]	115.0 [84]	159.9 [18]	870
$Pb(C_6H_4Cl-4)_4$	146.6 [506]	138.3 [71]	129.8 [87]	135.4 [23]	870, 1743
$Pb(C_6H_4Cl-4)_3Me$	146.8 [438]	138.0 [74]	129.4 [81]	134.7 [21]	870
$Pb(C_6H_4Me-3)_4$	150.1 [470]	137.9 [64]	138.7 [79]	129.1	1645
		134.4 [65]	129.1 [85]		
	150.1 [473]	138.0 [65]	138.7 [78]	129.2 [20]	1743
		134.5 [66]	129.2 [84]		
$Pb(C_6H_4OMe-3)_4$	151.0 [479]	123.0 [77]	160.4 [103]	113.9 [18]	1743
		130.0 [69]	130.2 [96]		
$Pb(C_6H_4Cl-3)_4$	150.5 [482]	136.6 [76]	136.4 [108]	129.5 [17]	1743
		135.1 [66]	130.9 [90]		
$Pb(C_6H_4Me-2)_4$	152.8 [461]	144.8 [54]	128.6 [64]	130.0 [18]	1743
		137.3 [78]	126.7 [82]		
$Pb(C_6H_4OMe-2)_4$	141.6 [591]	162.9 [16]	110.0 [43]	129.2 [16]	1743
		138.1 [55]	121.8 [86]		
$Pb\{C_6H_3(OMe)_2-2,4\}_4$	132.5 [598]	164.2 [25]	98.0 [49]	161.4 [18]	1743
		138.5 [65]	106.1 [92]		
$Pb\{C_6H_3(OMe)_2-2,5\}_4$	143.0 [556]	137.1 [16]	110.6 [53]	114.1 [16]	1743
		123.6 [64]	154.8 [106]		
$Pb(1-naphthyl)Me_3$	150.1 [388.0]	134.6 [60.0]	125.8 [78.0]	128.1 [16.0]	880
		C^9 134.3 [40.0]	C^{10} 139.2 [48.0]		

COMPOUND	$\delta(^{13}C)$/p.p.m., J/Hz				REFERENCE
	C^1	$C^{2,6}$	$C^{3,5}$	C^4	
$Pb(2\text{-naphthyl})Me_3$	136.2 [60.4]	146.3 [332.2]	133.4 [64.9]	127.4 [65.0]	880
		C^9 133.9 [69.5]	C^{10} 133.1 [12.1]		
$(2,2'\text{-biphenyl})PbMe_2$	C^2 149.3 344 , C^1 148.6 107				1646
$Pb(2\text{-pyridyl})Bu^n_3$	C^2 183.2, C^3 132.9, C^4 133.3, C^6 150.9, C^5 121.4				451
$Pb(3\text{-pyridyl})Bu^n_3$	C^2 156.9, C^3 145.5, C^4 144.4, C^6 148.5, C^5 124.6				451
$Pb(4\text{-pyridyl})Bu^n_3$	C^2 149.3, C^3 132.8, C^4 160.5				451
Phosphorus, (^{31}P)					
$[P(C_6H_5)_4]^+$		134.5 (10.5)	131.0 (12.9)	135.9 (3.1)	1236
$P(C_6H_5)_3$	137.1 to 138.3 (11.4 to 12.6)	133.6 to 134.4 (19.4 to 19.7)	128.4 to 129.2 (6.2 to 7.2)	128.5 to 129.3 (0.3)	15, 68, 443, 929, 1236, 1264, 1299
$P(C_6H_5)_3$*	134.9 (2.9)	134.1 (16.7)	129.5 (7.6)	130.2	929
$M\text{-}P(C_6H_5)_3$	121.1 to 140.1 (8 to 65.6)	130.9 to 136.1 (6.2 to 24.3)	127.4 to 131.3 (5.1 to 15.0)	126.9 to 135.5 (0 to 3.1)	15, 68, 193, 235, 269, 377, 443, 445, 482, 490, 554, 629, 646, 681, 722, 776, 830, 938, 1001, 1037, 1117, 1164, 1236, 1276, 1419, 1539
$[R\text{-}P(C_6H_5)_3]^+$	117.5 to 127.6 (80 to 93)	132.8 to 135.3 (8 to 12)	129.0 to 131.3 (10 to 13.2)	132.0 to 136.6 (2.1 to 3.7)	188, 626, 663, 860, 942, 1001, 1165, 1420, 1652, 1665

*In SO_2

COMPOUND	$\delta(^{13}C)$/p.p.m., J/Hz				REFERENCE
	C^1	$C^{2,6}$	$C^{3,5}$	C^4	
$RR'C{=}P(C_6H_5)_3$	124.1 to 133.6	132.0 to 133.8	128.3 to 129.1	130.1 to 132.6	472, 860, 1165, 1282,
	(81.2 to 94.2)	(8.5 to 12)	(11 to 13)	(2 to 3)	1665, 1682
$RN{=}P(C_6H_5)_3$	124.4 to 136.4	132.1 to 133.4	127.8 to 130.2	130.1 to 134.6	411, 923, 1604, 1655
	(93.6 to 104.2)	(8.3 to 11.1)	(11.0 to 13.2)	(2.4 to 4)	
$[R_2NP(C_6H_5)_3]^+$	118.6, 119.8	134.1, 135.5	130.0, 130.4	135.2, 135.7	923
	(102.5, 103.3)	(11.0, 11.6)	(13.4)	(2.7)	
$OP(C_6H_5)_3$	134.3	132.2, 132.4	128.4, 128.6	131.5, 132.0	1236, 1299
	(102.8)	(10.1, 9.6)	(11.6, 12.1)	(1.8)	
$SP(C_6H_5)_3$	136.2	132.4, 132.7	128.7, 128.4	131.7, 131.3	1236, 1299
	(85.0)	(10.7, 10.4)	(12.6, 10.8)	(3.1, 4.0)	
$SeP(C_6H_5)_3$		132.9	128.7	131.7	1236
		(10.9)	(12.7)	(3.2)	
Other PhP compounds					
					15, 193, 198, 235, 378, 411, 416, 443, 445, 525, 569, 582, 585, 627, 681, 858, 863, 864, 929, 942, 1053, 1078, 1233, 1275, 1299, 1539, 1604, 1648, 1649, 1657, 1665, 1666, 1667, 1673, 1687, 1738, 1777, 1798
Other aryl phosphorus compounds					
					569, 585, 698, 942, 1264, 1299, 1540, 1663
Arsenic					
$As(C_6H_5)_5$	131.3	134.6	131.5	132.5	837
$[As(C_6H_5)_4]^+$	122.4	134.5	132.5	135.9	119
		132.9	131.7	135.0	1236

COMPOUND	$\delta(^{13}C)$/p.p.m., J/Hz				REFERENCE
	C^1	$C^{2,6}$	$C^{3,5}$	C^4	
$As(C_6H_5)_3$	140.5	134.3	129.3	129.0	68
	139.3	133.3	128.1	128.3	837
	140.6	134.6	129.5	129.3	1236
	139.6				443
$As(C_6H_5)_3Cr(CO)_5$	136.1				443
$As(C_6H_5)_3Cr(CO)_2(arene)$	140.2 to 140.5				938
$As(C_6H_5)_3Mo(CO)_5$	135.1	132.7	129.3	130.2	1236
$As(C_6H_5)_3MoCp\{CN_2H-(p-tol)_2\}(CO)$	138.0				1748
$As(C_6H_5)_3W(CO)_5$	136.3	132.6	129.3	130.4	1236
$As(C_6H_5)_3WCp\{CN_2H-(p-tol)_2\}(CO)$	138.6				1748
$As(C_6H_5)_3Mn(CO)_4Br$		133.6	129.7	131.2	1236
$As(C_6H_5)_3Fe(CO)_4$		132.7	130.2	131.4	1236
$As(C_6H_5)_3Co(CO)_3SnCl_3$		132.3	130.1	132.1	1236
$As(C_6H_5)_3RhCp(1-MeC_3H_4)Cl$	133.1	134.7	130.8	132.5	1164
$As(C_6H_5)_3RhCp(2-MeC_3H_4)Cl$	132.8, 130.6, 129.5, 128.6				1164
$As(C_6H_5)_3Rh(C_5Me_5)(1-MeC_3H_4)Cl$	131.9	135.2	130.8	132.3	1164
$\{As(C_6H_5)_3\}_2Rh(CO)(Me_2N_3)CuCl$	133.9	133.8	128.8	130.3	554
$As(C_6H_5)_3Ni(CO)_3$	137.0				443
$As(C_6H_5)_3Pt(C_2H_4)_2$	136.9				1539
$[As(C_6H_5)_3PtMe_2(OAc)]_2$		128.3	129.4	134.2	1243
$[As(C_6H_5)_3CH_2COR]^+$	121.9, 122.2				1665
$As(C_6H_5)_3CHCOR$	128.9, 129.9				1665
$\{As(C_6H_5)_3CHCOMe\}_2PdCl_2$	127.8				1665
$As(C_6H_5)_3CHCOC_6H_4X-4$	131.6 to 132.0	132.3 to 132.5	129.4 to 129.6	127.9 to 128.9	784
$As(C_6H_5)_3F_2$	137.4 (J_{CF} = 18.3)				1782
$\{As(C_6H_5)_2Me\}_4Ir_4(CO)_8$	139.4				305
$As(C_6H_5)_2CH_2CH_2PPh_2$	140.9				1752
$\{As(C_6H_5)_2CH_2CH_2PPh_2\}V(C_3H_5)(CO)_3$	139.2				1752
$\{As(C_6H_5)_2CH_2CH_2PPh_2\}V(1-MeC_3H_4)(CO)_3$	138.8, 139.5				1752

COMPOUND	δ(^{13}C)/p.p.m., J/Hz				REFERENCE
	C^1	$C^{2,6}$	$C^{3,5}$	C^4	
$\{As(C_6H_5)_2CH_2CH_2PPh_2\}V(2\text{-}MeC_3H_4)(CO)_3$	139.6				1752
$As(C_6H_5)_2MeF_2$	136.9 (J_{CF} = 16.1)				1782
$\{As(C_6H_5)_2F_2\}_2CH_2$	136.7 (J_{CF} = 15.5)				1654
cis-$\{As(C_6H_5)Me_2\}_2PtMe_2$	139.5	131.8	128.9	129.5	30
fac-$\{As(C_6H_5)Me_2\}_2PtMe_3Cl$	136.8	131.9	129.5		30
fac-$\{As(C_6H_5)Me_2\}_2PtMe_3Br$	136.5	131.9	129.4	130.0	30
fac-$\{As(C_6H_5)Me_2\}_2PtMe_3I$	135.9	131.8	129.3	129.8	30
cis-$\{As(C_6H_5)Me_2\}_2PtMe_4$	137.5	132.6	129.7	130.5	30
$[o\text{-}\{Me(C_6H_5)As\}_2C_6H_4]Mo(CO)_4$	139.4		130.4	128.9	396
	139.4	131.5	130.3	128.9	
$As(C_6H_4Me\text{-}4)_5$	136.3	133.5	129.0	137.6	837
$As(C_6H_4Me\text{-}4)_3$	136.2	129.1	126.4	138.0	837
$\{o\text{-}(Me_2As)_2C_6H_4\}W(CO)_3N_2C_6H_4OMe$	$C^{1,2}$ 139.0, 139.5				1407
$\{o\text{-}(Me_2As)_2C_6H_4\}PtMeX$	$C^{1,2}$ 136.9 to 143.3; $C^{3,6}$ 129.8 to 131.1; $C^{4,5}$ 130.8, 133.8				402
$\{o\text{-}(MePhAs)_2C_6H_4\}Mo(CO)_4$	$C^{1,2}$ 142.0, $C^{3,6}$ 131.4, $C^{4,5}$ 129.4				396
	$C^{1,2}$ 141.9, $C^{3,6}$ 130.3, $C^{4,5}$ 129.4				
$\overline{AsC_6H_4NCH}$	C^1 142.7				1784
$Me_2\overline{AsC_6H_4NCH}$	C^1 143.0				1784
Antimony					
$Sb(C_6H_5)_5$		134.6	127.5	127.9	6
	146.3	134.4	127.2	127.7	837
$Sb(C_6H_5)_3$	139.3	136.8	129.4	129.2	68, 1264
	138.0	135.8	128.1	128.4	837
	140.0	137.1	129.8	129.5	1236
	138.4	136.1	128.8	128.6	1705
	139.7				470
	138.3				443

COMPOUND	$\delta(^{13}C)$/p.p.m., J/Hz				REFERENCE
	C^1	$C^{2,6}$	$C^{3,5}$	C^4	
$Sb(C_6H_5)_3Cr(CO)_5$	132.0				443
$Sb(C_6H_5)_3Mo(CO)_5$		135.2	129.5	130.3	1236
$Sb(C_6H_5)_3W(CO)_5$	132.1	135.2	129.7	130.6	1236
$Sb(C_6H_5)_3Mn(CO)_4Br$		131.1	130.1	131.3	1236
$Sb(C_6H_5)_3Fe(CO)_4$	131.0	135.4	130.2	131.4	1236
$Sb(C_6H_5)_3Co(CO)_3SnCl_3$		134.8	130.2	132.3	1236
$Sb(C_6H_5)_3Ni(CO)_3$	134.0				443
$\{Sb(C_6H_5)_3PtMe_2(OAc)\}_2$		128.3	129.5	136.3	1243
$Sb(C_6H_5)_3Pr_2$	140.2				470
$Sb(C_6H_5)_3(CH_2Ph)_2$	139.5				470
$Sb(C_6H_5)_3F_2$	134.3 (J_{CF} = 15.4)				1782
$Sb(C_6H_5)_3Cl_2$	141.3				470
$Sb(C_6H_5)_3Br_2$	142.4				470
$Sb(C_6H_5)_2MeF_2$	133.6 (J_{CF} = 13.8)				1782
$Sb(C_6H_5)_2OAc$	146.3	134.6	129.0	129.7	1705
$Sb(C_6H_5)_2(SC_6H_4F\text{-}4)$	140.1	135.3	129.0	129.3	1705
$Sb(C_6H_5)_2Cl$	144.4	134.3	129.1	130.0	1705
$Sb(C_6H_5)_2Br$	141.4	134.9	129.1	129.9	1705
$Sb(C_6H_5)(OCMe_2CMe_2O)_2$	129.6 to 136.0				990
$Sb(C_6H_4Me\text{-}4)_5$	144.6	135.6	129.1	138.5	837
$Sb(C_6H_4Me\text{-}4)_3$	134.6	135.8	129.3	137.8	837
$Sb(C_6H_4\text{-}2)_3CH$	C^1 141.2	C^2 146.3	C^3 128.6	C^4 127.2	1264
		C^6 135.2	C^5 125.1		
Bismuth					
$Bi(C_6H_5)_3$	131.1	138.1	131.0	128.3	68
	134.1	138.3	131.3	128.5	1236
$Bi(C_6H_5)_3(CH_2Ph)_2$	161.7				470
$Bi(C_6H_5)_3F_2$	153.7				1782
$Bi(C_6H_5)_3Br_2$	157.4				470

COMPOUND	$\delta(^{13}C)$/p.p.m., J/Hz				REFERENCE
	C^1	$C^{2,6}$	$C^{3,5}$	C^4	
Selenium, [^{77}*Se*]					
Se(C_6H_5)Me	131.9	129.9	128.7	125.6	1710
	130.0				1786
Se(C_6H_5)Et	130.4	132.3	128.6	128.6	1710
{Se(C_6H_5)Et}$_2$RuX$_3$(NO)		131.3	130.7	129.9	799
Se(C_6H_5)Pr^n	130.7	132.2	128.6	126.2	1710
Se(C_6H_5)Pr^i	129.5	134.5	128.4	126.8	1710
Se(C_6H_5)Bu^t	127.9	137.8	128.1	128.1	1710
Se(C_6H_5)$C_3H_2Cl_3$	129.2				1788
Se(C_6H_5)C_2H_3	129.0	132.6	128.9	127.0	766
	129.2	133.2	128.9	127.1	1220
Se(C_6H_5)CN	123.2				242
Se(C_6H_5)O_2SnMe$_3$	149.0				1632
Se$_2$(C_6H_5)$_2$	130.9	131.5	129.1	127.6	981
	130.1	130.8	129.3	127.7	1226
Se(C_6H_4Me-4)CH=CH$_2$	125.1	133.3	129.8	136.8	766
	125.2	136.9	129.7	136.9	1220
Se(C_6H_4Me-4)CN	119.4				242
Se(C_6H_4Me-4)(CH_2CH_2Cl)Cl$_2$	136.9 [98.8]	129.3 [12.4]	130.7 [4.3]	142.8	610
Se(C_6H_4Me-4)(CH_2CH_2)Cl (ring)	[13.3]				286
Se(C_6H_4Me-4)($C_3H_2Cl_3$)	125.3				1788
Se($C_6H_4CF_3$-4)CH=CH$_2$	135.1	131.7	125.7	129.7	1220
Se$C_6H_4CH_2$CHMe (ring)	C^1 [98.4]	C^6 [13.4]	C^5 [3.8]		743
Se$C_6H_4CH_2$CMe$_2$ (ring)	C^1 [100.0]	C^6 [12.3]	C^5 [2.9]		743
Se$C_6H_4CH_2$CO (ring)	C^1 134.0 or 136.6; $C^{2,6}$ 134.0 or 136.6, 124.7, 126.2, 127.2				743
SeC_6H_4NHCS (ring)	C^1 129.9, C^2 142.7				1226, 1713

COMPOUND	δ(^{13}C)/p.p.m., *J*/Hz				REFERENCE
$Se\overline{C_6H_4NCH}$	C^1 137.3	C^2 154.8			1226, 1713
$Se\overline{C_6H_4NMeCOC_6H_2(OMe)(NO_2)CH=CMe}$	C^1 146.6	C^2 119.4 C^6 125.3	C^3 109.3 C^5 120.2	C^4 126.4	1070
	C^1	$C^{2,6}$	$C^{3,5}$	C^4	
$Se(C_6H_4NH_2\text{-}4)CH=CH_2$	114.8	136.4	115.4	146.3	1220
$Se(C_6H_4NO_2\text{-}4)CH=CH_2$	139.9	130.6	123.8	146.9	1220
	141.3	129.9	123.6	146.0	766
$Se(C_6H_4NO_2\text{-}4)C_3H_2Cl_3$	141.0				1788
$Se(C_6H_4OMe\text{-}4)CH=CH_2$	118.2	135.6	114.8	159.4	766
	118.3	136.0	114.7	159.6	1220
$Se(C_6H_4OMe\text{-}4)C_3H_2Cl_3$	118.4				1788
$Se(C_6H_4F\text{-}4)CH=CH_2$	123.2	135.7	116.3	162.5	1220
$Se(C_6H_4Cl\text{-}4)CH=CH_2$	127.3	133.6	128.9	133.1	766
	127.2	134.3	129.2	133.8	1220
$Se(C_6H_4Cl\text{-}4)C_3H_2Cl_3$	127.4				1788
$Se(C_6H_4Br\text{-}4)CH=CH_2$	128.0	134.5	132.2	121.8	1220
	127.9	133.7	131.7	121.2	766
$Se(C_6H_4Br\text{-}4)C_3H_2Cl_3$	128.0				1788
$Se(C_6H_4Br\text{-}3)C_3H_2Cl_3$	131.3				1788
<u>*Tellurium*, [^{125}*Te*]</u>					
$Te(C_6H_5)Me$	112.5				1786
	112.6				1790
$[Te(C_6H_5)Me_2]^+$	140.9 [171.6]				1799
$Te(C_6H_5)Et$	112.0				1790
$Te(C_6H_5)Pr^i$	111.7				1790
$Te_2(C_6H_5)_2$	108.0	137.6	129.2	128.0	981

TABLE 2.5

^{13}C N.m.r. Data for Some Carbene Derivatives

COMPOUND	$\delta(^{13}C)$/p.p.m., J/Hz	REFERENCE
Niobium		
$Nb(CHBu^t)(CH_2Bu^t)_3$	246.0	1345
$Nb(CHBu^t)ClCp_2$	299	1346
Tantalum		
$Ta(CH_2)MeCp_2$	224.0, 228	591, 1344
$Ta(CHBu^t)_3(PMe_2Ph)_2$	246, 274	1341
$Ta(CHBu^t)_2Cp(PMe_3)$	235, 255	1341
$Ta(CHBu^t)(CH_2Bu^t)_3$	250, 250.1 {90}	309, 1345
$[Ta(CHBu^t)Cp_2(PMe_3)]^+$	294 (13.8)	1346
$Ta(CHBu^t)Cp_2(CHPh_2)$	267.4	1346
$Ta(CHBu^t)Cp_2Cl$	273, 274	591, 1346
$Ta(CHBu^t)(C_5H_4Me)_2Cl$	269.4	1346
$Ta(CHBu^t)CpCl_2$	246 {84}	1102
$Ta(CHPh)Cp_2(CH_2Ph)$	246	591, 1346
$Ta(CHSiMe_3)Cp_2Me$	236	1346
Chromium		
$Cr(CHNMe_2)(CO)_5$	246.2	1372
$Cr(CPh_2)(CO)_5$	399.4	1065
$Cr(CPhC_6H_4Me\text{-}4)(CO)_5$	397.0	1065
$Cr(CPhC_6H_4CF_3\text{-}4)(CO)_5$	394.0	1065
Cr(*C*Ph-2-furanyl)$(CO)_5$	334.5	1065
Cr(*C*Ph-2-thiophenyl)$(CO)_5$	355.6	1065
Cr(*C*Ph-2-*N*-methylpyrolle)$(CO)_5$	315.0	1065
Cr{*C*(2-furanyl)$_2$}$(CO)_5$	286.7	1065
Cr{*C*(2-thienyl)$_2$}$(CO)_5$	322.3	1065
Cr{*C*(2-furanyl)(2-thienyl)}$(CO)_5$	304.6	1065
$Cr(CMeNHMe)(CO)_5$	276, 284.8	35, 39, 170
$Cr(CMeNHPr^i)(CO)_5$	274	35, 170
$Cr(CMeNMe_2)(CO)_5$	270.1, 271 ($^1J_{NC}$ = 4.0 Hz)	35, 170, 175, 908
$Cr\{CMeN(CH_2)_4\}(CO)_5$	265.9	175
Cr{*C*Me(NCyCONHCy)}$(CO)_5$	259.4	369
$Cr(CEtNHCy)(CO)_5$	284.3	35, 39
Cr{*C*Ph(glyOMe)}$(CO)_5$	280.6, 283.1	811
Cr{*C*Ph(valOMe)}$(CO)_5$	283.3	811
Cr{$\overline{C\text{NCyC(=NCy)C}}H_2$}$(CO)_5$	291.2	369
Cr{$\overline{C(NEt_2)CH=C}$(OMe)Me}$(CO)_5$	267.0	436
Cr{*C*(NEt_2)CH=C(OMe)Ph}$(CO)_5$	271.0	436
Cr{*C*$(NEt_2)C(NEt_2)$=CPhOMe}$(CO)_4$	263.5	810

COMPOUND	$\delta(^{13}C)$/p.p.m., J/Hz	REFERENCE
$Cr(\mathit{C}PhNH_2)(CO)_5$	282.9	248
$Cr(\mathit{C}PhNH_2)(C_6H_6)(CO)_2$	302.4	935
$Cr(\mathit{C}PhNH_2)(C_6H_5Me)(CO)_2$	302.3	935
$Cr(\mathit{C}PhNH_2)$(*p*-xylene)$(CO)_2$	301.9	935
$Cr(\mathit{C}PhNH_2)$(mesitylene)$(CO)_2$	302.3	935
$Cr(\mathit{C}PhNHMe)(CO)_5$	289, 282.2	35, 170, 175, 811
$Cr(\mathit{C}PhNMe_2)(CO)_5$	277.5, 270.6	39, 175
$Cr(\mathit{C}PhNMe_2)(C_6H_6)(CO)_2$	295.1	935
$Cr(\mathit{C}PhNMe_2)(C_6H_5Me)(CO)_2$	295.5	935
$Cr(\mathit{C}PhNMe_2)$(*p*-xylene)$(CO)_2$	297.1	935
$Cr(\mathit{C}PhNMe_2)$(mesitylene)$(CO)_2$	296.3	935
$Cr\{\mathit{C}PhN(CH_2)_4\}(CO)_5$	271.3, 266.2	39, 175
$Cr\{\mathit{C}PhN(CH_2)_5\}(CO)_5$	270.5	39
$Cr\{\mathit{C}(C_6H_4Me\text{-}4)NH_2\}(CO)_5$	282.4	248
$Cr\{\mathit{C}(C_6H_4Br\text{-}4)NH_2\}(CO)_5$	281.8	248
$Cr\{\mathit{C}(NH_2)$(2-furanyl)$\}(CO)_5$	255.6	39
$Cr\{\mathit{C}(NH_2)$(2-thienyl)$\}(CO)_5$	271.8	39
$Cr\{\mathit{C}(NC_4H_8)$(2-furanyl)$\}(CO)_5$	253.7	39
$Cr\{\mathit{C}(NC_4H_8)$(2-thienyl)$\}(CO)_5$	266.9	39
$Cr(\mathit{C}PhPMe_3)(CO)_4Br$	311.1 (4.9)	435
$Cr\{\mathit{C}(CH_2Ph)(PMe_3)\}(CO)_4Br$	323.1 (9.8)	1363
$[Cr(\mathit{C}PhPMe_3)(C_6H_6)(CO)_2]^+$	307.3 (7.3)	906, 1048
$[Cr(\mathit{C}PhPMe_3)$(*p*-xylene)$(CO)_2]^+$	305.5 (9.8)	1048
$[Cr(\mathit{C}PhPMe_3)$(mesitylene)$(CO)_2]^+$	307.5 (9.8)	1048
$Cr\{\mathit{C}(C_6H_4Me\text{-}4)(PMe_3)\}(CO)_4Cl$	318.9	1363
$Cr\{\mathit{C}(C_6H_4Me\text{-}4)(PMe_3)\}(CO)_4Br$	311.1 (4.9)	1363
$Cr\{\mathit{C}(C_6H_4Me\text{-}4)(PMe_3)\}(CO)_4I$	319.3 (2.4)	1363
$Cr\{\mathit{C}(C_6H_2Me_3\text{-}2,4,6)(PMe_3)\}(CO)_4Br$	321.2 (7.3)	1363
$Cr\{\mathit{C}Me(OMe)\}(CO)_5$	360.2; 362.3	60, 89, 894
$Cr\{\mathit{C}Me(OEt)\}(CO)_5$	357	35
$Cr\{\mathit{C}Me(OSiMe_3)\}(CO)_5$	374.9	1052
trans-$Cr\{\mathit{C}Me(OMe)\}(PMe_3)(CO)_4$	352.1 (12.2)	440, 891, 894
cis-$Cr\{\mathit{C}Me(OMe)\}(PMe_3)(CO)_4$	360.3	894
trans-$Cr\{\mathit{C}Me(OMe)\}(AsMe_3)(CO)_4$	350.6	894
cis-$Cr\{\mathit{C}Me(OMe)\}(AsMe_3)(CO)_4$	361.3	892, 894
trans-$Cr\{\mathit{C}Me(OMe)\}(SbMe_3)(CO)_4$	352.6	894
cis-$Cr\{\mathit{C}Me(OMe)\}(SbMe_3)(CO)_4$	362.2	894
cis-$Cr\{\mathit{C}Me(OMe)\}(\overline{CNMeCH_2CH_2NMe})(CO)_4$	359.6	1226
$Cr\{\mathit{C}(OEt)C(OH)=CS(CH_2)_3S\}(CO)_4$	313.3	734
$Cr\{\mathit{C}(OMe)CH(OMe)Ph\}(CO)_5$	360.6	430
$Cr\{\mathit{C}(OMe)C(OMe)_2Ph\}(CO)_5$	363.8	429, 430
$Cr\{\mathit{C}(OMe)C(OEt)_2Ph\}(CO)_5$	363.8	430
$Cr\{\mathit{C}(OMe)CH=CH_2\}(CO)_5$	323.2	171
$Cr\{\mathit{C}(OSiMe_3)CH=PMe_3\}(CO)_5$	250.9 (10)	1359
$Cr\{\mathit{C}(OEt)C(OH)CPhSPh\}(CO)_4$	321.0	1365

COMPOUND	$\delta(^{13}C)$/p.p.m., *J*/Hz	REFERENCE
$Cr\{C(OEt)C(OH)C(CH_2)_3S\}(CO)_4$	313.0	1365
$Cr\{C(OEt)C(OEt)CPhSPh\}(CO)_4$	330.0	1365
$Cr\{C(OEt)C(OEt)C(CH_2)_3S\}(CO)_4$	312.0	1365
$Cr\{C(OH)Ph\}(CO)_5$	340.5	368
$Cr\{C(OMe)Ph\}(CO)_5$	350.9, 351.4, 354.5	35, 39, 89, 248, 368, 371, 429
$Cr\{C(OEt)Ph\}(CO)_5$	351	35
$Cr\{C(OSiMe_3)Ph\}(CO)_5$	369.1	1052
$\{(OC)_5CrCPh\}_2O$	354.1	368
$Cr\{C(O_2CMe)Ph\}(CO)_5$	357.6	1049
$Cr\{C(O_2CEt)Ph\}(CO)_5$	357.3	1049
$Cr\{C(O_2CPh)Ph\}(CO)_5$	356.5	1049
$Cr\{C(OMe)Ph\}(C_6H_6)(CO)_2$	347.7	935
$Cr\{C(OMe)Ph\}(C_6H_5Me)(CO)_2$	346.1	935
$Cr\{C(OMe)Ph\}$(*p*-xylene)$(CO)_2$	346.2	935
$Cr\{C(OMe)Ph\}$(mesitylene)$(CO)_2$	343.9	935
$Cr\{C(OH)p\text{-tol}\}(CO)_5$	340.9	944
$Cr\{C(OMe)p\text{-tol}\}(CO)_5$	348.0	248
$Cr\{C(OPh)p\text{-tol}\}(CO)_5$	350.0	1273
$Cr\{C(OSiMe_3)p\text{-tol}\}(CO)_5$	369.1	1052
$Cr\{C(O_2CMe)p\text{-tol}\}(CO)_5$	356.3	1049
$Cr\{C(O_2CEt)p\text{-tol}\}(CO)_5$	356.0	1049
$Cr\{C(O_2CPh)p\text{-tol}\}(CO)_5$	352.4	1049
$Cr\{C[OW(CO)_4CPh]p\text{-tol}\}(CO)_5$	353.1	941
$Cr\{C(OMe)C_6H_4CF_3\text{-}4\}(CO)_5$	350.2	248
$Cr\{C(O_2CEt)C_6H_4CF_3\text{-}4\}(CO)_5$	354.0	1049
$Cr\{C(OMe)C_6H_4OMe\text{-}4\}(CO)_5$	342.8, 340.5	39, 248
$Cr\{C(OMe)C_6H_4F\text{-}4\}(CO)_5$	346.8	248
$Cr\{C(OMe)C_6H_4Cl\text{-}4\}(CO)_5$	350.1, 347.9	39, 248
$Cr\{C(OMe)C_6H_4Br\text{-}4\}(CO)_5$	348.2	248
$Cr\{C(OMe)C_6H_4CF_3\text{-}3\}(CO)_5$	348.8	248
$Cr\{C(OMe)C_6H_4OMe\text{-}3\}(CO)_5$	350.5	248
$Cr\{C(OMe)C_6H_4Cl\text{-}3\}(CO)_5$	348.7	248
$Cr\{C(OEt)2\text{-furanyl}\}(CO)_5$	313.6	39
$Cr\{C(OSiMe_3)2\text{-furanyl}\}(CO)_5$	320.0	1052
$Cr\{C(OEt)2\text{-thienyl}\}(CO)_5$	319.8	39
$Cr\{C(OSiMe_3)2\text{-thienyl}\}(CO)_5$	330.6	1052
$Cr\{C(OEt)\text{ferrocenyl}\}(CO)_5$	332.0	39
$Cr\{C(NMe_2)SiPh_3\}(CO)_5$	305.0	907, 1042
$Cr\{C(OMe)SiPh_3\}(CO)_5$	432.4	1042
$Cr\{C(OEt)SiPh_3\}(CO)_5$	431.1	907, 1042
$Cr\{C(NEt_2)NCO\}(CO)_5$	229.9	962
$Cr(CNMeCH_2CH_2NMe)(CO)_5$	219.6	1223
cis-$Cr(CNMeCH_2CH_2NMe)_2(CO)_4$	228.4	1223
cis-$Cr(CNMeCH_2CH_2NMe)\{C(OMe)Me\}(CO)_4$	224.9	1223

COMPOUND	$\delta(^{13}C)$/p.p.m., J/Hz	REFERENCE
$\mathrm{Cr(\overline{\mathit{C}NEtCH_2CH_2N}Et)(CO)_5}$	217.8	1223
cis-$\mathrm{Cr(\overline{\mathit{C}NEtCH_2CH_2N}Et)_2(CO)_4}$	226.9	1223
$\mathrm{Cr\{\mathit{C}(NEt_2)(OEt)\}(CO)_5}$	239.7	652, 1376
$\mathrm{Cr\{\mathit{C}(NEt_2)F\}(CO)_5}$	245.7 ($^1J_{FC}$ = 393.1 Hz)	896
$\mathrm{Cr\{\mathit{C}(NEt_2)Cl\}(CO)_5}$	244.3	652
$\mathrm{Cr\{\mathit{C}(NEt_2)Br\}(CO)_5}$	241.3	962
$\mathrm{Cr\{\mathit{C}(OEt)_2\}(CO)_5}$	266.4	840
cis-$\mathrm{Cr\{\mathit{C}(SMe)_2\}_2(CO)_4}$	141.6	202
Molybdenum		
$\mathrm{[Mo\{\mathit{C}H(NMe_2)\}Cp(CO)_3]^+}$	231.4	1372
$\mathrm{Mo\{\mathit{C}{=}C(CN)_2\}Cp\{P(OMe)_3\}_2Cl}$	351.3	280
$\mathrm{Mo\{\mathit{C}(\mathit{p}\text{-}tol)PMe_3\}(PMe_3)(CO)_3Cl}$	244.6 (7.3)	1363
$\mathrm{Mo\{\mathit{C}(NMe_2)SiPh_3\}(CO)_5}$	300.4	1042
$\mathrm{Mo\{\mathit{C}(OMe)SiPh_3\}(CO)_5}$	426.7	1042
$\mathrm{Mo\{\mathit{C}(OEt)SiPh_3\}(CO)_5}$	424.3	1042
$\mathrm{[Mo(\mathit{C}F_2)Cp(CO)_3]^+}$	279.8	1378
$\mathrm{[Mo(\mathit{C}F_2)Cp(CO)_2PPh_3]^+}$	264.1	1378
Tungsten, [^{183}W]		
$\mathrm{W(\mathit{C}HBu^t)(CBu^t)(CH_2Bu^t)(PMe_3)_2}$	286 [120] (14)	1402
$\mathrm{W(\mathit{C}HBu^t)(CBu^t)(CH_2Bu^t)\text{-}(Me_2PCH_2CH_2PMe_2)}$	256 (13, 31)	1402
$\mathrm{[W\{\mathit{C}H(NMe_2)\}Cp(CO)_3]^+}$	215.5 [70]	1372
$\mathrm{W(\mathit{C}Ph_2)(CO)_5}$	358, 357.9 [92.8], 358.4	450, 933, 1065, 1107
$\mathrm{W(\mathit{C}Ph\mathit{p}\text{-}tol)(CO)_5}$	357.0	1065
$\mathrm{W\{\mathit{C}Ph(2\text{-}furanyl)\}(CO)_5}$	299.3	1065
$\mathrm{W\{\mathit{C}Ph(2\text{-}thienyl)\}(CO)_5}$	318.8	1065
$\mathrm{W\{\mathit{C}Ph(2\text{-}\mathit{N}\text{-}Me\text{-}pyrryl)\}(CO)_5}$	282.2	1065
$\mathrm{W\{\mathit{C}(2\text{-}furanyl)_2\}(CO)_5}$	254.0	1065
$\mathrm{W\{\mathit{C}(2\text{-}thienyl)_2\}(CO)_5}$	288.3	1065
$\mathrm{W\{\mathit{C}(2\text{-}furanyl)(2\text{-}thienyl)\}(CO)_5}$	271.0	1065
$\mathrm{W\{\mathit{C}(NHMe)Me\}(CO)_5}$	255.6, 255.7*, 258.7*	89, 178
$\mathrm{W\{\mathit{C}(NMe_2)Me\}(CO)_5}$	253.3 [92.8], 251.6 ($^1J_{NC}$ = 4.7 Hz)	175, 908, 933
$\mathrm{W\{\mathit{C}(NMeEt)Me\}(CO)_5}$	250*, 251.3*	175
$\mathrm{W\{\mathit{C}(CH{=}CPhNMe_2)NMe_2\}(CO)_5}$	243.7 [89.1]	933
$\mathrm{\overline{W\{\mathit{C}(NEt_2)C(N}Et_2){=}CPh(OMe)\}(CO)_4}$	245.2	810
$\mathrm{\overline{W\{\mathit{C}(NMe_2)CHMeCO}\}Cp(CO)_2}$	236.8 [95.2]*, 235.5 [97.7]*	881
$\mathrm{W\{\mathit{C}Ph(PMe_3)\}(CO)_3(PMe_3)Br}$	239.6 (7.3)	851
$\mathrm{W\{\mathit{C}(\mathit{p}\text{-}tol)(PMe_3)\}(CO)_3(PMe_3)Cl}$	241.4 (7.3)	1363
$\mathrm{W\{\mathit{C}(\mathit{p}\text{-}tol)(PMe_3)\}(CO)_3(PMe_3)Br}$	239.6 (7.3)	1363
$\mathrm{W\{\mathit{C}(SiMe_3)(PMe_3)\}(CO)_3(PMe_3)Br}$	235.8 (7.3)	1363
$\mathrm{W\{\mathit{C}(OH)Me\}(CO)_5}$	334.4	179, 941
$\mathrm{W\{\mathit{C}(OMe)Me\}(CO)_5}$	332.9 102.5	89, 178, 933, 941

*two isomers

COMPOUND	$\delta(^{13}C)$/p.p.m., J/Hz	REFERENCE
cis-W{*C*(OEt)Me}(CNMeCH$_2$CH$_2$NMe)(CO)$_4$	328.7	1210
W{*C*(OSiMe$_3$)Me}(CO)$_5$	346.1	1052
W{*C*[OW(CO)$_4$CMe]Me}(CO)$_5$	330.8	941
W{*C*(OEt)CH$_2$CH$_2$CH$_2$CHCPh$_2$}(CO)$_5$	334	1409
W{*C*(OEt)C$_5$H$_8$CHCPh$_3$}(CO)$_5$	336.0	1411
W{*C*(OEt)CH=CPhNMe$_2$}(CO)$_5$	269.7 [92.8]	933
W{*C*(OEt)C(OH)CS(CH$_2$)$_3$S}(CO)$_4$	295.0	1365
W{*C*(OEt)C≡CPh}(CO)$_5$	286.1 [102.5]	449, 933
W{*C*(OH)Ph}(CO)$_5$	314.6	179, 368
W{*C*(OMe)Ph}(CO)$_5$	321.9, 322.1, 322.6, 323.9	39, 248, 368, 1107
cis-W{*C*(OEt)Ph}(CNMeCH$_2$CH$_2$NMe)(CO)$_4$	313.9	1210
W{*C*[OW(CO)$_4$CPh]Ph}(CO)$_5$	298.3	368
W{*C*(OH)C$_6$H$_4$Me-4}(CO)$_5$	311.8	941
W{*C*(OMe)C$_6$H$_4$Me-4}(CO)$_5$	319.8	248
W{*C*(O$_2$CMe)C$_6$H$_4$Me-4}(CO)$_5$	316.5	1049
W{*C*[OW(CO)$_4$(C$_6$H$_4$Me-4)]C$_6$H$_4$Me-4}(CO)$_5$	329.9	941
W{*C*(OMe)C$_6$H$_4$CF$_3$-4}(CO)$_5$	322.2	248
W{*C*(OMe)C$_6$H$_4$OMe-4}(CO)$_5$	313.3	248
W{*C*(OMe)C$_6$H$_4$F-4}(CO)$_5$	318.5	248
W{*C*(OMe)C$_6$H$_4$Cl-4}(CO)$_5$	319.4	248
W{*C*(OMe)C$_6$H$_4$Br-4}(CO)$_5$	319.6	248
W{*C*(OEt)C$_5$H$_4$Mn(CO)$_3$}(CO)$_5$	305.4	1007
W{*C*(OSiMe$_3$)-2-furanyl}(CO)$_5$	291.8	1052
W{*C*(OSiMe$_3$)-2-thienyl}(CO)$_5$	299.6	1052
W{*C*(SMe)Me}(CO)$_5$	322.5	89, 178
W{*C*(SMe)Ph}(CO)$_5$	332.0	353, 438
W{*C*(SMe)C$_6$H$_4$Me-4}(CO)$_5$	332.8	353, 438
W{*C*(SMe)C$_6$H$_4$OH-4}(CO)$_5$	333.3	353, 438
W{*C*(SMe)C$_6$H$_4$OMe-4}(CO)$_5$	333.3	353, 438
W{*C*(SMe)C$_6$H$_4$F-4}(CO)$_5$	322.4	353, 438
W{*C*(SMe)C$_6$H$_4$Br-4}(CO)$_5$	331.1	353, 438
W{*C*(SeMe)Me}(CO)$_5$	355.5	178
W{*C*(NMe$_2$)SiPh$_3$}(CO)$_5$	285.9	1042
W{*C*(OMe)SiPh$_3$}(CO)$_5$	402.1 [101.1]	1042
W{*C*(OEt)SiPh$_3$}(CO)$_5$	399.8	1042
cis-W(*C*NMeCH$_2$CH$_2$NMe)$_2$(CO)$_4$	211.4	1210
trans-W(*C*NMeCH$_2$CH$_2$NMe)$_2$(CO)$_4$	215.4	1210
W(*C*NMeCH$_2$CH$_2$NMe)(CO)$_5$	206.6, 215.4	439, 1210
cis-W(*C*NMeCH$_2$CH$_2$NMe){*C*(OEt)Me}(CO)$_4$	212.7	1210
cis-W(*C*NMeCH$_2$CH$_2$NMe){*C*(OEt)Ph}(CO)$_4$	212.4	1210
cis-W(*C*NEtCH$_2$CH$_2$NEt)$_2$(CO)$_4$	213.6	1210
fac-W(*C*NEtCH$_2$CH$_2$NEt)$_2$py(CO)$_3$	226.5	1210
W(*C*NEtCH$_2$CH$_2$NEt)(CO)$_5$	205.4 [94.6]	1210
cis-W(*C*NEtCH$_2$CH$_2$NEt)(PEt$_3$)(CO)$_4$	213.2 (9.2)	1210

COMPOUND	δ(^{13}C)/p.p.m., *J*/Hz	REFERENCE
fac-W($\overline{C\text{NEtCH}_2\text{CH}_2\text{N}}$Et){P(OMe)$_3$}$_2(CO)_3$	214.9 (10.7)	1210
cis-W($\overline{C\text{NMeCH=CHN}}$Me)$_2(CO)_4$	187.1 [92.8]	933
trans-W($\overline{C\text{NMeCH=CHN}}$Me)$_2(CO)_4$	183.1 [87.9]	933
W($\overline{C\text{NMeCH=CHN}}$Me)(CO)$_5$	178.1 [100.1]	933
W($\overline{C\text{NMeN=CHN}}$Me)(CO)$_5$	180.5 [97.7]	933
W{*C*(OEt)NEt$_2$}(CO)$_5$	226.8	361, 1376
Manganese		
cis-Mn{*C*H(NMe$_2$)}(CO)$_4$	256.8	1372
Mn(*C*Me$_2$)Cp(CO)$_2$	372.8	756
Mn(*C*MePh)Cp(CO)$_2$	363.7	756
Mn(*C*=CHPh)Cp(CO)$_2$	379.5	970
Mn(*C*=C=CBu^t$_2$)Cp(CO)$_2$	331.2	752
Mn{*C*(PMe$_3$)Ph}Cp(CO)$_2$	318.0 (7.3)	1048
Mn{*C*(OMe)Me}(CO)$_2$(C$_5$H$_4$Me)	339.2	1417
Mn{*C*(OMe)Me}(PF$_2$NMePF$_2$)(C$_5$H$_4$Me)	330.9	1417
$\overline{\text{Mn}\{C\text{(OGe}}Me_2$)Me}(CO)$_4$	335.2*, 337.7*, 338.2*	182, 311
Mn{*C*(OMe)C$_{10}$H$_{19}$}Cp(CO)$_2$	351.3	1418
Mn{*C*(OR)Ph}Cp(CO)$_2$	332.5 to 334.8	756, 1058
Mn{*C*(SCN)Ph}Cp(CO)$_2$	285.1	946
Rhenium		
Re(*C*HPh)Cp(CO)$_2$	277.1	1423
Re(*C*MePh)Cp(CO)$_2$	306.0	832
Re{*C*(CN)Ph}Cp(CO)$_2$	229.2	946
Re(*C*=CHPh)Cp(CO)$_2$	329.5	969, 970
Re{*C*(PMe$_3$)Ph}Cp(CO)$_2$	262.2 (14.7)	1021, 1048
Re{*C*(OEt)Me}(GeMe$_3$)(CO)$_4$	314.8	717
Re{*C*(SCN)Ph}Cp(CO)$_2$	239.9	946
fac-[Re{*C*(OEt)SiPh$_3$}dppe(CO)$_3$]$^+$	393.6 (9)	1123
Iron		
trans-Fe$_2$(μ-*C*C$_5$H$_5$)(μ-CO)Cp$_2$(CO)$_2$	448.3	1451
cis-Fe$_2$(μ-*C*C$_5$H$_5$)(μ-CO)Cp$_2$(CO)$_2$	443.6	1451
Fe(*C*HPh)Cp(CO)$_2$	342.4 {146}	1095
Fe(*C*HPh)Cp(CO)PPh$_3$	341.2 (21) {136}	1095
[Fe(*C*=CHMe)Cp(dppe)]$^+$	358.3	1446
[Fe(*C*=CMe$_2$)Cp(dppe)]$^+$	363.3	1446
Fe{*C*(NHMe)$_2$}Cp(CO)CN	214.4	1470
[Fe{*C*(NHMe)$_2$}Cp(CO)(PMe$_3$)]$^+$	208.9	1470
[Fe{*C*(NHMe)$_2$}Cp(CO)(PPh$_3$)]$^+$	209.4	1470
[Fe{*C*(NHMe)$_2$}Cp(CO)(CNMe)]$^+$	206.4	1470
[Fe{*C*(NHMe)$_2$}Cp(CO)(CNPr^i)]$^+$	206.4	1470
[Fe{*C*(NHMe)$_2$}(indenyl)(CO)(CNMe)]$^+$	206.3	1470
[Fe{*C*(NHMe)$_2$}(indenyl)(CO)(CNEt)]$^+$	204.8, 205.5	1470
[Fe{*C*(NHMe)$_2$}(indenyl)(CO)(CNPr^i)]$^+$	206.5	1470

*in equilibrium with dimer

COMPOUND	$\delta(^{13}C)$/p.p.m., J/Hz	REFERENCE
$[Fe\{C(NHMe)_2\}(indenyl)(CO)(CNBu^t)]^+$	206.4	1470
$Fe\{C(NHMe)(NHEt)\}Cp(CO)CN$	214.2	1470
$[Fe\{C(NHMe)(NHEt)\}Cp(CO)(PMe_3)]^+$	206.8	1470
$[Fe\{C(NHMe)(NHEt)\}Cp(CO)(PPh_3)]^+$	207.2	1470
$[Fe\{C(NHMe)(NHEt)\}Cp(CO)(CNMe)]^+$	204.0, 204.9	1470
$[Fe\{C(NHMe)(NHEt)\}Cp(CO)(CNEt)]^+$	204.5, 205.3	1470
$[Fe\{C(NHMe)(NHEt)\}(indenyl)(CO)(CNMe)]^+$	204.3, 205.1	1470
$Fe\{C(NHMe)(NHPr^n)\}Cp(CO)CN$	215.8	1470
$Fe\{C(NHMe)(NHPr^i)\}Cp(CO)CN$	215.6	1470
$[Fe\{C(NHMe)(NHPr^i)\}Cp(CO)(PPh_3)]^+$	206.3	1470
$[Fe\{C(NHMe)(NHPr^i)\}Cp(CO)(CNMe)]^+$	203.3, 204.5	1470
$[Fe\{C(NHMe)(NHPr^i)\}Cp(CO)(CNPr^i)]^+$	202.5, 204.3	1470
$[Fe\{C(NHEt)_2\}Cp(CO)(CNEt)]^+$	203.9	1470
$[Fe\{C(NHPr^i)_2\}(indenyl)(CO)(CNPr^i)]^+$	201.6	1470
trans-$Fe(\overline{CNMeCH_2CH_2NMe})_2(CO)_3$	222.4	1224
$Fe(\overline{CNMeCH_2CH_2NMe})_2(NO)_2$	231.2	1224
$Fe(\overline{CNMeCH_2CH_2NMe})(CO)_4$	212.8, 213.2	202, 481, 1224
$Fe(\overline{CNMeCH_2CH_2NMe})(CO)(NO)_2$	220.4	1224
$Fe_2(\overline{CNEtCH_2CH_2NEt})(\mu\text{-}CO)_2Cp_2(CO)$	214	1224
$[\overline{Fe\{C(NHMe)NMeC}(NHMe)\}(CNMe)_4]^{2+}$	210.5	703
$[\overline{Fe\{C(NHMe)NMeC}(NHMe)\}(bipy)_2]^{2+}$	223.3	703
$\overline{Fe\{C(OMe)C(CO_2Me)CH_2}\}(CO)_3$	270.0	741
$\overline{Fe\{C(OMe)C(CO_2Me)CH}(CO_2Me)\}(CO)_3$	270.9	741
$Fe(\overline{COCMe_2CMe_2O})(CO)_4$	251.2	338, 481
$Fe(\overline{CSCH{=}CHS})(CO)_2\{P(OMe)_3\}_2$	250.7	1450
$Fe\{\overline{CSC(C_6H_4Cl\text{-}4){=}C(CHO)S}\}(CO)_2\{P(OMe)_3\}_2$	239.6	1450
$Fe\{\overline{CSC(CO_2Me){=}C(CO_2Me)S}\}(CO)_2\{P(OMe)_3\}_2$	239.4	1450
$Fe(CCl_2)$(tetraphenylporphyrin)	224.7	1186
Ruthenium		
$[Ru(C{=}CHPh)Cp(PPh_3)_2]^+$	350	1484
$Ru\{CH(NMe_2)\}(PEt_3)_2Cl_2(CO)$	248.5 (22)	1372
trans-$[Ru(\overline{CNMeCH_2CH_2NMe})_4(CO)Cl]^+$	208.1	736, 1477
trans-$[Ru(\overline{CNMeCH_2CH_2NMe})_4(PF_3)Cl]^+$	208.1 (16.8)	736, 1477
$[Ru(\overline{CNMeCH_2CH_2NMe})_4(NO)]^+$	220.4	1425
trans-$Ru(\overline{CNMeCH_2CH_2NMe})_4Cl_2$	227.8	1477
trans-$Ru(\overline{CNMeCH_2CH_2NMe})_3Cl_2$	216.3	1477

COMPOUND	δ(^{13}C)/p.p.m., *J*/Hz	REFERENCE
$Ru(\overline{CNMeCH_2CH_2NMe})_3I_2$	216.3	1477
$Ru(\overline{CNEtCH_2CH_2NEt})_3Cl_2$	216.5	1477
trans-$Ru(\overline{CNEtCH_2CH_2NEt})_3Cl_2(CO)$	216.5*, 212.7§	1477
$Ru(\overline{CNEtCH_2CH_2NEt})\{CN(C_6H_4Me\text{-}4\text{-}2)\text{-}CH_2CH_2N(p\text{-tol})\}Cl(PPh_3)$	213.4 (88.5)	1196
$Ru_3(\overline{CNEtCH_2CH_2NEt})(CO)_{11}$	198.6	1224
trans-$Ru\{\overline{CN(CH_2Ph)CH_2CH_2N(CH_2Ph)}\}_4\text{-}Cl_2$	228.1	1477
$[Ru\{\overline{CN(CH_2Ph)CH_2CH_2N(CH_2Ph)}\}_4(NO)]^+$	216.2	1484
$Ru\{\overline{CN(CH_2Ph)CH_2CH_2N(CH_2Ph)}\}_2(NO)\text{-}(CO)Cl$	207.3	1484
$Ru\{\overline{CN(p\text{-tol})CH_2CH_2N(p\text{-tol})}\}(CO)\text{-}(PPh_3)_2Cl$	216.9 (9.9)	1196
$Ru\{\overline{CN(p\text{-tol})CH_2CH_2N(p\text{-tol})}\}(CO)\text{-}(PEt_3)_2Cl$	219.3 (9.7)	1196
$Ru\{CN(C_6H_4\text{-}4\text{-}Me\text{-}2)CH_2CH_2N(p\text{-tol})\}\text{-}(PEt_3)_2Cl$	223.3 (9.9)	1196
$Ru\{CN(C_6H_4\text{-}4\text{-}Me\text{-}2)CH_2CH_2N(p\text{-tol})\}\text{-}(\overline{CNEtCH_2CH_2NEt})Cl(PPh_3)$	220.6 (9.1)	1196
Osmium		
trans-$[Os(\overline{CNMeCH_2CH_2NMe})_4Cl(NO)]^{2+}$	177.6	1484
trans-$[Os\{\overline{CN(CH_2Ph)CH_2CH_2N}\text{-}(CH_2Ph)\}_4Cl(NO)]^{2+}$	180.2	1484
$[Os\{\overline{CN(CH_2Ph)CH_2CH_2N(CH_2Ph)}\}_3Cl_2\text{-}(NO)]^+$	174.9*, 179.6§	1484
Cobalt		
$Co(\overline{CNEtCH_2CH_2NEt})_2(CO)(NO)$	228.4	1224
Rhodium, [^{103}*Rh*]		
$Rh_2(CH_2)Cp_2(CO)_2$	100.9 [29]	1513
$Rh\{C(NHBu^t)NBu^nC(NHBu^t)\}Bu^n{}_2\text{-}(CNBu^t)_2$	187 [35]	703
$Rh\{C(NHBu^t)NBu^nC(NHBu^t)\}IMe(CNBu^t)_2$	187.2 [35]	381
Iridium		
$[Ir\{CH(NMe_2)\}Cl_2(CO)(PPh_3)_2]^+$	192.7 (3)	1372

**cis*, §mutually *trans*

COMPOUND	$\delta(^{13}C)$/p.p.m., *J*/Hz	REFERENCE
Nickel		
cis-Ni($\overline{CNMeCH_2CH_2NMe}$)$_2Cl_2$	194.5	1224
cis-Ni($\overline{CNMeCH_2CH_2NMe}$)$_2I_2$	208.8	1224
cis-Ni($\overline{CNMeCH_2CH_2NMe}$)I$_2$(PPh$_3$)	206.3	1224
Ni($\overline{CNEtCH_2CH_2NEt}$)$_2(CO)_2$	223.4	1224
Ni($\overline{CNEtCH_2CH_2NEt}$)(CO)$_3$	214.4	1224
Ni($\overline{CNEtCH_2CH_2NEt}$)(NO)Cl	225.6	1484
Palladium		
trans-Pd($\overline{CNMeCH_2CH_2NMe}$)Cl$_2$(PBu$^n{}_3$)	200.5 (180.5)	202
cis-Pd($\overline{CNMeCH_2CH_2NMe}$)Cl$_2$(PBu$^n{}_3$)	195.1 (<2.5)	202
Pd($\overline{CNPhCH_2CH_2NPh}$)(C$_6H_4CH_2NMe_2$)Cl	201.6	1527
Platinum, [^{195}*Pt*]		
trans-[Pt{*C*Me(NH$_2$)}Me(AsMe$_3$)$_2$]$^+$	255.6 [666]	569
trans-[Pt{*C*Me(NHMe)}Me(AsMe$_3$)$_2$]$^+$	247.3 [687]	569
trans-[Pt{*C*Me(NHMe)}Cl(AsMe$_3$)$_2$]$^+$	215.1 [1047]	569
trans-[Pt{*C*Me(NHEt)}Me(AsMe$_3$)$_2$]$^+$	245.2 [684]	569
trans-[Pt{*C*Me(NHPh)}Me(AsMe$_3$)$_2$]$^+$	250.3 [686]	580
trans-[Pt{*C*Me(NHC$_6$H$_4$Me-4)}Me-(AsMe$_3$)$_2$]$^+$	248.4 [687]	580
trans-[Pt{*C*Me(NHC$_6$H$_4$OMe-4)}-Me(AsMe$_3$)$_2$]$^+$	246.2 [686]	580
trans-[Pt{*C*Me(NHC$_6$H$_4$Cl-4)}-Me(AsMe$_3$)$_2$]$^+$	252.0 [686]	580
trans-[Pt{*C*Me(NHC$_6$H$_4$Br-4)}-Me(AsMe$_3$)$_2$]$^+$	252.0 [686]	580
trans-[Pt{*C*Me(NMe$_2$)}Me(AsMe$_3$)$_2$]$^+$	245.1 [694]	569
trans-[Pt{*C*Me(NMe$_2$)}Cl(AsMe$_3$)$_2$]$^+$	210.3 [1070]	203, 569
trans-[Pt{*C*Me(OMe)}Me(AsMe$_3$)$_2$]$^+$	321.0 [759]	31, 569
trans-[Pt{*C*Me(OMe)}Cl(AsMe$_3$)$_2$]$^+$	278.3 [1125]	203, 569
trans-[Pt{*C*Me(OEt)}Me(AsMe$_3$)$_2$]$^+$	317.8 [756]	569
trans-[Pt($\overline{CCH_2CH_2CH_2O}$)Me(PMe$_2$Ph)$_2$]$^+$	299.2	203, 569
trans-[Pt($\overline{CCH_2CH_2CH_2O}$)Ph(AsMe$_3$)$_2$]$^+$	275.6 [764]	409
trans-Pt($\overline{CNMeCH_2CH_2NMe}$)Me$_2$(PEt$_3$)	218.1 (144.0)	202
cis-Pt($\overline{CNMeCH_2CH_2NMe}$)Me$_2$(PEt$_3$)	214.5 (14.4)	202
trans-Pt($\overline{CNMeCH_2CH_2NMe}$)Cl$_2$(PEt$_3$)	197.3 (156)	202
cis-Pt($\overline{CNMeCH_2CH_2NMe}$)Cl$_2$(PEt$_3$)	177.8 (*ca.* 50)	202
trans-Pt($\overline{CNMeCH_2CH_2NMe}$)Br$_2$(PEt$_3$)	196.2 (149)	202
trans-Pt($\overline{CNMeCH_2CH_2NMe}$)Cl$_2$(PBu$^n{}_3$)	196.5 (146.4)	202
cis-Pt($\overline{CNMeCH_2CH_2NMe}$)Cl$_2$(PBu$^n{}_3$)	178.0 (*ca.* 60)	202

COMPOUND	$\delta(^{13}C)$/p.p.m., J/Hz	REFERENCE
trans-Pt($\overline{CNMeCH_2CH_2NMe}$)$Cl_2(PPr^i_3)$	195.8 (151)	202
cis-Pt($\overline{CNMeCH_2CH_2NMe}$)$Cl_2(PPr^i_3)$	178.0 (7.2)	202
trans-Pt($\overline{CNMeCH_2CH_2NMe}$)$Cl_2(AsEt_3)$	188.6 [1074]	202
cis-Pt($\overline{CNMeCH_2CH_2NMe}$)$Cl_2(AsEt_3)$	175.1 [756]	202
trans-Pt($\overline{CNPhCH_2CH_2NPh}$)$Me_2(PEt_3)$	192.7 (170.6)	202
trans-[Pt{*C*($NHBu^n$)(NHC_6H_4OMe-4)}-($CH=CH_2$)$(PEt_2Ph)_2$]$^+$	202	1171
Phosphorus, (^{31}P)		
$Me_3P=CH_2$	-2.3 (90.5) {149}	523, 1063
	-2.5 (90) {150}	416
	-7.0 (+90.5)	143
	-1.5 (90.5)	1769
$Me_2(C_3H_5)P=CH_2$	-13.7 (100.1) {149.1}	1057
$Me_2Bu^tP=CH_2$	-9.6 (90.3)	1289
$MeBu^t_2P=CH_2$	-13.2 (83.0)	1289
$Me(C_4H_8)P=CH_2$	-1.4 (87.9)	1686
$Me(C_5H_{10})P=CH_2$	-5.4 (92.7)	1057
$Et_3P=CH_2$	-14.2 (86.8)	523
$Me_2P(=CH_2)(CH_2)_4P(=CH_2)(C_4H_8)$	-3.4, -7.5	1686
$Me(H_2C=)P\{(CH_2)_4\}_2P(=CH_2)Me$	-4.4	1685
$Bu^t_3P=CH_2$	-13.2 (68.4)	1289
$Ph_3P=CH_2$	4.1	472
	-4.1 (51.9)	1769
$Me_2P(=CH_2)N=PMe_3$	10.8 (89.7, 2.9)	1040
	10.0 (89.7, 2.9)	758
$C_5H_{10}P(=CH_2)N=PMe(C_5H_{10})$	-3.9 (87.9)	1057
$Et_3P=CHMe$	-7.4 (113.2)	523
$Ph_3P=CHMe$	3.2 (110.7)	860
$Ph_3P=CH(C_2H_3)$	28.7 (131.4)	860
$Ph_3P=CHPh$	28.0 (128.0)	860
$PhMe_2P=CHCOMe$	52.2 (108.0)	1665
$Ph_2MeP=CHCOMe$	51.2 (108.5)	1665
$Ph_3P=CHCOMe$	51.3 (108.0)	860, 1665
$PhMe_2P=CHCOPh$	51.2 (111.5)	1665
$Ph_2MeP=CHCOPh$	50.7 (112.5)	1665
$Ph_3P=CHCOAr$	50.0 to 53.6 (111.7)	782, 860, 1282, 1665
$Ph_3P=CHCO_2Me$	29.8 (126.7)	860
$Me_3P=CHC(OSiMe_3)Cr(CO)_5$	94.7 (67.6)	1359
$Me_3P=CHSiMe_3$	0.7 (88.2) {134.6}	1063
$\overline{Me_2P=CHSiMe_2CH_2SiMe_2CH_2}$	3.1 (90.3)	1605
$Me_3P=CHGeMe_3$	1.3 (87.9) {144.0}	1063
$[Ph_3P=CH-PBu^n_3]^+$	-6.2 (123.9, 108.6)	1779
$Ph_3P=CHCl$	25.7 (64.4)	1033
$Ph_3P=CMe_2$	9.0 (121.5)	860, 1769
$P(=CMe_2)(=NSiMe_3)N(SiMe_3)_2$	77.4 (215.9)	1612
$P(=CMeR)(=NSiMe_3)N(SiMe_3)_2$	84.1 to 88.9 (207.4 to 214.0)	1612

COMPOUND	$\delta(^{13}C)$/p.p.m., J/Hz	REFERENCE
$Ph_3P{=}\mathit{C}(CH_2)_3$ (ring)	14.6 (77.3)	860
$Ph_3P{=}\mathit{C}CH_2CH_2$ (ring)	4.3 (132.8)	860
2,4,6-$Me_3C_6H_2P{=}\mathit{C}Ph_2$	193.4 (43.5)	1669
$Ph_3P{=}\mathit{C}CH{=}CHCH{=}CH$ (ring)	78.3 (113.1)	234, 707, 860
	78.0 (112)	1165
Ph_3P(indene)	53.3 (128.7)	241
$Ph_3P{=}\mathit{C}(CO_2Me)(COCO_2Me)$	68.0 (111)	1682
$Et_2P{=}\mathit{C}MeSiEt(OMe)(CH_2)_3$ (ring)	-4.6 (82.5)	634
$Me_3P{=}\mathit{C}(SiMe_3)_2$	0.3 (63.3)	1063
$Me_3P{=}\mathit{C}(SiMe_2)_2CH_2$	13.2 (95)	759
	12.9 (95.2)	1605
$MeP{=}\mathit{C}(SiMe_2CH_2SiMe_2CH_2)_2$ (ring)	30.0 (65.9)	1605
$Me_3P{=}\mathit{C}(GeMe_3)_2$	1.9 (65.9)	1063
$Me_2P\{{=}\mathit{C}(GeMe_3)_2\}CH_2GeMe_3$	2.9 (65.9)	1063
$Me_3P{=}\mathit{C}(SnMe_3)_2$	-6.7 (51.3)	1063
$Me_3P{=}\mathit{C}(PbMe_3)_2$	14.7 (57.4)	1063
$PhP{=}\mathit{C}Bu^tNPhSiMe_3$	206 (57)	1685
$PhP{=}\mathit{C}PhNPhSiMe_3$	193 (47)*	1685
	205 (50)*	
$Me_3P{=}\mathit{C}{=}PMe_3$	10.8 (32)	527
$Bu^n{}_3P{=}\mathit{C}{=}PPh_3$	7.6 (127.7)	1779
$Ph_3P{=}\mathit{C}{=}PPh_3$	7.6 (127.7)	1779
$[Ph_3P\mathit{C}{=}PPh_2]_2$	18.3 (111.7, 76.2, 15.4)	1025
$RP{=}\mathit{C}Bu^t(OSiMe_3)$	212.0 to 227.0	1024, 1247
	(66.0 to 80.5)	1250
$Ph_3P{=}\mathit{C}Cl_2$	73.9 (72.3)	760
Arsenic		
$Ph_3As{=}\mathit{C}H_2$	1.6	472
$Ph_3As{=}\mathit{C}HCOMe$	56.9	1665
$Ph_3As{=}\mathit{C}HCOPh$	57.1	1665
$As{=}\mathit{C}HNC_6H_4$-2 (ring)	172.7	1784
$Me_2As{=}\mathit{C}HNC_6H_4$-2 (ring)	202.7	1784
$Ph_3As{=}\mathit{C}HCOC_6H_4X$-4	56.6 to 59.9	784
$PhAs{=}\mathit{C}Bu^tOSiMe_3$	239.0	1024
$Me\mathit{C}AsCo_2(CO)_6$	170.4	1767
$Ph\mathit{C}AsCo_2(CO)_6$	165.9	1767
Antimony		
$Me(Me_3SiCH_2)_2Sb{=}\mathit{C}(SiMe_3)_2$	9.8	1614

*isomers

TABLE 2.6

^{13}C N.m.r. Data for Some Carbyne Derivatives

COMPOUND	$\delta(^{13}C)$/p.p.m., *J*/Hz	REFERENCE
Niobium		
$Nb_2(\mathit{C}SiMe_3)_2(CH_2SiMe_3)_4$	406	897
Tantalum		
$Ta(\mathit{C}Ph)(C_5Me_5)(PMe_3)_2Cl$	345 (20)	1347
$Ta(\mathit{C}Bu^t)Cp(PMe_3)_2Cl$	348 (16)	1347
$Ta(\mathit{C}Bu^t)(C_5Me_5)(PMe_3)_2Cl$	354 (19)	1347
$Ta_2(\mathit{C}SiMe_3)_2(CH_2SiMe_3)_4$	406	897
Chromium		
trans-$[Cr(\mathit{C}Me)(CO)_4PMe_3]$	365.4, 364.2 (9.3)	440, 892
trans-$[Cr(\mathit{C}Me)(CO)_4SbMe_3]$	366.3	892
trans-$Cr(\mathit{C}Me)(CO)_4Br$	338.2	893
trans-$Cr(\mathit{C}Me)(CO)_4I$	340.8	893
$Cr(\mathit{C}Me)(CO)_3(PMe_3)Cl$	329.6 (29.3)	892, 904
$Cr(\mathit{C}Me)(CO)_3(PMe_3)Br$	331.1 (29.3)	892
$Cr(\mathit{C}Me)(CO)_3(AsMe_3)Br$	331.6	893
$Cr(\mathit{C}Me)(CO)_3(SbMe_3)Br$	338.2	893
$Cr(\mathit{C}Me)(CO)_3(PMe_3)I$	331.9 (26.9)	892
$Cr(\mathit{C}Me)(CO)_3(PMe_3)BF_4$	350.8	904
$Cr(\mathit{C}CH_2Ph)(CO)_4Br$	334.4	1363
$Cr(\mathit{C}Ph)(CO)_4Br$	318.2	893, 1067
$Cr(\mathit{C}Ph)(CO)_4Re(CO)_5$	321.1	1047
$Cr(\mathit{C}Ph)(CO)_3(PPh_3)Br$	313.0 (24.4)	1061
$Cr(\mathit{C}Ph)(CO)_3\{P(OPh)_3\}Br$	313.3 (34.2)	1061
$Cr(\mathit{C}Ph)(CO)_3(AsPh_3)Br$	313.0	1061
$Cr(\mathit{C}Ph)(CO)_3(SbPh_3)Br$	313.2	1061
$Cr(\mathit{C}Ph)(CO)_2(CNBu^t)_2Br$	310.4	1061
$Cr(\mathit{C}Ph)(CO)_2(py)_2Br$	303.2	1061
$Cr(\mathit{C}Ph)(CO)_2$(*o*-phen)Br	304.5	1061
$[Cr(\mathit{C}Ph)(CO)_2(C_6H_6)]^+$	349.1	906, 935
$[Cr(\mathit{C}Ph)(CO)_2(\text{p-xylene})]^+$	348.5	935
$[Cr(\mathit{C}Ph)(CO)_2(\text{mesitylene})]^+$	347.8	935
$Cr(\mathit{C}C_6H_4Me\text{-}4)(CO)_4Cl$	318.2	1363
$Cr(\mathit{C}C_6H_4Me\text{-}4)(CO)_4Br$	319.1	435, 1067, 1363
$Cr(\mathit{C}C_6H_4Me\text{-}4)(CO)_4I$	319.2	1363
$Cr(\mathit{C}C_6H_4NMe_2\text{-}4)(CO)_4Br$	296.9	1067
$Cr(\mathit{C}C_6H_4OMe\text{-}4)(CO)_4Br$	319.4	1067
$Cr(\mathit{C}C_6H_4CF_3\text{-}4)(CO)_4Br$	312.4	1067
$Cr(\mathit{C}C_6H_3Cl_2\text{-}2,6)(CO)_4Br$	302.2	1067
$Cr(\mathit{C}C_6H_3Ph\text{-}2)(CO)_4Br$	317.8	1067
$Cr(\mathit{C}C_6H_2Me_3\text{-}2,4,6)(CO)_4Br$	323.4	1067, 1363

COMPOUND	$\delta(^{13}C)$/p.p.m., *J*/Hz	REFERENCE
$Cr(\mathit{C}C_5H_4FeCp)(CO)_4Br$	322.0	1369
$Cr(\mathit{C}C_5H_4FeCp)(CO)_4I$	322.6	1369
$[Cr(\mathit{C}NEt_2)(CO)_5]^+$	282.2	896, 1376
$Cr(\mathit{C}NEt_2)(CO)_4Cl$	263.9	1376
$Cr(\mathit{C}NEt_2)(CO)_4Br$	264.1	437, 762, 1376
$Cr(\mathit{C}NEt_2)(CO)_4I$	268.2, 268.4	762, 1376
Molybdenum		
$Mo(\mathit{C}CH_2Bu^t)Cp\{P(OMe)_3\}_2$	299.8 (27)	1096
$[Mo(\mathit{C}CH_2Bu^t)CpH\{P(OMe)_3\}_2]^+$	346.7 (24)	1096
$Mo(\mathit{C}Bu^t)(CH_2Bu^t)_3$	324	1402
$Mo(\mathit{C}Ph)(CO)_4Re(CO)_5$	294.0	1047
$Mo(\mathit{C}\text{-}p\text{-}tol)(CO)_4Cl$	291.1	1363
$Mo(\mathit{C}C_5H_4FeCp)(CO)_4Br$	293.7	1369
Tungsten, $[^{183}W]$		
$W(\mathit{C}Me)(CO)_4BF_4$	329.6	1400
$W(\mathit{C}Me)(CO)_4Cl$	288.8, 289.0 [168.5]	176, 893, 941
$W(\mathit{C}Me)(CO)_4Br$	288.1, 288.0 [178.2]	176, 893, 933
$W(\mathit{C}Me)(CO)_4I$	286.3, 286.2 [185.5]	176, 893, 933
$W(\mathit{C}Me)(CO)_4NCS$	302.0	1400
$W(\mathit{C}Me)(CO)_4OCMeW(CO)_5$	294.1	941
$W(\mathit{C}Et)(CO)_4Cl$	296.4	893
$W(\mathit{C}Et)(CO)_4Br$	296.3	893
$W(\mathit{C}Bu^t)(CH_2Bu^t)_3$	317 [230]	1402
$W(\mathit{C}Bu^t)(CHBu^t)(CH_2Bu^t)(PMe_3)_2$	316 [210] (14)	1402
$W(\mathit{C}Bu^t)(CHBu^t)(CH_2Bu^t)\text{-}(Me_2PCH_2CH_2PMe_2)$	296 (12)	1402
$W\{\mathit{C}CH{=}CPh(NMe_2)\}(CO)_4Br$	283.7, 283.9 [168]	335, 449, 933
$W(\mathit{C}C{\equiv}CPh)(CO)_4Br$	230.5, 230.6 [185]	335, 449, 933
$W(\mathit{C}Ph)(CO)_4BF_4$	300.9	1400
$W(\mathit{C}Ph)(CO)_4Cl$	272.4, 272.6 [168.5]	368, 893, 933, 941
$W(\mathit{C}Ph)(CO)_4Br$	271.3, 271.0 [173.3]	368, 893, 933, 1067
$W(\mathit{C}Ph)(CO)_4I$	268.4, 268.3 [180.7]	893, 933
$W(\mathit{C}Ph)(CO)_4NCS$	283.8	1400
$W(\mathit{C}Ph)(CO)_4CN$	290.6	1400
$W(\mathit{C}Ph)(CO)_4MoCp(CO)_3$	318.7	1047
$W(\mathit{C}Ph)(CO)_4WCp(CO)_3$	295.7	1047
$W(\mathit{C}Ph)(CO)_4Mn(CO)_5$	278.6	1047
$W(\mathit{C}Ph)(CO)_4Re(CO)_5$	280.9	1047
$W(\mathit{C}Ph)(CO)_4OC(p\text{-}tol)Cr(CO)_5$	299.3	941
$W(\mathit{C}Ph)(CO)_4OCPhW(CO)_5$	291.0	368
$[W(\mathit{C}Ph)(CO)_4PPh_3]^+$	294.3	1400
$W(\mathit{C}Ph)(CO)_3(PPh_3)Br$	271.3 (9.8)	1061
$W(\mathit{C}Ph)(CO)_3(AsPh_3)Br$	270.0	1061
$W(\mathit{C}Ph)(CO)_3(SbPh_3)Br$	270.0	1061
$W(\mathit{C}Ph)(CO)_2(CNBu^t)_2Br$	266.9	1061
$W(\mathit{C}Ph)Cp(CO)_2$	299.3	637, 1043
$W(\mathit{C}\text{-}p\text{-}tol)(CO)_4Cl$	274.1	1363

COMPOUND	$\delta(^{13}C)$/p.p.m., *J*/Hz	REFERENCE
$W(C\text{-}p\text{-tol})(CO)_4Br$	271.4	941, 1067, 1363
$W(C\text{-}p\text{-tol})(CO)_4OC(p\text{-tol})W(CO)_5$	299.2	941
$W(C\text{-}p\text{-tol})Cp(CO)_2$	300.1	637, 1043
$W(CC_6H_4CF_3\text{-}4)(CO)_4Br$	266.2	1067
$W(CC_6H_4NMe_2\text{-}4)(CO)_4Br$	278.4	1067
$W(CC_6H_2Me_3\text{-}2,4,6)(CO)_4Br$	275.1	1067
$W(CC_6H_2Me_3\text{-}2,4,6)Cp(CO)_2$	300.6	1043
$W(CC_6H_4OMe\text{-}4)(CO)_4Br$	273.2	1067
$W(CC_6H_4OMe\text{-}4)Cp(CO)_2$	300.0	1043
$W\{CC_6H_3(OMe)Br\text{-}4,3\}Cp(CO)_2$	291.6	1043
$W(CC_6H_3Cl_2\text{-}2,6)(CO)_4Br$	256.4	1067
$W\{CC_5H_4Mn(CO)_3\}(CO)_4Br$	258.4	1007
$W(CC_5H_4FeCp)(CO)_4Cl$	276.2	1369
$W(CC_5H_4FeCp)(CO)_4Br$	275.1	751, 1369
$W(CC_5H_4FeCp)(CO)_4I$	272.7	1369
$W(CC_5H_4FeCp)Cp(CO)_2$	300.5	1043
$W_2(CSiMe_3)_2(CH_2SiMe_3)_4$	354	897
$W(CSiMe_3)(CO)_4Br$	337.1	1363
$W(CSiPh_3)(CO)_4Br$	337.1 [146.5]	1026
$W(CSiPh_3)Cp(CO)_2$	354.3 [178.2]	1026
$W(CNMe_2)(CO)_4Cl$	236.2	1376
$W(CNMe_2)(CO)_4Br$	235.2 [197.8], 235.5	933, 1376
$W(CNMe_2)(CO)_4I$	234.5	1376
$W(CNEt_2)(CO)_4Br$	235.6	361, 1376
$W(CNEt_2)(CO)_4I$	234.6	361, 1376
Manganese		
$[Mn(CPh)Cp(CO)_2]^+$	356.9	1058
Rhenium		
$[Re(CPh)Cp(CO)_2]^+$	317.7	832
Iron		
$Fe_3(CEt)(C_5H_2Me_2C_2H_3)(CO)_8$	345.6	1178
Cobalt		
$Co(CMe)(CO)_6As$	170.4	1767
$Co(CPh)(CO)_6As$	165.9	1767
$[Co_6C(CO)_{15}]^{2-}$	330.5	715
Phosphorus		
$P{\equiv}CH$	154 (54.0)	524

TABLE 2.7

^{13}C N.m.r. Data for Some Acyl, Thioacyl and Selenoacyl Derivatives

COMPOUND	$\delta(^{13}C)$/p.p.m., J/Hz	REFERENCE
Zirconium		
$Zr(\mathit{C}Op\text{-}tol)Cp_2p\text{-}tol$	300, 301*	1339
Chromium		
$[Cr(\mathit{C}O\text{-}2\text{-}thienyl)(CO)_5]^-$	276.2	39
$[Cr\{\mathit{C}OC(OMe)_2Ph\}(CO)_5]^-$	293.4	429, 430
$[Cr\{\mathit{C}OC(OEt)_2Ph\}(CO)_5]^-$	294.0	430
Molybdenum		
trans-$Mo(\mathit{C}OMe)Cp(CO)_2PPh_3$	263.2 (10.6)	1394
trans-$Mo(\mathit{C}OMe)Cp(CO)_2PMe_2Ph$	266.8 (12.5)	1394
trans-$Mo(\mathit{C}OMe)Cp(CO)_2CNCy$	267.6	1394
cis-$Mo(\mathit{C}OMe)Cp(CO)_2CNCy$	270.6	1394
Tungsten, $[^{183}W]$		
$W(\mathit{C}OMe)Cp(CO)_2PPh_3$	251.2 (11.6)	1394
$W(\mathit{C}OCH{=}PPh_3)(CO)_5$	200.0	1397
$W\{\mathit{C}OCHMeC(NMe_2)\}Cp(CO)_2$	242.7 [48.8]	881
$W\{\mathit{C}OCHMeC(NEt_2)\}Cp(CO)_2$	242.6 [51.3]	881
$W\{\mathit{C}OC(NEt_2){=}CH_2\}Cp(CO)_2$	248.7 [58.6]	881
$W\{\mathit{C}OC(NEt_2){=}CHCO_2Et\}Cp(CO)_2$	243.9 [53.7]	881
$W(\mathit{C}SMe)(CO)_4I$	252.7	674
$W(\mathit{C}SCOMe)(CO)_4I$	233.9	674
Manganese		
$Mn(\mathit{C}OMe)(CO)_5$	255.0	1744
$\{$*cis*-$Mn(\mathit{C}OMe)_2(CO)_4\}_3Al$	325.8	1744
$Mn\{\mathit{C}OC(NMe_2){=}CMe_2\}(CO)_4$	251.1	515
Rhenium		
$[Re_2(\mathit{C}HO)(CO)_9]^-$	264.2	1427
$Re(\mathit{C}OMe)(CO)_5$	244.0, 252.9	514, 717, 1744
cis-$Re\{(\mathit{C}OMe)_2H\}(CO)_4$	297.9	1744
cis-$Re\{(\mathit{C}OMe)_2BF_2\}(CO)_4$	314.0	1744
cis-$Re\{(\mathit{C}OMe)_2BCl_2\}(CO)_4$	314.2	1744
cis-$Re\{(\mathit{C}OMe)_2BPhCl\}(CO)_4$	310.5	1744
$\{$*cis*-$Re(CO)_4(\mathit{C}OMe)_2\}_3Al$	301	1744
cis-$Re\{(C^1OMe)(C^2OPr^i)H\}(CO)_4$	C^1 297.4, C^2 304.1	1744

*two isomers

COMPOUND	$\delta(^{13}C)$/p.p.m., *J*/Hz	REFERENCE
cis-Re{(C^1OMe)(C^2OPr^i)BF_2}$(CO)_4$	C^1 313.9, C^2 321.9	1744
cis-Re{(C^1OMe)(C^2OPr^i)BPhCl}$(CO)_4$	C^1 310.0, C^2 317.7	1744
{*cis*-Re$(CO)_4$(*C*OMe)(*C*OPr^i)}$_3$Al	300.8, 302.3, 303.0, 304.8, 306.0, 306.4, 306.6, 308.2	1744
[Re(*C*OMe)$GeMe_2(CO)_4]_2$	305.7	717
cis-Re{(*C*OMe)($COCH_2Ph$)BF_2}$(CO)_4$	314.4	1744
{*cis*-Re$(CO)_4$(*C*OMe)($COCH_2Ph$)}$_3$Al	299.3, 299.5, 299.8, 300.7, 301.1, 302.0, 303.6	1744
Re(*C*OPr^i)$(CO)_5$	254.0	1744
Re(*C*OCH_2Ph)$(CO)_5$	246.6	1744
Re(*C*OPh)$(CO)_5$	245.4	514, 717
fac-Re(*C*$OSiPh_3$)$(CO)_3$(dppe)	340.1 (10)	717, 1123
Iron		
$[Fe(CHO)(CO)_4]^-Na^+$	275.8	189
$[Fe(CHO)(CO)_4]^-[N(PPh_3)_2]^+$	260.1	189
$Fe_3H(\mu$-$COH)(CO)_{12}$	358.8	1441
$[Fe(CMeO)(CO)_4]^-$	277.2	1144
$[Fe(CEtO)(CO)_4]^-Na^+$	279.7	189
$[Fe(CEtO)(CO)_4]^-[N(PPh_3)_2]^+$	261.5	189
Fe(*C*MeO)$Cp(CO)_2$	254.4	60, 1394
Fe(*C*MeO)Cp(CO)PPh_3	277.0, 274.1 (21.4)	1276, 1394
Fe(*C*MeO)Cp(CO)(CNCy)	268.4	1394
$[Fe_2(CMeO)(\mu$-$PPh_2)_2(CO)_5]^-$	273.3	1144
$[Fe(CO_2Me)(CO)_4]^-$	207.5	657
$Fe(COC_7H_{10})(CO)_3$	242.2	816
Fe(COCH=CH*C*O)$(CO)_4$	245.1	956
$[Fe(COCH{=}CHCOMe)(CO)_3]^-$	246.1	741
$[Fe(COCH{=}CHCO_2Me)(CO)_3]^-$	242.7	741
Fe{COC(C_3H_5)=C(C_3H_5)*C*O}$(CO)_4$	239.8	945
Fe{COCMe=C(C_2Me)*C*O}$(CO)_4$	238.7, 242.4	847
$[Fe\{COC(CO_2Me)CH_2\}(CO)_3]^-$	246.0	741
$[Fe\{COC(CO_2Me)CHCO_2Me\}(CO)_3]^-$	235.7	741
Fe(S*C*=NMe_2)(S_2CNMe_2)$(CO)_4$	208.4 or 246.0	982
Osmium		
Os_3(*C*OMe)$H(CO)_{10}$	352.2	1475
Os_3(*C*OEt)$H(CO)_{10}$	349.7	1475
Cobalt		
Co{COC(CO_2Me)=C(OMe)O}Cp(CO)	205.6 or 214.1	1020
Co_3(*C*OPh)$(CO)_9$	207.3	1493
Palladium		
trans-Pd(*C*OCH_2Ph)Cl$(PEt_3)_2$	233.8	1532

COMPOUND	δ(^{13}C)/p.p.m., *J*/Hz	REFERENCE
trans-Pd(*C*OCH$_2$Ph)Br(PEt$_3$)$_2$	234.1	1532
[Pd(*C*OCH$_2$CH$_2$NEt$_2$)NHEt$_2$]$^+$	212	1125
Platinum, [^{195}*Pt*]		
Pt(*C*OMe)Me$_2$Cl(AsMe$_3$)$_2$	193.2 [851]	400
Pt(*C*OCOCPh=CH)(AsPh$_3$)$_2$	218.1	1176
Pt(*C*OCOCPh=CH)(PPh$_3$)$_2$	228.4 (90, 6)	1176
Pt(*C*OCOCPh=CH)(PEt$_2$Ph)$_2$	231.6 (108, 3)	1176
Pt(*C*OCOCPh=CH)(PEt$_3$)$_2$	235.9 (107, 7)	1176
Pt(*C*OCOCPh=CH)(PMePh$_2$)$_2$	231.5 (107, 9)	1176
Pt(*C*O$_2$Me)(PPh$_3$)$_2$Cl	168.5 [1346]	579
Pt(*C*O$_2$Me)(PPh$_3$)$_2$Br	169.3 [1345]	579
Pt(*C*O$_2$Me){C(CO$_2$Me)=C(CO$_2$Me)OMe}-(PPh$_3$)$_2$	192.1 (11.2)	1064
Pt(*C*O$_2$Me){C(CO$_2$Me)=C(CO$_2$Me)OH}-(PPh$_3$)$_2$	193.1 (11.0)	1064
Boron		
Me$_3$NBH$_2$*C*ONHEt	182.8	713
Silicon		
Si(*C*OMe)Me$_3$	254.1	649
Si(*C*OMe)Bu$^t{}_2$H	251.5	649
Si(*C*OMe)Bu$^t{}_2$F	252.4	649
Si(*C*OPh)Ph$_3$	230.7	717
Re(CO)$_3$(dppe)*C*OSiPh$_3$	340.1 (10)	717
Phosphorus, (^{31}*P*)		
P(*C*OPh)$_3$	205.5 (33.4)	1674
P(*C*OC$_6$H$_4$Me-4)$_3$	205.0 (33.1)	1674
P(*C*OC$_6$H$_4$Me-3)$_3$	205.8 (33.2)	1674
P(*C*OC$_{10}$H$_7$-1)$_3$	208.0 (34.3)	1674
P(*C*OC$_{10}$H$_7$-2)$_3$	205.5 (33.2)	1674
P(*C*OBu^t)$_3$	220.0 (49.5)	1247
P(*C*OBu^t)$_2$Bu^t	224.0 (60)	1777
P(*C*OBu^t)$_2$Ph	224.0 (48)	1777
P(*C*OBu^t)HPh	226.0 (48)	1777
P(*C*OBu^t)HBu^t	229.5 (50.0)	1777
P(*C*OBu^t)Me(SiMe$_3$)	229.0 (61.0)	1024, 1250
P(*C*OBu^t)CBu^t(OSiMe$_3$)	220.5 (74.0) or 227.0 (73.0)	1247
Arsenic		
As(*C*OBu^t)Ph(SiMe$_3$)	228.5	1024
Antimony		
Sb(*C*O$_2$Et)$_2$Ph$_3$	179.9	470

COMPOUND	$\delta(^{13}C)$/p.p.m., *J*/Hz	REFERENCE
Bismuth		
$Bi(\mathit{C}O_2Et)_2Ph_3$	176.6	470
Selenium		
$\overline{Se\mathit{C}OCH_2C_6H_4\text{-}2}$	204.9	781
$\overline{Se\mathit{C}OC(CHNMe_2)\mathit{C}_6H_4\text{-}2}$	186.7	781
$\overline{Se\mathit{C}SNHC_6H_4\text{-}2}$	193.5	1226, 1713
$Se\mathit{C}SeSCH{=}CHCH_2CHMeCH_2$ (ring)	221.1	1793
$Se\mathit{C}SeSeCH{=}CHCH_2CHMeCH_2$ (ring)	194.0	1793

TABLE 2.8

^{13}C Chemical Shift Data for Some Carbonyl Derivatives

COMPOUND	$\delta(^{13}C)$/p.p.m., *J*/Hz		REFERENCE
Vanadium, [^{51}V]			
$[V(CO)_6]^-$	225.7 [116, 146]		93, 388
Chromium			
$Cr(CO)_6$	211.2, 212.1, 212,3, 212.5, 214.6		15, 19, 69, 117, 388, 395
	cis	*trans*	
$Cr(CO)_5\{Me(MeO)C{=}CNC_6H_{11}\}$	214.8	221.2	1368
$[Cr(CO)_5C(OMe)MePh]^-$	222.6	227.4	1036
$[Cr(CO)_5C(OMe)Ph_2]^-$	219.7	225.3	1036, 1065
$Cr(CO)_5C(OMe)PhN(CH_2CH_2)_3N$	223.8	228.5	172, 371
$Cr(CO)_5C(OMe)PhNC_7H_{13}$	224.3	228.9	371
$[Cr(CO)_5COC_6H_4Me\text{-}4]^-$	224.1		941
$[Cr(CO)_5COC(OMe)_2Ph]^-$	222.8	228.1	429, 430
$[Cr(CO)_5COC(OEt)_2Ph]^-$	223.5	228.6	430
$Cr(CO)_5CH(NMe_2)$	217.5	224.0	1372
$Cr(CO)_5CPh(C_6H_4R\text{-}4)$	215.8 to 216.8	236.1 to 236.8	1065
$Cr(CO)_5CPh$(2-furanyl)	217.7	234.6	1065
$Cr(CO)_5CPh$(2-thienyl)	218.0	234.1	1065
$Cr(CO)_5CPh$(2-*N*-Me-pyrryl)	218.9	230.0	1065
$Cr(CO)_5C(\text{2-furanyl})_2$	218.6	233.5	1065
$Cr(CO)_5C(\text{2-thienyl})_2$	218.0	232.6	1065
$Cr(CO)_5C$(2-furanyl)(2-thienyl)	218.3	232.8	1065
$Cr(CO)_5CR^1(NR^2R^3)$	217 to 220	223 to 226	35, 39, 175, 248, 369, 436, 811, 1368
$Cr(CO)_5\{C(MeOC{=}CH_2)(NHC_6H_{11})\}$	218.3	235.2	1368
$Cr(CO)_5\overline{CCH_2C(=NCy)N}Cy$	222.9	223.3	369
$Cr(CO)_5CR^1(OR^2)$	215.0 to 218.7	223 to 227	35, 39, 60, 89, 171, 172, 248, 368, 371, 429, 430, 894, 941, 1052, 1273
$Cr(CO)_5CMe(O_2CR)$	216.1 to 216.5	227.7 to 228.5	1049
$Cr(CO)_5C(CH{=}PMe_3)OSiMe_3$	211.6	222.0	1359

COMPOUND	$\delta(^{13}C)$/p.p.m., J/Hz		REFERENCE
	cis	*trans*	
$Cr(\mathit{C}O)_5C(NMe_2)SiPh_3$	217.8	225.3	907, 1042
$Cr(\mathit{C}O)_5C(OMe)SiPh_3$	216.7	227.3	1042
$Cr(\mathit{C}O)_5C(OEt)SiPh_3$	216.8	227.7	907, 1042
$Cr(\mathit{C}O)_5CNPr^i$	216.3	218.2	39
$Cr(\mathit{C}O)_5CNCy$	214.4	216.7	574
$Cr(\mathit{C}O)_5CN\mathit{p}$-tol	216.5	218.7	39
$Cr(\mathit{C}O)_5CNSiMe_3$	214.5	216.2	967
$Cr(\mathit{C}O)_5CNGeMe_3$	215.4	217.6	967
$[Cr(\mathit{C}O)_5CNEt_2]^+$	207.7, 207.9	201.9	896, 1374
$Cr(\mathit{C}O)_5CNMeCH_2CH_2NMe$	218.7	221.1	1223
$Cr(\mathit{C}O)_5CNEtCH_2CH_2NEt$	218.4	222.5	1223
$Cr(\mathit{C}O)_5C(NCO)(NEt_2)$	216.9	222.9	962
$Cr(\mathit{C}O)_5C(NEt_2)OEt$	217.9	222.6	1376
$Cr(\mathit{C}O)_5C(NEt_2)F$	216.5 (J_{FC} = 9.8)	221.5	896
$Cr(\mathit{C}O)_5C(NEt_2)Cl$	216.5	223.1	652
$Cr(\mathit{C}O)_5C(NEt_2)Br$	217.2	223.5	962
$Cr(\mathit{C}O)_5C(OEt)_2$	218.2	223.0	840
$Cr(\mathit{C}O)_5COCMe_2CMe_2O$	216.5	223.2	338
$Cr(\mathit{C}O)_5CS$	211.4	211.4	682
	212.4	209.4	1745
$Cr(\mathit{C}O)_5(CSe)$	211.7	208.1	1745
$Cr(\mathit{C}O)_5$(N-ligand)	213.7 to 214.6	219.7 to 221.2	442, 640, 689, 1348, 1356
$Cr(\mathit{C}O)_5NSNC_6H_2Me_2$	197.5	201.1	961
$Cr(\mathit{C}O)_5PH_3$	215.2 (13.5)	219.7 (7)	442
$Cr(\mathit{C}O)_5PR_3$	216.5 to 217.8 (13 to 14.6)	221.2 to 222.5 (7 to 8)	388, 395, 442, 894
$[\{Cr(\mathit{C}O)_5PHCB_9H_9\text{-}1,7\}_2Fe]^-$	217.9 (12)	224.3	7
$Cr(\mathit{C}O)_5P(OMe)_3$	215.2 (21)	219.2 (4)	442
$Cr(\mathit{C}O)_5P(OPh)_3$	213.8 (20), 213.9	218.2 (0), 217.6	388, 395, 442
$Cr(\mathit{C}O)_5AsPh_3$	216.6	222.1	442
$Cr(\mathit{C}O)_5SbPh_3$	217.0	222.3	442
$Cr(\mathit{C}O)_5SR_2$	214.9 to 215.4	222.0 to 222.4	1352, 1353
$Cr(\mathit{C}O)_5S(O)R_2$	213.8	220.8, 221.2	1353
$Cr(\mathit{C}O)_5S{=}CSCH_2CH_2S$	214.8	227.5	829
$Cr(\mathit{C}O)_5S{=}CSCH{=}CHS$	215.3	216.8	829
$Cr(\mathit{C}O)_5SCMe_2$	214.5	223.0	995
$[Cr(\mathit{C}O)_5Cl]^-$	216.1, 217.6, 216.7	223.6, 223.8, 225.3	175, 442, 1394
$[Cr(\mathit{C}O)_5Br]^-$	216.1 to 217.7	224.7 to 225.7	175, 1394
$[Cr(\mathit{C}O)_5I]^-$	216.4	226.7	1394

COMPOUND	$\delta(^{13}C)$/p.p.m., J/Hz		REFERENCE
	cis	*trans*	
$Cr(CO)_4$(norbornadiene)	226.8	234.5	114
	225.5	228.6	117, 1115
$Cr(CO)_4(CPh)Re(CO)_5$	228.8		1047
cis-$Cr(CO)_4(CNCy)_2$	218.1	221.1	574
cis-$Cr(CO)_4(\overline{CNRCH_2CH_2NR})_2$	222.2, 222.6	228.7, 230.0	1223
cis-$Cr(CO)_4(\overline{CNMeCH_2CH_2NMe})$CMe-(OMe)	222.9	227.6, 233.0	1223
cis-$Cr(CO)_4\{C(SMe)_2\}_2$	215.9	277.0	202
$Cr(CO)_4(CH_2{=}CHCH_2C_3H_4N_2)$	223.5, 223.6, 224.3, 230.7		1375
$\overline{Cr(CO)_4C(NEt_2)C\{=C(OMe)Ph\}NEt_2}$	220.7	230.5, 232.7	810
trans-$[Cr(CO)_4CMe(PMe_3)]^+$	209.4 (17.1)		440, 891
cis-$Cr(CO)_4\{C(OMe)Me\}(PMe_3)$	222.3 (17.1)	226.4 to 228.4, 231.1 to 231.8	894
trans-$Cr(CO)_4\{C(OMe)Me\}(PMe_3)$	222.9 (12.2)		440, 891, 894
trans-$[Cr(CO)_4CMe(AsMe_3)]^+$	209.4		891
cis-$Cr(CO)_4\{C(OMe)Me\}(AsMe_3)$	222.2	228.1, 231.1	892, 894
trans-$Cr(CO)_4\{C(OMe)Me\}(AsMe_3)$	223.1		894
trans-$[Cr(CO)_4CMe(SbMe_3)]^+$	210.8		891
cis-$Cr(CO)_4\{C(OMe)Me\}(SbMe_3)$	223.1	229.6, 231.6	894
trans-$Cr(CO)_4\{C(OMe)Me\}(SbMe_3)$	223.5		894
$\overline{Cr(CO)_4C(OEt)C(OR)CRS}$	216 to 217	230 to 232	1365
$Cr(CO)_4(CR)X$, X = Cl, Br, I	206.6 to 209.0		893, 1067, 1363
$Cr(CO)_4(CNEt_2)X$, X = Cl, Br, I	211.2 to 212.9		437, 962, 1376
$Cr(CO)_4(CMePh)Br$	207.7		435
$Cr(CO)_4(CRPMe_3)X$, X = Cl, Br, I	218.7 to 221.7 (4.9)		435, 1363
$[Cr(CO)_4\{HC(NR)_2\}]^-$	217.9 to 220.4	231.3 to 231.9	1366
$[Cr(CO)_4\{HC(NMe)N(CO)Me\}]^-$	222.3, 232.4, 232,9, 245.4		1366
$Cr(CO)_4(\overline{NMeCH_2CH_2NMeC}{=}{-}\overline{CNMeCH_2CH_2NMe})$	213.9, 214.5		1223
$Cr(CO)_4$(glyoxalbis-t-butyl-amine	211.9	230.4	1367
$\overline{Cr(CO)_4\{N(C_6H_4Me_2\text{-}3,5)CHCHN}$-$(C_6H_4Me_2\text{-}3,5)\}$		225.8	448

COMPOUND	$\delta(^{13}C)$/p.p.m., *J*/Hz		REFERENCE
	cis	*trans*	
cis-$Cr(\mathit{C}O)_4(PH_3)_2$	210.3 (13.4)	218.5 (14)	442
trans-$Cr(\mathit{C}O)_4(PBu^n{}_3)_2$	223.8 (14)		442
trans-$Cr(\mathit{C}O)_4\{P(OMe)_3\}_2$	218.9 (21)		442
trans-$Cr(\mathit{C}O)_4\{P(OPh)_3\}_2$	216.7 (20)		388, 395, 442
$Cr(\mathit{C}O)_4SMeC(SMe)C(SMe)SMe$	227.0	215.9	439
$Cr(\mathit{C}O)_3$(arene)	229.8 to 235.5		60, 114, 117, 290, 370, 388, 395, 432, 434, 787, 927, 938, 952, 1088, 1294, 1305 1575, 1746
$Cr(\mathit{C}O)_3(C_6Me_6)$	236.3		117
$[Cr(\mathit{C}O)_3(C_6H_5CMe_2)]^+$	228.4		927
$[Cr(\mathit{C}O)_3(4\text{-}MeC_6H_4CHMe)]^+$	226, 226.5, 227.3		1294
$[Cr(\mathit{C}O)_3(4\text{-}MeC_6H_4CH\mathit{p}\text{-}tol)]^+$	226.5, 227.3, 229.4		1294
$[Cr(\mathit{C}O)_3(4\text{-}MeC_6H_4)_2CH]^+$	230.8		1361
$[Cr(\mathit{C}O)_3H(C_6H_5OMe)]^+$	220.4, 224.3		787
$Cr(\mathit{C}O)_3(C_7H_8)$	230.1*, 241.3§		114, 117, 173, 447, 927, 1765
$Cr(\mathit{C}O)_3(C_8H_8)$	226.5*, 245.0§		447
	225.8*, 244.4§		307
$Cr(\mathit{C}O)_3(C_8H_{10})$	228.8*, 241.4§		117, 447
$Cr(\mathit{C}O)_3(C_6H_6NCO_2Et)$	226.0*, 239.8§		1288
$Cr(\mathit{C}O)_3(C_{11}H_{14})$	233.0*, 236.4§		1351
$Cr(\mathit{C}O)_3(C_{11}H_{12}Me_2)$	232.7*, 236.2§		1351
$[Cr(\mathit{C}O)_3Cp]^-$	246.8		295, 388, 395
$Cr(\mathit{C}O)_3(C_5H_4SMe_2)$	241.9		1165
$Cr(\mathit{C}O)_3(C_5H_4PPh_3)$	241.0		1165
$Cr(\mathit{C}O)_3(C_2BN_2H_2Bu^t{}_2)$	234.6		1031
$Cr(\mathit{C}O)_3$(norbornadiene)(PPh_3)	235.8 (20.6), 240.0 (7.4)		433
$[Cr(\mathit{C}O)_3(CMe)(PMe_3)(OH_2)]^+$	211.5 (19.5)*, 212.9 (7.3)§		904
$Cr(\mathit{C}O)_3(CMe)(PMe_3)BF_4$	211.6 (19.5)*, 212.9 (7.3)§		904
$Cr(\mathit{C}O)_3(CMe)(PMe_3)Cl$	215.1 (22.0)*, 217.3§		892, 904
$Cr(\mathit{C}O)_3(CMe)(PMe_3)Br$	214.6 (19.5)*, 217.0§		892
$Cr(\mathit{C}O)_3(CPh)(PPh_3)Br$	213.2 (19.1)*, 217.8 (12.8)§		1061
$Cr(\mathit{C}O)_3(CPh)\{P(OPh)_3\}Br$	211.0 (26.9)*, 214.4 (14.6)§		1061
$Cr(\mathit{C}O)_3(CMe)(PMe_3)I$	214.1 (17.1)*, 216.0§		892
$Cr(\mathit{C}O)_3(CMe)(AsMe_3)Br$	214.7*, 218.1§		892

*intensity 2, §intensity 1

COMPOUND	δ(^{13}C)/p.p.m., *J*/Hz		REFERENCE
$Cr(CO)_3(CPh)(AsPh_3)Br$	213.2*, 218.4§		1061
$Cr(CO)_3(CMe)(SbMe_3)Br$	216.1*, 217.8§		890
$Cr(CO)_3(CPh)(SbPh_3)Br$	215.6*, 217.0§		1061
$Cr(CO)_2(C_{10}H_{10})$	251.1		432
$Cr(CO)_2(C_6H_3Me_2CH_2OCH_2CH{=}CH_2$	246.1, 248.1		1354
$[Cr(CO)_2(arene)CPh]^+$	224.9 to 226.6		906, 935
$Cr(CO)_2(arene)CPh(NR_2)$	238.5 to 240.0		935
$[Cr(CO)_2(arene)CPh(PMe_3)]^+$	248.2 to 250.5 (4.9, 7.3)		906, 1048
$Cr(CO)_2(arene)CPh(OMe)$	235.5 to 236.8		935
$Cr(CO)_2(CSe)(C_6H_6)$	229.0		1745
$Cr(CO)_2(NO)Cp_2$	237.6		1360
$Cr(CO)_2(NS)Cp_2$	239.4		1360
$Cr(CO)_2(arene)PR_3$	240.8 to 242.8 (20 to 23)		938, 1099
$[Cr(CO)_2(mesitylene)$-$H\{PMePh(CH_2Ph)\}]^+$	229.7 to 236.2 (42)†		1099
$Cr(CO)_2(arene)AsPh_3$	241.2, 242.4		938
$Cr(CO)_2(C_7H_8)PMe_3$	234.8 (19.5)		1765
$Cr(CO)_2(C_7H_8)P(OMe)_3$	235.6 (29.3)		1765
$Cr(CO)_2(C_7H_8)AsMe_3$	235.5		1765
$Cr(CO)_2(CPh)(CNBu^t)_2Br$	220.6		1061
$Cr(CO)_2(CPh)(py)_2Br$	230.4		1061
$Cr(CO)_2(CPh)(o\text{-phen})Br$	232.6		1061
$Cr(CO)_2\{(MeO)_2PCH_2CH_2P$-$(OMe)_2\}_2$	233.0		1371
$Cr(CO)(C_7H_8)\{P(OMe)_3\}_2$	239.0 (36.6)		1765
$Cr_2(CO)Cp_2C_4H_2Ph_2$	316.0		1349
<u>*Molybdenum*, [^{95}*Mo*]</u>			
$Mo(CO)_6$	202.0, 204.1 [68]		15, 69, 117
	cis	*trans*	
$Mo(CO)_5CNMeCH_2CH_2NMe$	207.4	212.5	439
$Mo(CO)_5C(NMe_2)OSiPh_3$	205.0	215.8	1042
$Mo(CO)_5C(OMe)SiPh_3$	205.8	217.3	1042
$Mo(CO)_5C(OEt)SiPh_3$	206.0	217.6	1042
$Mo(CO)_5CNCy$	203.8	207.9	574
$Mo(CO)_5CNBu^t$	203.6	206.8	574, 767
$Mo(CO)_5CNSiMe_3$	203.4	206.0	767
$Mo(CO)_5CNGeMe_3$	204.0	207.0	767
$Mo(CO)_5NH_2Cy$	204.1		442
$Mo(CO)_5(piperidine)$	203.9	206.9	442
$Mo(CO)_5PBu^n{}_3$	206.3 (9)	209.6 (21)	442
$Mo(CO)_5PPh_3$	205.7 (9)	210.2 (23)	442
	206.5	211.0	15, 68
$Mo(CO)_5PPh_2CH_2CH_2NMe_2$	199.4 (8.8)		359
$[Mo(CO)_5(PCHB_9H_{10}$-7,9$)]^-$	197.6 (18)	200.2 (26)	7, 155
$Mo(CO)_5P(OMe)_3$	204.3 (14)	208.2 (40)	442
	206.4 (7.3)	209.2 (18.3)	15
$Mo(CO)_5P(OEt)_3$	206.8 (6.7)	208.7 (18.9)	15

*intensity 2, §intensity 1, †several species present

COMPOUND	$\delta(^{13}C)$/p.p.m., *J*/Hz		REFERENCE
	cis	*trans*	
$Mo(CO)_5P(OPr^i)_3$	206.3 (14.6)	209.7 (12.2)	15, 60
$Mo(CO)_5P(OPh)_3$	203.1 (13)	206.9 (46)	442
$Mo(CO)_5PCl_3$	200.7 (11)	206.0 (66)	442
$Mo(CO)_5AsPh_3$	205.2	210.2	442
$Mo(CO)_5SbPh_3$	205.6	209.5	442
$[Mo(CO)_5Cl]^-$	205.3	213.8	1394
$[Mo(CO)_5Br]^-$	204.8	214.0	1394
$[Mo(CO)_5I]^-$	202.0	214.4	1394
$[Mo(CO)_4(BH_4)]^-$	209.8	226.2	1145
$Mo(CO)_4(CPh)Re(CO)_5$	215.4		1047
$Mo(CO)_4$(norbornadiene)	213.8 to 215.4	217.7 to 219.0	93, 114, 117, 1115
cis-$Mo(CO)_4(\overline{CNMeCH_2CH_2N}Me)$	211.5	219.4	439
cis-$Mo(CO)_4(CNCy)_2$	206.5	210.8	574
cis-$Mo(CO)_4(CNBu^t)_2$	206.3	210.6	574
$Mo(CO)_4$(1-allylpyrazole)	209.6, 210.3	217.9, 218.6	1375
$Mo(CO)_4(Cp\text{-tol})Cl$	199.3		1363
$Mo(CO)_4(CC_5H_4FeCp)Br$	198.9		1369
$Mo(CO)_4(NR_3)_2$	205.8 to 207.8	221.1 to 223.0	439, 448, 971, 1223 1367
$[Mo(CO)_4\{HC(NR)_2\}]^-$	208.1 to 209.8	225.6 to 226.3	1366
$[Mo(CO)_4\{HC(NMe)NCOMe\}]^-$	211.8, 224.9, 226.3, 239.9		1366
trans-$Mo(CO)_4(PBu^n_3)_2$	212.5 (9)		442
cis-$Mo(CO)_4$(dppe)	208.3, 210.6 (8.4)	217.4, 218.5 (24.0)	15, 1140
cis-$Mo(CO)_4\{(Ph_2PCH_2CH_2)_2NEt\}$	210.2 (9.5)		359
trans-$Mo(CO)_4\{P(OMe)_3\}_2$	210.3 (13.4)		15
cis-$Mo(CO)_4\{1,2\text{-}(MePhAs)_2\text{-}C_6H_4\}$racemic	208.6	217.7	396
meso	208.1, 208.8	217.9	396
$Mo_2(CO)_6C_{10}H_8$	218.1, 220.8, 224.4		671
$Mo_2(CO)_6C_{10}H_5Me_2Pr^i$	219.1, 222.4, 225.6		677
$Mo_2(CO)_6(PEt_3)C_{10}H_5Me_2Pr^i$	225.0, 230.2 (20.6), 235.0 (17.7)*		677
	227.2, 229.1 (19.3), 232.6 (17.7)*		677
$[Mo(CO)_3Cp]_2$	218.6		93
$Mo(CO)_3CpW(CO)_4CPh$	201.4	225.2	1047
$[Mo(CO)_3(EtO_2CCH{=}CHCO_2Et)_2]_2$	192, 203, 223		848
$Mo(CO)_3$(*m*-xylene)	223.1		60
$Mo(CO)_3$(mesitylene)	223.7		114, 117
$Mo(CO)_3$(durene)	224.4		114, 117
$Mo(CO)_3(C_6Me_6)$	225.9		117
$[Mo(CO)_3(C_7H_7)]^+$	206.3		927
$Mo(CO)_3C_7H_8$	217.9§, 229.2†		114, 117, 173, 445, 447, 927

*two isomers, §intensity 2, †intensity 1

COMPOUND	$\delta(^{13}C)$/p.p.m., *J*/Hz	REFERENCE
Mo(*C*O)$_3$C$_8$H$_{10}$	216.6*, 228.4§	447
Mo(*C*O)$_3$C$_8$H$_8$	213.5*; 214.0*, 228.2§; 228.9§	209, 447
Mo(*C*O)$_3$(C$_6$H$_6$NCO$_2$Et)	214.4, 226.0	1288
Mo(*C*O)$_3$C$_8$H$_8$O	218.7, 226.3	1461
Mo(*C*O)$_3$CpMe	226.1*; 227.2*, 239.2§; 240.3§	1389, 1396
Mo(*C*O)$_3$CpEt	227.9*, 239.9§	1396
Mo(*C*O)$_3$(C$_5$H$_4$Me)Me	227.7*, 241.0§	1396
Mo(*C*O)$_3$(C$_5$H$_4$CH$_2$CH$_2$)	228.9*, 236.3§	1396
Mo(*C*O)$_3$(C$_5$H$_4$CH$_2$CH$_2$CH$_2$)	227.9*, 239.9§	1396
Mo(*C*O)$_3$CpCH=C(CN)$_2$	223.4*, 232.7§	280
Mo(*C*O)$_3$CpCCl=C(CN)$_2$	227.8, 234.2	441
Mo(CO)$_3$Cp{C=CPhC(CF$_3$)$_2$OCH$_2$}	218.6	1312
Mo(*C*O)$_3$C$_5$H$_4$PPh$_3$	230.4	1165
Mo(*C*O)$_3$C$_5$H$_4$SMe$_2$	231.5	1165
Mo(*C*O)$_3$(C$_5$H$_5$SOMe)	224.9	1746
fac-Mo(*C*O)$_3$(CNCy)$_3$	213.8	574
[Mo(*C*O)$_3$Cp{CH(NMe$_2$)}]$^+$	225.0*, 231.4§	1372
Mo(*C*O)$_3$CpPNMeCH$_2$CH$_2$NMe	215.2	1381
Mo(*C*O)$_3$CpCl	224.5; 225.6*, 241.3§	1394
Mo(*C*O)$_3$CpBr	222.8*, 239.0§	1394
Mo(*C*O)$_3$CpI	220.8*, 236.1§	1394
Mo(*C*O)$_3${C(PMe$_3$)*p*-tol}-(PMe$_3$)Cl	201.9 (7.3), 221.8 (17.1)	1363
Mo$_2$(*C*O)$_6$(μ-NHPPh$_3$)$_3$	219.6	297
[Mo(*C*O)$_3$(py){HC(N*p*-tol)$_2$}]$^-$	229.1, 231.8	1366
[Mo(*C*O)$_3$(PPh$_3$){HC-(N*p*-tol)$_2$}]$^-$	222.2 (48), 229.9 (8)	1366
Mo$_2$(*C*O)$_4$Cp$_2$(H$_2$C=C=CH$_2$)	233.5, 237.2	1112, 1392
Mo$_2$(CO)$_4$Cp$_2$(MeHC=C=CH$_2$)	233.0, 233.7, 237.0, 239.6	1392
Mo$_2$(CO)$_4$Cp$_2$(MeHC=C=CHMe)	232.9, 239.3	1392
Mo$_2$(CO)$_4$Cp$_2$(MeC≡CMe)	230.8	1383
Mo$_2$(CO)$_4$Cp$_2$(EtC≡CEt)	224.6, 227.7, 230.8, 232.1, 233.2, 234.7	862
Mo$_2$(*C*O)$_4$Cp$_2$(HC≡CCF$_3$)	226.7, 230.0	1383
Mo$_2$(*C*O)$_4$Cp$_2$(MeO$_2$CC≡CCO$_2$Me)	231.1	1383
Mo$_2$(*C*O)$_4$Cp$_2$(NCNMe$_2$)	237.3, 241.9, 247.1, 248.3	1379
trans-Mo(*C*O)$_2$Cp(COMe)(CNCy)	233.9	1394
cis-Mo(*C*O)$_2$Cp(COMe)(CNCy)	238.7, 246.0	1394
[Mo(*C*O)$_2$CpCPhNCPhNMe]$^+$	280.1	1377
Mo(*C*O)$_2$CpCPhNMeCPHNMe	250.9, 260.4	1377
Mo(*C*O)$_2$Cp(MeCNMe)	250.0, 252.6	1385
Mo(*C*O)$_2$(C$_5$H$_4$Me)(MeCNMe)	250.9, 253.0	1385
Mo(*C*O)$_2$Cp{C$_3$(CN)$_2$NH$_2$}	224.3	441
Mo(*C*O)$_2$Cp{C$_3$H(CN)(OH)NH}	249.2	441

*intensity 2, §intensity 1

COMPOUND	$\delta(^{13}C)$/p.p.m., J/Hz	REFERENCE
$Mo(\mathit{CO})_2Cp\{C_3(CO_2Me)(CN)(OR)NH\}$	247.7, 247.5	441
$Mo(\mathit{CO})_2Cp\{Me_2NCC(CN)_2\}$	233.3	441
$Mo(\mathit{CO})_2Cp\{(CH_2)_5NCC(CN)_2\}$	231.1	441
$Mo(\mathit{CO})_2Cp(R_2C{=}NO)$	222.6 to 223.8, 228.8 to 229.6	1129
$Mo(\mathit{CO})_2(C_7H_8)PR_3$	224.1 to 224.8 (10 to 21)	445
$Mo(\mathit{CO})_2(C_7H_8)P(OPh)_3$	213.4 (17)	445
trans-$Mo(\mathit{CO})_2Cp(PPh_3)Me$	235.7 (23.0)	1394
trans-$Mo(\mathit{CO})_2Co(PPhMe_2)COMe$	236.9 (23.6)	1394
trans-$Mo(\mathit{CO})_2Cp(PPh_3)COMe$	238.1 (23.6)	1394
$\overline{Mo(\mathit{CO})_2CpCH{=}CHCMeC{=}O}$	248.4	749
$Mo(\mathit{CO})_2(C_7H_7)Cl$	216.6	1394
$Mo(\mathit{CO})_2(C_7H_7)Br$	214.9	1394
$Mo(\mathit{CO})_2(C_7H_7)I$	212.3	1394
$Mo(\mathit{CO})_2(NO)Cp$	227.3	93
$Mo(\mathit{CO})_2Cp(RNCHNR)$	260.1 to 266.6	1079
$Mo(\mathit{CO})_2Cp(PhN_3Ph)$	253.6	1121
$Mo(\mathit{CO})_2Cp(p\text{-}tolN_3p\text{-}tol)$	237.5	654
$Mo(\mathit{CO})_2Cp\{p\text{-}tolN_3C_6H_3(CF_3)_2\text{-}3,5\}$	239.9	654
$Mo(\mathit{CO})_2Cp\{N_3[C_6H_3(CF_3)_2\text{-}3,5]_2\}$	242.1	654
$Mo(\mathit{CO})_2Cp(Pr^iN_3C_6H_4Cl\text{-}4)$	254.1, 257.0	654
$Mo(\mathit{CO}_2)(C_5H_4Me)\{p\text{-}tolN_3C_6H_3\text{-}(CF_3)_2\text{-}3,5\}$	253.5	654
$Mo(\mathit{CO})_2Cp\{HC(NMe)N(COMe)\}$	247.4, 251.9	1079
$Mo(\mathit{CO})_2Cp(Me_2CH_2Ph)$	224.2, 232.6	1121
$[Mo(\mathit{CO})_2Cp(pzH)_2]^+$	249.9	1152
$[Mo(\mathit{CO})_2Cp(imH)_2]^+$	257.3	1152
$Mo(\mathit{CO})_2Cp(MeCNPh)\{P(OMe)_3\}$	237.3 (38.5)	1385
$Mo(\mathit{CO})_2Cp(py)Cl$	254.1, 265.2	1152
$Mo(\mathit{CO})_2Cp(pzH)Cl$	254.1, 262.7	1152
$Mo(\mathit{CO})_2Cp(ImH)Cl$	252.9, 259.0	1152
$Mo(\mathit{CO})_2(C_4H_6)(PBu^n_3)_2$	237.9	998
$Mo(\mathit{CO})_2(\eta^4\text{-}C_8H_8)(PBu^n_3)_2$	247.2	998
cis-$Mo(\mathit{CO})_2Cp(PPh_3)Cl$	242.8 (5.4), 255.6 (28.6)	1394
$Mo(\mathit{CO})_2Cp(PF_2NHMe)Cl$	232.5, 248.7	1390
$Mo(\mathit{CO})_2Cp\{(PF_2)_2NMe\}Cl$	228.6, 246.6 (41)	1390
cis-$Mo(\mathit{CO})_2Cp(PPh_3)Br$	240.6 (5.8), 253.6 (28.2)	1394
cis-$Mo(\mathit{CO})_2Cp(PPh_3)I$	236.8 (5.0), 250.3 (28.6)	444, 1394
trans-$Mo(\mathit{CO})_2Cp(PPh_3)I$	232.4 (27.4)	444, 1394
$[Mo(\mathit{CO})_2(\eta^3\text{-}C_3H_5)(bipy)(NH_3)]^+$	226.4	947
$[Mo(\mathit{CO})_2(\eta^3\text{-}C_3H_5)(bipy)py]^+$	225.6	947
$[Mo(\mathit{CO})_2(\eta^3\text{-}2\text{-}MeC_3H_4)(bipy)\text{-}py]^+$	226.0	947
$[Mo(\mathit{CO})_2(\eta^3\text{-}2\text{-}MeC_3H_4)(bipy)\text{-}P(OPh)_3]^+$	222.3	947
$[Mo(\mathit{CO})_2(\eta^3\text{-}2\text{-}MeC_3H_4)(bipy)\text{-}(OCMe_2)]^+$	226.4	947
$Mo(\mathit{CO})_2(\eta^3\text{-}C_3H_5)py(acac)$	227.9, 229*	1386

*two isomers

COMPOUND	$\delta(^{13}C)$/p.p.m., *J*/Hz		REFERENCE
Mo(*C*O)$_2${(3,5-Me$_2$C$_3$HN$_2$)$_3$-BH}(NO)	225.3		701
Mo$_2$(*C*O)$_2$Cp$_2$(HC≡CH)	229.7, 233.3		984
Mo(*C*O)(NO)Cp{(PF$_2$)$_2$NMe}	230.3 (12)		1390
trans-Mo(*C*O)Cp(PPh$_3$)I	267.0 (27.2)		1394
Tungsten, [^{183}*W*]			
W(*C*O)$_6$	191.0 [124.5], 191.9 [126], 192.1, 192.4, 193.7		15, 39, 60, 69, 117, 933, 1398
	cis	*trans*	
W(*C*O)$_5$HCl	190.1	194.7	1056
W(*C*O)$_5$HBr	188.3	192.6	1056
W(*C*O)$_5$HI	185.8	189.9	1056
W(*C*O)$_5$CS	192.4	189.3	682, 1398, 1745
W(*C*O)$_5${Me(MeO)C=CNC$_6$H$_{11}$}	199.2	202.4	1368
W(*C*O)$_5$CAr^{1}Ar^2	198.1 to 198.4 [127.0]	214.5 to 215.3 [102.5]	450, 933, 1065, 1107
W(*C*O)$_5$CR^{1}R^2	198.0 to 200.0	209.3 to 211.9	1065
(R^1 = Ph or R^2, R^2 = 2-furanyl, 2-thienyl, 2-N-Me-pyryl)	198.6 to 200.0 [127.0]	203.4 to 204.3 [129.4]	89, 175
W(*C*O)$_5$CR(NR$^1_2$)	198.6 to 200.0 [127.0]	203.4 to 204.3 [129.4]	89, 175
W(*C*O)$_5$CR(OR)	197.1 to 200.3 [127.0 to 125.0]	203.4 to 205.5 [117.2 to 127.0]	15, 39, 89, 178, 179, 248, 449, 933, 1107, 1411
W(*C*O)$_5$CMe(OSiMe$_3$)	199.3	206.9	1052
W(*C*O)$_5$C(*p*-tol)(O$_2$CMe)	197.5	198.3	1049
W(*C*O)$_5$C(2-furanyl)-(OSiMe$_3$)	199.5	206.0	1052
W(*C*O)$_5$C(*p*-tol)OW(CO)$_4$-Cptol	199.7	206.3	941
W(*C*O)$_5$CMeOW(CO)$_4$CMe	194.4	200.3	941
W(*C*O)$_5$CMe(SR)	196.8 to 198.0	207.4 to 207.8	89, 178, 353, 438
W(*C*O)$_5$CMe(SeMe)	197.9	205.0	178
W(*C*O)$_5$C(NMe$_2$)SiPh$_3$	199.3	204.8	1042
W(*C*O)$_5$C(OR)SiPh$_3$	198.0, 198.3 [127]	207.4, 207.7	1042
W(*C*O)$_5$C(NR$_2$)$_2$	197.5 to 198.7 [124.5, 126.7]	200.7 to 201.7 [131.8]	933, 1210
W(*C*O)$_5$C(OEt)NEt$_2$	198.4	201.3	361, 1376
W(*C*O)$_5$CNR	193.8 to 194.3 [128]	195.9 to 196.5 [137]	574, 718
W(*C*O)$_5$CNGeMe$_3$	194.8	196.9	967
W(*C*O)$_5$C(OMe)MeR	203.6	207.4	1036

COMPOUND	δ(^{13}C)/p.p.m., *J*/Hz		REFERENCE
	cis	*trans*	
$W(\mathit{C}O)_5C(PMe_3)R^1R^2$	200.6 to 201.5	202.1 to 203.6	450, 1062
$W(\mathit{C}O)_5C(O)CH_2PPh_3$	205.9 [136]	209.0	1397
$[W(\mathit{C}O)_5C(O)\mathit{p}\text{-tol}]^-$	204.6	208.5	941
$W(\mathit{C}O)_5$(nitrogen ligand)	197.9 to 199.5 [132]	201.4 to 202.5	15, 69, 180, 442, 640, 662, 1398
$W(\mathit{C}O)_5(NCMe)$	196.1	199.3	967
$W(\mathit{C}O)_5(NCSR)$	196.8	200.2	689
$W(\mathit{C}O)_5(N_2SC_6H_2Me_2)$	214.3		961
$W(\mathit{C}O)_5(PR_3)$	196.4 to 198.6 [124 to 129] (6, 7)	197.9 to 200.4 [140, 142] (19 to 22)	15, 39, 69, 117, 442, 718, 1398
$W(\mathit{C}O)_5\{P(OR)_3\}$	193.7 to 197.2 [123, 125.1] (9 to 11.6)	196.2 to 199.6 [135 to 139] (36 to 45.4)	15, 69, 117, 442, 1398
$W(\mathit{C}O)_5PPh_2NHR$	194.9 to 196.9 (7.5)	199.4 (25.5)	1168
$W(\mathit{C}O)_5AsPh_3$	197.5 [124]	199.7	15, 69
$W(\mathit{C}O)_5SbPh_3$	197.0	199.1	69
$W(\mathit{C}O)_5BiPh_3$	197.8	198.3	69
$W(\mathit{C}O)_5SR_2$	196.2 to 198.6	199.2 to 202.1	673, 993, 1352
$W(\mathit{C}O)_5SCX_2$	196.9 to 197.5	200.8 to 202.0	1747
$[W(\mathit{C}O)_5Cl]^-$	198.7 [129], 199.4	201.6, 201.9	718, 1394, 1398
$[W(\mathit{C}O)_5Br]^-$	197.1 to 199.0	201.6 to 202.0	175, 1393, 1398
$[W(\mathit{C}O)_5I]^-$	196.3 to 197.1 [127.0]	201.6 [175.8] to 202.0	933, 1393, 1398
$W(\mathit{C}O)_4(CPh)Mo(CO)_3Cp$	198.7		1047
$W(\mathit{C}O)_4(CPh)W(CO)_3Cp$	198.3		1047
$W(\mathit{C}O)_4(CPh)Mn(CO)_5$	205.3		1047
$W(\mathit{C}O)_4(CPh)Re(CO)_5$	207.3		1047
$W(\mathit{C}O)_4(\mu\text{-}CO)(\mu\text{-}CPhPMe_3)\text{-}Re(CO)_4$	from 187.4, 191.5, 193.0, 197.9, 199.5		1029
cis-$W(\mathit{C}O)_4(CNCy)_2$	197.3	200.6	574
$W(\mathit{C}O)_4$(norbornadiene)	203.3 [117.9]	209.3	1115
trans-$W(\mathit{C}O)_4(CS)(CNCy)$	195.4		1398
cis-$W(\mathit{C}O)_4(CS)(CNCy)$	191.4	195.8, 198.2	1398
trans-$W(\mathit{C}O)_4(CN)CPh$	193.1		1400
cis-$W(\mathit{C}O)_4(CNMeCH_2CH_2NMe)_2$	205.3	214.0	1210
trans-$W(\mathit{C}O)_4(CNMeCH_2CH_2N\text{-}Me)_2$	210.8		1210
cis-$W(\mathit{C}O)_4(CNEtCH_2CH_2NEt)_2$	204.8 [128.2]	211.1	1210
cis-$W(\mathit{C}O)_4(CNMeCH{=}CHNMe)_2$	205.4 [127.0]	211.5 [136.7]	933

COMPOUND	$\delta(^{13}C)$/p.p.m., *J*/Hz		REFERENCE
	cis	*trans*	
trans-W(*C*O)$_4$(CNMeCH=CHNMe)$_2$	210.6 [129.4]		933
cis-W(*C*O)$_4$(CNMeCH$_2$CH$_2$NMe)-{C(OEt)R}	205.0	209.5, 215.3	1210
trans-W(*C*O)$_4$(CS)(NC$_5$H$_4$Me-4)	199.6		1398
trans-W(*C*O)$_4$(CR)SCN	192.8, 193.2		1400
W(*C*O)$_4$C(NEt$_2$)C{CPh(OMe)}N-Et$_2$	208.3	217.1, 219.3	810
trans-[W(*C*O)$_4$(CPh)PPh$_3$]$^+$	194.1		1400
trans-W(*C*O)$_4$(CS)(PR$_3$)	198.1 (5.9)		1398
trans-W(*C*O)$_4$(CS){P(OMe)$_3$}	196.4 (9.8)		1398
trans-W(*C*O)$_4$(CS){P(OPh)$_3$}	195.2 (9.8)		1398
cis-W(*C*O)$_4$(CS)(PBu$^n{}_3$)	193.9 (8.9)	198.4 (6.9), 199.6 (20.6)	1398
cis-W(*C*O)$_4$(CS)(PPh$_3$)	193.8 (9.8)	198.6 (7.9), 200.6 (23.6)	1398
cis-W(*C*O)$_4$(CS){P(*p*-tol)$_3$}	193.9 (9.8)	198.9 (5.9) 200.9 (23.6)	1398
cis-W(*C*O)$_4$(CS){P(OMe)$_3$}	192.0 (12.8)	196.8 (10.8) 199.1 (40.4)	1398
cis-W(*C*O)$_4$(CS){P(OPh)$_3$}	191.1 (12.8)	195.4 (9.8) 197.8 (47.3)	1398
cis-W(*C*O)$_4$(CNEtCH$_2$CH$_2$NEt)-(PEt$_3$)	204.0 (9.1)	207.8 (22.9), 208.4	1210
W(*C*O)$_4$(CMe)OCMeW(CO)$_5$	194.4		941
W(*C*O)$_4$(CAr)OCArW(CO)$_5$	192.9, 193.4		368, 941
W(*C*O)$_4$(C*p*-tol)OC(*p*-tol)-W(CO)$_5$	194.1		941
W(*C*O)$_4$C(OEt)CHOH-CSCH$_2$CH$_2$CH$_2$S	200.0	211.0, 214.0	1365
trans-W(*C*O)$_4$CR(BF$_4$)	196.1, 196.3		1400
trans-W(*C*O)$_4$(CR)Cl	193.4 to 194.7 [129.4]		176, 368, 893, 933, 941, 1363, 1369, 1394
trans-W(*C*O)$_4$(CNMe$_2$)Cl	195.4		1376
trans-[W(*C*O)$_4$(CS)Cl]$^-$	199.2		682, 1394, 1398
trans-W(*C*O)$_4$(CR)Br	191.0 to 194.4 [127.0 to 131.8]		176, 355, 361, 368, 449, 751, 893, 933, 941, 1007, 1026, 1067, 1363, 1369, 1376, 1394
trans-[W(*C*O)$_4$(CS)Br]$^-$	198.5		682, 1394, 1398

COMPOUND	$\delta(^{13}C)$/p.p.m., J/Hz		REFERENCE
	cis	*trans*	
trans-$W(\mathit{C}O)_4(CR)I$	189.5 to 192.3	129.4	176, 361, 933, 1369, 1376, 1394
trans-$[W(\mathit{C}O)_4(CS)I]^-$	196.5 128		682, 1394, 1398
trans-$W(\mathit{C}O)_4(CSMe)I$	188.8		674
trans-$W(\mathit{C}O)_4(CSCOMe)I$	188.1		674
$[W(\mathit{C}O)_4\{HC(NR)_2\}]^-$	205.1 to 206.2	217.9 to 219.1	1366
$[W(\mathit{C}O)_4\{HC(NMe)(NCOMe)\}]^-$	206.2, 218.8, 219.0, 235.7		1366
$W(\mathit{C}O)_4$(1-allyl-pyrazole)	203.0, 204.0, 208.6, 212.8		1375
$W(\mathit{C}O)_4(Bu^tNSNBu^t)$	211.8		640
$W(\mathit{C}O)_4(NR{=}CR{-}CR{=}NR)$	210.8		448
cis-$W(\mathit{C}O)_4(PEt_3)_2$	204.7 (6.3)	204.4 (15.2)*	117
trans-$W(\mathit{C}O)_4(PEt_3)_2$	204.7 (5)		117
$W(\mathit{C}O)_4$(dppe)	204.6 (9)		15
$W(\mathit{C}O)_3CpW(CO)_4CPh$	201.1, 214.3		1047
$W_2(\mathit{C}O)_6C_{10}H_6MePr^i$	208.6, 209.4, 210.4, 212.5, 220.5		671
$W(\mathit{C}O)_3$(toluene)	210.1		1745
$W(\mathit{C}O)_3$(mesitylene)	212.6		114, 117
$W(\mathit{C}O)_3$(durene)	213.7		114, 117
$W(\mathit{C}O)_3(C_6Me_6)$	215.7		117
$W(\mathit{C}O)_3(C_7H_8)$	208.5§, 218.9†		114, 117, 447, 927
$W(\mathit{C}O)_3(C_8H_8)$	193.8, 207		117, 307
$W(\mathit{C}O)_3(C_6H_6NCO_2Et)$	205.3§, 216.9†		1288
$W(\mathit{C}O)_3CpMe$	215.5, 216.8, 217.8§	228.6, 229.8, 239.2†	104, 1060, 1394, 1396
$W(\mathit{C}O)_3(C_5H_4Me)Me$	217.6§	230.4†	1396
$W(\mathit{C}O)_3(C_5H_4CH_2CH_2)$	218.7§	227.5†	1396
$W(\mathit{C}O)_3(C_5H_4CH_2CH_2CH_2$	218.0§	229.2†	1396
$W(\mathit{C}O)_3Cp(C_3H_5)$	217.7 [161.0]§	229.7 [129.5]†	449
$W(\mathit{C}O)_3Cp(CH_2C_2H)$	217.5 [158.5]§	230.2 [129.5]†	449
$[W(\mathit{C}O)_3Cp\{CH(NMe_2)\}]^+$	214.8 [138]§	220.4 [125]†	1372
$W(\mathit{C}O)_3CpCH{=}C(CN)_2$	213.0§	220.9†	280
$W(\mathit{C}O)_3CpC(CN){=}C(CN)_2$	214.7§	219.7†	280
$W(\mathit{C}O)_3CpCCl{=}C(CN)_2$	217.7§	222.5†	441
$W(\mathit{C}O)_3C_5H_4PPh_3$	220.2		1165
$W(\mathit{C}O)_3C_5H_4SMe_2$	221.8		1165
$W(\mathit{C}O)_3(C_5H_5SOMe)$	216.0		1746
fac-$W(\mathit{C}O)_3(CNEtCH_2CH_2N$-$Et)_2py$	224.1§	230.7†	1210
fac-$W(\mathit{C}O)_3(CNEtCH_2CH_2N$-$Et)\{P(OMe)_3\}_2$	210.6 (9.9)§	210.7 (19.1, 103.7)†	1210
$W(\mathit{C}O)_3(PMe_3)_2\{H_2C{=}C$-$(CO_2Me)_2\}$	201.1 (4.9)§	219.3 (13.5)†	1408

* $|^2J_{PC(cis)} + {}^2J_{PC(trans)}|$, § intensity 2, † intensity 1

COMPOUND	δ(^{13}C)/p.p.m., *J*/Hz		REFERENCE
	cis	*trans*	
$W(\mathit{CO})_3\{C(PMe_3)\mathit{p}\text{-tol}\}$-$(PMe_3)Cl$	206.9 (7.3)§	213.8 (12.2)†	1363
$W(\mathit{CO})_3(CPh)(PPh_3)Br$	199.9 (7.3)§	201.7†	1061
$W(\mathit{CO})_3\{C(PMe_3)\mathit{p}\text{-tol}\}$-$(PMe_3)Br$	205.9 (7.3)§	212.3 (12.2)†	851, 1363
$W(\mathit{CO})_3\{C(PMe_3)SiMe_3\}$-$(PMe_3)Br$	205.2 (9.8)§	209.5 (12.2)†	1363
$W(\mathit{CO})_3(CS)(diars)$	201.9§	210.2†	682
$W(\mathit{CO})_3(CPh)(AsPh_3)Br$	199.4§	200.9†	1061
$W_2(\mathit{CO})_6(\mu\text{-}AsPh_2CH_2AsPh_2)$-$(\mu\text{-}Br)(MeC_2Me)$	205.5		1023
$W(\mathit{CO})_3(CPh)(SbPh_3)Br$	199.5§	199.7†	1061
$[W(\mathit{CO})_3\{HC(N\text{-}\mathit{p}\text{-tol})_2py]^-$	224.0§	228.1†	1366
$[W(\mathit{CO})_3\{HC(N\text{-}\mathit{p}\text{-tol})_2\}PPh_3]^-$	217.7 (47)§	224.0 (6)†	1366
$W(\mathit{CO})_3(diars)(NH_2C_6H_4OMe\text{-}4)$	195.5§	207†	1407
fac-$W(\mathit{CO})_3(PEt_3)_3$	212.4 [135] (14, 5.5)		15, 117
$W(\mathit{CO})_2CpCR$	221.3 to 222.6 [202.6]		637, 1026, 1043
$\overline{W(\mathit{CO})_2CpCOCHMeCN}Me_2$	234.5 [141.6]	228.7 [156.2]	881
$\overline{W(\mathit{CO})_2CpCO(NEt_2)}$=CHR	228.2 [146.5], 228.3 [151.4] 225.0 [170.9], 226.6 [166.0]		881
$W(\mathit{CO})_2CpN_2Me$	191.4		1060
$W(\mathit{CO})_2CpNH_2C_3H(CN)$	214.1		441
$W(\mathit{CO})_2Cp\{(CH_2)_5NCC(CN)_2\}$	220.8		441
$W(\mathit{CO})_2Cp(COMe)PPh_3$	230.0 (17.1)		1394
$W(\mathit{CO})_2Cp(PMe_3)C_2(\mathit{p}\text{-tol})$	232.6 [142.7] (65.9) (23.7), 247.7 [139.7]		1403
trans-$W(\mathit{CO})_2Cp(PMe_3)$-$C(\mathit{p}\text{-tol})CO$	224.1 (17.1)		1404
$\overline{W(\mathit{CO})_2CpCH=CHCMe=O}$	239.2 [154]		749
$W(\mathit{CO})_2(CNBu^t)_2(CPh)Br$	207.1		1061
$W(\mathit{CO})_2Cp(SnMe_3)(PR_3)$	223.0 to 223.3 (17.5)		71, 688
$W(\mathit{CO})_2Cp(SnMe_3)\{P(OR)_3\}$	220.5, 220.6 (25)		71, 688
$W(\mathit{CO})_2Cp(RNCHNR)$	253.4 to 260.8		1079, 1748, 1749
$W(\mathit{CO})_2Cp(RN_3R)$	240.3 to 248.2		654, 1749
$W(\mathit{CO})_2Cp(Me_2CNNPh)$	232.3, 238.2		1121
$W(\mathit{CO})_2Cp(R_2CNO)$	230.1 to 231.6	235.9 to 237.2	1129
$[W(\mathit{CO})_2Cp(pzH)_2]^+$	248.4		1152
$[W(\mathit{CO})_2Cp(imH)_2]^+$	252.6		1152
$W(\mathit{CO})_2Cp\{HC(NR)NCOR\}$	238.9 to 244.6		1079, 1749
$W(\mathit{CO})_2Cp(pzH)Cl$	252.2, 259.0		1152
$W(\mathit{CO})_2Cp(imH)Cl$	254.9, 261.0		1152
$W(\mathit{CO})_2Cp\{PF(OMe)_2\}Cl$	227.6 (11), 241.2 (31)		1390
$W(\mathit{CO})_2Cp\{PF(OEt)_2\}Cl$	229.4 (11), 242.4 (34)		1390
$W(\mathit{CO})_2Cp\{(PF_2)_2NMe\}Cl$	219.7, 236.5 (29)		1390

§intensity 2, †intensity 1

COMPOUND	$\delta(^{13}C)$/p.p.m., J/Hz		REFERENCE
	cis	*trans*	
$W(CO)_2\{(MeO)_2PCH_2CH_2P(OMe)_2\}_2$	214.0		1371
$W(CO)Cp(PMe_3)_2CRCO$	245.2 to 251.8 (17.1 to 19.5)		1037
$W(CO)LCpHC(N\text{-}p\text{-}tol)_2$	257.0 to 262.5		1748
$W(CO)Cp(PR_3)OC_2R$	228.1 to 230.1 [166] (7.3 to 8.8)		1037
$W_2(CO)_2\{\mu\text{-}HC(NC_6H_3Me_2\text{-}3,5)\}_4H$	262.6, 275.7		1410
Manganese			
$Mn(CO)_5H$	211.4 {14}	210.8 {7}	986
$Mn(CO)_5W(CO)_4CPh$	216.6, 224.3		1047
$Mn(CO)_5Me$	214		67
$Mn(CO)_5CH_2CH{=}CH_2$	213.9	211.8	1414
$Mn(CO)_5CH_2Ph$	211.1	209.1	336
$Mn(CO)_5C{=}CPhC(CN)XC(CN)XCH_2$	208.2, 208.4		1312
$Mn(CO)_5COMe$	210.0		1744
$Mn(CO)_5C_5Cl_5$	205		534
$Mn(CO)_5I$	205.4		986
$[MnFe_2(CO)_{12}]^-$	223		372
$Mn_2(CO)_8(MeC_2NEt_2)$	211.0, 232.9		684
$Mn(CO)_4(\eta^3\text{-}C_3H_5)$	215.6*, 216.4*, 222.6§		1414
$Mn(CO)_4(\eta^3\text{-}1\text{-}EtC_3H_4)$	215.6, 217.0, 220.8, 221.5		1414
$Mn(CO)_4(\eta^3\text{-}1\text{-}Pr^iC_3H_4)$	215.9, 217.4, 221.4, 222.1		1414
$Mn(CO)_4COC(NMe_2)CMe_2$	212.5*, 214.0§, 217.2*		515
$Mn(CO)_4(Me_2C{=}C{=}NMe_2)$	212.8		515
$\{cis\text{-}Mn(CO)_4(COMe)_2\}_3Al$	211.9, 217.2		1744
$Mn(CO)_4\{CH(NMe_2)\}Cl$	210.4§, 212.9*, 218.3*		1372
$MnFe(CO)_6C_7H_7$	224.3		716
$Mn(CO)_3Cp$	224.5 to 225.8		181, 295, 388, 534, 778, 885, 1712, 1745
$Mn(CO)_3(C_5H_4R)$	223.2 to 225.7		67, 181, 878, 885, 1007, 1702, 1712, 1745
$[Mn(CO)_3(C_5H_4CR^1R^2)]^+$	214.2 to 222.2		878, 1702
$Mn(CO)_3(C_5H_4O_2CMe)$	222.8		181
$Mn(CO)_3(C_5H_4BX_2)$	221.6 to 223.2		885
$[Mn(CO)_3(C_5H_4PPh_3)]^+$	222.4		1420
$Mn(CO)_3(C_5H_4MeBI_2)$	221.8		885
$Mn(CO)_3\{C_5H_2(CO_2Me)_3\}$	220.2, 221.9, 225.9		535, 1116
$Mn(CO)_3(C_5H_2I_3)$	223.5		1089
$Mn(CO)_3(C_5Cl_5)$	219.8		534
$Mn(CO)_3(C_5Cl_4Br)$	220.8		534
$Mn(CO)_3(C_5Br_5)$	220.9		1089
$Mn(CO)_3(C_{13}H_9)$	225.2		1131

*intensity 1, §intensity 2

COMPOUND	$\delta(^{13}C)$/p.p.m., J/Hz *cis*	*trans*	REFERENCE
$[Mn(CO)_3C_{13}H_{10})]^+$	216.5		1131
$Mn(CO)_3(C_4H_4N)$	221.1		958
$Mn(CO)_3(C_4H_{4-n}R_nP)$	222.5 to 223.8		1412
$Mn(CO)_3(2,5\text{-}Me_2C_4H_2As)$	224.2		1416
$Mn(CO)_3(NC_4H_3\text{-}2\text{-}CMeO)$	221.1		958
$Mn_2(CO)_4(C_5H_4Me)_2CH_2$	233.0		533
$Mn(CO)_2(CS)(C_5H_4R)$	223.9, 224.0		295, 1712, 1745
$Mn(CO)_2(CSe)(C_5H_4R)$	223.0, 223.1		1712
$Mn(CO)_2(CRR^1)Cp$	232.9, 233.4		756
$Mn(CO)_2\{CR(OR^1)\}Cp$	231.4 to 235.3		756
$Mn(CO)_2Cp(C{=}C{=}CBu^t_2)$	229.8		752
$Mn(CO)_2(C_5H_4R)(PhC{=}C{=}O)$	238.0		754
$Mn(CO)_2Cp(C{=}CHPh)$	226.7		970
$Mn_2(CO)_4Cp_2(C{=}CHPh)$	229.2, 232.4, 234.1, 235.5		970
$[Mn(CO)_2CpCPh]^+$	209.4		1058
$[Mn(CO)_2CpCPhPMe_3]^+$	232.5 (4.9)		1048
$Mn(CO)_2CpCPhSCN$	204.3		946
$Mn(CO)_2Cp(C_8H_{11})$	234.5		295
$[Mn(CO)_2Cp(\eta^2\text{-olefin})]^+$	233.9 to 235.5		1189
$Mn(CO)_2Cp(CS_2)$	228.4, 230.2		1032
$\{Mn(CO)_2Cp\}_2CS_2$	228.1*, 229.0*, 234.4§		1032
$Mn(CO)_2Cp(NHC_5H_{10})$	236.2		295
$Mn(CO)_2(C_5H_4R)(PR_3)$	230.7 to 235.0 (23, 26)		181, 295
$Mn(CO)_2CpP(OMe)_3$	229.5 (34)		295
$Mn(CO)_2CpP(OPh)_3$	226.4, 228.8 (36)		181, 295
$[Mn(CO)_2(C_5H_4CR^1R^2)(PR_3)]^+$	222.7 to 225.8		1699
$Mn(CO)_2(C_4H_2Me_2P)PPh_3$	231.2		1419
$Mn(CO)_2CpAsPh_3$	232.4		181
$Mn(CO)_2CpSbPh_3$	230.8		181
$Mn(CO)Cp(PMe_3)(p\text{-}tolC_2O)$	229.6 (7.4)		963
$Mn(CO)Cp(dppe)$	220.5		181
Rhenium			
$Re(CO)_5H$	183.2 {7}	182.7 {8}	986
$Re(CO)_5Cr(CO)_4CPh$	184.2, 196.5		1047
$Re(CO)_5Mo(CO)_4CPh$	199.4		1047
$Re(CO)_5W(CO)_4CPh$	197.1		1047
$Re_2(CO)_{10}$	192.8	183.7	336
$[Re_2(CO)_9H]^-$	189.9, 195.4, 199.0, 201.0, 202.0		1427
$[Re_2(CO)_9CHO]^-$	189.1, 194.5, 199.4, 200.1, 203.6		1426, 1427
$Re(CO)_5Me$	185.2	181.2	514
$Re(CO)_5Ph$	183.6	181.7	514
$Re(CO)_5C(NMe_2){=}CMe_2$	193.4, 195.0		515
$Re(CO)_5COMe$	183.0	181.2	514
	188.0	186.0	1744
$Re(CO)_5COPr^i$	184.5	182.2	1744

*intensity 1, §intensity 2

COMPOUND	$\delta(^{13}C)$/p.p.m., J/Hz		REFERENCE
	cis	*trans*	
$Re(\mathit{C}O)_5COCH_2Ph$	183.8	181.9	1744
$Re(\mathit{C}O)_5COPh$	182.9	181.0	514
$Re(\mathit{C}O)_5SiMe_3$	187.9	182.9	514
$Re(\mathit{C}O)_5SiCl_3$	180.5	178.5	514
$Re(\mathit{C}O)_5GeMe_3$	187.5	182.7	514
$Re(\mathit{C}O)_5SnMe_3$	187.7	182.9	514
$Re(\mathit{C}O)_5PbMe_3$	*ca* 189.3		514
$[Re(\mathit{C}O)_5NCMe]^+$	178.6	179.6	514
$Re(\mathit{C}O)_5Br$	177.8, 179.3	176.3, 177.9	336, 514
$Re(\mathit{C}O)_5I$	176.5	176.1	986
$ReW(\mathit{C}O)_9CPhPMe_3$	from 187.4, 191.5, 193.0, 197.9, 199.5		1029
$[Re_2(\mathit{C}O)_8Br(CPh)]^+$	185.9, 186.1, 186.9		762
$Re(\mathit{C}O)_4COCMe_2CNMe_2$	191.5*, 192.5§, 193.9*		515
$Re(\mathit{C}O)_4COCMe_2CNC_5H_{10}$	191.4*, 192.8§, 194.0*		515
cis-$Re(\mathit{C}O)_4(COR)_2X$	185.2 to 191.9		1744
$[Re(\mathit{C}O)_4(dppe)]^+$	182.9 (8)	184.1 (8, 41)	717, 1123
$ReFe(\mathit{C}O)_6C_7H_7$	192.4		716
$Re(\mathit{C}O)_3Cp$	195.0		1712
$Re(\mathit{C}O)_3(C_5H_4X)$, (X = Cl, Br)	193.0		1425
fac-$[Re(\mathit{C}O)_3(dppe)C(OEt)SiPh_3]^+$	191.9 (7)	192.5 (9, 50)	1123
fac-$Re(\mathit{C}O)_3(dppe)COSiPh_3$	194.6 (6)	196.1 (9, 50)	717, 1123
mer-$Re(\mathit{C}O)_3(dppe)SiPh_3$	198.3	198.9	1123
mer-$Re(\mathit{C}O)_3\{P(OPh)_3\}_2Cl$	186.7§	183.9*	1422
$[Re(\mathit{C}O)_3(OH)]_4$	196.7		1424
$[Re(\mathit{C}O)_2HCp(PPh_3)]^+$	189.0		648
$Re(\mathit{C}O)_2CpCHPh$	206.5		1423
$Re(\mathit{C}O)_2CpCMePh$	206.5		832
$Re(\mathit{C}O)_2CpC(CN)Ph$	203.6		946
$Re(\mathit{C}O)_2CpCCHPh$	198.6		969, 970
$[Re(\mathit{C}O)_2CpCPh]^+$	186.8		832
$[Re(\mathit{C}O)_2CpCPhPMe_3]^+$	202.5		1048
$[Re(\mathit{C}O)_2CpC(PMe_3)_2Ph]^+$	208.0		1019
$Re(\mathit{C}O)_2CpC(SCN)Ph$	204.3		946
$Re(\mathit{C}O)_2Cp(CS)$	196.4		1712
$Re(\mathit{C}O)_2Cp(CSe)$	196.4		1712

Iron, [^{57}Fe]

A. Mononuclear Compounds

i) Iron pentacarbonyl

$Fe(CO)_5$	209.6 to 213.5 [23.4]		13, 14, 41, 93, 115
	206.4†	224.1#	276

ii) Iron tetracarbonyl derivatives

$Fe(\mathit{C}O)_4H_2$	205.9 {9}		986
$[Fe(\mathit{C}O)_4Me]^-$	223.0		1144

*intensity 1; §intensity 2; †equatorial, basal; #apical

COMPOUND	$\delta(^{13}C)$/p.p.m., J/Hz	REFERENCE
$[Fe(CO)_4COMe]^-$	220.5	1144
$\overline{Fe(CO)_4COCR{=}CR^1CO}$	198.8 to 203.2	847, 945, 956
$[Fe(CO)_4Pr(CO_2Et)]^-$	221.6	1454
$\overline{Fe(CO)_4CH_2CH_2CH_2SiMe_2}$	206.1*, 207.5*, 211.5§	1436
$Fe(CO)_4(C_3F_7)I$	198.8	93
$Fe(CO)_4\overline{CNMeCH_2CH_2NMe}$	217.2, 217.4	202, 481, 1224
$Fe(CO)_4\overline{COCMe_2CMe_2O}$	215.3	338, 481
$Fe(CO)_4(\eta^2$-olefin)	204 to 212.8	320, 355, 597, 618, 725, 993, 1456, 1459, 1461
$[Fe(CO)_4(C_3H_{5-n}R_n)]^+$	196.1 to 200.0	679, 1469
$Fe(CO)_4CNBu^t$	212.4	574
$Fe(CO)_4CS$	212.2	1472
cis-$Fe(CO)_4(SiMe_3)_2$	207.6†, 208.5# [25 averaged] ($^2J_{SiC}$ = 6 Hz)	727, 1130
$\overline{Fe(CO)_4SiMe_2CH_2CH_2SiMe_2}$	207.9†, 208.9#	393
cis-$Fe(CO)_4(SiMe_2Cl)_2$	203.3†, 205.7#	727
cis-$Fe(CO)_4(SiMeCl_2)_2$	200.1†, 202.5#	727
trans-$Fe(CO)_4(SiMeCl_2)_2$	203.6	727
cis-$Fe(CO)_4(SiCl_3)_2$	197.3†, 199.4#	727
trans-$Fe(CO)_4(SiCl_3)_2$	199.4	727
cis-$Fe(CO)_4(GeMe_2H)_2$	205.5†, 207.1#	1438
cis-$Fe(CO)_4(GeMe_3)_2$	207.1†, 208.9#	727
cis-$Fe(CO)_4(SnMe_3)_2$	208.1†, 207.9# {$^2J_{SnC}$ = 33 (*cis*); 65 (*trans*)} ($^2J_{SnC}$ = 101)	727, 1128
cis-$Fe(CO)_4(SnR_3)_2$	206.5 to 208.7† ($^2J_{SnC}$ = 61 to 80) 206.7 to 207.8# ($^2J_{SnC}$ = 96 to 112)	1128 1128
cis-$Fe(CO)_4(SnCl_3)_2$	195.2 ($^2J_{SnC}$ = 170) 195.3 ($^2J_{SnC}$= 235)	1128
$Fe(CO)_4NMe_3$	217.5	1184
$Fe(CO)_4PR_3$	214.8 to 215.3 (19)	330
$Fe(CO)_4\overline{PRNMeCH_2CH_2NMe}$	214.7	1381
$\{Fe(CO)_4PF_2\}_2NMe$	209.7, 210.4, 211.1	1443
$Fe(CO)_{5-n}(PF_3)_n$	(237.6_{trans}, 58.8_{cis}, -96_{vic})×	276
$Fe(CO)_4O{=}\overline{SCH_2CH{=}CHCH_2}$	212.7	656
$Fe(CO)_4SR_2$	214.2 to 215.6	1429
$Fe(CO)_4I_2$	198.5†, 204.6#	986

*intensity 1; §intensity 2; †equatorial, basal; #apical; ϕ1 signal; Ψ2 signals; ×assumed independent of n and obtained by a fit as a function of n

COMPOUND	$\delta(^{13}C)$/p.p.m., J/Hz	REFERENCE
iii) Iron tricarbonyl derivatives		
$Fe(CO)_3(B_5H_9)$	206.6, 207.7, 210.5	710
$[Fe(CO)_3H(C_4H_6)]^+$	196.0, 199.2, 203.1	679, 925, 1225
$[Fe(CO)_3H(C_4H_4)]^+$	200.2, 202.7	785, 1225
$[Fe(CO)_3H(C_6H_8)]^+$	196.2, 200.3, 203.0	679
$[Fe(CO)_3H$(norbornadiene)$]^+$	197.3, 205.6	924, 1225
$Fe(CO)_3(B_2SMe_2C_2$-heterocycle)	210.4, 211.1	479
$Fe(CO)_3$(σ-π compound)	201.5 to 205.8†, 206.5 to 216.4#	816, 1198, 1435, 1440, 1460, 1473, 1716
$Fe(CO)_3(C_4H_4)$	209.0, 214.9, 215.6 [28.7]	132, 481, 877, 1766
$Fe(CO)_3(C_4H_6)$	210.1 to 211.3*, 216.6 to 217.4§ [27.9 averaged]	326, 358, 865
$Fe(CO)_3$(diene)	206.7 to 216.8 (averaged) 202.5 to 212.4*, 212.9 to 217.5§	132, 135, 137, 188, 303, 312, 326, 358, 488, 513, 684, 705, 721, 725, 726, 738, 740, 802, 816, 846, 865, 869, 874, 918, 925, 939, 957, 959, 993, 1008, 1082, 1206, 1225, 1280, 1295, 1463, 1456, 1461, 1466, 1716, 1766
$Fe(CO)_3\{C_4(CF_3)_4CO$	201.0, 203.8	865
$Fe(CO)_3$(1,5-diene)	217.2 to 218.8	865, 924, 1225
$[Fe(CO)_3(C_4H_3CR^1R^2)]^+$	201.0 to 202.8	869
$Fe(CO)_3$(trimethylenemethene)	212, 211.6	55, 618
$Fe(CO)_3(C_4H_5R)$	209.7 to 214.8	10, 845
$[Fe(CO)_3Cp]^+$	202.7, 203.0	287, 295, 388

*equatorial, basal; §apical; †1 signal; #2 signals

COMPOUND	$\delta(^{13}C)$/p.p.m., J/Hz	REFERENCE
$[Fe(\mathit{C}O)_3(dienyl)]^+$	196.3 to 199.1*, 206.0 to 210.0§	258, 675, 787, 924, 940, 1225, 1261, 1432
trans-$Fe(\mathit{C}O)_3(\overline{CNMeCH_2CH_2N}Me)_2$	225.9	1224
$Fe(\mathit{C}O)_3(CNBu^t)_2$	214.9	574
$Fe(\mathit{C}O)_3(C_2H_3SiMe_2)(SiMe_3)$	212	708
$Fe(\mathit{C}O)_3\{MeO_2CCH{=}C(CO_2Me)_2\}py$	207.3, 209.9	1435
$Fe(\mathit{C}O)_3(NRCHCHNR)$	215.5, 216.3	874
$Fe(\mathit{C}O)_3(NPhCHCHNPh)$	206.1, 209.5, 212.5	874
$Fe(\mathit{C}O)_3(NPh{=}CHCHPh)$	209.1, 212.0, 214.1	874
$Fe(\mathit{C}O)_3\{C_{12}H_{12}(OMe)_2\}$	213.8	843
$Fe(\mathit{C}O)_3SO_2C_7H_7$	206.9	1198
$Fe(\mathit{C}O)_3(CHPh{=}CHCH{=}O)$	203.8, 209.3, 210.8	874
$Fe(\mathit{C}O)_3(allyl)X$	202.8 to 209.9	174, 568, 679, 869, 925, 1225
$Fe(\mathit{C}O)_3\{Ph_2P(CH_2)_3PPh_2\}$	219.9 (8.9)	483
$Fe(\mathit{C}O)_3(PR_3)_2$	215.5, 216.7	330
$Fe(\mathit{C}O)_3(Me_2PCH_2CH_2PMe_2)$	221.5	1
$Fe(\mathit{C}O)_3(Me_2PCF_2CH_2PMe_2)$	220.0 (5.7, 10.1)	483
$Fe(\mathit{C}O)_3(diars)$	221.4	483
iv). Iron dicarbonyl derivatives		
$Fe(\mathit{C}O)_2(C_6H_8)(CNEt)$	217.2	1465
$Fe(\mathit{C}O)_2(C_{12}H_{16})$	215.7	725
$Fe(\mathit{C}O)_2CpMe$	218.3, 218.4	60, 70, 763, 1394
$Fe(\mathit{C}O)_2CpR^*$	214.0 to 219.2	60, 70, 356, 763, 856, 980, 1001, 1232, 1312, 1394
$Fe(\mathit{C}O)_2(C_5IX_4)C_3F_7$	209.3 to 212.7	1447
$Fe(\mathit{C}O)_2Cp$(substituted vinyl)	211.2 to 214.3	280, 1213, 1312
$[Fe(\mathit{C}O)_2Cp(\eta^2\text{-olefin})]^+$	201.3 to 211.0	287, 314, 484, 485, 489, 659
$[Fe(\mathit{C}O)_2(CS)Cp]^+$	203.3	295, 661
$[Fe(\mathit{C}O)_2(CHPh)Cp]^+$	206.5	1095
$Fe(\mathit{C}O)_2CpCN$	211.1, 211.4	60, 70
$Fe(\mathit{C}O)_2CpMR_3$ (M = Si, Ge, Sn)	213.2 to 216.7	70, 1455, 1729
$Fe(\mathit{C}O)_2(NHMe_2)(C_4H_2Me_2SO_2)$	209.0	735
$[Fe(\mathit{C}O)_2CpNH_3]^+$	211.8	295
$[Fe(\mathit{C}O)_2Cp(NCMe)]^+$	209.1	287
$[Fe(\mathit{C}O)_2Cp(py)]^+$	210.8	287
$Fe(\mathit{C}O)_2\{P(OMe)_3\}\overline{CSCR^1CR^2S}$	219.6 to 222.8 (35.9 to 38.1)	1450
$Fe(\mathit{C}O)_2\{CS_2C(CO_2Me)C(CO_2Me)\}$-$(PMe_3)_2$	207.9 (19), 212.0 (15)	1433
$Fe(\mathit{C}O)_2(C_3H_4S)PPh_3$	214.3, 215	721
$[Fe(\mathit{C}O)_2Cp(PR_3)]^+$	209.7, 210.5 (21.8)	295, 485
$[Fe(\mathit{C}O)_2Cp(thf)]^+$	208.9	1003

*basal, §apical

COMPOUND	δ(^{13}C)/p.p.m., *J*/Hz	REFERENCE
$Fe(\mathit{C}O)_2(S_2CNMe_2)(CSNMe_2)$	214.4	982
$Fe(\mathit{C}O)_2CpCl$	212.9, 213.3	60, 70
$Fe(\mathit{C}O)_2CpBr$	213.2, 213.5	60, 70
$Fe(\mathit{C}O)_2(C_5H_4X)I$	212.6 to 213.8	60, 70, 1447
$Fe(\mathit{C}O)_2(NO)_2$	207	13, 14, 93
v) Iron monocarbonyl derivatives		
$Fe(\mathit{C}O)HCp(PPh_3)$	210.8	722
$[Fe(\mathit{C}O)(C_6H_8)(C_7H_9)]^+$	224.3	331
$\overline{Fe(\mathit{C}O)CpC(CF_3)=C(CF_3)CH_2CH}=CHMe$	219.6	1116
$[Fe(\mathit{C}O)Cp(CN)_2]^-$	219.2	60
$[Fe(\mathit{C}O)Cp(CNR)_2]^+$	212.1, 212.6	60, 487, 574
$[Fe(\mathit{C}O)Cp(CNR)\{C(NHR)(NHR^1)\}]^+$	216.2 to 217.5	1470
$Fe(\mathit{C}O)Cp(CN)\{C(NHR)(NHR^1)\}$	220.3 to 220.6	1470
$Fe(\mathit{C}O)$(1-allylpyrazole)	213.1	1375
$Fe(\mathit{C}O)\{C_4H_4(CO_2Et)_2\}$(bipy)	214.5	972
$Fe(\mathit{C}O)(\overline{CNMeCH_2CH_2N}Me)(NO)_2$	225.8	1224
$Fe(\mathit{C}O)Cp(PPh_3)R$	219.6 to 226.1 (30.2 to 48.0)	482, 776, 980 1001, 1117, 1276, 1394
$[Fe(\mathit{C}O)Cp(PPh_3)(CH_2PPh_3)]^+$	209 (24.1)	1001
$Fe(\mathit{C}O)Cp\{P(OR)_3\}R^1$	216.2 to 219.5 (38 to 50)	856, 1109
$[Fe(\mathit{C}O)Cp(PPh_3)(CHPh)]^+$	215.4 (29)	1095
$Fe(\mathit{C}O)Cp(PR_3)CN$	216.9 to 218.2 (27.2 to 29.4)	482, 489
$[Fe(\mathit{C}O)Cp(PPh_3)(CNEt)]^+$	213.8 (26)	482
$[Fe(\mathit{C}O)Cp(PR_3)\{C(NHR)(NHR^1)\}]^+$	217.9 to 219.6	1470
$[Fe(\mathit{C}O)Cp(CS)PR_3]^+$	209.0 to 211.9 (23 to 35)	661
$Fe(\mathit{C}O)Cp(PF_2NMeR)Cl$	213.9, 215.1 (44)	1390
haemoglobin-CO	206.0 to 208.2	124, 412-414
$Fe(\mathit{C}O)(PF_2C_6H_{10}PF_2)_2$	189	1442
<u>B. Binuclear Compounds</u>		
$Fe_2(\mathit{C}O)_6C_{14}H_{20}$	211.8*, 215.4§	489
$Fe_2(\mathit{C}O)_6(C_4H_6)$	209.5	597
$Fe_2(\mathit{C}O)_6C_4H_2(CBu^t{}_2)_2$	207.0*, 211.8§	597
$Fe_2(\mathit{C}O)_6C_4HR^1R^2R^3$	207.3 to 210.4*	635, 847
	205.2 to 207.2§	684
	209.5 to 213.7†	949, 1219
$Fe_2(\mathit{C}O)_6C_4H_2Bu^t{}_2$	211.8*, 211.6*,	859, 1219
	209.1§, 208.9§	859, 1219
	214.1§, 222.2*	
$Fe_2(\mathit{C}O)_6C_4Ph_4$	212.3*, 210.6*, 204.5§, 203.2§ 216.2†, 214.7†	635, 859
$Fe_2(\mathit{C}O)_6C_4H_2Ph_2$	210.0*, 209.2*, 206.4§, 204.5§, 214.1†, 212.3†	859

*intensity 1, §intensity 2, †intensity 3

COMPOUND		$\delta(^{13}C)$/p.p.m., J/Hz	REFERENCE
$Fe_2(\mathit{C}O)_6C_6H_4CHCPh$		210.5^{*}, 207.8^{*}, 206.6^{*}, $216.5^{\dagger}$	859
$Fe_2(\mathit{C}O)_5(PPh_3)C_4Ph_4$		209.5 (13.1), 217.7	635
$Fe_2(\mathit{C}O)_5(PBu^n{}_3)C_4Ph_4$		208.3 (15.5), 220.9	635
$Fe_2(\mathit{C}O)_6(PhC_2NMe_2)$		211, 228.3	684
$Fe_2(\mathit{C}O)_6C_4R^1{}_2R^2{}_2$		209.6 to 211.6	1462
$Fe_2(\mathit{C}O)_6(C_{10}H_{12})$		211.5, 215.0	486
$Fe(\mathit{C}O)_3M(CO)_nC_7H_7$		211.4 to 213.0	716
$Fe(\mathit{C}O)_6(C_8H_{10})_2CO$		211.0	945
$Fe_2(\mathit{C}O)_6(Ph_4C_4CO)$		206.5, 207.3, 210.3	691
		206.3^{*}, 206.9^{*}, 210.0^{*}, $207.7^{\dagger}$	
$Fe_2(\mathit{C}O)_6(CBu^tCHCOCBu^tCH)$		206.9^{*}, 210.5^{*}, 212.2^{*}, $208.6^{\dagger}$	859
$Fe_2(\mathit{C}O)_6(CHCBu^tCOCBu^tCH)$		205.8^{*}, 209.5^{*}, 209.9^{*}, $208.4^{\dagger}$	859
$Fe_2(\mathit{C}O)_6(C_9H_{11}O)$		209.6, 214.1	945
$Fe_2(\mathit{C}O)_6(C_7H_8)$		212.6, 215.0	186
$Fe_2(\mathit{C}O)_6(C_8H_{10})$		211.7	186
$Fe_2(\mathit{C}O)_6(C_9H_{12})$		206.5, 209.5, 210.6, 216.6, 217.2, 218.8	186, 997
$Fe_2(\mathit{C}O)_6(C_9H_{11}Cl)$		211.9	186
$Fe_2(\mathit{C}O)_6(C_{10}H_{14})$		211.4	186
$Fe_2(\mathit{C}O)_5(C_9H_8)$		215.2^{*}, 223.0^{*}, $217.1^{\dagger}$	639, 1126
$Fe_2(\mathit{C}O)_4(PEt_3)(C_9H_8)$		216.5, 223.5, 215.7 (20.6), 231.2 (12.0)	1126
$Fe_2(\mathit{C}O)_5(C_9H_7Me)$		$214.4^{§}$, $217.7^{\dagger}$	639
$Fe_2(\mathit{C}O)_5(C_{10}H_8)$		$208.2^{\#}$, $210.1^{\#}$, $229.8^{\#}$, 214.5^{ϕ}, 220.0^{ϕ}	639, 1134
$Fe_2(\mathit{C}O)_5(C_{10}H_5Me_2Pr^i)$	Isomer 1	214.8^{ϕ}, 220.5^{ϕ}, $208.7^{\#}$, $210.9^{\#}$, $229.7^{\#}$	1101
	Isomer 2	$219.8^{§}$, $206.6^{\#}$, $213.7^{\#}$, $219.0^{\#}$	1101
$Fe_2(\mathit{C}O)_4(PEt_3)(C_{10}H_5Me_2Pr^i)$		214.2, 217.2, 222.3, 234.7	1101
cis-$Fe_2(\mathit{C}O)_4Cp_2$		210.0^{Ψ}, 210.9^{Ψ}, $274.1^{\times}$, $285.1^{\times}$	13, 14, 93, 712
trans-$Fe_2(\mathit{C}O)_4Cp_2$		$242.3^{@}$	712
cis-$Fe_2(\mathit{C}O)_4(MeC_5H_4)_2$		210.8^{Ψ}, $276.0^{\times}$	712
trans-$Fe_2(\mathit{C}O)_4(MeC_5H_4)_2$		$243.4^{@}$	712
cis-$Fe_2(\mathit{C}O)_4\{(CH_2)_4C_5H_3\}_2$		211.4^{Ψ}, $277.0^{\times}$	712
trans-$Fe_2(\mathit{C}O)_4\{(CH_2)_4C_5H_3\}_2$		$244.0^{@}$	712
cis-$Fe_2(\mathit{C}O)_4(indenyl)_2$		210.3^{Ψ}, $271.9^{\times}$	712

*intensity 1, §intensity 2, †intensity 3, $^{\#}Fe(CO)_3$, $^{\phi}Fe(CO)_2$, $^{\Psi}$terminal, $^{\times}$bridge, $^{@}$averaged

COMPOUND	$\delta(^{13}C)$/p.p.m., J/Hz	REFERENCE
trans-$Fe_2(CO)_4(indenyl)_2$	211.6*, 273.2§	712
cis-$FeNi(CO)_3Cp_2$	209.4*, 254.5§	712
$Fe_2(CO)_3(CNBu^t)Cp_2$	214.0*, 281.8§	506
$Fe_2(CO)_3(CNPh)Cp_2$	215.1*, 275.9§	506
$Fe_2(CO)_3(\overline{CNMeCH_2CH_2NMe})Cp_2$	214	1224
cis-$[Fe_2(CO)_3(CC_5H_5)Cp_2]^+$	208.7*, 255.2§	1451
trans-$[Fe_2(CO)_3(CC_5H_5)Cp_2]^+$	207.4*, 253.5§	1451
$Fe_2(CO)_3(GeMe_2)Cp_2$	217.5*, 282.2§	389
$Fe_2(CO)_3\{P(OEt)_3\}Cp_2$	216.2*, 279.8 (22)§	397
$Fe_2(CO)_3\{P(OPh)_3\}Cp_2$	215.1*, 277.6 (20)§	387
$Fe_2(CO)_4(diazulenyl)$	211.0*, 269.0§	686
$Fe_2(CO)_4(C_5H_4CMe_2CMe_2C_5H_4)$	211.7*, 274.3§	686
$Fe_2(CO)_3\{P(OMe)_3\}(C_5H_4CMe_2CMe_2C_5H_4)$	214.1*, 278.5§	686
$Fe_2(CO)_3(C_4H_4)_2$	238.8	480
$Fe_2(CO)_6(C_6H_4CH_2NMe)$	209.7, 211.5, 211.9	1457
$Fe_2(CO)_6(C_6H_4CHPhNMe_2)$	212.3, 212.7	792
$Fe_2(CO)_6(C_6H_4CH_2NPh_2)$	209.4, 210.9, 211.0	1457
$Fe_2(CO)_6(C_3H_2COS)$	189.3, 209.8	960
$Fe_2(CO)_6(PhC{=}CPhS)$	204.6†, 209.6#	1457
$Fe_2(CO)_5(PPh_3)(PhC{=}CPhS)$	204.5†, 210.3ϕ, 220.2 (26.7)†, 221.3 (17.0)†	1457
$Fe_2(CO)_6(ArCHSAr)$	209.3 to 210.7	792
$Fe_2(CO)_7(N_2C_4H_4)$	211.9ϕ, 214.5Ψ, 283.4§	1103
$Fe_2(CO)_6(PhC_2H_2N)_2CO$	207.3	1431
$Fe_2(CO)_6(\mu\text{-}NHPr^i)(\mu\text{-}Me_2CNO)$	204.8, 205.8, 212.2, 212.8, 214.2, 214.8	1437
$Fe_2(CO)_7\{P(OEt)_2\}_2O$	214.3*, 260.3§	1444
$Fe_2(CO)_5(\{P(OEt)_2\}_2O)_2$	215.7 (23.3)*, 221.3 (14.8)*, 280.8 (18.9)§	1444
$Fe_2(CO)_7\{(PF_2)_2NMe\}$	215.2	1444
$Fe_2(CO)_6(PPhCH_2CH_2CH_2PPh)$	212.0 (4.7)	1445
$Fe_2(CO)_5(PMePh_2)(PPhCH_2CH_2CH_2PPh)$	215.4 (4.3)×, 217.4 (4.5, 12.9)ϕ	1445
$Fe_2(CO)_5\{PMePh(CH_2Ph)\}(PPhCH_2\text{-}CH_2CH_2PPh_2)$	215.4 (4.1)×, 218.0ϕ	1445
$[Fe_2(CO)_5(COMe)(\mu\text{-}PPh_2)_2]^-$	219.3, 221.7	1144
$Fe_2(CO)_5(\mu\text{-}AsMe_2)_2(PMe_2Ph)$	212.7×, 215.1 (9)ϕ	1445
$Fe_2(CO)_5(\mu\text{-}AsMe_2)_2\{PMePh(CH_2Ph)\}$	214.0×, 214.9ϕ	1445
$Fe_2(CO)_4(\mu\text{-}AsMe_2)_2(PMe_2Ph)_2$	217.0×, 219.9 (19)ϕ	1445
C. Trinuclear Compounds		
$Fe_3(CO)_{12}$	212.9 [8.3]	372, 1280
$[Fe_2Mn(CO)_{12}]^-$	223.3	372
$Fe_2Ru(CO)_{12}$	206.4	372
$[Fe_3(CO)_{11}H]^-$	208.9, 210.4, 215.8, 219.3, 221.3, 223.8, 285.7; 214.8@, 259.8†	834; 372
$Fe_2Co(CO)_9Cp$	216.6	555

*terminal, §bridge, †intensity 1, #intensity 5, ϕintensity 2, Ψintensity 4, ×intensity 3, @intensity 10

COMPOUND	$\delta(^{13}C)$/p.p.m., J/Hz	REFERENCE
$FeRh_2(CO)_6Cp_2$	190.0 (J_{RhC} = 2 Hz); 193.3 (J_{RhC} = 2 Hz)	555
$Fe_3(CO)_{10}H(\mu\text{-COH})$	203.4, 204.9, 210.6, 214.0, 215.4, 217.2	1441
$Fe_3(CO)_{10}\{C(OR)\}H$	204.1*, 207.3*, 210.2*, 212.7*, 214.1§, 215.8§	1475
$Fe_3(CO)_8$(1,3,6-trimethylhexa-1,3,5-triene-1,5-diyl)	205.4, 208.5, 209.5, 214.7, 216.5, 217.4	696
$Fe_3(CO)_8(C_4Ph)_4$	204.2, 205.8, 255.8†	665
$Fe_3(CO)_8CEt\{C_5H_2Me_2(C_2H_3)\}$	205.0, 212.6, 210.8, 214.1, 252.8†	1178
$[Rh\{Fe(CO)_2(PPh_2)Cp\}_2]^+$	208.5 (8), 236.6 (6, 20) (J_{RhC} = 32 Hz)†	1190
$Fe_3(CO)_9(CO)_2S$	196.2, 207.7	658
$FeCo_2(CO)_9S$	209.8*, 215.1§	1214
$FeCo_2(CO)_8S\{P(OPh)_3\}$	210.6, 211.8, 217.0	1214
$FeCo_2(CO)_8S(PBu^n_3)$	210.2, 214.8, 217.2	1214
$FeCo_2(CO)_7S(PBu^n_3)_2$	215.6§, 219.9	1214
$FeCo_2(CO)_6S(PBu^n_3)_3$	219.1, 224.6	1214
$FeCo_2(CO)_7S$(dppe)	212.2 (2), 221.7	1214
D. Tetranuclear Complexes		
$Fe_4(CO)_{10}$(diazulenyl)	210.4#, 211.2§, 272.2§	686
$FeCo_3(CO)_{12}H$	210.2	1458
$FeCo_3(CO)_{11}H(PPh_3)$	208.2 (9.5)	1458
$FeCo_3(CO)_{10}H(PPh_3)_2$	210.9 (9.8)	1458
$FeCo_3(CO)_{10}H\{P(OPh)_3\}_2$	209.9 (17.0)	1458
$FeCo_3(CO)_9H\{P(OPh)_3\}_3$	211.0 (14.3)	1458
Ruthenium		
A. Mononuclear Compounds		
$Ru(CO)_4H_2$	192.6$^\phi$, 190.4$^\Psi$	986
cis-$Ru(CO)_4(SiMe_3)_2$	198.5$^\phi$, 191.6$^\Psi$	727
$Ru(CO)_4(SiMe_2CH_2CH_2SiMe_2)$	198.2$^\phi$, 193.6$^\Psi$	393
cis-$Ru(CO)_4(SiMe_2Cl)_2$	194.5$^\phi$, 188.1$^\Psi$	727
trans-$Ru(CO)_4(SiMe_2Cl)_2$	191.1	727
cis-$Ru(CO)_4(SiMeCl_2)_2$	190.6$^\phi$, 185.4$^\Psi$	727
trans-$Ru(CO)_4(SiMeCl_2)_2$	187.0	727
cis-$Ru(CO)_4(SiCl_3)_2$	187.5$^\phi$, 183.6$^\Psi$	727
cis-$Ru(CO)_4(GeMe_3)_2$	198.0$^\phi$, 191.3$^\Psi$	727
cis-$Ru(CO)_4(SnMe_3)_2$	197.4 (J_{SnC} = 61 Hz)$^\phi$ 192.3 ($J_{SnC\,trans}$ = 143 Hz, $J_{SnC\,cis}$ = 36 Hz)	727
cis-$Ru(CO)_4(PbMe_3)_2$	197.0 (J_{PbC} = 94 Hz)$^\phi$ 192.2 ($J_{PbC\,trans}$ = 252 Hz, $J_{PbC\,cis}$ = 66 Hz)	727
$Ru(CO)_4I_2$	178.2$^\phi$, 177.8 ($^2J_{CC}$ = 4 Hz)	986
$Ru(CO)_3(C_4H_6)$	197.5	1156

*intensity 2, §intensity 1, †bridge, #intensity 4, $^\phi$axial, $^\Psi$equatorial or basal

COMPOUND	$\delta(^{13}C)$/p.p.m., J/Hz	REFERENCE
$Ru(\mathit{C}O)_3(C_6H_8)$	197.9§, 202.8*	865
$Ru(\mathit{C}O)_3(C_8H_8)$	198.2, 198.9	726
$Ru(\mathit{C}O)_3(Ph_4C_4CO)$	194.1	865
$Ru(\mathit{C}O)_3\{(CF_3)_4C_4CO\}$	189.6§, 188.0*	865
$Ru(\mathit{C}O)_2(C_5Me_4Et)Cl$	198.7	1499
cis-$Ru(\mathit{C}O)_2Cl_2(PEt_3)_2$	195.4 (10.6)	193
cis-$Ru(\mathit{C}O)_2Cl_2(PEt_2Ph)_2$	193.9 (11.0)	193
cis-$Ru(\mathit{C}O)_2Cl_2(PEtPh_2)_2$	193.7 (10.6)	193
cis-$Ru(\mathit{C}O)_2Cl_2(PEt_2Bu^t)_2$	197.4 (10.9)	193
cis-$Ru(\mathit{C}O)_2Cl_2(PBu^n{}_2Bu^t)_2$	197.3 (11.0)	193
cis-$Ru(\mathit{C}O)_2Cl_2(PMeBu^t{}_2)_2$	198.3 (10.7)	193
trans-$[Ru(\mathit{C}O)(\overline{CNMeCH_2CH_2NMe})_4Cl]^+$	203.9	1477
trans-$Ru(\mathit{C}O)(\overline{CNMeCH_2CH_2NMe})_3Cl_2$	207.6	1477
$Ru(\mathit{C}O)(C_5Me_4Et)(AsPh_3)Cl$	205.6	1499
$Ru(\mathit{C}O)\{CN(C_6H_3Me\text{-}4\text{-}2)CH_2CH_2N\mathit{p}\text{-tol}\}$-$Cl(CO)(PPh_3)_2$	202.4 (9.9)	1196
$Ru(\mathit{C}O)\{CN(C_6H_3Me\text{-}4\text{-}2)CH_2CH_2N\mathit{p}\text{-tol}\}$-$Cl(CO)(PEt_3)_2$	201.2 (10.7)	1196
$Ru(\mathit{C}O)(C_2H_4)Cl_2(PMe_2Ph)$	194.0	1480
$Ru(\mathit{C}O)\{\overline{CN(CH_2Ph)CH_2CH_2N}(CH_2Ph)\}_2$-$(NO)Cl$	240.7 or 207.3	1484
$Ru(\mathit{C}O)\{CH(NMe_2)\}(PEt_3)_2Cl_2$	203.0 (26.9)	1372
$Ru(\mathit{C}O)$(tetraarylporphyrin)py	180.1, 180.3	690
$Ru(\mathit{C}O)$(octaethylporphyrin)py	183	690
B. Binuclear Compounds		
$Ru_2(\mathit{C}O)_6(C_{10}H_{12})$	193.0, 198.8, 204.6 195.7, 203.5, 206.6	1215
$Ru_2(\mathit{C}O)_5(C_{10}H_{10})$	203.8†, 210.2† 191.7#, 193.1#, 213.0#	1094
$Ru_2(\mathit{C}O)_5(C_{10}H_7Me_2Pr^i)$ Isomer 1	201.3†, 207.4† 188.9#, 190.0#, 210.6#	1094
Isomer 2	202.6†, 209.1† 194.5#, 214.3#	1094
$Ru_2(\mathit{C}O)_4Cp_2$	199.3$^{\phi}$, 251.7$^{\Psi}$	712
$Ru_2(\mathit{C}O)_4(C_5H_4Me)$	199$^{\phi}$, 255.6$^{\Psi}$	712
$Ru_2(\mathit{C}O)_4\{C_5H_3(CH_2)_4\}_2$	202.2$^{\phi}$, 252.9$^{\Psi}$	712
$Ru_2(\mathit{C}O)_4(indenyl)_2$	198.5$^{\phi}$, 250.0$^{\Psi}$	712
C. Trinuclear Compounds		
$Ru_3(\mathit{C}O)_{12}$	199.7	372
$RuFe_2(\mathit{C}O)_{12}$	206.4	372
$Ru_3(\mathit{C}O)_{10}(\overline{CNEtCH_2CH_2NEt})$	204.9	1224
$Ru_3(\mathit{C}O)_{10}(C_2Ph_2)$	185.1, 187.1, 197.6, 198.3	1209

*axial, §equatorial or basal, †$Ru(CO)_2$, #$Ru(CO)_3$, $^{\phi}$terminal, $^{\Psi}$bridge

COMPOUND	$\delta(^{13}C)$/p.p.m., J/Hz	REFERENCE
$Ru_3(CO)_{10}(NO)_2$	$198.4^{*\dagger}$	372
	$196.6^{\S\dagger}$, $182.8^{\S\#}$, $192.2^{\#}$, $196.2^{\#}$, $196.3^{\#}$, $198.4^{\dagger}$	1486
$Ru_3(CO)_{10}(N_2C_4H_4)$	196.2, 198.6, 200.4, 201.6, 202.4, 256.7, 262.3	1094
$Ru_3(CO)_{10}(Ph_2PCH_2PPh_2)$	192.5, 196.9, 214.3, 218.3	1153
$Ru_3(CO)_9H_3CMe$	189.3, 190.1,	25
	189.7, 190.5 {12}	658
$[Ru_3(CO)_9H_2(HC_2Bu^t)]^{2+}$	222.5	1482
$Ru_3(CO)_9H(C_6H_9)$	192.1, 196.6, 198.5, 199.3, 199.7	58
$Ru_3(CO)_9H(MeCCHCMe)$	190.4, 192.3, 193.2, 195.9, 197.7, 206.0	868
$Ru_3(CO)_9H(HCCMeCMe)$	188.9, 190.2, 191.6, 192.7, 195.8, 196.1, 197.6	868
$Ru_3(CO)_9H(EtCCHMe)$	188.1, 190.8, 191.4, 192.1, 192.3, 192.8, 195.7, 197.2, 203.9	868
$Ru_3(CO)_9H(EtCCHCMe)$	190.5, 192.0, 192.3, 196.1, 197.7	868
$Ru_3(CO)_9H_2S$	181.8 {13}, 186.1, 187.9, 193.3 {8}, 197.7	658
$Ru_3(CO)_9(PEt_3)_3$	202.4, 221.6	700
$Ru_3(CO)_8(NO)(PPh_3)_2$	197.7 (8.3)$^{\#}$, $200.4^{\#}$, $206.1^{\dagger}$	1486
D. Tetranuclear Complexes		
$Ru_4(CO)_{13}H_2$	185.7^{Ψ}, 187.0^{Ψ}, $189.5^{\#}$, $191.1^{\#}$, 201.2^{ϕ}	1481
$[Ru_4(CO)_{12}H_2]^{2-}$	$199.0^{\#}$, 200.0^{Ψ}, 200.3^{Ψ}, $203.3^{\#}$, 205.1^{Ψ}, $280.9^{\#}$, 281.1^{Ψ}	1479
Osmium		
A. Mononuclear Complexes		
$Os(CO)_4H_2$	$173.5^{\times}$, $171.6^{@}$	986
$Os(CO)_4Me_2$	$177.7^{\times}$, $170.4^{@}$	986
cis-$Os(CO)_4(SiMe_3)_2$	$181.5^{\times}$, $172.6^{@}$	727
trans-$Os(CO)_4(SiMe_3)_2$	186.2	727
$Os(CO)_4(SiMe_2CH_2CH_2SiMe_2)$	181.0, 175.5	393
cis-$Os(CO)_4(SiMe_2Cl)_2$	$177.5^{\times}$, $169.6^{@}$	727
trans-$Os(CO)_4(SiMe_2Cl)_2$	179.5	727
cis-$Os(CO)_4(SiMeCl_2)_2$	$173.7^{\times}$, $167.2^{@}$	727
trans-$Os(CO)_4(SiMeCl_2)_2$	174.4	727
trans-$Os(CO)_4(SiCl_3)_2$	170.3	727
cis-$Os(CO)_4(GeMe_3)_2$	$180.4^{\times}$, $171.9^{@}$	727
trans-$Os(CO)_4(GeMe_3)_2$	185.0	727
cis-$Os(CO)_4(SnMe_3)_2$	179.1 (J_{SnC} = 54 Hz)$^{\times}$ 172.3 (J_{SnCcis} = 40 Hz, $J_{SnCtrans}$ = 118 Hz)$^{@}$	727

*Ru(CO)$_4$, §Ru(CO)$_3$, †intensity 4, $^{\#}$intensity 2, $^{\phi}$intensity 7, $^{\Psi}$intensity 1, $^{\times}$axial, $^{@}$equatorial

COMPOUND	$\delta(^{13}C)$/p.p.m., J/Hz	REFERENCE
trans-$Os(CO)_4(SnMe_3)_2$	185.5	727
cis-$Os(CO)_4(PbMe_3)_2$	179.3 (J_{PbC} = 78 Hz)$^{\times}$, 172.3 (J_{PbCcis} = 53 Hz, $J_{PbCtrans}$ = 22 Hz)$^{@}$	727
$Os(CO)_4I_2$	159.2$^{\times}$, 156.4$^{@}$	986
$[Os(CO)_3(C_7H_{10})]_2$	177.9, 180.3	823
cis-$Os(CO)_2(PPr^n{}_2Bu^t)_2Cl_2$	177.6 (7.3)	72, 193
$[Os(CO)(NO)(C_2H_4)(PPh_3)_2]^+$	182.9	86
B. Trinuclear Compounds		
$Os_3(CO)_{12}$	170.4, 182.3	372
$Os_3(CO)_{11}PEt_3$	170.4, 172.8, 173.8, 176.8, 178.1, 184.4, 186.3, 194.1	700
$Os_3(CO)_{10}(COMe)H$	168.7, 169.4, 173,9, 174.5, 174.6, 175.4, 178.6, 178.7, 179.1, 180.6	1475
$Os_3(CO)_{10}(COEt)H$	169.3, 170.0, 174.4, 175.0, 175.2, 176.0, 179.3, 179.7, 181.2	1475
$Os_3(CO)_{10}(CO_2Me)H$	177.0†, 178.7§, 184.6*, 186.2*	1486
$Os_3(CO)_{10}(CO_2CF_3)H$	175.3§, 175.8§, 176.1§, 177.1§, 183.4*, 185.0*	1486
$Os_3(CO)_{10}H(NHBu^n)$	172.8§, 174.9§, 178.2§, 179.4§, 183.4*, 191.3*	1486
$Os_3(CO)_{10}H(OH)$	169.6§, 172.3§, 176.4§, 180.6*, 182.1*	1486
$Os_3(CO)_{10}H(SEt)$	169.5§, 170.4§, 173.7§ 176.3§, 179.8*, 180.0*	1486
$Os_3(CO)_{10}H(SPh)$	169.3§, 171.8§, 173.8§, 176.0§, 180.1*, 180.7*	1486
$Os_3(CO)_{10}HCl$	172.5§, 172.9§, 176.9§, 177.5§, 183.3*, 183.8*	1486
$Os_3(CO)_{10}HBr$	172.3§, 173.3§, 177.1§, 178.1§, 184.8*, 185.4*	1486
$Os_3(CO)_{10}HI$	168.5§, 170.5§, 173.8§, 175.3§, 183.8*, 184.9§	1486
$Os_3(CO)_{10}(C_7H_8)$	$^2J_{CC}$ = 35 Hz	963
$Os_3(CO)_{10}(C_6H_8)$	166.9*, 170.9*, 174.1*, 174.9*, 180.8*, 182.8*, 184.0*, 184.8*, 185.6§	1151
$Os_3(CO)_{10}(C_8H_{14})_2$	163.8*, 165.5*, 176.2§, 177.8*, 183.0*, 183.8§, 193.4§	1142
$Os_3(CO)_{10}(N_2C_4H_4)$	178.7§, 180.1§, 184.6§, 185.9* (J_{CC} = 33 Hz), 186.3§, 191.1$^{\#}$ (J_{CC} = 33 Hz)*	1120
$Os_3(CO)_{10}(PEt_3)_2$	174.2, 178.9, 180.9, 183.8, 187.8, 196.2, 197.1	700
$Os_3(CO)_{10}Cl_2$	177.0§, 177.7$^{\#}$, 178.3§, 182.6§	1192

$^{\times}$axial, $^{@}$equatorial, *intensity 1, §intensity 2, †intensity 6, $^{\#}$intensity 4

COMPOUND	$\delta(^{13}C)$/p.p.m., J/Hz	REFERENCE
$Os_3(CO)_9H_3CMe$	166.7, 167.6 {8}	658
$Os_3(CO)_9H_2S$	156.1, 165.1, 166.5, 169.5 {6}, 176.1	658
$[Os_3(CO)_9HPEt_3]^+$	177.1§, 180.1*, 183.4§, 184.6*, 185.3*, 186.7§	644
$Os_3(CO)_9H(COMe)(CNBu^t)$	168.4, 174.6, 176.4, 179.7, 180.9	1475
$Os_3(CO)_9(PEt_3)_3$	186.4, 199.4	700
C. Tetranuclear and Higher Compounds		
$Os_3Co(CO)_{12}H$	169.0†, 176.0$^\#$	1490
$[Os_5(CO)_{15}H]^-$	169.3, 180.3, 182.3, 184.8, 189.2	731
$[Os_6(CO)_{18}]^{2-}$	195.0	745
$[Os_6(CO)_{18}H]^-$	184.6, 191.5	745
Cobalt, [^{59}Co]		
A. Mononuclear Compounds		
$[Co(CO)_4]^-$	[287]	98
$Co(CO)_4CF_3$	199.2 (J_{FC} = 14.4 Hz)$^\phi$, 191.7 (J_{FC} = 10.4 Hz)$^\Psi$	1139
$Co(CO)_4SiPh_3$	203.6$^\phi$, 197.6$^\Psi$	1139
$Co(CO)_4SiF_3$	199.6$^\phi$, 191.4$^\Psi$	1139
$Co(CO)_4SiCl_3$	199.3 (J_{SiC} = 106 Hz)$^\phi$, 192.0$^\Psi$	1139
$Co(CO)_4GePh_3$	204.3$^\phi$, 197.7$^\Psi$	1139
$Co(CO)_4GeCl_3$	197.4$^\phi$, 189.7$^\Psi$	1139
$Co(CO)_4SnMe_3$	205.8 (J_{SnC} = 68 Hz)$^\phi$, 198.6 (J_{SnC} = 72 Hz)$^\Psi$	1139
$Co(CO)_4SnBu^n{}_3$	206.2$^\phi$, 198.8 (J_{SnC} = 84 Hz)$^\Psi$	1139
$Co(CO)_4Sn(CH_2Ph)_3$	205.3$^\phi$, 196.6 (J_{SnC} = 99 Hz)$^\Psi$	1139
$Co(CO)_4SnPh_3$	204.9 (J_{SnC} = 95 Hz)$^\phi$, 197.1 (J_{SnC} = 100 Hz)$^\Psi$	1139
$Co(CO)_4SnCl_3$	191.1$^\times$	1139
$Co(CO)_4PbPh_3$	206.4$^\phi$, 197.6 (J_{PbC} = 148 Hz)	
$Co(CO)_2(C_5Me_4Et)$	208.0	1499
$\overline{Co(CO)OC(OMe)=C(CO_2Me)CO}$	200.0	1020
$Co(CO)(\overline{CNEtCH_2CH_2NEt})_2(NO)$	221.3	1224
B. Binuclear Compounds		
$Co_2(CO)_8$	204.0	1501
$Co_2(CO)_6(RC_2R)$	195.6 to 201.7	336, 1306
$Co_2(CO)_5L(RC_2R)$	202.9 to 208.0	778, 1306
$Co_2(CO)_4Cp_2\{CH(CO_2Et)\}$	207.2	1494
$Co_2(CO)_4Cp_2\{C(CO_2Et)_2\}$	206.7	1494
$Co_2(CO)_6PCMe$	199.9	1497

*intensity 1, §intensity 2, †intensity 8, $^\#$intensity 4, $^\phi$axial, $^\Psi$equatorial, $^\times$averaged

COMPOUND	$\delta(^{13}C)$/p.p.m., J/Hz	REFERENCE
C. Trinuclear Compounds		
$Co_3(CO)_9CX$	198.2 to 200.9	60, 329, 664
$[Co_3(CO)_9CCHR]^+$	192.4 to 193.2	329
$Co_2Fe(CO)_9S$	198.5	1214
$Co_3(CO)_8(PMePh_2)CH$	206 (11)	538
$Co_3(CO)_8(PPh_3)CMe$	204 (11)	535
$Co_3(CO)_8(PBu^n{}_3)CMe$	209.2	535
$Co_3(CO)_8(PCy_3)CMe$	210.9	535
$Co_2Fe(CO)_8S\{P(OPh)_3\}$	199.9§, 200.8*, 206.1*	1214
$Co_2Fe(CO)_8S(PBu^n{}_3)$	200.1§, 202.8*, 211.6*	1214
$Co_2Fe(CO)_7S(PBu^n{}_3)_2$	206.7†, 213.5†	1214
$Co_2Fe(CO)_7S(dppe)$	200.6†, 210.9†	1214
$Co_2Fe(CO)_6S(PBu^n{}_3)_3$	211.3*, 219.1*, 214.9†	1214
D. Tetranuclear Compounds		
$Co_4(CO)_{12}$	191.9$^\phi$, 195.9$^\phi$, 243.1	540, 594, 1501
$Co_3Rh(CO)_{12}$	195.5$^{†\phi}$, 200.1$^{†\phi}$, 201.1$^{†\phi}$, 238.3 (J_{RhC} = 38 Hz)$^\#$, 251.2$^\#$	379, 1501
$CoFe(CO)_{12}H$	203.1	1458
$Co_4(CO)_{11}P(OMe)_3$	193.7$^{†\phi}$, 196.1$^{§\phi}$, 197.2$^\phi$, 198.0$^\phi$, 245.6$^{*\#}$, 248.2$^{†\#}$	540
$Co_3Fe(CO)_{11}H(PPh_3)$	191.0$^{*\phi}$, 203.5$^{†\phi}$, 210.8$^{†\phi}$, 234.8$^{*\#}$, 238.4$^{*\#}$	1458
$Co_4(CO)_{10}(Ph_2C_2)$	193.7†, 198.9†, 206.8$^×$, 192.2, 197.2, 211.6, 193.5$^\Psi$, 198.7†, 213.6$^\Psi$	594, 665, 1501
$Co_4(CO)_{10}\{C_2(CO_2Me)_2\}$	191.3$^\Psi$, 198.2†, 211.2$^\Psi$, 190.1†, 197.7†, 203.2$^×$	594, 1501
$Co_3Fe(CO)_{10}H(PPh_3)_2$	195.0*, 207.8†, 211.8*, 243.1$^{*\#}$, 245.2$^{*\#}$	1458
$Co_3Fe(CO)_{10}H\{P(OPh)_3\}_2$	196.9*, 202.8†, 211.1*, 237.0$^{†\#}$	1458
$Co_4(CO)_9(C_6H_5Me)$	180.9§, 198.7§, 247.3$^{§\#}$	1501
$Co_3Fe(CO)_9H\{P(OPh)_3\}_3$	203.9, 240.0$^\#$	1458
E. Hexanuclear Compound		
$[Co_6(CO)_{15}C]^{2-}$	224.9, 234.9, 252	715
Rhodium, [^{103}Rh]		
A. Mononuclear Compounds		
$[Rh(CO)_4]^-$	206.3 [74.7]	306
$Rh(CO)_2(PhCH{=}CHCHNC_3H_7)$	182.1	545
$[Rh(CO)_2Cl_2]^-$	183.1 [72]	1202
$[Rh(CO)_2Br_2]^-$	183.4 [72]	1202
$Rh(CO)_2$(piperidine)Cl	238.6	723
$Rh(CO)_2$(pyridine)Cl	181.3 [73], 185.6 [67]	723

*intensity 1, §intensity 3, †intensity 2, $^\#$bridge, $^\phi$terminal, $^\Psi$intensity 4, $^×$intensity 6

COMPOUND	$\delta(^{13}C)$/p.p.m., J/Hz	REFERENCE
B. Binuclear Compounds		
$RhFe(CO)_5(C_7H_7)$	194.3 [72.4]	716
$Rh_2(CO)_4(N_2C_3H_3)_2$	186.3 [65], 185.0 [70]	557
$Rh_2(CO)_4Cl_2$	180.4 [76.4], [68.8],	42, 383,
	177.3 [76.6]	723
$Rh_2(CO)_3Cp_2$	191.8 [83]*, 231.8 [45]§	57, 1501
$Rh_2(CO)_2Cp_2\{P(OPh)_3\}$	190.4 [85]*,	1501
	239.4 [41, 50] (19)§	1501
$Rh_2(CO)_2Cp_2(PMe_2Ph)$	235.4 [26, 60] (8)§	1501
$Rh_2(CO)_2Cp_2(CF_3C_2CF_3)$	189.8 [39.4]†, 189.0 [80]$^{\#}$	556
$Rh_2(CO)(CPh_2)_2py_2Cl_2$	206.5 [47.3]§	382
$Rh_2(CO)(CPh_2)_2Cp_2$	223.4 [42.4]§	882
C. Trinuclear Compounds		
$RhFe_2(CO)_{10}$	218.8 [9]	555
$RhFe_2(CO)_5Cp_2$	209.2 [41.0]$^{\S\phi}$, 234.5 [50]$^{\S\Psi}$	555
$Rh_3(CO)_3Cp_3$	232.5 [49]§	709
	216.8 [28], 237.0 [31]§	1508
$[Rh\{Fe(CO)_2PPh_2Cp\}_2]^+$	236.6 [32] (20, 6)§	1190
$Rh_3(CO)(PhC_2Ph)Cp_3$	241.6 [43.7, 28.4]§	206
$Rh_3(CO)(C_6F_5C{=}CC_6F_5)Cp_3$	217.8 [48.5]§	206
D. Tetranuclear Compounds		
$Rh_4(CO)_{12}$	175.5 [62]$^{\times}$, 181.8 [64]$^{\times}$,	42, 207
	183.4 [75]$^{\times}$, 228.8 [35]$^{\times @}$	1501
$Rh_4(CO)_{11}(PMePh_2)$	188.9$^{\varepsilon}$, 242.6$^{@\theta}$, 249.0$^{@\infty}$	1501
$Rh_4(CO)_{11}(PPh_3)$	178.9 [67]$^{\infty}$, 181.7 [63]$^{\infty}$, 184.4$^{\Omega}$,	1501
	234.8 [33]$^{@\theta}$, 235.9 [30]$^{@\infty}$	
$RhCo_3(CO)_{12}$	183.1 [51]$^{\times}$, 188.2 [78]$^{\times}$,	379, 1501
	238.3 [38]$^{@\theta}$	
$Rh_4(CO)_2Cp_4$	235.6 [49, 35]$^{@}$	709
E. Hexanuclear and Higher Complexes		
$[Rh_6(CO)_{15}C]^{2-}$	198.1 [77.1, 3.9], 225.2 [30.8]$^{@}$,	306
	236.3 [51.8]$^{@}$	
$[Rh_6(CO)_{15}I]^-$	183.3, 232.9 [28]$^{@}$, 239.2 [28]$^{@}$,	1501
	245.3$^{@}$	
$[Rh_{12}(CO)_{30}]^{2-}$	186.3 [73], 186.1 [73],	383
	183.4 [87.7], 211.5 [36.6]$^{@}$,	
	237.4 [*ca* 27.5]$^{\partial}$	
$[Rh_{13}(CO)_{24}H_3]^{2-}$	209.9, 229.6 [35.1]$^{@}$	1197
$[Rh_{13}(CO)_{24}H_2]^{3-}$	209.8, 235.4 [38.4]$^{@}$	1197
$[Rh_{17}(CO)_{32}S_2]^{3-}$	194.1 [97.3], 195.8 [83.9],	1506
	231.9 [44.5]$^{@}$, 252.7 [40.9]$^{@}$	

*terminal, §bridge, †30°C, $^{\#}$-66°C, $^{\phi}$room temperature, $^{\Psi}$-70°C, $^{\times}$intensity 3, $^{@}$bridging, $^{\varepsilon}$intensity 8, $^{\theta}$intensity 2, $^{\infty}$intensity 1, $^{\Omega}$intensity 6, $^{\partial}$face bridge

COMPOUND	$\delta(^{13}C)$/p.p.m., *J*/Hz	REFERENCE
Iridium		
$Ir(CO)_2Cp$	173.8	336
$[Ir(CO)_2Cl_2]^-$	169.6	1202
$[Ir(CO)\{CH(NMe_2)\}Cl_2(PPh_3)_2]^+$	158.8 (3)	1372
$Ir_4(CO)_{11}PMePh$	155.9 (26), 157.1, 158.3, 171.1, 173.2 (4), 196.2, 206.8	1234
$Ir_4(CO)_{10}(cod)$	153.2*, 153.7§, 159.2§, 172.5§, 198.0*, 210.0§	1522
$Ir_4(CO)_{10}(PMePh_2)_2$	156.3, 158.7 (6), 159.7, 159.8 (24), 160.9, 172.6 (15), 175.0, 205.6, 206.9, 217.6	1234
$Ir_4(CO)_9(PMePh_2)_3$	159.9, 160.7 (3), 165.7 (25), 176.6 (9, 14), 217.1, 217.8	1234
$Ir_4(CO)_8(cod)_2$	155.0*, 157.1§, 162.5* 176.8*, 212.1§, 220.7*	1522
$Ir_4(CO)_8(PMePh_2)_4$	164.5 (8, 33), 171.1 (4, 32), 178.2 (9, 15), 218.4, 223.4	1234
$Ir_4(CO)_8(AsMe_2Ph)_4$	169.4, 176.6, 221.2	305
$Ir_4(CO)_8(AsMePh_2)_4$	169.1, 175.5, 220.3, 223.7	305
$Ir_4(CO)_6(cod)_3$	160.0, 221.7	1522
Nickel		
$Ni(CO)_4$	191.6, 193	13, 14, 93, 536
$Ni(CO)_3(\overline{CNEtCH_2CH_2NEt})$	199.1	1224
$Ni(CO)_3PBu^n{}_3$	197.3	536
$Ni(CO)_3PPh_3$	195.9; 199.4	536
$Ni(CO)_3PPhCl_2$	192.3	536
$Ni(CO)_3P(OEt)_3$	195.2	536
$Ni(CO)_3PCl_3$	190.2 (12)	536
$Ni(CO)_3AsPh_3$	195.8	536
$Ni(CO)_3SbPh_3$	196.5	536
$Ni(CO)_2(\overline{CNEtCH_2CH_2NEt})_2$	205.4	1224
$Ni(CO)_2(PBu^n{}_3)(PPh_3)$	200.8 (3)	536
$Ni(CO)_2(PBu^n{}_3)\{P(OMe)_3\}$	200.1 (5)	536
$Ni(CO)_2(PBu^n{}_3)\{P(OPh)_3\}$	198.4 (6)	536
$Ni(CO)_2(PPh_3)_2$	199.4	536
$Ni(CO)_2(PPh_3)\{P(OMe)_3\}$	198.7 (3)	536
$Ni(CO)_2(PPh_3)\{P(OPh)_3\}$	194.2	536
$Ni(CO)_2\{P(OEt)_3\}_2$	198.3	536
$Ni(CO)_2\{P(OMe)_3\}\{P(OPh)_3\}$	196.5	536
$\overline{Ni(CO)(C_5H_4CH_2CH_2CO)}$	190.8 or 235.0	809
$Ni(CO)\{P(OEt)_3\}_3$	201.2	536
$[Ni(CO)Cp]_2$	226.8†	712
$NiFe(CO)_3Cp_2$	254.5	712
$\{Ni(CO)(C_5H_4CH_2CH_2)\}_2CO$	207.1, 226.6	809
$Ni_2(CO)_3\{(PF_2)_2NMe\}_2$	221.6	1443
$Ni_2(CO)_2\{(PF_2)_2NMe\}_3$	189.2	1443

*intensity 1, §intensity 2, †bridge

COMPOUND	$\delta(^{13}C)$/p.p.m., J/Hz	REFERENCE
Palladium		
$[Pd(CO)Cl_3]^-$	163.4	1202
$[Pd(CO)Br_3]^-$	164.9	1202
Platinum, $[^{195}Pt]$		
$Pt(CO)_2Cl_2$	151.6 [1576]	1202
$Pt(CO)_2Br_2$	152.0 [1566]	1202
trans-$[PtH(CO)(PEt_3)_2]^+$	182.8 [990]	579
$Pt(CO)MeCp$	163.4 [2459]	790
$Pt(CO)Ph_2(SEt_2)$	182.1	1244
$Pt(CO)(p\text{-tol})_2(SEt_2)$	180.3	1244
trans-$[Pt(CO)Me(PMe_2Ph)_2]^+$	177.6 [986] (10)	203
trans-$[Pt(CO)(CH_2Ph)(PEt_3)_2]^+$	175.7 [960]	579
trans-$[Pt(CO)(C_6H_4Cl)(PEt_3)_2]^+$	176.8 [978]	579
trans-$[Pt(CO)Me(AsMe_3)_2]^+$	178.7 [1007], [1000]	31, 203
trans-$[Pt(CO)Ph(AsMe_3)_2]^+$	177.6 [884]	409
$[Pt(CO)MeCl_2]^-$	162.5 [2013]	1562
$[Pt(CO)EtCl_2]^-$	161.9 [2155]	1562
$[Pt(CO)Pr^nCl_2]^-$	161.8 [2141]	1562
$[Pt(CO)Bu^nCl_2]^-$	161.8 [2144]	1562
$[Pt(CO)Pr^iCl_2]^-$	162.3 [2237]	1562
$[Pt(CO)PhCl_2]^-$	159.4 [2042]	1562
trans-$[Pt(CO)(NCS)(PEt_3)_2]^+$	164.9 [1817]	579
trans-$Pt(CO)Cl_2(NC_5H_5)$	151.1 [1669.7]	1313
trans-$Pt(CO)Cl_2(NC_5H_4Me\text{-}4)$	151.3 [1656.6]	1313
trans-$Pt(CO)Cl_2(NC_5H_4COMe\text{-}4)$	150.5 [1692.2]	1313
trans-$Pt(CO)Cl_2(NC_5H_4NC\text{-}4)$	149.7 [1726.0]	1313
trans-$Pt(CO)Cl_2(NC_5H_4OBu^n\text{-}4)$	151.9 [1649.5]	1313
trans-$[Pt(CO)(NO_3)(PEt_3)_2]^+$	160.5 [1817]	579
trans-$[Pt(CO)Cl(PMe_3)_2]^+$	160.4 [1813] (9.9)	1202
trans-$[Pt(CO)Cl(PEt_3)_2]^+$	161.7 [1777] (8.8)	1750
trans-$[Pt(CO)Cl(PPh_3)_2]^+$	158.6 [1788]	579
trans-$[Pt(CO)Br(PPh_3)_2]^+$	159.0 [1772]	579
trans-$[Pt(CO)I(PPh_3)_2]^+$	157.4 [1658]	579
$Pt(CO)Cl_2(PMe_3)$	158.4 [1808] (7.6)	1202
trans-$[Pt(CO)Cl(AsMe_3)_2]^+$	159.2 [1747]	203
trans-$[Pt(CO)Cl(AsEt_3)_2]^+$	158.8 [1740]	579
trans-$[Pt(CO)Cl(AsPh_3)_2]^+$	158.2 [1724]	579
cis-$[Pt(CO)Cl_2(AsEt_3)]$	155.8 [1725]	579
$[Pt(CO)Cl_3]^-$	151.8 [1730]	579
	151.4 [1757]	1708
	152.0 [1732]	1202
	[1720]	1537
$[Pt(CO)Br_5]^-$	[1225]	1537
$[Pt(CO)Br_3]^-$	153.0 [1701]	1202
$[Pt(CO)I_3]^-$	156.0 [1636]	1202
Copper		
$[Cu(CO)_n]^+$	168.4 to 171.1	683, 1319

COMPOUND	$\delta(^{13}C)$/p.p.m., *J*/Hz	REFERENCE
Silver		
$[Ag(\mathit{C}O)_2]^+$	170.0 to 173.4	683, 1319
Gold		
Au(*C*O)Cl	170.8	1202
Boron, [^{11}*B*]		
$BH_3\mathit{C}O$	[30.2]	549

TABLE 2.9

^{13}C Chemical Shift Data for Some CS and CSe Derivatives

COMPOUND	δ(^{13}C)/p.p.m., *J*/Hz	REFERENCE
Chromium		
Cr(*C*S)(CO)$_5$	331.1	682
Tungsten		
W(*C*S)(CO)$_5$	298.7	682, 1398
trans-W(*C*S)(CO)$_4$CNCy	298.6	1398
cis-W(*C*S)(CO)$_4$CNCy	302.7	1398
trans-W(*C*S)(CO)$_4$(NC$_5$H$_4$Me-4)	302.7	1398
trans-W(*C*S)(CO)$_4$(PBu$^n{}_3$)	300.0 (15.8)	1398
cis-W(*C*S)(CO)$_4$(PBu$^n{}_3$)	305.1 (6.9)	1398
trans-W(*C*S)(CO)$_4$(PPh$_3$)	299.6 (21.7)	1398
cis-W(*C*S)(CO)$_4$(PPh$_3$)	306.0 (6.9)	1398
trans-W(*C*S)(CO)$_4${P(*p*-tol)$_3$}	299.8 (19.7)	1398
cis-W(*C*S)(CO)$_4${P(*p*-tol)$_3$}	306.7 (5.9)	1398
trans-W(*C*S)(CO)$_4${P(OMe)$_3$}	302.8 (35.4)	1398
cis-W(*C*S)(CO)$_4${P(OMe)$_3$}	302.5 (10.8)	1398
trans-W(*C*S)(CO)$_4${P(OPh)$_3$}	301.4 (42.3)	1398
cis-W(*C*S)(CO)$_4${P(OPh)$_3$}	301.5 (9.8)	1398
trans-[W(*C*S)(CO)$_4$Cl]$^-$	287.3	682, 1398
trans-[W(*C*S)(CO)$_4$Br]$^-$	287.4	682, 1398
trans-[W(*C*S)(CO)$_4$I]$^-$	285.7	682, 1398
W(*C*S)(CO)$_3$(diars)	306.9	682
Manganese		
Mn(*C*S)(CO)$_2$Cp	442.6	295
	335.4	1712, 1745
Mn(*C*S)(CO)$_2$(C$_5$H$_4$Me)	336.6	1712
Mn(*C*Se)(CO)$_2$Cp	357.2	1712
Mn(*C*Se)(CO)$_2$(C$_5$H$_4$Me)	358.8	1712
Rhenium		
Re(*C*S)(CO)$_2$Cp	288.7	1712
Re(*C*Se)(CO)$_2$Cp	308.0	1712
Iron		
Fe(*C*S)(CO)$_4$	327.5	1472
Fe(*C*S)(CO)$_2$Cp	307.9	295
[Fe(*C*S)(CO)(PPh$_3$)Cp]$^+$	317.4 (26)	661
[Fe(*C*S)(CO){P(C$_6$H$_4$F-4)$_3$}Cp]$^+$	316.4 (27)	661
[Fe(*C*S)(CO)(PCy$_3$)Cp]$^+$	320.0 (27)	661
[Fe(*C*S)(CNPh)$_2$Cp]$^+$	320.1	661
Fe(*C*S)(porphyrin)(EtOH)	313.5	1449

TABLE 2.10

Chemical Shift Data for Some Olefins and Acetylenes π-Bonded to Metals

COMPOUND	$\delta(^{13}C)$/p.p.m., *J*/Hz	REFERENCE
Niobium		
$Nb(C_2H_4)HCp_2$	8.0 {156}, 13.4 {153}	313, 1389
$Nb(C_2H_4)EtCp_2$	27.6 {154.5}, 29.4 {153}	313, 1389
Tantalum		
$Ta(C_2H_4)MeCp_2$	20.2, 20.9	1344
Chromium		
$Cr(C^1H_2{=}C^2HCH_2C_6H_3Me_2\text{-}3,5)(CO)_2$	C^1 38.6 or 41.5, C^2 67.0	1350
$Cr(C^1H_2{=}C^2HOC_6H_3Me_2\text{-}3,5)(CO)_2$	C^1 34.7, C^2 59.2	1350
$Cr(norbornadiene)(CO)_4$	74.7, 76.0	114, 117, 1115
$Cr(norbornadiene)(CO)_3(PPh_3)$	59.8, 97.5	433
$Cr(benzonorbornadiene)(CO)_2$	27.5	432
$Cr_2(C^1Ph{=}C^2HCPhC)Cp_2(CO)$	C^1 194 or 207, C^2 107	1349
$Cr(cod)(CO)_4$	92.5	1751
$Cr(cod)(CO)_3\{P(OMe)_3\}$	81.7, 93.4	1751
Molybdenum		
$Mo(C_2H_4)_2H(cis\text{-}Ph_2PCH{=}CHPPh_2)$	22.9, 36.6, 41.2	1185
$Mo(C_2H_4)Cp_2$	11.8	1389
$Mo_2(C^1H_2{=}C^2{=}C^1H_2)Cp_2(CO)_4$	C^1 36.7, C^2 197.0	1392
$Mo_2(C^1HMe{=}C^2{=}C^3H_2)Cp_2(CO)_4$	C^1 56.2, C^2 196.8, C^3 35.1	1392
$Mo_2(C^1HMe{=}C^2{=}C^1HMe)Cp_2(CO)_4$	C^1 54.9, C^2 195.9	1392
$Mo(C_8H_8O)(CO)_3$	49.5	1461
$Mo(norbornadiene)(CO)_4$	78.5, 79.6	114, 117, 1115
$Mo_2(C_8H_8)Cp_2(CO)_2$	65.8, 100.5	739
$[Mo(MeO_2CCH{=}CHCO_2Me)_2(CO)_3]_2$	129, 130, 136	848
$Mo\{(NC)_2C^1{=}C^2{=}C^3{=}NH_2\}(CO)_2Cp$	C^1 216.3, C^2 188.7, C^3 65.2	441
$Mo\{(NC)HC^1{=}C^2{=}C^3{=}NHOH\}(CO)_2Cp$	C^1 176.0, C^2 101.2	441
$Mo\{(NC)(MeO_2C)C^1{=}C^2{=}C^3{=}NHOMe\}(CO)_2Cp$	C^1 233.6, C^2 176.1, C^3 102.1	441
$Mo\{(NC)(MeO_2C)C^1{=}C^2{=}C^3{=}NHOEt\}(CO)_2Cp$	C^1 233.5, C^2 175.8, C^3 102.5	441
$Mo\{(NC)_2C{=}C{=}NMe_2\}(CO)_2Cp$	224.1	441
$Mo\{(NC)_2C{=}C{=}N(CH_2)_5\}(CO)_2Cp$	219.6	441
$Mo(HC{\equiv}CH)Cp_2$	117.7	1389

COMPOUND	δ(^{13}C)/p.p.m., J/Hz	REFERENCE
$Mo(HC^1{\equiv}C^2Me)Cp_2$	C^1 106.0, C^2 126.1	1389
$Mo_2(Me\mathit{C}{\equiv}\mathit{C}Me)Cp_2(CO)_4$	82.6	1383
$Mo_2(HC{\equiv}CCH_2CH_2OH)Cp_2(CO)_4$	87.2	1383
$Mo_2(MeO_2C\mathit{C}{\equiv}\mathit{C}CO_2Me)Cp_2(CO)_4$	62.0	1383
Tungsten, [^{183}W]		
$[W(\mathit{C}_2H_4)HCp_2]^+$	7.7, 15.4	343
$[W(\mathit{C}_2H_4)MeCp_2]^+$	27.7, 28.0	343
$W_2(H_2\mathit{C}{=}C{=}\mathit{C}H_2)Cp_2(CO)_4$	31.6	1392
trans-$W\{\mathit{C}H_2{=}C(CO_2Me)_2\}(CO)_3(PMe_3)_2$	39.6 (9.8)	1408
trans-$W\{$*cis*-$MeO_2CCH{=}CHCO_2Me\}(CO)_3(PMe_3)_2$	42.2 (4.9)	1408
W(norbornadiene)$(CO)_4$	68.9	1115
$W\{(NC)_2C{=}C{=}N(CH_2)_5\}(CO)_2Cp$	208.9	441
$W\{(NC)_2C{=}C{=}NH_2\}(CO)_2Cp$	60, 177.3, 201.0	441
$W\{C^1H_2{=}C^2(NEt_2)CO\}(CO)_2Cp$	C^1 11.5 [24.4] {153.8} C^2 109.4 [9.8]	881
$W\{MeO_2CC^1H{=}C^2(NEt_2)CO\}(CO)_2Cp$	C^1 30.5 [29.3] {151.4} C^2 107.9 [14.6]	881
$W(PhC^1C^2O)(CO)(PMe_3)Cp$	C^2 203.1, C^2 198.2 (4.9)	1037
$W(4\text{-}MeC_6H_4C^1C^2O)(CO)(PMe_3)Cp$	C^1 203.7, C^2 197.6 (5.9)	1037
$W(4\text{-}MeOC_6H_4C^1C^2O)(CO)(PMe_3)Cp$	C^1 204.4, C^2 197.4 (7.3)	1037
$W(CpFeC_5H_4C^1C^2O)(CO)(PMe_3)Cp$	C^1 203.6, C^2 197.6 (7.3)	1037
$W(2,4,6\text{-}Me_3C_6H_2C^1C^2O)(CO)(PMe_3)Cp$	C^1 199.9, C^2 200.0 (8.8)	1037
$W(4\text{-}MeC_6H_4C^1C^2O)(CO)(PPh_3)Cp$	C^1 202.5, C^2 201.8 (7.3)	1037
$W_2(Me\mathit{C}{\equiv}CMe)(Ph_2AsCH_2AsPh_2)(CO)_5Br_2$	99.7, 100.4	1023
Manganese		
$Mn(C_8H_8)(CO)_2Cp$	56.9	1189
$Mn(C_7H_8)(CO)_2Cp$	52.6, 53.6	1189
$[Mn(4\text{-}MeC_6H_4C^1C^2O)(CO)PMe_3)Cp]^+$	C^1 197.6 (5.9), C^2 203.7	1034
$Mn(\mathit{C}S_2)(CO)_2Cp$	254.1	1032
$Mn\{\mathit{C}S_2Mn(CO)_2Cp\}(CO)_2Cp$	251.5	1032
Iron		
$[Fe(\mathit{C}_2H_4)(CO)_2Cp]^+$	56.8, 56.9	287, 489, 659, 1003
$[Fe(\mathit{C}_2H_4)(CO)_2(indenyl)]^+$	63.5	489
$[Fe(C^1H_2{=}C^2HMe)(CO)_2Cp]^+$	C^1 58.0, C^2 88.8 C^1 53.9, C^2 83.9	4, 489
$[Fe(C^1H_2{=}C^2HMe)(CO)_2(indenyl)]^+$	C^1 61.3, C^2 89.7	489
$[Fe(C^1H_2{=}C^2HEt)(CO)_2Cp]^+$	C^1 52.3, C^2 89.3	489
$[Fe(C^1H_2{=}C^2HPh)(CO)_2Cp]^+$	C^1 48.7, C^2 85.6	659
$[Fe(C^1H_2{=}C^2HCO_2Me)(CO)_2Cp]^+$	C^1 34.5, C^2 44.8	320, 618
$[Fe\{C^1H_2{=}C^2HP(O)Ph_2\}(CO)_2Cp]^+$	C^1 117.0, C^2 120.8 (86.9)	828
$[Fe(C^1H_2{=}C^2Me_2)(CO)_2Cp]^+$	C^1 59.0, C^2 118.8	489
cis-$Fe\{C^1H_2C^2CH(CO_2R)CH(CO_2R)\}(CO)_4$	C^1 23, C^2 50	618
trans-$Fe\{C^1H_2C^2CH(CO_2R)CH(CO_2R)\}(CO)_4$	C^1 24, C^2 55	618

COMPOUND	δ(^{13}C)/p.p.m, *J*/Hz	REFERENCE
$Fe\{C^1H_2{=}C^2CHCOOCOCH\}(CO)_4$	C^1 19, C^2 56	618
$Fe\{C^1H_2{=}C^2CH{=}CHC{=}CH_2\}(CO)_4$	C^1 25.2, C^2 81.7	725
$Fe(C^1H_2{=}C^2COC_4H_4)(CO)_4$	C^1 31.5, C^2 85.8	1462
$Fe_2(C^1H_2{=}C^2{=}C^2{=}CH_2)(CO)_6$	C^1 70.6, C^2 123.3	597, 1462
$Fe_2(C^1HMe{=}C^2{=}C^3{=}C^4H_2)(CO)_6$ *	C^1 C^2 121.8, C^4 69.6, C^3 124.2, C^1 92.4, C^2 123.1, C^4 71.1, C^3 124.6	1462
$Fe_2(C^1HPh{=}C^2{=}C^3{=}C^4H_2)(CO)_6$ *	C^1 97.1, C^2 116.6 C^4 70.7, C^3 124.2 C^1 96.6, C^2 120.1 C^4 71.8, C^3 125.6	1462
$Fe_2(C^1Ph_2{=}C^2{=}C^3{=}C^4H_2)(CO)_6$	C^1 93.3, C^2 122.8 C^4 119.2, C^3 120.8	1462
$Fe_2(\mathit{C}Me_2{=}\mathit{C}{=}\mathit{C}{=}CH_2)(CO)_6$	120.1, 120.3	1462
$[Fe(\mathit{trans}\text{-}\mathit{C}HMe{=}\mathit{C}HMe)(CO)_2Cp]^+$	89.2	489
$Fe(\mathit{trans}\text{-}MeO_2C\mathit{C}H{=}\mathit{C}HCO_2Me)(CO)_4$	44.9	320
$Fe(\mathit{cis}\text{-}MeO_2C\mathit{C}H{=}\mathit{C}HCO_2Me)(CO)_4$	46.6	320
$Fe\{\mathit{C}HMe{=}\mathit{C}HCH_2C(CF_3){=}C(CF_3)\}(CO)Cp$	71.7, 72.6	1116
$Fe(C^1HPh{=}C^2HCH{=}NPh)(CO)_3$	C^1 74.9, C^2 62.6	874
$Fe(PhCO\mathit{C}H{=}\mathit{C}HC_5H_4FeCp)(CO)_4$	50.0, 59.0	355
$Fe(Ph\mathit{C}H{=}\mathit{C}HCOC_5H_4FeCp)(CO)_4$	51.1, 58.3	355
$Fe(Ph\mathit{C}H{=}\mathit{C}HCOC_5H_4FeCp)(CO)_3$	61.2, 67.7	355
$[\{Fe(CO)_2Cp\}_2\mathit{C}_4H_4]^{2+}$	74.8 {182}	314, 659
$Fe\{\mathit{C}H{=}\mathit{C}HC({=}CH_2)C({=}CH_2)\}(CO)_4$	61.0	725
Fe(norbornadiene)$(CO)_3$	38.3, 38.7 {151.6}	865, 924
[Fe(norbornadiene)H$(CO)_3]^+$	39.8 {177.1}, 55.8 {181.4}	924
[Fe(norbornadiene)$(CO)_2Cp]^+$	86.9	1003
$Fe(C_8H_8O)(CO)_4$	54.9 {171}	993, 1461
$Fe\{C_8H_8OFe(CO)_3\}(CO)_4$	58.6	1461
$Fe(C_9H_{10})(CO)_4$	57.6, 63.9	1456
$Fe(C_{10}H_{12})(CO)_3$	from 65.3, 61.9, 57.7, 55.2, 55.1, 35.0, 34.2	1716
$Fe\{C_{16}H_{16}Fe(CO)_3\}(CO)_3$	from 64.3, 59.1, 57.2, 54.9, 52.2, 49.8, 41.1, 38.7, 35.8, 35.7	1716
$Fe\{C_{12}H_{12}(OMe)_2\}(CO)_3$	117.1	843
$Fe(C_{12}H_6F_4)(CO)_3$	43.8	865
[Fe(cyclohexene)$(CO)_2Cp]^+$	82.7	1003
[Fe(1,4-cyclohexadiene)$(CO)_2Cp]^+$	80.2	1003
[Fe(cycloheptene)$(CO)_2Cp]^+$	80.0	1003
[Fe(cod)(CO)Cp]$^+$	82.3	1003
Fe(cod)$(CO)_3$	83.7	865
Fe(cod)$(CNBu^t)_3$	74.1	745
trans-$Fe\{C_8H_{10}Fe(CO)_3\}(CO)_4$	from 54.7, 57.0, 61.2, 62.7	1456

* two isomers

COMPOUND	$\delta(^{13}C)$p.p.m., J/Hz	REFERENCE
cis-$Fe\{C_8H_{10}Fe(CO)_3\}(CO)_4$	from 60.8, 61.5, 91.6	1456
$Fe\{C_8H_{10}Fe(CO)_3\}(CO)_4$	from 56.6, 61.1, 66.9, 70.4, 88.7, 89.8	1456
$Fe_2(\mathit{C}H=\mathit{C}HCOS)(CO)_6$	77.2, 87.6	960
$Fe\{MeO_2CC^1H=C^2(CO_2Me)_2\}(CO)_3py$	C^1 52.1, C^2 65.2	1435
$Fe(Bu^tN\mathit{C}HCH=NBu^t)(CO)_3$	106.5	874
$Fe_2(C^1HMe=C^2C^3=C^4HMe)(CO)_6$*	C^1 94.4, C^2 121.1 C^4 71.1 C^3 123.3	1462
	$C^{1,4}$ 92.1, $C^{2,3}$ 123.2	1462
$Fe_2(C^1HPh=C^2C^3=C^4HPh)(CO)_6$*	C^1 96.8, C^2 120.5 C^4 98.7, C^3 118.4	1462
	$C^{1,4}$ 97.3, $C^{2,3}$ 117.3	1462
$Fe_2(C^1HMe=C^2C^3=C^4Me_2)(CO)_6$*	C^1 96.0, C^2 121.4, C^4 118.9, C^3 118.9	1462
	C^1 118.5, C^2 121.1 C^4 71.9, C^3 124.1	1462
$Fe_2\{C^1HPh=C^2C^3=C(CH_2)_5\}(CO)_6$	C^1 98.4, C^2 119.9, C^3 117.0	1462
$Fe_2\{(C^1H=C^2Me)_2CO\}(CO)_6$	C^1 171.1 {161}, C^2 100.7	859
$Fe_2\{C^1H=C^2Bu^t)_2CO\}(CO)_6$	C^1 163.1 {160}, C^2 122.3	859
$Fe_2(C^1Me=C^2HCOC^3Me=C^4H)(CO)_6$	C^1 200.9, C^2 83.5 {161} C^4 169.7 {166} C^3 101.4	859
$Fe_2(C^1Bu^t=C^2HCOC^3Bu^t=C^4H)(CO)_6$	C^1 216.3, C^2 78.2 {166} C^4 161.9 {161}, C^3 123.9	859
$Fe(Bu^t{}_2C^1=C^2=C^3=C^2=C^1Bu^t{}_2)(CO)_4$	C^1 161.7, C^2 83.1, C^3 139.7	597
$Fe_2(Bu^t{}_2C^1=C^2=C^3=C^2=C^1Bu^t{}_2)(CO)_6$	C^1 168.0, C^2 126.5, C^3 91.6	597
$Fe_2\{(CH_2)_5C^1=C^2C^2=C^1(CH_2)_5\}(CO)_6$	C^1 129.8, C^2 117.8	1462
$Fe_2\{(C^1Ph=C^2Ph)_2CO\}(CO)_6$	C^1 192.4 or 195.5, C^2 94.2	859
$Fe_2\{\mathit{C}(CO_2Me)=C(CO)MeNR\}(CO)_6$	112.0, 125.8	1431
$Fe_2(C_{16}H_{15}O)(CO)_6$	63.3, 82.9	945

Ruthenium

COMPOUND	$\delta(^{13}C)$p.p.m., J/Hz	REFERENCE
$Ru(\mathit{C}_2H_4)(CO)(PMe_2Ph)_2Cl_2$	58.6	1480
Ru(norbornadiene)$(CO)_2Cl_2$	64.8, 77.1	301
Ru(norbornadiene)$(NH_2Cy)_2Cl_2$	63.3, 71.0*	849
Ru(norbornadiene)$(NH_2Ph)_2Cl_2$	70.2	849
Ru(norbornadiene)$(NHC_5H_{10})_2Cl_2$	71.4	849
Ru(norbornadiene)$(py)_2Cl_2$	75.5	849
Ru(cod)$(H_2NCy)_2Cl_2$	84.8, 87.3*	849
Ru(cod)$(H_2NC_{12}H_{23})_2$	84.7, 87.3*	849
Ru(1-3,5,6-η^5-C_8H_{11})(η^6-$C_6H_5BF_3$)	from 69.6, 77.3, 80.7	1180
Ru(1-3,5,6-η^5-8-$Me_3SiC_8H_8$)$(CO)_2$ $SiMe_3$	from 40.2, 71.8, 102.7, 115.6	1439
$Ru_2(C_{10}H_{12})(CO)_6$	73.6 and 72.4 or 75.0	1215

* two isomers

COMPOUND	$\delta(^{13}C)$p.p.m., J/Hz	REFERENCE
Osmium		
$[Os(C_2H_4)(CO)(NO)(PPh_3)_2]^+$	54.4	86
Cobalt		
$Co(C_{10}H_{12})Cp$	65.6	539
$Co(C_6H_4O_2)Cp$	91.3	562
$Co(cod)_2Li(thf)_2$	50.7, 71.5, 74.3, 79.6	748
$Co_2(HC_2H)(CO)_6$	70.8	1306
$Co_2(HC_2H)(CO)_5PEt_2Ph$	69.6	1306
$Co_2(HC_2H)(CO)_5AsPh_3$	70.2	1306
$Co_2(HC^1{\equiv}C^2Me)(CO)_6$	C^1 73.0, C^2 90.8	1306
$Co_2(HC^1{\equiv}C^2C_5H_{11}{}^i)(CO)_6$	C^1 74.0, C^2 99.0	1306
$Co_2(HC^1{\equiv}C^2Bu^t)(CO)_6$	C^1 73.4, C^2 112.0	1306
$Co_2(HC^1{\equiv}C^2Bu^t)(CO)_5PEt_2Ph$	C^1 71.2, C^2 108.7	1306
$Co_2(HC^1{\equiv}C^2CO_2Me)(CO)_6$	C^1 73.5, C^2 76.9	1306
$Co_2(MeC{\equiv}CMe)(CO)_6$	94.4	1306
$Co_2(MeC^1{\equiv}C^2Ph)(CO)_6$	C^1 94.2, C^2 91.6	1306
$Co_2(PhC^1{\equiv}C^2Ph)(CO)_6$	89.6, 91.0	336,1306
$Co_2(PhC^1{\equiv}C^2Ph)(CO)_5PBu^n{}_3$	89.2	1306
$Co_2(CF_3C{\equiv}CCF_3)(CO)_6$	29.4	1306
$Co_2(HO_2CC{\equiv}CCO_2H)(CO)_6$	81.0	1306
$Co_2(MeO_2CC{\equiv}CCO_2Me)(CO)_6$	79.0	1306
Rhodium, [^{103}*Rh*]		
$Rh(C_2H_4)_2Cp$	60.0 [14]	11, 244 1389
$Rh(C_2H_4)_2$(2,6-lutidine)Cl	58.0 [13.2], 67.5 [11.0]	1511
$Rh(C_2H_4)_2$(acac)	36.3, 59.4	11, 1389, 1428
$Rh(C_2H_4)_2$(dbm)	59.9	1510
$Rh(C_2H_4)_2$(dpm)	59.3	1510
$Rh(C_2H_4)_2$(tfac)	60.6	1510
$Rh(C_2H_4)\{P(p\text{-tol})_3\}_2Cl$	44.6 [16]	323
$Rh(C^1H_2{=}C^2HMe)$(acac)	C^1 61.4, C^2 73.0	1428
	C^1 63.8 [15.5]* C^2 75.8 [13.7], C^1 65.0 [15.5]*	4
$Rh(C^1H_2{=}C^2HMe)$(dbm)	C^1 61.6, C^2 74.0	1510
$Rh(C^1H_2{=}C^2HMe)$dpm)	C^1 62.0, C^2 72.4	1510
$Rh(C^1H_2{=}C^2HMe)$(tfac)	C^1 61.9, C^2 75.0	1510
$Rh\{(C^1H_2{=}C^2HCH_2CH_2CH_2)_3P\}Cl$	C^1 55.4 [5.5], C^2 78.5 [5.5]	824
$Rh\{(C^1H_2{=}C^2HCH_2CH_2CH_2CH_2)_3P\}Br$	C^1 54.4 [6.4], C^2 77.6 [5.5]	824
$Rh\{(C^1H_2{=}C^2HCH_2CH_2)_3P\}Cl$	C^1 59.4 [6.4] C^2 81.0 [6.0] (6.0)	824
$Rh\{(C^1H_2{=}C^2HCH_2CH_2)_3P\}Br$	C^1 58.9 [5.5] C^2 81.7 [5.5] (5.7)	824

*two isomers

COMPOUND	$\delta(^{13}C)$p.p.m., J/Hz	REFERENCE
Rh{($C^1H_2=C^2HCH_2)_2C(CO_2Me)_2$}(acac)	C^1 61.9 [14.9] C^2 70.0 [13.2]	1507
Rh{($C^1H_2=C^2HCH_2)_2$NMe}(acac)	C^1 55.4 [13.3] C^2 70.7 [14.8]	1507
Rh{($C^1H_2=C^2HCH_2)_2NCC_3H_5O$}(acac)	C^1 55.3 [13.2] C^2 58.8 [16.2]	1507
Rh{$C^1H_2=C^2H)_2C_3H_4$}(thfac)	C^1 81.8, C^2 63.1	560
Rh($C^1H_2=C^2HPh)_2$(acac)	C^1 54.2, C^2 73.9	1428
Rh($C^1H_2=C^2HPh)_2$(dbm)	C^1 55.3, C^2 74.4	1510
Rh($C^1H_2=C^2HPh)_2$(dpm)	C^1 50.0, C^2 73.4	1510
Rh($C^1H_2=C^2HPh)_2$(tfac)	C^1 54.4, C^2 75.6	1510
Rh($C^1H_2=C^2HCO_2Me)_2$(acac)	C^1 61.4, C^2 61.4	1428
Rh($C^1H_2=C^2HCO_2Me)_2$(dbm)	C^1 65.2, C^2 62.5	1510
Rh($C^1H_2=C^2HCO_2Me$)(dpm)	C^1 61.4, C^2 61.4	1510
Rh($C^1H_2=C^2HCO_2Me$)(tfac)	C^1 65.2, C^2 62.5	1510
Rh($C^1H_2=C^2HOAc$)(acac)	C^1 42.4, C^2 97.9	1428
Rh($C^1H_2=C^2HOAc$)(dmb)	C^1 44.1, C^2 98.5	1510
Rh($C^1H_2=C^2HOAc$)(dpm)	C^1 43.7, C^2 97.5	1510
Rh($C^1H_2=C^2HOAc$)(tfac)	C^1 44.1, C^2 98.5	1510
Rh($C^1H_2=C^2HCl$)(acac)	C^1 60.5, C^2 81.8	1428
Rh($C^1H_2=C^2HCl$)(dbm)	C^1 60.0, C^2 82.6	1510
Rh($C^1H_2=C^2HCl$)(dpm)	C^1 59.5, C^2 79.2	1510
Rh($C^1H_2=C^2HCl$)(tfac)	C^1 61.0, C^2 83.2	1510
Rh{($C^1HPh=C^2HCH_2)_2O$}(acac)	C^1 64.6, [14.7], C^2 70.9, [11.8]	1507
Rh($Ph_2PCH_2CH_2$*C*H=*C*$HCH_2CH_2PPh_2$)Cl	62.3 [16.5]	642, 966
Rh($C_6H_9PCy_2$)(PCy_3)Cl	53.0 [14.3], 59.2 [13.9]	1514
Rh(norbornadiene)Cp	27.8 [10], 28.7 [10]	11,229
[Rh(norbornadiene)(SC_4Me_4)]$^+$	54.4 [9.2]	1517
[Rh(norbornadiene)(C_5Me_5)(NCMe)]$^{2+}$	47.0, 51.2, 94.9 [4.6], 105.7 [4.6]	1518
[Rh{2,3-($MeO_2C)_2C_7H_6$}Cl]$_2$	$C^{2,3}$ 50.3 [12.1], $C^{5,6}$ 55.9 [10.1]	1314
[Rh{2,3-($MeO_2C)_2C_7H_6$}Br]$_2$	$C^{2,3}$ 50.5 [11.4], $C^{5,6}$ 55.6 [10.1]	1314
[Rh{2,3-($MeO_2C)_2C_7H_6$}I]$_2$	$C^{2,3}$ 38.3 [11.5], $C^{5,6}$ 52.8	1314
Rh{2,3-($MeO_2C)_2C_7H_6$}(acac)	$C^{2,3}$ 50.9 [12.5], $C^{5,6}$ 57.1 [11.0]	1314
Rh{2,3-($MeO_2C)_2C_7H_6$}(acac)py	$C^{2,3}$ 48.0 [12.5], $C^{5,6}$ 63.9 [8.0]	1314
Rh{2,3-($MeO_2C)_2C_7H_6$}(hfac)	$C^{2,3}$ 52, $C^{5,6}$ 58.4 [10.4]	1314
Rh{2,3-($MeO_2C)_2C_7H_6$}(hfac)py	$C^{2,3}$ 29.0 [14.0] $C^{5,6}$ 62.3 [8.0]	1314
[Rh($C_{10}H_{11}$)(C_5Me_5)]$^+$	93.4 [9.2] or 104.1 [10.7]	1518
[Rh(C_9H_{12})Cl]$_2$	72.9 [13], 79.1 [12]	1502
Rh(C_6H_4O)Cp	93.5	561
[Rh(1,2,5-η^3-C_8H_{13})Cp]$^+$	75.5, 84.0 [15]	1172
[Rh(1,2,4-η^3-C_8H_{13})Cp]$^+$	80.3 [15], 87.0 [5]	1172
Rh(cod){$(CH_2)_2PBu^t{}_2$}	78.5 [9.8]	1512
Rh(cod)Cp	62.4 [14]	11

COMPOUND	$\delta(^{13}C)$p.p.m., J/Hz	REFERENCE
$Rh(cod)(C_5Cl_5)$	80.4 [13]	534
$[Rh(cod)(C_4Me_2H_2S)]^+$	83.1 [12.2], 86.9 [10.7]	1517,1504
$[Rh(cod)(C_4Me_4S)]^+$	85.1 [12.2]	1517
$Rh(cod)(RNCHC_6H_4O)$	84.3 to 85.7, 69.6 to 74.3	1516
Rh(cod)(acac)	75, 76.0 [14]	11,871
Rh(cod)(tfac)	76.3 [14]	11,871
Rh(cod)(hfac)	78.2 [16]	11,871
$[Rh(cod)Cl]_2$	78.3 to 78.9 [14.6]	131,871
	79.2 [14]	
Rh(cot)Cp	68 [12]	298
$Rh(C_8H_9)Cp$	C^6 77.8 [10] or 89.1 [10]	298
	C^7 38.6 [5]	
$[Rh(C_{10}H_{12})Cl]_2$	88.0 [15]	1509
$[Rh\{C_{10}H_{10}(O_2CMe)_2\}Cl]_2$	71.9 [5]	1509
$Rh_2(C_4Ph_4)Cp_2$	75.4 [11.8], 96.9 [5.8],	556
	101.5 [6.3], 161.5	
	[31.5,21.5]	
$Rh_2(C_7H_8)Cp_2$	C^5 58.0 [5], C^6 25.1 [7]	299
$Rh_2(C_8H_{10})Cp_2$	C^5 53.5, C^6 62.5	299
$[Rh_2(C_8H_{10})HCp_2]^+$	C^5 56.6 [9], or 59.5 [6],	299
	or 63.8 [8], C^6 75.1 [10]	
$Rh_3(PhC{\equiv}CPh)Cp_3$	301.6 [36.6]	556
$Rh_3(PhC{\equiv}CPh)(CO)Cp_3$	149.8 [25.2, 15.4, 7.0]	556
$Rh_2(CF_3C{\equiv}CCF_3)(CO)Cp_2$	115.1	556
Iridium		
$Ir(C_2H_4)(acac)$	40.9	1515
$Ir(C^1H_2{=}C^2HMe)(acac)$	C^1 46.7, C^2 54.6	1515
$[Ir(C^1H_2{=}C^2HMe)(C_3H_5)(C_5Me_5)]^+$	C^1 51.0, C^2 64.6	1518
$Ir(C^1H_2{=}C^2HPh)(acac)$	C^1 38.3, C^2 55.1	1515
$Ir(C^1H_2{=}C^2HCO_2Me)(acac)$	C^1 42.0, C^2 42.0	1515
$Ir(C^1H_2{=}C^2HOAc)(acac)$	C^1 27.9, C^2 78.2	1515
$Ir(C^1H_2{=}C^2HCl)(acac)$	C^1 41.0, C^2 60.7	1515
$Ir(Ph_2PCH_2CH_2CH{=}CHCH_2CH_2PPh_2)$	43.8	968
$Ir(C_6H_4O_2)Cp$	83.0	561
$Ir(C_6H_{10}PCy_2)(PCy_3)Cl$	36.0, 40.2	1514
$Ir(cod)(RNCHC_6H_4O)$	68.8 to 70.7, 52.3 to 56.1	1516
Ir(cod)(2-picoline)Cl	56.6, 60.0, 68.6, 69.8	979
$Ir(cod)(PCy_3)Cl$	90.0 (13), 51.5	979
$Ir(cod)(PPh_3)Cl$	94.0 (13), 53.6	979
$Ir(cod)\{P(OPh)_3\}Cl$	105.9 (20.5), 54.5	979
Ir(cod)(acac)	59.3	871
Ir(cod)(tfac)	60.1, 60.9	871
Ir(cod)(hfac)	62.4	871
Ir(cod)(dbm)	60.1	871
Ir(cod)(dpm)	59.2	871
Ir(cod)(MeCOCHCOPh)	59.6, 59.8	871
$Ir(cod)(F_3CCOCHCOPh)$	60.6, 61.3	871
$Ir(cod)(MeCOCHCO_2Me)$	58.6, 59.0	871
$[Ir(cod)Cl]_2$	62.1	131,871

COMPOUND	$\delta(^{13}C)$p.p.m., J/Hz	REFERENCE
$[Ir(1\text{-}3,5,6\text{-}\eta^5\text{-}C_8H_{11})(C_5Me_5)]^+$	36.8, 81.4	1518
Nickel		
$Ni(C_2H_4)\{P(OC_6H_4Me\text{-}2)_3\}_2$	47.4	603
$Ni(C^1H_2{=}C^2HCH_2CH_2CO)Cp$	C^1 48.8, C^2 83.3	284
$Ni(C^1H_2{=}C^2HCN)\{P(OC_6H_4Me\text{-}2)_3\}_2$	C^1 42.7, C^2 25.9	603
$Ni(C^1H_2{=}C^2HCO_2Me)\{P(OC_6H_4Me\text{-}2)_3\}_2$	C^1 44.7, C^2 52.4	603
$Ni(NC\mathit{C}H{=}\mathit{C}HCN)\{P(OC_6H_4Me\text{-}2)_3\}_2$	24.8	603
$Ni(MeO_2C\mathit{C}H{=}\mathit{C}HCO_2Me)\{P(OC_6H_4Me\text{-}2)_3\}_2$	51.1	603
$Ni(O_2C\mathit{C}H{=}\mathit{C}HCO)\{P(OC_6H_4Me\text{-}2)_3\}_2$	47.8	603
$Ni(C_6H_4O_2)(cod)$	100.5*, 112.2§	562
$Ni(cod)_2$	89.6, 89.7	1084
$Ni(cod)(C_4Ph_4AlPhOEt)$	86.2, 91.3	1084
$Ni(norbornene)(PPh_3)_2$	66.3 (N = 14.6 Hz)	890
$Ni(norbornene)_2PMe_3$	64.8, 65.7, 69.1, 70.9	890
$Ni(norbornene)_3$	73.1, 76.2, 76.5	890
Ni(*trans,trans,trans*-1,5,9-cyclododecatriene)PMe_3	88.0 (7.0), 89.9 (1.9)	890
Ni(cis,cis,cis-1,5,9-cyclododecatriene)	89.0	890
Ni(*cis,cis,cis*-1,5,9-cyclododecatriene)$P(OC_6H_4OPh\text{-}2)_3$	99.5	890
Palladium		
$Pd(C_2H_4)_3$	63.5	1175
$[Pd(C^1H_2{=}C^2HCH_2CH_2C^2H{=}C^1H_2)(2\text{-}MeC_3H_5)]^+$	C^1 129.5, C^2 96.2	1173
$Pd(C^1H_2{=}CHCH_2C_7H_6)(hfac)$	C^1 76.1, C^2 108.1	199
$Pd(C^1H_2{=}C^2ClCHCO_2MeCHCO_2Me)Cp$	C^1 62.5, C^2 77.8	862
$Pd(C^1H_2{=}C^2ClCHCO_2MeCHCO_2Me)(acac)$	C^1 78.6, C^2 91.0	862
$Pd(C^1H_2{=}C^2ClCHCO_2MeCHCO_2Me)(hfac)$	C^1 80.2, C^2 92.0	862
$[Pd(C^1H_2{=}C^2ClCHCO_2MeCHCO_2Me)Cl]_2$	C^1 78.3, C^2 89.8	862
$[Pd(MeOC^1H{=}C^2HCHCO_2MeCHCO_2Me)Cl]_2^{\dagger}$	C^1 143.3, C^2 43.1 C^1 141.9, C^2 41.3	862
$[Pd(EtOC^1H{=}C^2HCHCO_2MeCHCO_2Me)Cl]_2^{\dagger}$	C^1 142.7, C^2 43.1 C^1 141.1, C^2 41.1	862 862
$Pd(CBu^t{=}CMe\mathit{C}Me{=}\mathit{C}Bu^tCl)(acac)$	from 84.0, 124.9, 127.3, 141.3	1194
$Pd\{1,3,4\text{-}\eta^3\text{-}C_4(p\text{-}tol)_4Ph\}(S_2CNR_2)$	C^3 123.0 to 123.9, C^4 103.4 to 108.5	1529
$Pd(C_6H_4O_2)(PBu^n_3)_2$	104	589
$Pd_2(C_6H_4O_2)(PBu^n_3)_4$	65	589
$Pd(norbornene)_3$	82.5	602
$Pd(2,3,5\text{-}\eta^3\text{-}norbornenyl)(hfac)$	C^3 76.4, C^2 105.7	199
$[Pd(2,3,5\text{-}\eta^3\text{-}6\text{-}Ph\text{-}norbornenyl)Cl]_2$	C^2 112.5, C^3 71.1	1013

*cod, §duroquinone, †two isomers

COMPOUND	δ(^{13}C)p.p.m., *J*/Hz	REFERENCE
[Pd{7,7-(CHCO$_2$Et)Clnorbornene}Cl]$_2$	103.5, 105.1	300
[Pd{7,7-(CHCO$_2$Et)(MeO)norbornene}Cl]$_2$	101.1, 103.4	300
[Pd(C$_5$Me$_5$CHPhCH$_2$)Cl]$_2$	119.9 and 134.6 or 148.2	1177
Pd(C$_5$Me$_5$CHPhCH$_2$)(acac)	from 122.1, 128.9, 134.1, 147.0	1177
[Pd(C$_6$H$_7$MeBu^{t}CH$_2$)Cl]$_2$	95.6, 131.3	1086
[Pd{C$_6$H$_7$Me(CO$_2$Me)Bu^{t}CH$_2$}Cl]$_2$	95.3, 130.8	1086
[Pd(1,2,5-η^3-C$_8$H$_{13}$)Cl]$_2$	C^1 105.4 {155.6}, C^2 101.0 {159.3}	643
Pd(1,2,5-η^3-6-MeO$_2$CC$_8$H$_{12}$)(hfac)	97.1, 103.5	199
[Pd(cod)(2-MeC$_3$H$_5$)]$^+$	113.8	1173
[Pd(cod)(C$_8$H$_{13}$)]$^+$	113.4	1200
Pd(cod)Cl$_2$	117.2	131
[Pd(cot)(C$_3$H$_5$)]$^+$	114.9 (averaged)	1173
[Pd(cot)(2-MeC$_3$H$_4$)]$^+$	114.6 (averaged)	1173
[Pd(cot)(2-PhC$_3$H$_4$)]$^+$	115.0 (averaged)	1173
Pd(C$_9$H$_{12}$)Cl$_2$	117.7	1502
Pd(C$_{10}$H$_{12}$)Cl$_2$	113.3	131

Platinum [^{195}*Pt*]

A. Acyclic Olefins

COMPOUND	δ(^{13}C)p.p.m., *J*/Hz	REFERENCE
Pt(C_2H$_4$)$_3$	48.4 [113.0]	1175
Pt(C_2H$_4$)$_2$(C$_2$F$_4$)$_2$	65.9 [38]	578
Pt(C_2H$_4$)$_2$PMe$_3$	33.0 [158] (16), 37.9 [139] (-7)	1539
Pt(C_2H$_4$)$_2$PMe$_2$Ph	34.4 [159] (15) 39.5 [139] (-6)	1539
Pt(C_2H$_4$)$_2$PMePh$_2$	36.3 [156] (15), 42.0 [137] (-6)	1539
Pt(C_2H$_4$)$_2$PPh$_3$	38.4 [154] (12), 45.2 [137] (-5)	1539
Pt(C_2H$_4$)$_2$PCy$_3$	36.9 [146] (2) (averaged)	1539
Pt(C_2H$_4$)$_2$AsPh$_3$	37.1 [193], 41.6 [136]	1539
Pt(C_2H$_4$)Me(HBpz$_3$)	24.7 [384]	866
trans-[Pt(C_2H$_4$)Me(PMe$_2$Ph)$_2$]$^+$	84.4 [50]	32, 203
trans-Pt(C_2H$_4$)(NH$_3$)Cl$_2$	73.9	666
trans-Pt(C_2H$_4$)(NH$_2$Me)Cl$_2$	73.9 [160.3]	1296
trans-Pt(C_2H$_4$)(NH$_2$R)Cl$_2$	75.1 to 75.6 [160.3, 163.3]	666, 1296
trans-Pt(C_2H$_4$)(NHMe$_2$)Cl$_2$	74.8 [156.7]	1296
	67.7	666
	74.6 [156]	576
trans-Pt(C_2H$_4$)(NHR$_2$)Cl$_2$	73.1 to 74.7 [153.8 to 162.0]	1296
trans-Pt(C_2H$_4$)(HNC$_5$H$_{10}$)Cl$_2$	79.3 [153.8]	628
	74.4 [153.8]	1296

COMPOUND	$\delta(^{13}C)$p.p.m., *J*/Hz	REFERENCE
trans-Pt(C_2H_4)(py)Cl_2	74.8	576
	75.5	666
	75.3 [165]	577,660 1230
	75.2 [167.9]	1313
trans-Pt(C_2H_4)(substituted pyridine)Cl_2	73.4 to 76.0 [162.5 to 170]	576, 577, 666, 867, 1230, 1313
trans-Pt(C_2H_4)(NCMe)Cl_2	74.9 [198], 70.7	1230, 666
Pt(C_2H_4)(PPh_3)$_2$	39.6 [194]	32, 351 1389
trans-Pt(C_2H_4)(ONC_5H_5)Cl_2	61.0 [217],	576
	63.9 [218.0]	625
trans-Pt(C_2H_4)(ONC_5H_4Me-4)Cl_2	61.1 [215]	1230
	60.9 [215]	576
	60.0	666
trans-Pt(C_2H_4)(dmf)Cl_2	64.0	666
trans-Pt(C_2H_4){OC(NH_2)$_2$}Cl_2	67.1	666
[Pt(C_2H_4)Cl_3]$^-$	67.1 [195]	351, 1389
	67.2 [+192]	666, 1708
	66.9 [195.2]	425, 576, 867
	67.3 [194]	1230
	67.1 to 75.1 [188, 195]	32, 60, 131
	71.1 [188]	625
trans-Pt(C^1H_2=C^2HMe)(py)Cl_2	C^1 71.4 [159], C^2 99.5 [152]	660, 1230
trans-Pt(C^1H_2=C^2HMe)(NC_5H_4R)Cl_2	C^1 71.2 to 71.6 [159 to 165] C^2 98.2 to 101.1 [151 to 156]	867, 1230
trans-Pt(C^1H_2=C^2HMe)(NCMe)Cl_2	C^1 71.6, C^2 94.6	1230
trans-Pt(C^1H_2=C^2HMe)(ONC_5H_4Me-4)Cl_2	C^1 59.1 [213], C^2 83.4 [202]	1230
[Pt(C^1H_2=C^2HMe)Cl_3]$^-$	C^1 64.3 [190], C^2 87.5 [186]	425
	C^1 66.4 [190], C^2 86.9 [188]	1230
trans-Pt(C^1H_2=C^2HEt)(NC_5H_4Me-4)Cl_2	C^1 69.5 [160.0], C^2 105.1 [157.9]	867
[Pt(C^1H_2=C^2HEt)Cl_3]$^-$	C^1 63.9 [192], C^2 91.8 [195]	425
trans-Pt(C^1H_2=C^2HBu^n)(NC_5H_4Me-4)Cl_2	C^1 69.6 [159.9], C^2 103.4 [157.0]	867
trans-Pt(C^1H_2=C^2HCH$_2$Pr^i)(NC_5H_4Me-4)Cl_2	C^1 70.5 [159.6], C^2 102.2 [157.4]	867, 910
trans-Pt(C^1H_2=C^2HCH$_2$CH$_2$Pr^i)(NC_5H_4Me-4)	C^1 69.8 [160.1], C^2 104.0 [156.4]	867
trans-[Pt{C^1H_2=C^2H(CH$_2$)$_2$Pr^i}(NH_3)$_2$Cl]$^+$	C^1 67.8 [177.6], C^2 95.7 [174.2]	867

COMPOUND	δ(^{13}C)p.p.m., *J*/Hz	REFERENCE
$[Pt\{C^1H_2{=}C^2H(CH_2)_2Pr^i\}Cl_3]^-$	C^1 65.1 [190.9], C^2 91.5 [192.3]	867
$Pt(C^1H_2{=}C^2HCH_2$norbornane)(hfac)	C^1 56.2 [305], C^2 87.0 [277]	199
trans-$Pt(C^1H_2{=}C^2HCH_2Bu^t)(NC_5H_4Me\text{-}4)Cl_2$	C^1 71.3 [159.8], C^2 100.7 [158.4]	867
trans-$Pt(C^1H_2{=}C^2HCH_2Ph)(NC_5H_4Me\text{-}4)Cl_2$	C^1 69.2 [159.2], C^2 99.8 [162.7]	867
trans-$Pt\{C^1H_2{=}C^2H(CH_2)_4OH\}(NC_5H_4Me\text{-}4)Cl_2$	C^1 69.8 [159.2], C^2 102.8 [158.6]	867
trans-$Pt\{C^1H_2{=}C^2H(CH_2)_3OH\}(NC_5H_4Me\text{-}4)Cl_2$	C^1 69.5 [159.3], C^2 102.3 [157.3]	867
trans-$Pt\{C^1H_2{=}C^2H(CH_2)_2OH\}(NC_5H_4Me\text{-}4)Cl_2$	C^1 71.5 [158.6], C^2 99.1 [160.5]	867
trans-$Pt(C^1H_2{=}C^2HCH_2OH)(NC_5H_4Me\text{-}4)Cl_2$	C^1 67.4 [155.5], C^2 100.3 [159.3]	867
trans-$Pt\{C^1H_2{=}C^2H(CH_2)_4OAc\}(NC_5H_4Me\text{-}4)Cl_2$	C^1 70.1 [160.1], C^2 102.2 [160.2]	867
trans-$Pt\{C^1H_2{=}C^2H(CH_2)_4OMe\}(NC_5H_4Me\text{-}4)Cl_2$	C^1 70.0 [159.6], C^2 102.9 [158.0]	867
trans-$Pt\{C^1H_2{=}C^2H(CH_2)_3OAc\}(NC_5H_4Me\text{-}4)Cl_2$	C^1 70.4 [160.0], C^2 101.5 [161.9]	867
trans-$Pt\{C^1H_2{=}C^2H(CH_2)_3OMe\}(NC_5H_4Me\text{-}4)Cl_2$	C^1 70.0 [159.4], C^2 102.6 [158.7]	867
trans-$Pt\{C^1H_2{=}C^2H(CH_2)_2OAc\}(NC_5H_4Me\text{-}4)Cl_2$	C^1 71.5 [159.8], C^2 96.6 [165.9]	867
trans-$Pt\{C^1H_2{=}C^2H(CH_2)_2OMe\}(NC_5H_4Me\text{-}4)Cl_2$	C^1 71.2 [158.7], C^2 98.9 [161.7]	867
trans-$Pt(C^1H_2{=}C^2HCH_2CH{=}CMe_2)(NC_5H_4Me\text{-}4)Cl_2$	C^1 69.3 [160.1] C^2 101.4 [157.7]	867
trans-$Pt(C^1H_2{=}C^2HCH_2OAc)(NC_5H_4Me\text{-}4)Cl_2$	C^1 71.3 [154.8] C^2 89.3 [181.6]	867
trans-$Pt(C^1H_2{=}C^2HCH_2OMe)(NC_5H_4Me\text{-}4)Cl_2$	C^1 70.6 [157.8] C^2 93.0 [175.2]	867
trans-$Pt(C^1H_2{=}C^2HCH_2OPh)(NC_5H_4Me\text{-}4)Cl_2$	C^1 70.9 [156.0] C^2 90.5 [180.0]	867
trans-$[Pt(C^1H_2{=}C^2HCH_2OPh)(NH_3)_2Cl]^+$	C^1 64.9 [170.6] C^2 87.2 [188.7]	867
$[Pt(C^1H_2{=}C^2HCH_2OPh)Cl_3]^-$	C^1 65.2 [184.9] C^2 80.3 [212.2]	867
trans-$Pt(C^1H_2{=}C^2HCH_2Cl)(NC_5H_4Me\text{-}4)Cl_2$	C^1 66.1 [145.6] C^2 90.3 [175.1]	867
trans-$Pt(C^1H_2{=}C^2HBu^t)(NC_5H_4Me\text{-}4)Cl_2$	C^1 64.2 [146.8] C^2 116.6 [152.0]	867
trans-$Pt(C^1H_2{=}C^2HCH{=}CMe_2)(NC_5H_4Me\text{-}4)Cl_2$	C^1 61.6 [172.2] C^2 98.4 [119.8]	867, 910
trans-$[Pt(C^1H_2{=}C^2HPh)(NH_3)_2Cl]^+$	C^1 60.4 [190.8] C^2 89.9 [154.6]	867
trans-$Pt(C^1H_2{=}C^2HPh)(py)Cl_2$	C^1 62.1 [167.2] C^2 98.2 [139.9]	660, 1230 1313

COMPOUND	δ(^{13}C)p.p.m., J/Hz	REFERENCE
trans-Pt(C^1H_2=C^2HPh)(NC_5H_4X-4)Cl_2	C^1 61.7 to 62.3 [163.9 to 174] C^2 95.4 to 100.1 [136.7 to 140.9]	867, 1230 1313
trans-Pt(C^1H_2=C^2HPh)(NCMe)Cl_2	C^1 62.6, C^2 93.6	1230
trans-Pt(C^1H_2=C^2HPh)(ONC_5H_5)Cl_2	C^1 50.5, C^2 82.3	1230
[Pt(C^1H_2=C^2HPh)Cl_3]$^-$	C^1 57.4 [196], C^2 85.6	1230
	C^1 58.0 [195.4], C^2 86.0 [173.7]	867
trans-Pt(C^1H_2=$C^2HC_6H_4NMe_2$-4)(NC_5H_4Me-4)Cl_2	C^1 54.4 [178.9] C^2 106.4 [104.6]	867, 910
trans-[Pt(C^1H_2=$C^2HC_6H_4$OEt-4)$(NH_3)_2$Cl]$^+$	C^1 58.5 [201.0] C^2 92.1 [138.4]	867
trans-Pt(C^1H_2=$C^2HC_6H_4$OEt-4)(NC_5H_4Me-4)Cl_2	C^1 59.1 [171.1] C^2 100.7 [123.4]	867
[Pt(C^1H_2=$C^2HC_6H_4$OEt-4)Cl_3]$^-$	C^1 56.6 [201.0] C^2 88.0 [162.2]	867
trans-[Pt(C^1H_2=$C^2HC_6H_4$OPh-4)$(NH_3)_2$Cl]$^+$	C^1 59.2 [197.6] C^2 90.3 [144.0]	867
trans-Pt(C^1H_2=$C^2HC_6H_4$OPh-4)(NC_5H_4Me-4)Cl_2	C^1 60.5 [168.4] C^2 98.6 [130.0]	867
[Pt(C^1H_2=$C^2HC_6H_4$OPh-4)Cl_3]$^-$	C^1 57.5 [198.9] C^2 86.2 [168.0]	867
trans-Pt(C^1H_2=$C^2HC_6H_4$Me-4)(NC_5H_4Me-4)Cl_2	C^1 61.0 [167.2] C^2 98.8 [132.9]	867
[Pt(C^1H_2=$C^2HC_6H_4$Me-4)Cl_3]$^-$	C^1 57.9 [196.0] C^2 86.5 [171.9]	867
trans-Pt(C^1H_2=$C^2HC_6H_4$COMe-4)(NC_5H_4Me-4)Cl_2	C^1 63.9 [162.8] C^2 93.7 [144.4]	867
trans-[Pt(C^1H_2=$C^2HC_6H_4NO_2$-4)$(NH_3)_2$Cl]$^+$	C^1 62.5 [182.8] C^2 85.4 [165.3]	867
trans-Pt(C^1H_2=$C^2HC_6H_4NO_2$-4)(NC_5H_4Me-4)Cl_2	C^1 65.0 [162.7] C^2 90.9 [150.9]	867
[Pt(C^1H_2=$C^2HC_6H_4NO_2$-4)Cl_3]$^-$	C^1 59.8 [192.7] C^2 80.9 [181.0]	867
trans-[Pt(C^1H_2=$C^2HC_6H_4$Cl-4)$(NH_3)_2$Cl]$^+$	C^1 60.1 [187.7] C^2 87.9 [153.5]	867
trans-Pt(C^1H_2=$C^2HC_6H_4$Cl-4)(NC_5H_4Me-4)Cl_2	C^1 62.4 [166.6] C^2 95.6 [138.6]	867
[Pt(C^1H_2=$C^2HC_6H_4$Cl-4)Cl_3]$^-$	C^1 58.5 [197.4] C^2 84.0 [174.4]	867
Pt(C^1H_2=CHCN)Me($HBpz_3$)*	C^1 23.4 [396] C^2 8.0 [412]	866
	C^1 23.8 [397] C^2 7.4 [403]	
Pt(C^1H_2=C^2HCN)(acac)Cl	C^1 44.7 [254.9] C^2 65.7 [217.3]	1536
Pt(C^1H_2=C^2HCO_2Me)($HBpz_3$)	C^1 31.3 [356] C^2 21.6 [377]	866
trans-Pt(C^1H_2=C^2HCO_2Me)(py)Cl_2	C^1 71.6 [150] C^2 73.1 [183]	660, 1230

* two isomers

COMPOUND	$\delta(^{13}C)$p.p.m., J/Hz	REFERENCE
trans-$Pt(C^1H_2{=}C^2HCO_2Me)(NC_5H_4Me\text{-}4)Cl_2$	C^1 71.0, C^2 72.7 [185]	1230
trans-$Pt(C^1H_2{=}C^2HCO_2Me)$(collidine)Cl_2	C^1 71.8 [152] C^2 76.5 [187]	1230
trans-$Pt(C^1H_2{=}C^2HCO_2Me)(NCMe)Cl_2$	C^1 72.4, C^2 75.9	1230
trans-$Pt(C^1H_2{=}C^2HCO_2Me)(ONC_5H_4Me\text{-}4)Cl_2$	C^1 58.2, C^2 61.3	1230
$[Pt(C^1H_2{=}C^2HCO_2Me)Cl_3]^-$	C^1 62.8 [182] C^2 67.9 [212]	1230
trans-$Pt(C^1H_2{=}C^2HOMe)(NC_5H_4Me\text{-}4)Cl$	C^1 44.4 [193.1] C^2 144.2 [88.4]	867
trans-$Pt(C^1H_2{=}C^2HOAc)(py)Cl_2$	C^1 52.9 [166] C^2 111.6 [171]	660,1230
trans-$Pt(C^1H_2{=}C^2HOAc)(NC_5H_4X\text{-}4)Cl_2$	C^1 52.9 to 53.7 [165 to 169] C^2 110.0 to 112.4 [171 to 174]	867,1230
trans-$Pt(C^1H_2{=}C^2HOAc)(ONC_5H_4Me\text{-}4)Cl_2$	C^1 41.7, C^2 96.9	1230
$[Pt(C^1H_2{=}C^2HOAc)Cl_3]$	C^1 50.4 [191] C^2 100.4 [216]	1230
trans-$Pt(H_2C^1{=}C^2{=}CMe_2)(NC_5H_4Me\text{-}4)Cl_2$	C^1 35.8 [112.2] C^2 157.7 [277.5]	867
trans-Pt(*trans*-Me*C*H=*C*HMe)$(NC_5H_{10})Cl_2$	92.1 [145.9]	625
trans-Pt(*trans*-Me*C*H=*C*HMe)(py)Cl_2	91.0 [150]	1230
trans-Pt(*trans*-Me*C*H=*C*HMe)$(NC_5H_4Me\text{-}4)Cl_2$	90.6 [150], 92.7 [152.0]	1230, 867
trans-Pt(*trans*-Me*C*H=*C*HMe)$(NC_5H_4CN\text{-}4)Cl_2$	92.2 [154]	1230
trans-Pt(*trans*-Me*C*H=*C*HMe)(collidine)Cl_2	89.9 [153]	1230
[Pt(*trans*-Me*C*H=*C*HMe)$Cl_3]^-$	89.3 [178.2] 87.8 [178.1] 82.4 [182]	625 131 425
trans-Pt(*cis*-Me*C*H=*C*HMe)$(NC_5H_{10})Cl_2$	89.6 [142.8]	628
trans-Pt(*cis*-Me*C*H=*C*HMe)(py)Cl_2	91.0 [150]	660
trans-Pt(*cis*-Me*C*H=*C*HMe)$(NC_5H_4Me\text{-}4)Cl_2$	90.8 [149.8]	867
trans-Pt(*cis*-Me*C*H=*C*HMe)(NCMe)Cl_2	90.6 [189]	1230
trans-Pt(*cis*-Me*C*H=*C*HMe)$(ONC_5H_4Me\text{-}4)Cl_2$	77.1 [204]	1230
[Pt(*cis*-Me*C*H=*C*HMe)$Cl_3]^-$	81.3 [184] 85.5 [180] 86.9 [175.2] 89.3 [174.5]	1230 425 628 131
trans-Pt(*trans*-Et*C*H=*C*HEt)$(NC_5H_4Me\text{-}4)Cl_2$	97.1 [161.0]	867
trans-Pt(*cis*-Et*C*H=*C*HEt)$(NC_5H_4Me\text{-}4)Cl_2$	95.7 [155.8]	867
trans-Pt(*trans*-Me*C*H=*C*HPh)$(NC_5H_4Me\text{-}4)Cl_2$	84.0 [158.1] 93.2 [135.2]	867

COMPOUND	$\delta(^{13}C)$p.p.m., J/Hz	REFERENCE
Pt(*trans*-PhC^{1}H=C^{2}HCOCHCHPh)$(PEt_3)_2$	C^1 61.0 [180] (5, 5) C^2 49.7 [235] (0, 37)	302
Pt(*trans*-PhC^{1}H=C^{2}HCOCHCHPh)$(AsPh_3)_2$	C^1 58.0 [185], C^2 46.3 [275]	302
Pt(*trans*-NC*C*H=*C*HCN)Me$(HBpz_3)$	7.8 [406], 9.0 [416]	866
Pt(*trans*-$MeO_2CCH=CHCO_2Me)_2(PBu^t{}_2Ph)$	55.7 [212.2] (10.6) 56.2 [133.0] (12.0)	1106
Pt(*trans*-$EtO_2CCH=CHCO_2Me$)Me$(HBpz_3)$	30.9 [354], 31.0 [347]	866
$Pt(C_2F_4)Me(HBpz_3)$	79.6 [606]	866
B. Cyclic Olefins		
Pt(maleic anhydride)Me$(HBpz_3)$	28.8 [390]	866
trans-Pt(Cyclopentene)$(NC_5H_4Me\text{-}4)Cl_2$	101.0 [150.0]	867
$Pt(C_{10}H_{12}OAc)$hfac	85.4 [259], 87.9 [250]	199
$Pt(C_{10}H_{12})Cl_2$	116.6 [144]	131
Pt(norbornene)$_3$	68.0 [189]	578,1175
Pt(norbornenyl)(hfac)	61.3 [132], 79.2 [296]	199
Pt(norbornadiene)Me_2	88.3 [46]	403
trans-Pt(cyclohexene)$(NC_5H_4Me\text{-}4)Cl_2$	95.0 [146.0]	867
trans-Pt(cycloheptene)$(NC_5H_4Me\text{-}4)Cl_2$	95.6 [162.7]	867
Pt(*trans*-cyclooctene)$_3$	72.9 [154]	1175
trans-Pt(*cis*-cyclooctene)$(NC_5H_4Me\text{-}4)Cl_2$	93.8 [161.8]	867
$Pt(C_9H_{12})Cl_2$	93.9 [161], 102.5 [121]	1502
Pt(1,2,5-η^3-C_8H_{13})(hfac)	77.7 [269], 80.5 [266]	1200
Pt(1,2,5-η^3-C_8H_{13})(cod)	108.7 [154.3], 110.7 [157.2]	1200
Pt(1,2,5-η^3-6-$MeO_2CC_8H_{12}$)(hfac)	77.9 [260], 82.2 [255]	199
Pt(1,2,5-η^3-6-$MeOC_8H_{12}$)(py)Cl*	78.9, 82.9, 85.7, 88.5	1313
Pt(1,2,5-η^3-6-$MeOC_8H_{12})(NC_5H_4X\text{-}4)$Cl*	78.5 to 79.5 [240.5] 83.7 [243.5], 86.0 to 89.0 [208.2], 83.5 to 86.5 [203.4]	1313
Pt(cod)$_2$	73.3 [143]	1175
Pt(cod)Me_2	98.8 [55], 99.0 [55]	32,351, 403,573
Pt(cod)MeEt	97.4 [62.5], 99.8 [40.4]	1538
Pt(cod)Me(σ-Cp)	92.3 [104.3], 106.5 [55.0]	790
Pt(cod)Et_2	98.9 [47], 99.0 [47]	351,583
$\overline{Pt(cod)CH_2CMe=CMeCH_2}$	96.8 [51.9]	700
Pt(cod)$(CH_2Ph)_2$	100.1 [65]	573
Pt(cod)(σ-Cp)	101.3 [110.5]	790
$\overline{Pt(cod)CH(C_2H_3)(CH_2)_2CHC_2H_3}$	100.3 [64], 101.0 [63]	1593
$\overline{Pt(cod)\{C_8H_{12}C(CF_3)_2C}(CN)_2\}$	98.9 [134.3], 101.0 [137.3]	1174
Pt(cod)$(CF_3)_2$	110.8 [56], 111.0 [56]	32,573
Pt(cod)$(C_6H_4Br\text{-}p)_2$	105.1 [49]	1541

*two isomers

COMPOUND	$\delta(^{13}C)$p.p.m., J/Hz	REFERENCE
$Pt(cod)(C_6H_4Cl\text{-}m)_2$	105.3 [49]	1541
$Pt(cod)(C_6H_4OMe\text{-}o)_2$	102.2 [57]	1541
$Pt(cod)(2\text{-furanyl})_2$	103.2 [66]	1541
$Pt(cod)(2\text{-thienyl})_2$	104.4 [66]	1541
$Pt(cod)(2\text{-}C_8H_5O)_2$	105.2 [64]	1541
$[Pt(cod)(2\text{-}MeC_3H_4)]^+$	101.7 [105], 103.9 [127]	1173
$[Pt(cod)(C_7H_{11})]^+$	102.3 [103], 104.3 [121]	1200
$[Pt(cod)(C_7H_7)]^+$	109.3 [112]	1200
$[Pt(cod)(C_7H_6Ph)]^+$	109.2 [113]	1200
$[Pt(cod)(C_8H_{13})]^+$	115.2 [108.4], 116.3 [16.6] 121.2 [16.6], 126.9 [118.2]	1200
$Pt_2(cod)_2\{\mu\text{-}(PhC)_2CO\}$	91.6 [99], 94.2 [111]	1546
$[Pt_2(cod)_2\{\mu\text{-}(PhC)_2COH\}]^+$	96.7 [114], 97.5 [116]	1546
$[Pt(cod)Me(CNEt)]^+$	110.0 [126], 110.2 [35]	573
$[Pt(cod)Me(NCC_6H_4OMe\text{-}4)]^+$	91.8 [215], 113.8 [35]	573
$[Pt(cod)Me(NC_5H_4Me\text{-}4)]^+$	92.5 [178], 113.0 [36]	573
$[Pt(cod)Me(PPh_3)]^+$	108.6 [46], 116.2 [75] (12)	573
$[Pt(cod)Me(AsPh_3)]^+$	107.9 [44], 111.5 [104]	573
$\overline{Pt(cod)\{C_8H_{12}C(CF_3)_2O}\}$	79.8 [224.3], 79.4 [228.9] 112.7 [27.5], 112.9 [24.4]	1174
$\overline{Pt(cod)\{C(CF_3)_2OC(CF_3)_2O}\}$	87.0 [178.2], 113.2 [70.8]	1174
$\overline{Pt(cod)\{C(OH)(CF_3)_2CH{=}C(CF_3)O}\}$	79.7 [210], 84.8 [203] 114.8 [38], 115.0 [33]	1200
$Pt(cod)MeCl$	84.1 [214], 113.2 [30]	573
$Pt(cod)(CH_2Ph)Cl$	86.2 [230], 113.3 [30]	573
$Pt(cod)PhCl$	87.1 [208], 115.2 [28]	573,1541
$Pt(cod)(C_6H_4Me\text{-}4)Cl$	87.1 [210], 115.5 [27]	1541
$Pt(cod)(2\text{-furanyl})Cl$	89.6 [188], 112.1 [50]	1541
$Pt(cod)(2\text{-thienyl})Cl$	90.0 [188], 113.3 [50]	1541
$Pt(cod)(2\text{-}C_8H_5O)Cl$	90.4 [186], 113.5 [50]	1541
$Pt(cod)3\text{-}C_8H_7)Cl$	86.2 [206], 115.8 [30]	1541
$Pt(cod)Cl_2$	100.6 [153]	131
$Pt(cod)I_2$	103.2 [124]	32,573
$[Pt(C_8H_8)(2\text{-}MeC_3H_4)]^+$	102.3 [79], 104.8 [101]	1173
$Pt(C_2F_4)_2(C_4H_2)$	108.8 [470]	578

C. Acetylenes

COMPOUND	$\delta(^{13}C)$p.p.m., J/Hz	REFERENCE
$Pt(MeC{\equiv}CMe)(PPh_3)_2$	112.8 [52]	32
trans-$Pt(MeC{\equiv}CMe)Me(PMe_2Ph)_2$	69.5 [18]	32, 203
trans-$Pt(MeC{\equiv}CMe)(NC_5H_4X\text{-}4)Cl_2$	76.2 to 76.4 [182.2 to 183.5]	1313
$Pt(PhC{\equiv}CPh)_2$	124.8 [311]	733
$Pt_2\{\mu_2\text{-}(\eta^2\text{-}PhC{\equiv}CPh)\}(PMe_3)_4$	80.9 [278]	1181

COMPOUND	δ(^{13}C)p.p.m., J/Hz	REFERENCE
Copper		
Cu(norbornene)triflate	107.6	364
Cu($C_{10}H_{12}$)triflate	123.0, 128.4, 130.0, 130.6	364
Cu($C_{10}H_{12}$)(O_2CCF_3)	106.9, 110.5, 127.5, 128.4	850
Cu_3(C_7H_8)(O_2CCF_3)	113.6, 118.5, 125.7	850
Cu(cyclooctene)triflate	120.2	364
Cu(cod)(O_2CCF_3)	116.5	850
Cu(cod)triflate	123.1	364
Cu_2(cod)$(O_2CCF_3)_2$	109.7	850
Cu_2(C_8H_8)$(O_2CCF_3)_2$	116.8	850
Cu_4(C_8H_8)$(O_2CCF_3)_4$	90.5	850
Cu(*Z*,*Z*,*Z*-1,5,9-cyclododecatriene) triflate	124.9	364
Cu(*E*,*E*,*E*-1,5,9-cyclododecatriene) triflate	126.0	364
Cu(*Z*,*E*,*E*-1,5,9-cyclododecatriene) (O_2CCF_3)	117.4, 125.7, 130.7	850
Silver		
$[Ag(C^1H_2{=}C^2HMe)_2]^+$	C^1 109.4, C^2 138.5	4
$[Ag(CH_2{=}CC_9H_{14})]^+$	141.0, 156.0	321
[Ag(cyclopentene)]$^+$	126.2	130
[Ag(cyclohexene)]$^+$	122.8	130
Mercury		
[Hg(cyclohexene)]$^{2+}$	157.7	129
[Hg(norbornadiene)Me]$^+$	152.7	544
[Hg(cod)Me]$^+$	139.7	544
$[Hg(C^1H_2{=}C^2Ph_2)(O_2CCF_3)]^+$	C^1 141.9, C^2 226.6	424

TABLE 2.11

^{13}C Chemical Shift Data for Some Allyl Groups π-Bonded to Metals

COMPOUND	δ(^{13}C)/p.p.m., *J*/Hz			REFERENCE
Titanium	C^1	C^2	C^3	
Ti(1-MeC_3H_4)(C_4H_6)Cp	80.9	124.4	57.7	522
Vanadium				
V(C_3H_5)(CO)$_3$($Ph_2AsCH_2CH_2PPh_2$)	53.1	88.0	53.1	1752
V(1-MeC_3H_4)(CO)$_3$($Ph_2AsCH_2CH_2PPh_2$)	65.9	89.7	50.5	1752
V(1-MeC_3H_4)(CO)$_3${$C_6H_4(AsMe_2)_2$}	64.0	88.5	46.6	1752
V(2-MeC_3H_4)(CO)$_3$($Ph_2AsCH_2CH_2PPh_2$)	55.4	105.8	55.4	1752
V(2-MeC_3H_4)(CO)$_3${$C_6H_4(AsMe_2)_2$}	52.9	105.2	52.9	1752
V(1-PhC_3H_4)(CO)$_3$($Ph_2AsCH_2CH_2PPh_2$)	70.9	88.0	57.1	1752
V(1,1-$Me_2C_3H_3$)(CO)$_3$($Ph_2AsCH_2CH_2$-PPh_2)	81.2	90.6	51.1	1752
Molybdenum				
[Mo(C_3H_5)(CO)$_2$(NH_3)bipy]$^+$	59.1	74.2	59.1	947
[Mo(C_3H_5)(CO)$_2$(py)bipy]$^+$	60.8	74.6	60.8	947
Mo(C_3H_5)(CO)$_2$(py)(acac)*	61.2	100.2		1386
	57.0	100.1	60.5	1386
[Mo(2-MeC_3H_4)(CO)$_2$(py)bipy]$^+$	58.8	86.2	58.8	947
[Mo(2-MeC_3H_4)(CO)$_2${P(OPh)$_3$}bipy]$^+$	64.1	94.9 (7)	64.1	947
[Mo(2-MeC_3H_4)(CO)$_2$(OCMe$_2$)bipy]$^+$	57.4	85.1	57.4	947
$Mo_2(CH_2CCH_2)Cp_2(CO)_4$	36	196	36	1112
$Mo_2(C_8H_8)Cp_2(CO)_2$	from 59.8, 78.1, 91.8, 107.9			739
Manganese				
Mn(C_3H_5)(CO)$_4$	43.2	93.8	43.8	1414
Mn(1-EtC_3H_4)(CO)$_4$	71.0	92.7	37.0	1414
Mn(1-Pr^iC_3H_4)(CO)$_4$	80.1	91.2	36.9	1414
Iron				
A. Acyclic Allyl Compounds				
[Fe(C_3H_5)(CO)$_4$]$^+$	56.8	99.2	56.8	1469
	58.5	100.8	58.5	679
Fe(C_3H_5)(CO)$_3$$O_2$CMe	68.1	104.1	68.1	568
Fe(C_3H_5)(CO)$_3$$O_2CCF_3$	69.1	104.3	69.1	568
	70.4	107.1	70.4	679
Fe(C_3H_5)(CO)$_3$(ONO_2)	68.8	104.6	68.8	568
Fe(C_3H_5)(CO)$_3$Cl	68.2	102.4	68.2	568
Fe(C_3H_5)(CO)$_3$Br	65.0§	101.7	65.0	568

*Two isomers, §Isomers

COMPOUND	δ(^{13}C)/p.p.m., J/Hz			REFERENCE
	C^1	C^2	C^3	
	65.6§	102.5	65.6	869
	54.6†	112.8	54.6	568
	55.3†	113.5	55.3	869
$Fe(C_3H_5)(CO)_3I$	59.7§	100.8	59.7	174
	59.5§	100.6	59.5	568
	52.6†	106.6	52.6	174
	52.4†	106.5	52.4	568
$[Fe(anti\text{-}1\text{-}MeC_3H_4)(CO)_4]^+$	84.5	96.8	57.0	679
	{168}	{163}	{164.0}	
	82.4	94.3	56.5	1469
$[Fe(anti\text{-}1\text{-}MeC_3H_4)(CO)_3SO_2]^+$	101.4	83.2	70.9	925
	{162.5}	{165.0}	{168.0}	
$Fe(anti\text{-}1\text{-}MeC_3H_4)(CO)_3O_2CCF_3$	82.8	102.8	70.3	679
	{161}	{155}	{160}	
$[Fe(syn\text{-}1\text{-}MeC_3H_4)(CO)_4]^+$	83.6	98.2	49.1	1469
$Fe(1\text{-}MeC_3H_4)(CO)_3O_2CCF_3$	90.9	105.2	63.3	568
$Fe(1\text{-}MeC_3H_4)(CO)_3ONO_2$	91.1	105.2	62.8	568
$Fe(1\text{-}MeC_3H_4)(CO)_3Cl$	99.0	77.7	67.6	925
	{162.5}	{170.0}	{166.2}	
	89.8	103.0	62.2	568
$Fe(1\text{-}MeC_3H_4)(CO)_3Br$	76.1	113.6	48.0	568
	86.7	102.6	59.2	568
$[Fe(2\text{-}MeC_3H_4)(CO)_4]^+$	55.4	124.0	55.4	1469
$[\overline{Fe(1\text{-}C}H_2C_3H_4)H(CO)_3]^+$	81.9	101.9	54.0	679
	{179}	{176}	{161}	
	80.9	100.3	53.1	925
$[Fe(syn\text{-}1\text{-}EtC_3H_4)(CO)_4]^+$	91.0	96.7	49.3	1469
$Fe(1\text{-}PhC_3H_4)(CO)_3O_2CCF_3$	93.2	98.9	62.8	568
$Fe(1\text{-}PhC_3H_4)(CO)_3Cl$	91.8	97.1	61.8	568
$Fe(1\text{-}PhC_3H_4)(CO)_3Br$	88.6	96.5	58.5	568
	82.1	107.5	45.9	568
$\overline{Fe\{1\text{-}C}(CF_3)CHC_3H_2\}(CO)_3$	80.9	96.7	64.4	1211
$[Fe(1,1\text{-}Me_2C_3H_3)(CO)_4]^+$	115.7	93.7	47.6	1469
$[Fe(syn,anti\text{-}1,3\text{-}Me_2C_3H_3)(CO)_4]^+$	82.9	96.2	73.8	1469
$[Fe(syn,syn\text{-}1,3\text{-}Me_2C_3H_3)(CO)_4]^+$	76.4	100.7	76.4	1469
$[Fe(syn\text{-}1,2\text{-}Pr^iMeC_3H_3)(CO)_4]^+$	95.9	118.2	50.9	1469
$\overline{Fe[syn\text{-}anti\text{-}1,3\text{-}Me\{C_5H_4Fe}(CO)_2\}\text{-}C_3H_3](CO)_3$	50.5, 63.0, 71.7			697
$\overline{Fe\{1\text{-}C}(CF_3)CH\text{-}1\text{-}MeC_3H_3\}(CO)_3$		97.1	59.2	1211
$\overline{Fe\{1\text{-}C}(CF_3)CH\text{-}2\text{-}MeC_3H_3\}(CO)_3$	80.2	113.7	65.3	1211
$Fe(CH_2CHCHS)(CO)_2PPh_3$	48.9	95.2	97.6	721
$Fe(CH_2CHSiMe_2)(CO)_3SiMe_3$	43.0, 46.0			708
$\overline{Fe\{1\text{-}C}(CF_3)CH\text{-}1,2\text{-}Me_2C_3H_2\}(CO)_3$	91.0	112.0	62.1	1211
$Fe(MeO_2CC^1HC^2HC^3O)(CO)_3$	42.4	17.8	246.0	1435
$Fe(C^1H_2C^2MeC^3O)(CO)_3$	31.6	52.4	270.0	1435

§,† Isomers

COMPOUND	$\delta(^{13}C)$/p.p.m., J/Hz			REFERENCE
	C^1	C^2	C^3	
$Fe\{C^1H_2C^2(CHO)C^3O\}(CO)_3$	23.5	35.9	254.4	1435
$Fe\{C^1H_2C^2(COMe)C^3O\}(CO)_3$	23.2	25.4	246.1	1435
$Fe\{C^1H_2C^2(CO_2Me)C^3O\}(CO)_3$	24.6	23.3	242.7	1435
$Fe(MeO_2CC^1HC^2MeC^3O)(CO)_3$	41.8	51.4	270.9	1435
$Fe\{MeO_2CC^1HC^2(CO_2Me)C^3O\}(CO)_3$	31.7	25.1	235.7	1435
$Fe_2(CR^1R^2{=}CR^3CR^4{=}CR^5R^6)(CO)_6$	see olefin tables			

B. Cyclic Allyl Compounds

COMPOUND	$\delta(^{13}C)$/p.p.m., J/Hz	REFERENCE
$Fe_2(C_{10}H_{12})(CO)_6$	83.0	486
$Fe(C_{12}H_{16})(CO)_2$	104.2, 113.8	725
$[Fe(CHC^1HC^2HC^1H)(CO)_3H]^+$	C^1 65.8 {202.9}, C^2 117.3 {191.2}	785
$[Fe(CHC^1HC^2HC^3HCH_2CH_2)(CO)_3H]^+$	$C^{1,3}$ 81.2; 83.6, C^3 96.4	679
$Fe(CHCH_2\mathit{CHCHC}HCH_2CH_2)(CO)_3$	60.9, 78.6, 97.0	816
$Fe\{CH(CO)CH_2\mathit{CHCHC}HCH_2CH_2\}(CO)_3$	73.2, 84.3, 91.5	816
$Fe(CHCO\mathit{CHCHC}HCH_2CH_2)(CO)_3$	78.4, 89.0, 100.2	1473
$Fe_2\{\mathit{C}^1\mathit{C}^2(CH_2)_4CH\}_2(CO)_6$	C^1 84.4; 85.4, C^2 66.6	491, 823
$Fe(C_6H_7NCO_2Et)(CO)_3$*	64.2, 78.1, 83.0 66.1, 80.4, 83.4 56.2, 75.9, 84.9 59.9, 60.1, 88.6	1400
$Fe(C_9H_8Ph_2O)(CO)_3$	66.4, 72.3, 98.9	1460
$Fe\{C_9H_6(CN)_4O\}(CO)_3$§	45.9, 52.6, 57.1, 96.3 57.2, 65.1, 73.2, 99.7	1149
$Fe(C_8H_{12})(CO)_3$	88.6, 93.8	1198
$Fe(C_8H_{10}CO)(CO)_3$	77.4, 84.8	816
$Fe_2(C_7H_8)(CO)_6$	63.3, 64.0, 74.9	186
$Fe_2(C_8H_{10})(CO)_6$	48.6, 69.7, 74.0	186
$Fe_2(C_9H_8)(CO)_5$	51.3, 79.5, 59.8	639
$Fe_2(C_9H_2Me)(CO)_5$	50.5, 78.5, 58.4	639
$Fe_2(C_9H_{10})(CO)_6$	45.8, 68.5, 77.9	186
$Fe_2(C_9H_9Cl)(CO)_6$	48.6, 69.6, 74.1	186
$Fe_2(C_9H_{10}O)(CO)_6$	from 61.0, 81.9, 147.5, 182.6	945
$Fe_2(C_{10}H_{12})(CO)_6$	53.9, 70.2, 72.9	186
$Fe_2(C_{10}H_8)(CO)_5$	57.9, 80.6, 58.6	639
$Fe_2(C_{16}H_{15}O)(CO)_6$	from 69.1, 106.0, 147.9, 179.7	945
Ruthenium		
$Ru(1\text{-}3,5,6\text{-}\eta^5\text{-}C_8H_{11})$	24.6, 29.6 and 69.6 or 77.3 or 80.7	1180
$Ru(C_8H_8SiMe_3)(CO)_2SiMe_3$	from 40.2, 67.0, 71.8, 102.7, 115.6	1478

*Four isomers, §two isomers

COMPOUND	$\delta(^{13}C)$/p.p.m., J/Hz			REFERENCE
$Ru_2(C_{10}H_{12})(CO)_6$	from 50.6, 72.4, 75.0, 84.5			1215
	C^1	C^2	C^3	
$Ru_3(MeC^1C^2HC^3Me)H(CO)_9$	190.0	120.3 {157}	190.0	868
$Ru_3(MeC^1C^2MeC^3H)H(CO)_9$	186.8	130.4	161.6 {150}	868
$Ru_3(MeC^1C^2HC^3Et)H(CO)_9$	191.0	119.8 {162}	198.2	58, 868
$Ru_3(HMeC^1C^2C^3Et)H(CO)_9$	43.0 {166}	149.3	176.7	868
Osmium				
$[Os\{C^1H(C^2HCH_2CH_2)_2\}(CO)_3]_2$	145.0	111.4		823
Rhodium, [^{103}*Rh*]				
$[Rh(1\text{-}MeC_3H_4)CpCl]_2$	81.8 [9]	88.3 [6]	54.4 [10]	1164
$[Rh(1\text{-}MeC_3H_4)Cp(PPh_3)]^+$	76.5 [10]	84.6 [4]	48.1 [10]	1164
$[Rh(1\text{-}MeC_3H_4)Cp(AsPh_3)]^+$	75.2 [9] {161}	84.6 [6] {156}	46.8 [10] {156}	1164
$[Rh(1\text{-}MeC_3H_4)(C_5Me_5)Cl]_2$	73.1 [9]	96.6 [6]	58.5 [10]	1164
$[Rh(1\text{-}MeC_3H_4)(C_5Me_5)py]^+$	77.3 [9]	98.6 [4]	60.5 [12]	1164
$[Rh(1\text{-}MeC_3H_4)(C_5Me_5)PPh_3]^+$	71.3 [9]	92.4 [6]	52.9 [12]	1164
$[Rh(1\text{-}MeC_3H_4)(C_5Me_5)AsPh_3]^+$	70.0 [7]	92.5 [6]	51.5 [10]	1164
$[Rh(2\text{-}MeC_3H_4)CpCl]_2$	59.1 [10]	107.9 [6]	59.1 [10]	1164
$[Rh(2\text{-}MeC_3H_4)Cp(py)]^+$	60.1 [10]	111.8 [6]	60.1 [10]	1164
$[Rh(2\text{-}MeC_3H_4)Cp(PPh_3)]^+$	51.8 [10]	108.2 [6]	51.8 [10]	1164
$[Rh(2\text{-}MeC_3H_4)Cp(AsPh_3)]^+$	48.7 [10]	106.7 [6]	48.7 [10]	1164
$Rh\{CH_2(C^1HC^2HC^3H_2)_2\}hfac$	56.4 [9]	115.3 [6]	30.9 [7]	560
$[Rh(1\text{-}3,6,7\text{-}\eta^5\text{-}C_8H_9)Cp]^+$	46.4 [5]	82.9 [5]	77.8 [10] or 89.1 [10]	298
$Rh_2(C_7H_8)Cp_2$*	51.7 [9]	67.5 [7]	50.0 [10]	299
	56.9 [N=3]	67.0 [N=9]	69.3 [N=7]	299

*two isomers

COMPOUND	δ(^{13}C)/p.p.m., *J*/Hz			REFERENCE
	C^1	C^2	C^3	
$Rh_2(C_8H_{10})Cp_2$	48.4 [7]	72.2 [6]	49.5	299
$[Rh_2(C_8H_{10})HCp_2]^+$	from 56.6 [9], 59.5 [6], 63.2 [8], 74.9 [6]			299
$[Rh(C_{10}H_{11})C_5Me_5)]^+$	93.4 [9.2]	99.7 [5.6]	104.1 [10.7]	1518
Iridium				
$[Ir(C_3H_5)(CH_2{=}CHMe)(C_5Me_5)]^+$	43.3	86.0	49.1	1518
$[Ir(1\text{-}3,5,6\text{-}\eta^5\text{-}C_8H_{11})(C_5Me_5)]^+$	29.4	84.2	34.6	1518
Nickel				
cis-$Ni(C_3H_5)_2$	53.1	112.5	53.1	410
trans-$Ni(C_3H_5)_2$	52.6	112.1	52.6	410
$Ni(C_3H_5)Cp$	40.3	91.4	40.3	244, 410
$[Ni(C_3H_5)Cl]_2$	54.2	106.8	54.2	410
$[Ni(C_3H_5)Br]_2$	55.9	105.1	55.9	410
$[Ni(C_3H_5)I]_2$	58.9	102.3	58.9	410
$[Ni(1\text{-}Me\text{-}C_3H_4)Cl]_2$	70.0	106.9	48.0	33
$[Ni(1\text{-}Me\text{-}C_3H_4)Br]_2$	71.2	105.6	49.6	33
$[Ni(1\text{-}Me\text{-}C_3H_4)I]_2$	76.3	105.5	52.4	33
cis-$Ni(2\text{-}Me\text{-}C_3H_4)_2$	51.8	110.2	51.8	410
trans-$Ni(2\text{-}Me\text{-}C_3H_4)_2$	51.1	122.8	51.1	410
$Ni(2\text{-}Me\text{-}C_3H_4)COCOCHCMe{=}CH_2(PCy_3)$	40.5	123.4	71.2 (19)	1525
$[Ni(2\text{-}Me\text{-}C_3H_4)(NCMe)_2]^+$	57.8		57.8	987
$[Ni(2\text{-}Me\text{-}C_3H_4)\{SC(NMe_2)_2\}_2]^+$	56.3	120.6	56.3	1254
$Ni(indenyl)_2$	67.6	106.6	67.6	370
Palladium				
A. Acyclic Allyls				
cis-$Pr(C_3H_5)_2$	53.9	115.2	53.9	410
trans-$Pd(C_3H_5)_2$	54.7	115.2	54.7	410
$Pd(C_3H_5)Cp$	45.8	95.0	45.8	118, 194
	45.9	94.3	45.9	244
$[Pd(C_3H_5)(C_8H_8)]^+$	80.6	125.9	80.6	1173
$[Pd(C_3H_5)(PEt_2Ph)_2]^+$	71.6 (15)	122.6 (7)	71.6 (15)	235
$[Pd(C_3H_5)(PPh_3)_2]^+$	78.9 (15)	127.3 (8)	78.9 (15)	235
$Pd(C_3H_5)(PPh_3)Cl$	62.0	118.7	79.4	118, 194
$Pd(C_3H_5)(acac)$	55.8	113.5	55.8	118, 194
	60.7	113.2	60.7	1588
$[Pd(C_3H_5)(OAc)]_2$	54.8	110.5	54.8	1150
$[Pd(C_3H_5)Cl]_2$	63.2	111.9	63.2	118, 194
	62.8	111.3	62.8	60
	62.8	111.0	62.8	195
	62.9	111.1	62.9	1292
$[Pd(C_3H_5)Br]_2$	65.2	111.3	65.2	118, 194

COMPOUND	$\delta(^{13}C)$/p.p.m.,			REFERENCE
	C^1	C^2	C^3	
$[Pd(C_3H_5)I]_2$	68.0	110.2	68.0	118, 194
$Pd(1\text{-}MeC_3H_4)(acac)$	73.6	113.3	51.6	119, 194
	75.0	108.8		194
$[Pd(1\text{-}MeC_3H_4)Cl]_2$	81.6	111.6	58.4	195
	81.4	111.4	58.3	1292
$Pd(2\text{-}MeC_3H_4)Cp$	47.0	112.2	47.0	118, 194
$[Pd(2\text{-}MeC_3H_4)(cod)]^+$	76.3	142.2	76.3	1173
$[Pd(2\text{-}MeC_3H_4)(cot)]^+$	79.4	141.7	79.4	1173
$[Pd(2\text{-}MeC_3H_4)(C_6H_{10})]^+$	75.8	138.4	75.8	1173
$[Pd(2\text{-}MeC_3H_4)(C_6Me_6)]^+$	62.9	131.0	62.9	1173
$[Pd(2\text{-}MeC_3H_4)(C_4Me_2H_2S)]^+$	68.7	137.9	68.7	1504
$[Pd(2\text{-}MeC_3H_4)(C_4Me_4S)]^+$	64.1	135.4	64.1	1517
$[Pd(2\text{-}MeC_3H_4)(NCMe)_2]^+$	62.9	122.3	62.9	1173
$[Pd(2\text{-}MeC_3H_4)(PEt_2Ph)_2]^+$	71.0	136.3	71.0	235
	(15.9)	(7.5)	(15.9)	
$Pd(2\text{-}MeC_3H_4)(acac)$	54.8	129.2	54.8	118, 194
$[Pd(2\text{-}MeC_3H_4)Cl]_2$	61.7	127.9	61.7	118, 194
	61.9	126.9	61.9	1292
	62.1	127.7	62.1	1147
$[Pd(2\text{-}MeC_3H_4)Br]_2$	64.4	127.2	64.4	118, 194
$[Pd(2\text{-}MeC_3H_4)I]_2$	67.6	125.6	67.6	118, 194
$Pd(1\text{-}PhCH_2C_3H_4)(acac)$	77.6	112.0	52.6	119
$[Pd(1\text{-}PhC_3H_4)(C_4H_2Me_2S)]^+$	90.9		66.3	1504
$Pd(1\text{-}PhC_3H_4)(acac)$	75.0	108.8		119
	77.6	112.0	52.6	194
$[Pd(1\text{-}PhC_3H_4)Cl]_2$	86.6	112.6	64.2	1292
$[Pd(2\text{-}PhC_3H_4)(cot)]^+$	75.9	139.9	75.9	1173
$[Pd(2\text{-}PhC_3H_4)Cl]_2$	63.9	132.9	63.9	1292
$Pd_2(2\text{-}C_3H_4)_2(acac)_2$	55.4	122.6	55.4	196
$[Pd(1,2\text{-}Me_2C_3H_3)(dppe)]^+$	92.3	87.5	67.1	692
	(32)			
$[Pd(1,3\text{-}Me_2C_3H_3)(dppe)]^+$	85.9	123.8	85.9	692
	(34)	(6)	(34)	
$[Pd(1,3\text{-}Me_2C_3H_3)(O_2CMe)]_2$	70.7	113.0	70.7	1150
$[Pd(1\text{-}Me\text{-}3\text{-}EtC_3H_3)(O_2CMe)]_2$	71.3	111.0	77.6	1150
$[Pd(1\text{-}Me\text{-}3\text{-}MeCO_2CH_2C_3H_3)Cl]_2$	61.0	125.6	72.1	1533
$[Pd(1\text{-}Me\text{-}3\text{-}MeCO_2CHMeC_3H_3)(O_2CMe)]_2$	73.0	110.3	70.3	1150
$Pd(1\text{-}Me\text{-}3\text{-}PhC_3H_3)(acac)$	70.6	109.5	71.4	119, 194
$Pd\{1\text{-}Ph\text{-}3\text{-}(PhCH{=}CH)\text{-}3\text{-}AcOC_3H_2\}(acac)$	75.8	98.2	101.9	590
$Pd\{Ph(4\text{-}MeC_6H_4)CC_5HMe_2\text{-}(C_6H_4Me\text{-}4)_2\}(acac)$	66.1	115.7	82.6	1534
	67.1			
$[Pd\{Ph(4\text{-}MeC_6H_4)CC_5HMe_2\text{-}(C_6H_4Me\text{-}4)_2\}Cl]_2$	67.2	120.5	90.2	1534
	70.7			
$Pd_3\{C_3(C_6H_4OMe\text{-}4)_3\}_2(acac)_2$	129.8	134.4	142.5	1535
$Pd_3\{C_3(C_6H_4OMe\text{-}4)_2Ph\}_2(acac)_2$	from signals 127.4 to 142.7			1535
$Pd_3\{C_3(C_6H_4OMe\text{-}4)_2Ph\}_2(hfac)_2$	from signals 125.8 to 143.4			1535

COMPOUND	$\delta(^{13}C)$/p.p.m., J/Hz			REFERENCE
	C^1	C^2	C^3	
B. Cyclic Allyls				
$Pd\{Bu^tClC^1C^2C^3HCBu^t{=}CHCHBu^t\}(acac)$	96.3	125.7	74.7	377
$[Pd\{Bu^tClC^1C^2C^3HCBu^t{=}CHCHBu^t\}Cl]_2$	102.4	126.4	80.9	377
$[Pd\{C^1H_2{=}C^2C^3H(CH_2)_4\}Cl]_2$	58.2	126.0	68.8	937, 1292
$[Pd\{C^1H_2{=}C^2C^3HCH_2CHMe(CH_2)_2\}Cl]_2$	55.9	123.1	71.4	937, 1292
$[Pd\{C^1H_2{=}C^2C^3H(CH_2)_3CHMe\}Cl]_2$	57.3	121.8	71.7	937, 1292
$[Pd(steroidal\ allyl)Cl]_2$	65.2	134.2	83.6	1252
	65.3	134.1	83.1	1252
	65.3	123.9	83.4	1252
$[Pd\{C^1(p\text{-}tol)C^2(p\text{-}tol)C^3(p\text{-}tol)CPh\text{-}(p\text{-}tol)\}(cod)]^+$	109.8	125.0	109.8	1529
$[Pd\{C^1(p\text{-}tol)C^2(p\text{-}tol)C^3(p\text{-}tol)CPh\text{-}(p\text{-}tol)\}(bipy)]^+$	92.1	126.7	92.1	1529
$[Pd\{C^1(p\text{-}tol)C^2(p\text{-}tol)C^3(p\text{-}tol)CPh\text{-}(p\text{-}tol)\}(PMe_2Ph)_2]^+$	107.1 (36)	134.2 (18)	107.1 (36)	1529
$[Pd\{C^1(p\text{-}tol)C^2(p\text{-}tol)C^3(p\text{-}tol)CPh\text{-}(p\text{-}tol)\}\{P(OMe)_3\}_2]^+$	107.1 (62)	134.1 (13)	107.1 (62)	1529
$Pd\{C^1(p\text{-}tol)C^2(p\text{-}tol)C^3(p\text{-}tol)CPh\text{-}(p\text{-}tol)\}(PMe_2Ph)Cl$	113.1 (29)		85.5 (9)	1529
$Pd\{C^1(p\text{-}tol)C^2(p\text{-}tol)C^3(p\text{-}tol)CPh\text{-}(p\text{-}tol)\}\{P(OMe)_3\}Cl$	111.3 (46)		82.7 (14)	1529
$Pd\{C^1(p\text{-}tol)C^2(p\text{-}tol)C^3(p\text{-}tol)CPh\text{-}(p\text{-}tol)\}(acac)$	83.5	121.3	83.5	1529
$Pd\{C^1(p\text{-}tol)C^2(p\text{-}tol)C^3(p\text{-}tol)CPh\text{-}(p\text{-}tol)\}(hfac)$	88.5	122.7	88.5	1529
$Pd\{C^1(p\text{-}tol)C^2(p\text{-}tol)C^3(p\text{-}tol)CPh\text{-}(p\text{-}tol)\}(S_2CNR_2)$	90.8	125.6	90.8	1529
$[Pd\{C^1(p\text{-}tol)C^2(p\text{-}tol)C^3(p\text{-}tol)CPh\text{-}(p\text{-}tol)\}Cl]_2$	89.6	119.4	89.6	1529
$Pd\{C^1(p\text{-}tol)C^2(p\text{-}tol)C^3(p\text{-}tol)C(OMe)\text{-}(p\text{-}tol)\}S_2CNMe_2$*	89.6	125.5	89.6	1529
$[Pd\{C^1(p\text{-}tol)C^2(p\text{-}tol)C^3(p\text{-}tol)C(OMe)\text{-}(p\text{-}tol)\}Cl]_2$	89.7§	115.9	89.7	1529
	89.8*	119.2	89.8	1529
$[Pd\{C^1HC^2HC^3H(CH_2)_2\}(dppe)]^+$	92.3 (32)	116.3 (7)	92.3 (32)	692
$[Pd\{C^1HC^2HC^3HCHCH(OMe)CHCHCH\}Cl]_2$	84.2	100.7	84.2	197
$Pd(C^1HC^2Bu^tC^3HCHBu^tC{=}CBu^tCl)(PPh_3)Cl$	96.8 (24.5)	132.4 (6)	77.6 (3)	377

* *exo* OMe, § *endo* OMe

COMPOUND	$\delta(^{13}C)$/p.p.m., J/Hz			REFERENCE
	C^1	C^2	C^3	
$[Pd(C^1HC^2Bu^tC^3HCHBu^tC{=}CBu^tCl)Cl]_2$	80.9	132.6	78.8	377
$[Pd(C^1MeC^2MeC^3MeCMeCH_2CHPhCMe)$-(cod)$]^{+*}$	115.6	130.4	122.1	1177
	119.5	129.7	124.3	1177
$[Pd(C^1MeC^2MeC^3MeCMeCH_2CHPhCMe)$-(cot)$]^+$		130.8		1177
$[Pd(C^1MeC^2MeC^3MeCMeCH_2CHPhCMe)$-$(NCMe)_2]^+$	103.0	119.5	106.7	1177
$[Pd(C^1MeC^2MeC^3MeCMeCH_2CHPhCMe)$-(acac)*	87.7	113.8	88.2	1177
	90.0	112.6	92.9	1177
$[Pd(C^1MeC^2MeC^3MeCMeCH_2CHPhCMe)Cl_2]^*$	96.2	113.3	96.6	1177
	99.6	113.4	103.1	1177
$[Pd\{C^1HC^2HC^3H(CH_2)_3\}(dppe)]^+$	86.9 (35)	113.6 (7)	86.9	692
$[Pd\{C^1HC^2HC^3H(CH_2)_3\}(dpae)]^+$	86.9	111.6	86.9	692
$[Pd\{C^1HC^2HC^3H(CH_2)_3\}Cl]_2$	78.8	101.7	78.8	937, 1182
$[Pd\{C^1HC^2MeC^3H(CH_2)_3\}Cl]_2$	77.9	116.2	77.9	937, 1182
$[Pd(C^1HC^2MeC^3HCH_2CHMeCH_2)Cl]_2$	78.2	115.0	78.2	937, 1182
$[Pd\{C^1MeC^2MeC^3H(CH_2)_3\}Cl]_2$	90.7	113.3	75.7	937, 1182
$[Pd\{C^1HC^2PhC^3H(CH_2)_3\}Cl]_2$	76.0	117.4	76.0	937, 1182
$Pd\{C^1(COMe)C^2HC^3HCHBu^t(CH_2)_2\}$(acac)	79.6	102.9	51.5	1588
[Pd(1-3-η^3-cyclohepta-1-3-dienyl)-(cod)]$^+$			93.7	728
[Pd(1-3-η^3-cyclohepta-1-3-dienyl)-(cot)]$^+$			95.7	728
[Pd(1-3-η^3-cyclohepta-1,3-dienyl)-$(C_6Me_6)]^§$			89.2	728
[Pd(1-3-η^3-cyclohepta-1,3-dienyl)-$(Me_2NCH_2CH_2NMe_2)]^+$			76.1	728
[Pd(1-3-η^3-cyclohepta-1,3-dienyl)-(bipy)]$^+$			78.9	728
[Pd(1-3-η^3-cyclohepta-1,3-dienyl)-$(PEt_3)_2]^+$			87.7	728
[Pd(1-3-η^3-cyclohepta-1,3-dienyl)-(dppe)]$^+$			88.5	692, 728
[Pd(1-3-η^3-cyclohepta-1,3-dienyl)-$\{P(OMe)_3\}_2]^+$			91.6	728
[Pd(1-3-η^3-cyclohepta-1,3-dienyl)-$(AsEt_3)_2]^+$	87.3	117.7	86.5	728
[Pd(1-3-η^3-cyclohepta-1,3-dienyl)-(dpae)]$^+$			87.9	728

*two isomers, §hexamethyl-Dewar-benzene

COMPOUND	$\delta(^{13}C)$/p.p.m., J/Hz			REFERENCE
	C^1	C^2	C^3	
Pd(1-3-η^3-cyclohepta-1,3-dienyl)(acac)			71.7	728
Pd(1-3-η^3-cyclohepta-1,3-dienyl)(hfac)			75.2	728
Pd(1-3-η^3-cyclohepta-1,3-dienyl)(S_2CNMe_2)			77.1	728
[Pd(1-3-η^3-cyclohepta-1,3-dienyl)($MeSCH_2CH_2SMe$)]$^+$			85.7	728
[Pd(1-3-η^3-cyclohepta-1,3-dienyl)Cl]$_2$			78.8	728
[Pd(1-3-η^3-cyclohepta-1,3-dienyl)Br]$_2$			80.9	728
[Pd(1-3-η^3-cyclohepta-1,3-dienyl)I]$_2$			84.3, 84.8	728
[Pd(C_7H_{11})(dppe)]$^+$	86.6 (37)	115.9 (6)	86.6 (37)	692
[Pd(C_7H_{11})(dpae)]$^+$	86.4	113.7	86.4	692
[Pd(C_7H_9)(dppe)]$^+$	86.4 (37)	113.0 (6)	86.4 (37)	692
Pd(C_8H_{11})Cp	62.2	91.1	67.1	119, 194
Pd(C_8H_{11})(acac)	68.0	103.6	73.6	119, 194

Platinum, [^{195}Pt]

COMPOUND	C^1	C^2	C^3	REFERENCE
trans-Pt(C_3H_5)$_2$	46.7 [225.8]	102.1 [61.0]	46.7 [225.8]	410
cis-Pt(C_3H_5)$_2$	48.5 [225.0]	104.1 [55.2]	48.5 [225.0]	410
[Pt(1-MeC_3H_4)(PEt_3)$_2$]$^+$	80.0 [74] (25.8)	111.5 (6)	54.8 [86] (27.6)	1545
[Pt(2-MeC_3H_4)(cod)]$^+$	67.6 [154]	140.4 [47]	67.6 [154]	1173
[Pt(2-MeC_3H_4)(C_8H_8)]$^+$	70.4 [142]	140.2 [43]	70.4 [142]	1173
Pt($C^1H_2C^2HC^3HCH_2CH_2CH{=}CH$)$PMe_3$	49.9 [55.7]	108.7 [32.3] (2.0)	64.4 [44.0] (36.1)	720
Pt(Ph_3C)(acac)	56.8 [375]	104.0 [66]	104.0 [96]	538
[Pt(C_9H_8OEt)(PPh_3)$_2$]$^+$	87.2	97.0	81.8	1544
[Pt(C_7H_{11})(cod)]$^+$	82.3 [161]	111.9 [46.4]	82.3 [161]	1200
[Pt(C_7H_7)(cod)]$^+$	108.5 [44] averaged			1200
[Pt(C_7H_6Ph)(cod)]$^+$	103.3, 114.2, 127.7, 129.9			1200
Pt_2{μ-$(PhC)_2CO$}(cod)$_2$	108.8 [429]	155.9 [144]	108.8 [429]	1546
Pt_2{μ-$(PhC)_2CO$}($CNBu^t$)$_4$	86.6 [342]	168.5 [125]	86.6 [342]	1017, 1546

COMPOUND	$\delta(^{13}C)$/p.p.m., J/Hz			REFERENCE
	C^1	C^2	C^3	
$[Pt_2\{\mu-(PhC)_2COH\}(cod)_2]^+$	100.5 [394]	144.5 [119]	100.5 [394]	1546
$[Pt_2\{\mu-(PhC)_2COH\}(CNBu^t)_4]^+$	80.4 [317]	155.6 [107]	80.4 [317]	1546
$Pt_3\{C_3(C_6H_4OMe\text{-}4)_3\}_2(acac)_2$*	127.6, 132.1, 137.8 [20]			1535

*A number of related compounds were reported but the C_3 signals are mixed with those of the aromatic substituents.

TABLE 2.12

^{13}C Chemical Shift Data for Cyclobutadienes π-Bonded to Metals

COMPOUND	δ(^{13}C)/p.p.m., *J*/Hz				REFERENCE
Molybdenum					
$Mo(C_4H_4)(CO)_4$	60.7				244
Iron, [^{57}Fe]					
$Fe(C_4H_4)(CO)_3$	64.5				480
	63.7				1766
	63.3				244
	61.0	{191}			18, 19, 132
	65.2	[3.6]			877
$Fe_2(C_4H_4)_2(CO)_3$	77.8				480
		C^1	$C^{2,4}$	C^3	
$Fe(C_4H_3CH_2OH)(CO)_3$		84.5	64.6 {192}	63.1 {193}	869
$Fe(C_4H_3CHMeOH)(CO)_3$		90.2	62.6 {191} 62.7 {191}	63.1 {189}	869
$Fe(C_4H_3CHPhOH)(CO)_3$		88.0	63.4 {188} 63.6 {188}	64.2 {190}	869
$[Fe(C_4H_3CH_2)(CO)_3]^+$		98.0	84.6	109.0	869
$[Fe(C_4H_3CHMe)(CO)_3]^+$		96.6	78.8 80.4	106.5	869
$[Fe(C_4H_3CHPh)(CO)_3]^+$		90.9	80.0	106.2	869
$Fe(C_4H_3Ph)(CO)_3$		82.8	61.7	61.4	1766
$Fe(C_4H_3C_6H_4F\text{-}4)(CO)_3$		82.2	61.3	61.2	1766
$Fe(C_4H_3C_6H_4Cl\text{-}4)(CO)_3$		81.2	62.0	61.4	1766
$Fe(C_4H_3C_6H_4Br\text{-}4)(CO)_3$		81.0	62.1	61.4	1766
$Fe(\overline{C_4H_2CH_2CH{=}C}H)(CO)_3$	$C^{1,4}$	88.6 97.2	$C^{2,3}$ 55.4 52.9		725
Cobalt					
$Co(C_4Ph_4)Cp$	75.0				1495
$Co(C_4Ph_4)(C_5H_4CHPh_2)$	74.8				1495
$[Co(C_4Ph_4)(C_5H_4CPh_2)]^+$	88.1				1495
Nickel					
$Ni(C_4Me_4)(bipy)$	82.2				1523

TABLE 2.13

^{13}C N.m.r. Chemical Shift Data for Some Dienes and Trimethylenemethane π-Bonded to Metals

COMPOUND	δ(^{13}C)/p.p.m., *J*/Hz		REFERENCE
	$C^{1,4}$	$C^{2,3}$	
Titanium			
Ti(C_4H_6)(1-MeC_3H_4)Cp	58.5, 61.4	114.2	522
Chromium			
Cr(C_4H_6)$(CO)_4$	56.5	86.4	1751
Cr(C_4H_6)$(CO)_3$P$(OMe)_3$	51.9	85.1	1751
Cr{PhB($C^1H{=}C^2HCH_2)_2$}$(CO)_4$	112.8	91.3	1364
$Cr_2(C^1HC^2HC^3HC^4H)_2Cp$	178.6	112.1	1350
	57.0	93.7	
Cr($C_{11}H_{12}$)$(CO)_3$	97.2	98.2	1351
Cr($C_{11}H_{10}Me_2$)$(CO)_3$	95.2	98.0	1351
Cr(tetramethylthiophene)-$(CO)_3$	100.3	100.7	1517
Molybdenum			
cis-Mo$(C_4H_6)_2(CO)_2$	43.4, 45.0	69.4, 92.9	1751
$[Mo(C_4H_6)Cp(dppe)]^+$	44.0	77.2	732
	48.1	102.3	
Mo(C_4H_6)$(CO)_2(PBu^n_3)_2$	43.2	83.2	998
Mo(C_8H_8O)$(CO)_3$	58.8	85.4	1461
Mo{PhB($C^1H{=}C^2HCH_2)_2$}$(CO)_4$	93.3	108.4	1364
Mo(C_8H_8)$(CO)_2(PBu^n_3)_2$	95.5 (averaged)		998
Tungsten			
W{PhB($C^1H{=}C^2HCH_2)_2$}$(CO)_4$	89.3	100.5	1364
Iron			
A. Acyclic Butadienes			
Fe(C_4H_6)$(CO)_3$	36	81	132
	41.1	85.4	135, 185
	40.3	85.1	244
	40.4 {160.6}	85.2 {168.7}	939, 1304
	41.0 {158.7}	85.8 {170.0}	925
	40.5	85.5	918
	{$^1J_{CH}$ + 161.5, + 158.0}	{$^1J_{CH}$ + 169.1}	
	{$^2J_{CH}$ + 3.4,}	{$^2J_{CH}$ − 0.02, − 0.90, + 2.3}	
	{$^3J_{CH}$ + 7.8}		

COMPOUND	$\delta(^{13}C)$/p.p.m., J/Hz		REFERENCE
	$C^{1,4}$ {$^{4}J_{CH}$ + 0.4, + 1.7}	$C^{2,3}$ {$^{3}J_{CH}$ + 4.1, + 9.4}	
	40.7	85.5	1469
	41.1	86.2	358
	41.1	85.8	721
	41.5	86.4	865
$Fe(C_4H_6)_2PPh_3$	43.0 (10.4)	82.8	269
$Fe(\mathit{anti}\text{-}1\text{-}MeC_4H_5)(CO)_3$	C^1 53.9, C^4 40.9,	C^2 88.2, C^3 90.4	1469
$Fe(\mathit{syn}\text{-}1\text{-}MeC_4H_5)(CO)_3$	C^1 58.6, C^4 39.6,	C^2 93.2, C^3 81.2	1469
$Fe(2\text{-}MeC_4H_5)(CO)_3$	C^1 44.1, C^4 38.2,	C^2 103.5, C^3 85.1	1469
	C^1 43.9 {158.2}, C^4 38.0 {158.2},	C^2 102.9, C^3 84.7 {162.0}	939
	C^1 44.2, C^4 38.6	C^2 103.6, C^3 85.0	874
$Fe(2\text{-}MeC_4H_5)_2PPh_3$	C^1 41.2 (10.0), C^4 42.0 (12.5),	C^2 94.1, C^3 88.1	269
$Fe(1\text{-}CH_2{=}CHC_4H_5)(CO)_3$	C^1 62.1, C^4 39.2,	C^2 81.3, C^3 85.8	185
	C^1 62.1, C^4 39.6,	C^2 81.4, C^3 85.9	1170
$Fe\{1\text{-}CH(OH)MeC_4H_5\}(CO)_3$	C^1 69.1 or 71.1 C^4 40.0,	C^2 81.4, C^3 84.7	185
$(OC)_3Fe(C_4H_5\text{-}1\text{-}CHMeCH{=}CH\text{-}1\text{-}C_4H_5)Fe(CO)_3$	C^1 62.4, 69.5, C^4 39.5, 40.0,	C^2 81.3, 81.6, C^3 85.6, 87.3	185
$Fe(2\text{-}MeOC_4H_5)(CO)_3$	C^1 34.0, C^4 31.2,	C^2 140.9, C^3 66.0	1304
$Fe(1,1\text{-}Me_2C_4H_4)(CO)_3$	C^1 71.5, C^4 41.2,	C^2 92.3, C^3 85.6	1469
$Fe(1,2\text{-}Me_2C_4H_4)(CO)_3$	C^1 58.8 {158.2}, C^4 36.4 {159},	C^2 103.4, C^3 81.8 {170.0}	939, 1469
$Fe(1,3\text{-}Me_2C_4H_4)(CO)_3$	C^1 56.2, C^4 43.4,	C^2 90.1, C^3 99.4	1469
$Fe(1\text{-}Me\text{-}4\text{-}MeCOC_4H_4)(CO)_3$	C^1 54.8, C^4 60.7,	C^2 81.9, C^3 89.8	358
$Fe\{1\text{-}Me\text{-}4\text{-}MeCH(OH)C_4H_4\}\text{-}(CO)_3$	C^1 57.6, C^4 68.1,	C^2 81.2, C^3 85.4	358
$[Fe\{1\text{-}Me\text{-}4\text{-}\mathit{exo}\text{-}(Et_3NCH_2)\text{-}C_4H_4\}(CO)_3]^+$	C^1 59.1, C^4 41.7,	C^2 96.3, C^3 82.9	740, 1753
$[Fe\{1\text{-}Me\text{-}4\text{-}\mathit{exo}\text{-}(Ph_3PCH_2)\text{-}C_4H_4\}(CO)_3]^+$	C^1 60.7, C^4 41.0 (9.8),	C^2 96.4, C^3 80.8 (2.9)	740, 1753

COMPOUND	$\delta(^{13}C)$/p.p.m., J/Hz		REFERENCE
[Fe{1-Me-4-*endo*-(Ph_3PCH_2)C_4H_4}-	C^1 59.3,	C^2 86.8,	740, 1753
$(CO)_3]^+$	C^4 43.6	C^3 83.6	
	(8.8),	(2.9)	
[Fe{1-Me-4-*endo*-(Ph_3AsCH_2)C_4H_4}-	C^1 59.2,	C^2 86.9,	740, 1753
$(CO)_3]^+$	C^4 44.8,	C^3 82.9	
[Fe(1-Me-4-Me-4-$Ph_3PCH_2C_4H_3$)-	C^1 53.0	C^2 79.9	1753
$(CO)_3]^+$	(9.8),	(2.9),	
	C^4 60.5,	C^3 96.6	
	C^1 56.5	C^2 83.4	1753
	(6.0),	(2.0)	
	C^4 59.6,	C^3 85.3	
[Fe(1-Et-4-Me-4-$Ph_3PCH_2C_4H_3$)-	C^1 51.7	C^2 79.5	1753
$(CO)_3]^+$	(11),		
	C^4 69.6,	C^3 95.6	
	C^1 54.7	C^2 82.8,	1753
	(6.0),		
	C^4 68.6,	C^3 84.5	
Fe{1,4-$(EtO_2C)_2C_4H_4$}$(CO)_3$	C^1 25.1	C^2 85.8	972
	{162},	{171},	
	C^4 49.2	C^3 82.3	
	{162},	{171}	
Fe($Bu^tN=C^2HC^3H=C^4HMe$)$(CO)_3$		C^2 110.2,	874
	C^4 57.9,	C^3 74.7	
Fe($O=C^2HC^3H=C^4HPh$)$(CO)_3$		C^2 123.4,	871
	C^4 83.3,	C^3 63.7	
Fe($C^1H_2=C^2CH_2CH_2C^2=C^1H_2$)$(CO)_3$	C^1 32.7 or	C^2 114.2	725
	36.0		
Fe{$C^1H_2=C^2(CH_2)_4C^2=C^1H_2$}$(CO)_3$	C^1 {160.2},	C^2 {156.0}	918
Fe{$C^1H_2=C^2(CH)_4C^2=C^1H_2$}$(CO)_3$	C^1 36.7,	C^2 100.5	918
	C^1 35.3	C^2 99.5	993
	{160}		
Fe{$C^1H_2=C^2CH(CH_2)_2CH(CH_2)C^2=C^1H_2$}-	C^1 {161.3},	C^2 {157.5}	918
$(CO)_3$			
Fe{$C^1H=C^2CH(CH_2CH_2)_2CHC^2=C^1H_2$}-	C^1 {161.3},	C^2 {156.0}	918
$(CO)_3$			
Fe{$C^1H=C^2CH[CH=CHFe(CO)_4]CH(O)C^2$-	C^1 34.7,	C^2 113.0	1461
$=C^1H_2$}$(CO)_3$			
Fe($C_8H_4Me_4$)$(CO)_3$	C^1 66.8,	C^2 116.7	802
Fe($C_9H_7Me_3$)$(CO)_3$		C^2 99.8,	802
		100.8	
Fe{$C^1H_2=C^2HC^3=C^4HCOCH(CMe_2Cl)(CH_2)_2$}-		C^2 82.5,	738
$(CO)_3$	C^4 56.7,	C^3 110.5	
Fe{$C^1H_2=C^2HC^3=C^4HCOC(CMe_2)(CH_2)_2$}-	C^1 37.7,	C^2 82.3,	738
$(CO)_3$	C^4 59.6,	C^3 110.6	
Fe(1,1,3-$Me_3C_4H_3$)$(CO)_3$	C^1 67.0,	C^2 92.9,	1469
	C^4 43.7,	C^3 102.9	

COMPOUND	$\delta(^{13}C)$/p.p.m., J/Hz		REFERENCE
$Fe_3(CO)_8$(1,3,6-Me_3-hexa-1,3,5-triene-1,5-diyl)	C^1 61.9 {162}, C^3 99.6 {166}, C^5 76.5 {156}	C^2 152.4, C^4 121.7,	696

B. Cyclic Dienes

COMPOUND	$\delta(^{13}C)$/p.p.m., J/Hz		REFERENCE
$Fe(C^2HC^1HCOC^1HC^2H)(CO)_3$	C^1 69.0,	C^2 82.9	959
$Fe(PhC^2C^1PhCOC^1PhC^2Ph)(CO)_3$	C^1 82.5,	C^2 103.7	865
$Fe\{C^2(CF_3)C^1(CF_3)COC^1(CF_3)C^2(CF_3)\}(CO)_3$	C^1 65.2, (J_{FC} = 35 Hz)	C^2 90.1	865
$Fe\{C_4Me_2(NEt_2)_2CO\}(CO)_3$	74.5 80.5	110.2 124.3	684
$Fe\{C_4(NEt_2)_4CO\}(CO)_3$	95.9	108.4	684
$Fe(C^2MeC^1HSO_2C^1HC^2Me)(CO)_3$	C^1 68.8	C^2 96.5	957
$Fe(C^2HC^1MeSO_2C^1MeC^2H)(CO)_2NHMe_2$	C^1 69.0	C^2 73.4 {188.7}	735

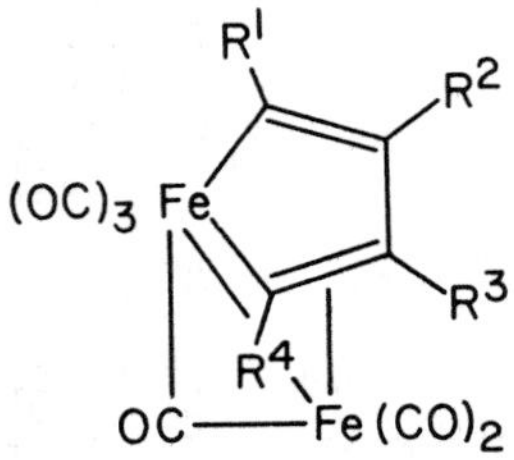

R^1	R^2	R^3	R^4			
H	H	H	H	$C^{1,4}$ 154.9,	$C^{2,3}$ 110.6	635, 1219
				$C^{1,4}$ 153.0,	$C^{2,3}$ 108.7	949
Me	H	H	H	C^1 179.1, C^4 151.1,	C^2 108.7, C^3 108.7	949
				C^1 180.8, C^4 152.9,	C^2 110.6, C^3 110.3	1219
H	Me	H	H	C^1 149.8, C^4 152.7,	C^2 129.3, C^3 110.4	949
				C^1 151.8, C^4 154.7,	C^2 131.1, C^3 112.3	1219
Me	H	H	Me	$C^{1,4}$ 178.3,	$C^{2,3}$ 110.4	1219
				$C^{1,4}$ 179.0,	$C^{2,3}$ 110.8 {162}	859
H	Me	Me	H	$C^{1,4}$ 153.6 {160},	$C^{2,3}$ 130.3	859
Me	H	Me	H	C^1 181.1,	C^2 113.2 {166},	859
				C^4 150.1 {161},	C^3 131.6	
Bu^t	H	H	Bu^t	$C^{1,4}$ 197.0,	$C^{2,3}$ 109.7 {166}	859, 1219
Ph	H	H	Ph	$C^{1,4}$ 176.5,	$C^{2,3}$ 112.6 {165}	859
				$C^{1,4}$ 174.9,	$C^{2,3}$ 111.4	635

COMPOUND				$\delta(^{13}C)$/p.p.m., J/Hz		REFERENCE
CH=CH-CH=CH	H	Ph		C^{1} 149.7, C^{4} 171.8,	C^{3} 113.6 {166}	859
NMe_2	H	H	NMe_2	$C^{1,4}$ 147.7,	$C^{2,3}$ 123.0	684
Ph	Ph	Ph	Ph	$C^{1,4}$ 172.3,	$C^{2,3}$ 147.6	635
				$C^{1,4}$ 173.8,	$C^{2,3}$ 129.2	859
$Fe_2(C_4Ph_4)(CO)_5PBu^n{}_3$				$C^{1,4}$ 172.1 (15.8),	$C^{2,3}$ 148.3	635
$Fe_2(C_4Ph_4)(CO)_5PPh_3$				$C^{1,4}$ 176.6 (12.4)	$C^{2,3}$ 149.3	635
$Fe_2\{C_4Me_2(C_2Me)_2\}(CO)_6$				84.9, 131.5, 140.0, 176.4		847
$Fe_2\{C_4Ph_2(NMe_2)_2\}(CO)_6$				from 102.6, 126.3, 128.3, 129.1, 146.3		684
Fe(cyclohexa-1,3-diene)$(CO)_3$				$C^{1,4}$ 62.5 {158.2},	$C^{2,3}$ 85.4 {172.8}	939, 1304
				$C^{1,4}$ 61.1,	$C^{2,3}$ 84.3	865
Fe(cyclohexa-1,3-diene)$(CO)_2$CNEt				$C^{1,4}$ 58.9,	$C^{2,3}$ 85.1	1465
Fe(cyclohexa-1,3-diene)(C_7H_9)(CO)				$C^{1,4}$ 77.9 or 88.8 or 89.0	$C^{2,3}$ 97.7	331
Fe(1-Me-cyclohexa-1,3-diene)$(CO)_3$				C^{4} 61.0,	C^{2} 87.8, C^{3} 81.3	939
Fe(2-Me-cyclohexa-1,3-diene)$(CO)_3$				C^{1} 65.3 {156.3}, C^{4} 58.7 {156.3},	C^{2} 102.9, C^{3} 85.7 {178.8}	939
Fe(2-MeO_2C-cyclohexa-1,3-diene)-$(CO)_3$				C^{1} 61.8, C^{4} 64.7,	C^{2} 86.5, C^{3} 88.7	1304
Fe(1-MeO-cyclohexa-1,3-diene)-$(CO)_3$				C^{1} 116.9, C^{4} 58.0,	C^{2} 78.2, C^{3} 77.2	488
Fe(2-MeO-cyclohexa-1,3-diene)-$(CO)_3$				C^{1} 55.1, C^{4} 50.9,	C^{2} 139.9, C^{3} 67.7	1304
Fe(5-Ph_3P-cyclohexa-1,3-diene)-$(CO)_3$				C^{1} 53.1, C^{4} 60.5 (7),	C^{2} 86.4, C^{3} 86.4	188
Fe(*trans*-AcO-cyclohexa-1,3-diene)$(CO)_3$				C^{1} 58.7, C^{4} 60.7,	C^{2} 84.1 C^{3} 86.8	1082
Fe(*cis*-AcO-cyclohexa-1,3-diene)-$(CO)_3$				C^{1} 59.0, C^{4} 62.0,	C^{2} 83.1, C^{3} 86.2	1082
Fe(5-F-cyclohexa-1,3-diene)$(CO)_3$				C^{1} 55.5, C^{4} 60.7,	C^{2} 84.7, C^{3} 84.1	1466
Fe(1,5,5-Me_3-3-NC_4H_8-cyclohexa-1,3-diene)$(CO)_3$				C^{1} 67.3, C^{4} 65.5 {154},	C^{2} 68.8 {164}, C^{3} 125.5	1206
Fe(2,6,6-Me_3-4-NC_4H_8-cyclohexa-1,3-diene)$(CO)_3$				C^{1} 70.5, C^{4} 112.6,	C^{2} 85.0, C^{3} 70.5	1206
Fe(5,6-disubstitued-cyclohexa-1,3-diene)$(CO)_3$				$C^{1,4}$ 61.9 to 71.6 {157.7 to 165.5} $C^{2,3}$ 82.3 to 87.7 {172.2 to 174.8}		312, 513, 705, 865, 918, 1295, 1456, 1716
Fe(steriodal diene)$(CO)_2P(OPh)_3$					$C^{2,3}$ 78.0, 80.7	1439

COMPOUND	$\delta(^{13}C)$/p.p.m., J/Hz		REFERENCE
Fe(cycloheptatriene)$(CO)_3$	$C^{1,4}$ 56.2; 60.7,	$C^{2,3}$ 88.6; 93.7	868
Fe(cyclohepta-1,3-diene-5-one)-$(CO)_3$	$C^{1,4}$ 58.0; 62.4,	$C^{2,3}$ 91.8; 92.2	865
Fe(cycloheptatriene-one)$(CO)_3$	$C^{1,4}$ 51.8; 62.1,	$C^{2,3}$ 91.1; 95.4	303
	$C^{1,4}$ 52.9; 62.3,	$C^{2,3}$ 92.5; 97.5	865
	$C^{1,4}$ 50.8; 60.4,	$C^{2,3}$ 91.3; 96.0	1008
[Fe(cycloheptatriene-ol)$(CO)_3]^+$	$C^{1,4}$ 52.0 {156}, 58.9 {156}	$C^{2,3}$ 95.0 {178} 99.3 {176}	1008
$[Fe(C_7H_8)(CO)_3]_n$	$C^{1,4}$ 61.3 ,	$C^{2,3}$ 88.8	846
$Fe(C_8H_{10})(CO)_3$	$C^{1,4}$ 59.5; 64.2,	$C^{2,3}$ 85.5; 88.1	816
$Fe(C_7H_8SO_2)(CO)_3$	77.0, 79.4, 82.2, 90.5		1198
Fe(cycloocta-1,3-diene)$(CO)_3$	$C^{1,4}$ 63.2,	$C^{2,3}$ 91.8	865
Fe(cycloocta-1,3-diene)$\{P(OMe)_3\}_3$		$C^{2,3}$ 86.3	678
	$C^{1,4}$ 50.4 (1.0)	$C^{2,3}$ 86.0	745
Fe(cycloocta-1,3,5-triene)$(CO)_3$	$C^{1,4}$ 59.0; 60.5,	$C^{2,3}$ 89.0; 89.3	865
	$C^{1,4}$ 59.4; 60.3,	$C^{2,3}$ 88.5; 88.9	1456
Fe(5-Ph_3C-cycloocta-1,3,5-triene)-$(CO)_3$	$C^{1,4}$ 74.9; 95.5,	$C^{2,3}$ 98.7; 106.2	513
Fe(7-Ph_3C-cycloocta-1,3,5-triene)-$(CO)_3$	$C^{1,4}$ 69.6; 71.0	$C^{2,3}$ 96.4; 99.6	513
Fe{cycloocta-1,3,5-triene-$Fe(CO)_4\}(CO)_3$	from 56.6, 61.1, 66.9, 70.4, 88.6, 89.8		1456
Fe{cycloocta-1,3,6-triene-$Fe(CO)_4\}(CO)_3$*	$C^{1,4}$ 54.7 or 57.0,	$C^{2,3}$ 61.2 or 62.7	1456
	$C^{1,4}$ 60.8 or 61.5,	$C^{2,3}$ 91.6 or 91.6	1456
$Fe(C_9H_{10})(CO)_3$	77.5, 89.3		1456
Fe(cycloocta-1,3,5-triene-7-one)-$(CO)_3$	$C^{1,4}$ 53.5; 60.1,	$C^{2,3}$ 90.9; 92.1	705
$Fe(C_8H_8)(CO)_3$	$C^{1,4}$ 62.9,	$C^{2,3}$ 89.2	137
	$C^{1,4}$ 60.6,	$C^{2,3}$ 89.6	1456
	$C^{1,4}$ 64.1,	$C^{2,3}$ 92.5	1230
	$C^{1,4}$ 73.7,	$C^{2,3}$ 92.5	726
Fe(5-Ph_3C-C_8H_7)$(CO)_3$	80.8, 85.4, 96.7, 117.9		513

*two isomers

COMPOUND	$\delta(^{13}C)$/p.p.m., J/Hz		REFERENCE

C. Trimethylene Methane

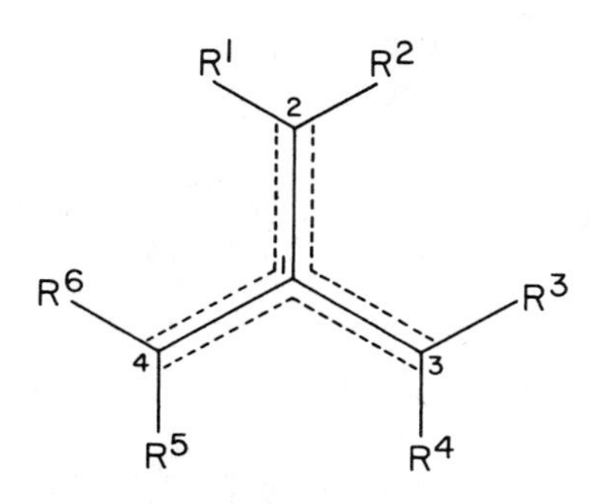

COMPOUND	$\delta(^{13}C)$/p.p.m., J/Hz		REFERENCE
R^1 to R^6 = H	C^1 105.0,	$C^{2,3,4}$ 53.0	55, 618
	C^1 106.1,	$C^{2,3,4}$ 54.9	845
R^1 = Et, R^2 to R^6 = H	C^1 103.2,	C^2 82.7 $C^{3,4}$ 39.9, 50.8	845
R^1 = $CH{=}CH_2$, R^2 to R^6 = H	C^1 = 105.2,	C^2 = 80.2, $C^{3,4}$ = 51.9, 52.5	10
R^1 = CHMe(COH), R^2 to R^6 = H	C^1 102.6	C^2 85.9 {147}, $C^{3,4}$ 51.4 {172} 52.7 {172}	845
R^1 = CHMe(OAc), R^2 to R^6 = H	C^1 103.6,	C^2 79.9, $C^{3,4}$ 51.7, 52.7	845
R^1 = CO_2Me, R^2 to R^6 = H	C^1 107.1,	C^2 69.4 {157}, $C^{3,4}$ 53.7 {162} 58.6 {162}	845
R^1, R^2 = $(CH_2)_2CC_3H_4Fe(CO)_3$, R^3 to R^6 = H	$C^{1,2}$ 96.1, 100.4	$C^{3,4}$ 37.5	512
Ruthenium			
$Ru(butadiene)(CO)_3$	$C^{1,4}$ 32.7, {$^1J_{CH}$ = 159.6; 156.2} {$^2J_{CH}$ = 3.6} {$^3J_{CH}$ = 7.8} {$^4J_{CH}$ = 0.7, 1.6}	$C^{2,3}$ 86.3 {$^1J_{CH}$ = 168.2} {$^2J_{CH}$ = -0.02, -0.8, 2.7} {$^3J_{CH}$ = 3.9, 8.9}	1156
Ru(cyclohexa-1,3-diene)-$(CO)_3$	$C^{1,4}$ 55.7,	$C^{2,3}$ 87.3	865
Ru(5,6-$C_{10}H_{10}$-cyclohexa-1,3-diene)$(CO)_3$	33.9, 41.2, 43.5, 62.9, 87.7		608
Ru(cyclohepta-1,3-diene)-$(CO)_3$	$C^{1,4}$ 52.9,	$C^{2,3}$ 90.3	518

COMPOUND	δ(^{13}C)/p.p.m., J/Hz		REFERENCE
Ru(PhC^{2}C^{1}PhCOC^{1}PhC^{2}Ph)-(CO)$_3$	C^{1,4} 81.9,	C^{2,3} 107.1	865
Ru{(CF$_3$)C^{2}C^1(CF$_3$)COC^1-(CF$_3$)C^2(CF$_3$)}(CO)$_3$	C^{1,4} 66.7 (J_{CF} = 35),	C^{2,3} 92.9	865
Ru(C_8H_8)(CO)$_3$	C^{1,4} 52.7,	C^{2,3} 92.1	726
Ru(C_8H_8)(C_6H_6)	91.6 (averaged)		827
Ru_2($C_8H_7SiMe_3$)(CO)$_4$SiMe$_3$	38.3, 42.6, 73.4, 79.7, 96.4, 99.9, 115.4, 148.4		607
Cobalt			
Co(C^{2}EtC^{1}EtCOC^{1}EtC^{2}Et)Cp	C^1 82.9,	C^2 93.3	798
Rhodium, [^{103}Rh]			
Rh(hexa-2,3-diene-1-ol)Cp	C^{2,5} 53.9 [15], 56.0 [15]	C^{3,4} 77.7 [9] 82.1 [7]	1519
Rh(hepta-3,5-diene-2-one)Cp	C^{3,6} 51.0 [15], 54.4 [16]	C^{4,5} 75.8 [9], 80.7 [6]	1519
Rh(5-Me-6-Ph-hexa-3,5-diene-2-one)Cp	C^{3,6} 50.6 [16], 60.2 [13]	C^{4,5} 77.6 [8], 93.4 [7]	1519
Rh(1,5-Ph$_2$-penta-2,4-diene-1-one)Cp	C^{2,5} 47.1 [16], 57.4 [13]	C^{3,4} 76.6 [8] 79.0 [7]	1519
Rh(EtC^{2}C^{1}EtCOC^{1}EtC^{2}Et)Cp	C^1 88.5 [9],	C^2 97.5 [12]	798
Rh{C_4(NEt$_2$)$_4$CO}Cp	from 84.5 [6], 97.4 [9], 100.9 [10]		684
Iridium			
Ir(hexa-2,4-diene-1-ol)Cp	C^{2,5} 41.0; 46.6,	C^{3,4} 66.8, 72.3	1519
Ir(hepta-3,5-diene-2-one)Cp	C^{3,6} 40.0; 43.6,	C^{4,5} 67.7, 70.7	1519

TABLE 2.14

^{13}C N.m.r. Chemical Shift Data for Some Cyclopentadienyl Groups π-Bonded to Metals

COMPOUND	$\delta(^{13}C)$/p.p.m., J/Hz	REFERENCE
Lithium		
$Li(C_5H_5)$	103.6 {159.1, 5.7, 8.2}	934
$Li(C_5H_4Me)$	C^1 113.1, $C^{2,5}$ 103.1, $C^{3,4}$ 101.7	1722
Sodium		
$Na(C_5H_5)$	103.4 {156.5, 7.0}	934
	102.1	93
Potassium		
$K(C_5H_5)$	104.8 {155.9, 6.9}	934
$K(C_5H_4Me)$	C^1 113.1, $C^{2,5}$ 102.2, $C^{3,4}$ 103.7	1722
$K(C_5H_4CHMePh)$	C^1 125.9, $C^{2,5}$ 102.2, $C^{3,4}$ 103.7	1723
Beryllium, [^9Be]		
$Be(C_5H_5)Ph$	104.7 [1.1] {177.4, 6.7}	934
$Be(C_5H_5)Br$	105.5 {179.1, 6.3}	934
Magnesium		
$Mg(C_5H_5)_2$	108.0 {167.5, 6.9}	934
	108.6	93
$Mg(C_5H_5)Cl$	103.6	64
$Mg(C_5H_4Me)Cl$	C^1 116.1, $C^{2,5}$ 104.1, $C^{3,4}$ 101.6	64
$Mg(C_5H_4SiMe_3)Cl$	C^1 112.6, $C^{2,5}$ 112.5, $C^{3,4}$ 105.7	64
$Mg(1,3\text{-}Me_2C_5H_3)Cl$	$C^{1,3}$ 114.4, C^2 105.2, $C^{4,5}$ 101.4	64
Scandium		
$Sc(C_5H_5)_2Me_2AlMe_2$	113.2	911
Yttrium		
$Y(C_5H_5)_2Me_2AlMe_2$	112.2	911
$[Y(\sigma\text{-}\pi\text{-}C_5H_4Me)(\eta^5\text{-}C_5H_4Me)]_2$	112.2, 116.6, 121.9	1335
Titanium		
$Ti(C_5H_5)_2Ph_2$	115.6, 116.8	60, 184, 932, 955

COMPOUND	$\delta(^{13}C)$/p.p.m., J/Hz			REFERENCE
$Ti(C_5H_5)_2p\text{-}tol_2$	115.5			339
$Ti(C_5H_5)_2N(SiMe_3)SiMe_2CH_2$	113.5			386
$Ti(C_5H_5)_2(NEt_2)_2$	109.9			184
$Ti_2(C_5H_5)_4N_2$	100			1338
$Ti(C_5H_5)_2(OC_6H_4F\text{-}4)_2$	115.6			932
$Ti(C_5H_5)(C_5H_4Me)(OC_6H_3Me_2\text{-}2,6)Cl$	117.2			628
$Ti(C_5H_5)_2F_2$	118.8			932
$Ti(C_5H_5)_2Cl_2$	121.3			60, 184, 955
	119.7			932
$Ti(C_5H_5)(C_5H_4CHMePh)Cl_2$	119.6			628
$Ti(C_5H_5)(C_5Me_5)Cl_2$	119.6			932
$Ti(C_5H_5)_2Br_2$	120			628
	119.5			932
$Ti(C_5H_5)_2I_2$	118.0			932
$Ti_2(C_5H_5)_2H_2(C_{10}H_8)$	104.6			322
$Ti(C_5H_5)(C_7H_7)$	96.5 {172}			334, 418
$Ti(C_5H_5)(C_7H_6Me)$	97.2 {172}			334, 418
$Ti(C_5H_5)Me(OPr^i)_2$	111.5 {172}			955
$Ti(C_5H_5)(NMe_2)_3$	110.6			521
$Ti(C_5H_5)(NEt_2)_3$	110.1			521
$Ti(C_5H_5)(OEt)_3$	112.3			244, 932, 955
$Ti(C_5H_5)(O_2CMe)_3$	122.7			932
$Ti(C_5H_5)(O_2CCCl_3)_3$	119.5			932
$Ti(C_5H_5)(OPh)_3$	118.2			932
$Ti(C_5H_5)(OC_6H_4Me\text{-}4)_3$	117.8			932
$Ti(C_5H_5)(OC_6H_3Me_2\text{-}2,4)_3$	115.9			932
$Ti(C_5H_5)(OC_6H_4F\text{-}4)_3$	117.3			932
$Ti(C_5H_5)(OC_6H_4Cl\text{-}4)_3$	119.4			932
$Ti(C_5H_5)(OEt)_2Cl$	114.7			932
$Ti(C_5H_5)(OEt)Cl_2$	118.7			932
$Ti(C_5H_5)Cl_3$	123.1			244, 932
$Ti(C_5H_5)Br_3$	122.4			932
$Ti(C_5H_4Me)_2F_2$	114.9,	117.8		932
	C^1	$C^{2,5}$	$C^{3,4}$	
$Ti(C_5H_4Me)(C_7H_7)$	107.3	99.0 {170}	96.0 {170}	334, 418
$Ti(C_5H_4Me)(NMe_2)_3$	122.2	110.0	111.4	507
$Ti(C_5H_4Me)Cp(OC_6H_3Me_2\text{-}2,6)Cl$	132.6	121.1, 115.6,	121.8, 109.6	628
$Ti(C_5H_4Me)Cl_3$	140.2	124.1	123.8	932
$Ti_2(C_5H_4C_5H_4)Cp_2H_2$	122.6	102.7	100.4	322
$Ti(C_5H_4Et)(NMe_2)_3$	134.2	112.5	111.7	521
$Ti(C_5H_4Pr^i)(NMe_2)_3$	135.2	109.0	108.3	521
$Ti(C_5H_4Bu^t)(NMe_2)_3$	140.3	108.4	109.0	521
$Ti(C_5H_4CHMePh)CpCl_2$	115.4, 122.0	117.9,	119.2,	628
$Ti\{C_5H_4(CH_2)_3C_5H_4\}Cl_2$	127.2,	120.3,	123.0	628
$Ti\{C_5H_4(CH_2)_2CHMeC_5H_4\}Cl_2$	132.2, 116.8, 123.5,	127.9, 121.3, 123.7,	115.4, 121.6, 123.4	628

COMPOUND	$\delta(^{13}C)$/p.p.m., J/Hz					REFERENCE
	C^1	$C^{2,5}$	$C^{3,4}$			
$Ti(C_5H_4Bu^t{}_2\text{-}1,3)(NMe_3)_2$	140.6	108.2,	109.0,	109.6,	111.9	521
$Ti\{C_5H_4Me(SiMe_3)\}(NMe_3)_2$	126.3	112.4,	113.4,	114.2,	117.6	521
$Ti(C_5H_4SiMe_3)(NMe_2)_2$	120.1	103.0	118.2			521
$Ti(C_5H_4GeMe_3)(NMe_2)_2$	121.4	112.7	117.5			521
$Ti(C_5Me_5)_2Me_2$	119.6					316, 328
$Ti(C_5Me_5)_2N_2$	108, 109.4					316, 328
$Ti(C_5Me_5)CpCl_2$	130.1					932
$Ti(C_5Me_5)(OEt)_3$	120.5					932
$Ti(C_5Me_5)(OPh)_3$	126.5					932
$Ti(C_5Me_5)Cl_3$	137.2					932
$Ti(C_5Me_5)Br_3$	138.2					932
Zirconium						
$Zr(C_5H_5)(C_7H_7)$	100.6 {172}					334, 418
$Zr(C_5H_5)_2Cl_2$	115.7					244
$Zr(C_5H_5)(acac)_2Cl$	116.5					244
Hafnium						
$Hf(C_5H_5)_2Cl_2$	114.4					244
Vanadium						
$V(C_5H_5)(CO)_4$	90.7					244
Niobium						
$Nb(C_5H_5)_2H(C_2H_4)$	91.1 {176, 7}					313
$Nb(C_5H_5)_2Et(C_2H_4)$	97.6 {194, 7}					313
$Nb(C_5H_5)_2(CHBu^t)Cl$	104.3, 106.3					1346
$Nb(C_5H_5)_2Me(S_2)$	105.8					1342
$[Nb(C_5H_5)(CO)(PhC_2Ph)]_2$	98.7					244
Tantalum						
$Ta(C_5H_5)_2Me(CH_2)$	100					591, 1344
$Ta(C_5H_5)_2Me$	97.6					1344
$Ta(C_5H_5)_2(CHBu^t)CHPh_2$	99.7, 102.9					1346
$Ta(C_5H_5)_2(CHPh)CH_2Ph$	101.0, 101.6					1346
$Ta(C_5H_5)_2(CHSiMe_3)Me$	101.3, 102.8					1346
$[Ta(C_5H_5)_2(CHBu^t)PMe_3]^+$	98.8, 102.1					1346
$Ta(C_5H_5)_2(CHBu^t)Cl$	103.2, 104.4					1346
$Ta(C_5H_5)(PMe_3)_2(CBu^t)Cl$	97					1347
$Ta(C_5H_5)(CHBu^t)Cl_2$	107 {180}					1102
$Ta(C_5H_5)(C_4H_8)Cl_2$	112.8 {181}					1102
$Ta(C_5H_4Me)_2(CHBu^t)Cl$	118.2, 122.7					1346
Chromium						
$Cr(C_5H_5)(C_7H_7)^-$	75.4 {170}					334, 1109
$[Cr(C_5H_5)(CO)_3]^-$	81.9					295
$Cr_2(C_5H_5)_2(C_8H_8)$	90.3					1350
$Cr_2(C_5H_5)(CO)(C_4H_2Ph_2)$	96.3, 98.3					1349
$Cr(C_5H_5)(CO)_2NO$	90.8					1360
$Cr(C_5H_5)(CO)_2NS$	92.8					1360

COMPOUND	δ(^{13}C)/p.p.m., J/Hz	REFERENCE
$Cr(C_5H_5)(CO)(NO)(N_2C_5H_8)$	94.3, 94.5	779
$Cr(C_5H_5)\{HC(NPh)_2\}(CO)_2$	94.2	1749
$Cr(C_5H_5)\{HC(NPh)NCOPh\}(CO)_2$	91.1	1749
$Cr(C_5H_4CPh_2)(CO)_3$	C^1 107.5, C^{2-5} 89.4, 93.9	1728
$Cr(C_5H_4CH{=}NMe_2)(CO)_3$	88.8, 89.9	1728
$Cr(C_5H_4PPh_3)(CO)_3$	C^1 67.0 (112), $C^{2,5}$ 89.7 (13), $C^{3,4}$ 87.3 (13)	1165, 1728
$Cr(C_5H_4SMe_2)(CO)_3$	C^1 78.6, $C^{2,5}$ 86.4, $C^{3,4}$ 84.3	1165, 1728
$Cr(C_2BN_2Bu^t{}_2)(CO)_3$	89.8	1031
$Cr(C_4Me_4S)(CO)_3$	100.3, 100.7	1517
Molybdenum		
$Mo(C_5H_5)_2(HC_2H)$	84.8	1389
$Mo(C_5H_5)_2(HC_2Me)$	84.1	1389
$Mo(C_5H_5)(C_7H_7)$	84.4	446
	83.8 {172}	334, 418
$Mo(C_5H_5)(CO)_3Me$	92.1, 92.4	1394, 1396
$Mo(C_5H_5)(CO)_3Et$	91.7, 92.7	93, 1396
$Mo(C_5H_5)(CO)_3CH_2Ph$	94.0	244
$Mo(C_5H_5)(CO)_3CH_2SMe$	93.2	93
$Mo(C_5H_5)(CO)_3CF_3$	91.9	93
$Mo(C_5H_5)(CO)_3CH{=}C(CN)_2$	95.2	280
$Mo(C_5H_5)(CO)_3\{C_4Ph(CF_3)_2O\}$	99.3	1312
$Mo(C_5H_5)(CO)_3CCl{=}C(CN)_2$	98.3	441
$Mo(C_5H_5)(CO)_3COC_3F_7$	94.9	93
$[Mo(C_5H_5)(CO)_3\{CH(NMe_2)\}]^+$	96.3	1372
$Mo_2(C_5H_5)_2(CO)_4(HC_2CF_3)$	91.0	1383
$Mo_2(C_5H_5)_2(CO)_4(MeC_2Me)$	91.5	1383
$Mo_2(C_5H_5)_2(CO)_4(HC_2CH_2CH_2OH)$	91.3	1383
$Mo_2(C_5H_5)_2(CO)_4(MeO_2CC_2CO_2Me)$	90.8	1383
$Mo_2(C_5H_5)_2(CO)_4(H_2C{=}C{=}CH_2)$	93.1	1112, 1392
$Mo_2(C_5H_5)_2(CO)_4(MeHC{=}C{=}CH_2)$	93.0, 93.3	1392
$Mo_2(C_5H_5)_2(CO)_4(MeHC{=}C{=}CHMe)$	93.1	1392
trans-$Mo(C_5H_5)(CO)_2(CNCy)COMe$	94.5	1394
cis-$Mo(C_5H_5)(CO)_2(CNCy)COMe$	94.9	1394
$Mo_2(C_5H_5)_2(CO)_2(C_8H_8)$	88.8, 92.5	739
	89.1, 92.3	735
$Mo(C_5H_5)(\mu\text{-}\eta^5\text{-}\eta^1\text{-}C_5H_4)Mn(CO)_4$	82.9	322
$Mo(C_5H_5)(CO)_3SnPh_3$	92.5	244
$\{Mo(C_5H_5)(CO)_3\}_3SnPh$	92.8	244
$Mo(C_5H_5)(CO)_2(Me_2CN_2Ph)$	95.5	1121
$Mo(C_5H_5)(CO)_2(Me_2CCNMe_2)$	92.1	515
$Mo(C_5H_5)(CO)_2(Me_2CCNC_5H_{10})$	92.4	515
$Mo(C_5H_5)(CO)_2(PhHCHMeCPhNMe)$	96.6	1377
$[Mo(C_5H_5)(CO)_2(PhHCNCPhNMe)]^+$	96.1	1377
$Mo_2(C_5H_5)_2(CO)_4(NCNMe_2)$	94.0, 95.1	1379
$Mo(C_5H_5)(CO)_2(MeCNMe)$	93.9	1385

COMPOUND	$\delta(^{13}C)$/p.p.m., J/Hz	REFERENCE
$Mo(C_5H_5)(CO)_2(RN_3R)$	94.7 to 95.6	654, 1079, 1121, 1749
$Mo(C_5H_5)(CO)_2HC(NR)_2$	95.4 to 95.6	1748, 1749
$Mo(C_5H_5)(CO)LHC(NR)_2$	92.0 to 94.3	1748
$Mo(C_5H_5)(CO)_2\{HC(NR)NCOR\}$	92.5 to 93.3	1079, 1749
$Mo(C_5H_5)(CO)_2(NO)$	93.6	93
$[Mo(C_5H_5)(CO)_2(pzH)_2]^+$	96.3	1152
$[Mo(C_5H_5)(CO)_2(imH)_2]^+$	96.8	1152
$Mo(C_5H_5)(CO)_2\{C_3(CN)_2NH_2\}$	95.0	441
$Mo(C_5H_5)(CO)_2\{C_3H(CN)(OH)NH\}$	94.3	441
$Mo(C_5H_5)(CO)_2\{C_3(CO_2Me)(CN)(OR)NH\}$	94.1	441
$Mo(C_5H_5)(CO)_2\{Me_2NCC(CN)_2\}$	95.4	441
$Mo(C_5H_5)(CO)_2\{C_5H_{10}NCC(CN)_2\}$	94.4	441
$Mo(C_5H_5)(CO)_2(R_2CNO)$	96.7 to 96.9	1129
$Mo(C_5H_5)(CO)_2(pzH)Cl$	96.4	1152
$Mo(C_5H_5)(CO)_2(imH)Cl$	96.7	1152
$Mo(C_5H_5)(CO)_2(py)Cl$	97.4	1152
trans-$Mo(C_5H_5)(CO)_2Me(PPh_3)$	91.8	1394
trans-$Mo(C_5H_5)(CO)_2COMe(PMe_2Ph)$	95.4	1394
trans-$Mo(C_5H_5)(CO)_2COMe(PPh_3)$	96.2	1394
$Mo(C_5H_5)(CO)_2(MeCNPh)\{P(OMe)_3\}$	94.6	1385
$Mo(C_5H_5)(CO)_3Cl$	94.9	93
$Mo(C_5H_5)(CO)_2\{(PF_2)_2NMe\}Cl$	94.7	1390
$Mo(C_5H_5)(CO)_2(PF_2NHMe)Cl$	94.6	1390
$Mo(C_5H_5)(CO)(NO)\{(PF_2)_2NMe\}$	93.4	1390
$Mo(C_5H_5)(C_3H_5)(NO)(S_2CNMe_2)$	102.8	376
$Mo(C_5H_5)(\sigma\text{-}C_5H_5)(NO)(S_2CNBu^n{}_2)$	105.5	1387
$Mo(C_5H_5)\{(PF_2)_2NMe\}_2Cl$	92.4	1390
$[Mo(C_5H_5)H\{P(OMe)_3\}_2CCH_2Bu^t]^+$	96.2	1096
$Mo(C_5H_5)\{P(OMe)_3\}_3CH{=}CHBu^t$	87.4	1096
$Mo(C_5H_5)\{P(OMe)_3\}_2CCH_2Bu^t$	89.3	1096
$\overline{Mo(C_5H_5)(CO)_2CHCHCMeO}$	93.8	749
$Mo(C_5H_4Me)(CO)_3Me$	111.3, 93.2, 89.2	1396
$Mo(C_5H_4Me)(CO)_2(MeCNMe)$	114.6, 94.5, 90.4	1385
$Mo(C_5H_4Me)(CO)_2\{p\text{-}tolN_3C_6H_3\text{-}(CF_3)_2\text{-}3,5\}$	88.7, 89.6, 95.5, 96.2	654
$\overline{Mo(C_5H_4CH_2CH_2)}(CO)_3$	67.0, 88.0, 89.6	1396
$\overline{Mo(C_5H_4CH_2CH_2CH_2)}(CO)_3$	110.9, 91.6, 87.8	1396
$Mo(\mu\text{-}\eta^5\text{-}\eta^1\text{-}C_5H_4)(C_5H_5)(CO)Mn(CO)_4$	104.5, 86.0	322
$Mo(C_5H_4PPh_3)(CO)_3$	71.8 (113), 94.9 (12), 91.8 (15)	1165
$Mo(C_5H_4SMe_2)(CO)_3$	83.4, 91.4, 89.4	1165
Tungsten		
$[W(C_5H_5)_2H(C_2H_4)]^+$	88.8	343
$[W(C_5H_5)_2Me(C_2H_4)]^+$	96.4	343
$W(C_5H_5)(CO)_3Me$	90.9, 91.0, 92.4	60, 1060, 1394, 1396
$W(C_5H_5)(CO)_3CH_2CH{=}CH_2$	92.6	449
$W(C_5H_5)(CO)_3CH_2C{\equiv}CH$	92.7	449

COMPOUND	$\delta(^{13}C)$/p.p.m., J/Hz	REFERENCE
$W(\mathit{C}_5H_5)(CO)_3CH_2Ph$	92.4	244
$W(\mathit{C}_5H_5)(CO)_3Ph$	92.2	244
$[W(\mathit{C}_5H_5)(CO)_3(CHNMe_2)]^+$	95.1	1372
$W(\mathit{C}_5H_5)(CO)_3CH{=}C(CN)_2$	93.7	280
$W(\mathit{C}_5H_5)(CO)_3C(CN){=}C(CN)_2$	96.4	280
$W(\mathit{C}_5H_5)(CO)_3CCl{=}C(CN)_2$	97.1	441
$W(\mathit{C}_5H_5)(CO)_3COCF_3$	94.1	93
$W(\mathit{C}_5H_5)(CO)_3Mn(CO)_5$	90.0	244
$\{W(\mathit{C}_5H_5)(CO)_3\}_2Sn(SEt)_2$	90.2	244
$W(\mathit{C}_5H_5)(CO)_2CR$	91.9 to 92.6	637, 1026, 1043
$\overline{W(\mathit{C}_5H_5)(CO)_2COCHMeC}(NR_2)$	90.8 {175.8}	881
$\overline{W(\mathit{C}_5H_5)(CO)_2COC(NEt_2)}{=}CH_2$	89.7 {178.2}	881
$\overline{W(\mathit{C}_5H_5)(CO)_2COC(NEt_2)}{=}CHCO_2Et$	92.3 {180.7}	881
$W_2(\mathit{C}_5H_5)_2(CO)_4(H_2C{=}C{=}CH_2)$	91.8	1392
$W(\mathit{C}_5H_5)(CO)_2(PMe_3)C_2\mathit{p}\text{-tol}$	91.9	1393
$W(\mathit{C}_5H_5)(CO)_2(PMe_3)C(\mathit{p}\text{-tol})CO$	91.0	1404
$W(\mathit{C}_5H_5)(CO)_2(PPh_3)COMe$	94.7	1394
$W(\mathit{C}_5H_5)(CO)_2(PR_3)SnMe_3$	86 to 87	71
$W(\mathit{C}_5H_5)(CO)_2(Me_2CCNMe_2)$	90.4, 91.7	515
$W(\mathit{C}_5H_5)(CO)_2\{C_3(CN)_2NH_2\}$	93.8	441
$W(\mathit{C}_5H_5)(CO)_2\{C_5H_{10}NCC(CN)_2\}$	93.0	441
$W(\mathit{C}_5H_5)(CO)_2(Me_2CN_2Ph)$	94.7	1121
$\overline{W(\mathit{C}_5H_5)(CO)_2CH{=}CHCMeO}$	92.9	749
$W(\mathit{C}_5H_5)(CO)_2N_2Me$	93.3	1060
$W(\mathit{C}_5H_5)(CO)_2(RNCHNR)$	93.8 to 94.3	1079, 1748, 1749
$W(\mathit{C}_5H_5)(CO)L(RNCHNR)$	90.0 to 92.3	1748
$W(\mathit{C}_5H_5)(CO)_2(RN_3R)$	93.3 to 93.9	654, 1121, 1749
$W(\mathit{C}_5H_5)(CO)_2\{HC(NR)N(CO)R\}$	91.2 to 91.9	1079, 1749
$[W(\mathit{C}_5H_5)(CO)_2(pzH)_2]^+$	95.3	1152
$[W(\mathit{C}_5H_5)(CO)_2(imH)_2]^+$	95.4	1152
$W(\mathit{C}_5H_5)(CO)_2(R_2CNO)$	95.8, 95.9	1129
$W(\mathit{C}_5H_5)(CO)_2(pzH)Cl$	95.2	1152
$W(\mathit{C}_5H_5)(CO)_2(imH)Cl$	95.3	1152
$W(\mathit{C}_5H_5)(CO)_2(PMe_3)CRCO$	93.3 to 94.5	1037
$W(\mathit{C}_5H_5)(CO)_2(PPh_3)(C\mathit{p}\text{-tol}CO)$	94.9	1037
$W(\mathit{C}_5H_5)(CO)_2\{(PF_2)_2NMe\}Cl$	92.8	1390
$W(\mathit{C}_5H_5)(CO)_2\{PF(OMe)_2\}Cl$	92.8	1390
$W(\mathit{C}_5H_5)(CO)_2\{PF(OEt)_2\}Cl$	93.1	1390
$W(\mathit{C}_5H_5)(CO)(\mu\text{-}\eta^5\text{-}\eta^1\text{-}C_5H_4)Mn(CO)_4$	78.7	322
$W(\mathit{C}_5H_5)(CO)(PMe_3)_2(CRCO)$	87.2 to 87.9	1037
$W(\mathit{C}_5H_5)\{(PF_2)_2NMe\}_2Cl$	89.5	1390
$W(\mathit{C}_5H_4Me)(CO)_3Me$	109.9, 920, 87.9	1396
$W(\mathit{C}_5H_4Me)(CO)_2\{\mathit{p}\text{-tol}N_3C_6H_3\text{-}(CF_3)_2\text{-}3,5\}$	87.9, 88.8, 95.0, 96.0	654
$\overline{W(\mathit{C}_5H_4CH_2C}H_2)(CO)_3$	70.8, 86.5, 87.4	1396

COMPOUND	$\delta(^{13}C)$/p.p.m., J/Hz		REFERENCE
$W(C_5H_4CH_2CH_2CH_2)(CO)_3$	110.0, 90.0, 86.9		1396
$W(\mu-\eta^5-\eta^1-C_5H_4)Cp(CO)Mn(CO)_4$	100.8, 81.3		322
$W(C_5H_4PPh_3)(CO)_3$	70.3 (109), 91.5 (12), 89.3 (13)		1165
$W(C_5H_4SMe_2)(CO)_3$	77.4, 89.0, 87.3		1165
Manganese			
A. Unsubstituted Cyclopentadienyls			
$Mn(C_5H_5)(CO)_3$	75 {175}		93, 534, 685
	83.1		181, 244
	82.7		295
	83.0		885, 1712
$Mn(C_5H_5)(CO)_2CMe_2$	90.0		756
$Mn(C_5H_5)(CO)_2CMePh$	90.6		756
$Mn(C_5H_5)(CO)_2C{=}CHPh$	88.2 {179.4}		970
$Mn_2(C_5H_5)_2(CO)_4C{=}CHPh$	89.0 {179.4}, 88.7 {179.4}		970
$Mn(C_5H_5)(CO)_2C{=}C{=}CBu^t_2$	88.4		752
$Mn(C_5H_5)(CO)_2CPh{=}C{=}O$	89		754
$[Mn(C_5H_5)(CO)_2CPh]^+$	93.9		1058
$Mn(C_5H_5)(CO)_2(octene)$	83.9		295
$Mn(C_5H_5)(CO)_2(\eta^2-C_7H_8)$	84.0		1189
$Mn(C_5H_5)(CO)_2(\eta^2-C_8H_8)$	86.3		1189
$Mn(C_5H_5)(CO)_2CPh(PMe_3)$	96.3		1046
$Mn(C_5H_5)(CO)_2C(OMe)Ph$	88.4		756, 1058
$Mn(C_5H_5)(CO)_2C(OR)Ph$	88.6 to 89.1		1058
$Mn(C_5H_5)(CO)_2C(OMe)(C_{10}H_{19})$	87.8		1418
$Mn(C_5H_5)(CO)_2CS$	85.7, 86.3		295, 1712
$Mn(C_5H_5)(CO)_2C(SCN)Ph$	94.5		946
$Mn(C_5H_5)(CO)_2CSe$	86.9		1712
$Mn(C_5H_5)(CO)_2(NHC_5H_{10})$	81.9		295
$[Mn(C_5H_5)(CO)_2H(PR_3)]^+$	87.4		347
$Mn(C_5H_5)(CO)_2PR_3$	79.0 to 82.4		181, 295, 347
$Mn(C_5H_5)(CO)_2AsPh_3$	80.8		181
$Mn(C_5H_5)(CO)_2SbPh_3$	79.1		181
$[Mn(C_5H_5)(CO)H\{Ph_2P(CH_2)_nPPh_2\}]^+$	86.8		347
$[Mn(C_5H_5)(CO)H(PPh_3)_2]^+$	90.0		347
$[Mn(C_5H_5)(CO)(NO)(PPh_3)]^+$	95.8		347
$Mn(C_5H_5)(CO)(PPh_3)_2$	82.2, 82.0		181, 347
$Mn(C_5H_5)(CO)(Ph_2PCH_2PPh_2)$	76.7		181, 244
$Mn(C_5H_5)(CO)(dppe)$	79.6		181, 347
$Mn(C_5H_5)(CO)(Ph_2PCH_2CH_2CH_2PPh_2)$	81.8, 80.6		181, 347
$Mn(C_5H_5)(CO)\{(PF_2)_2NMe\}$	78.8		1390
$Mn(C_5H_5)\{(PF_2)_2NMe\}_2$	78.1		1390
B. Substituted Cyclopentadienyls			
	C^1	$C^{2,3,4,5}$	
$Mn(C_5H_4Me)(CO)_3$	100.9	80.1, 79.4	67
	102.4	82.4, 82.1	885
	103.0	83.1, 82.5	181

COMPOUND	$\delta(^{13}C)$/p.p.m., J/Hz		REFERENCE
	C^1	$C^{2,3,4,5}$	
$Mn(C_5H_4Me)(CO)_2(CPh{=}C{=}O)$	105.4	89.3, 89.4	754
$Mn(C_5H_4Me)(CO)_2C(OMe)Me$	104.3	88.4, 86.9	1417
$Mn_2(C_5H_4Me)_2(CO)_4CH_2$	102.0	85.7, 86.5, 88.2, 89.9	533
$Mn(C_5H_4Me)(CO)_2CS$	104.4	85.5, 84.4	1712
$Mn(C_5H_4Me)(CO)_2CSe$	105.5	86.8, 85.5	1712
$Mn(C_5H_4Me)\{C(OMe)Me\}$-$\{(PF_2)_2NMe\}$	99.8	88.4, 83.6	1417
$Mn(C_5H_4Et)(CO)_3$	109.0	81.7, 81.4	181
$Mn(C_5H_4Pr^i)(CO)_3$	112.5	81.2, 82.1	181
$Mn(C_5H_4Bu^t)(CO)_3$	118.0	80.9, 83.4	181
$Mn(C_5H_4COMe)(CO)_3$	92.6	84.2, 87.4	181
$Mn(C_5H_4CO_2Me)(CO)_3$	84.5	83.1, 87.1	181
$Mn\{C_5H_4C(OEt)W(CO)_5\}(CO)_3$	110.3	90.8, 86.7	1007
$Mn\{C_5H_4CW(CO)_4Br\}(CO)_3$	101.0	89.5, 82.6	1007
$Mn\{C_5H_4C(OH)HPh\}(CO)_3$	108.2	82.1, 80.0 82.5, 81.2	878
$Mn\{C_5H_4C(OH)MeEt\}(CO)_3$	114.1	83.3, 78.7 82.5, 80.7	878
$Mn\{C_5H_4C(OH)MePh\}(CO)_3$	114.0	84.4, 81.2 83.7, 79.4	878
$Mn\{C_5H_4C(OH)Ph_2\}(CO)_3$	111.5	86.9, 80.0	878
$Mn\{C_5H_4C(OH)Ph(C_6H_4OMe\text{-}4)\}$-$(CO)_3$	112.1	89.7, 80.1, 79.9	878
$\{Mn(C_5H_4)(CO)_3\}_2CHOH$	107.4	81.1, 82.2, 83.1	1702
$\{Mn(C_5H_4)(CO)_3\}_2CMeOH$	112.3	79.8, 80.9, 83.5, 84.0	1702
$\{Mn(C_5H_4)(CO)_3\}_2CPhOH$	112.0	80.1, 81.1, 86.0, 86.6	1702
$\{Mn(C_5H_4)(CO)_3\}\{Mn(C_5H_4)$-$(CO)_2PPh_3\}CHOH$	107.8 105.0	80.2, 80.9, 81.6 83.3, 85.5, 87.9	1702
$\{Mn(C_5H_4)(CO)_3\}\{Fe(C_5H_4)Cp\}$	108.7	81.0, 81.6, 82.7	1702
$Mn\{C_5H_4C(OH)HMe\}(CO)_2PPh_3$	108.2	84.1, 83.6, 80.6, 80.0	1699
$Mn\{C_5H_4C(OH)Me_2\}(CO)_2PPh_3$	112.9	80.9, 83.7	1699
$Mn\{C_5H_4C(OH)Me_2\}(CO)_2PPr^i_3$	112.2	78.8, 81.3	1699
$Mn\{C_5H_4C(OH)Et_2\}(CO)_2PPh_3$	111.9	82.0, 83.4	1699
$Mn\{C_5H_4C(OH)HPh\}(CO)_2PPh_3$	107.0	81.2, 84.9, 80.0, 83.1	1699
$Mn\{C_5H_4C(OH)MePh\}(CO)_2PPh_3$	112.6	81.5, 82.8, 85.2	1699
$Mn\{C_5H_4C(OH)Ph_2\}(CO)_2PPh_3$	110.2	84.1, 84.8	1699
$[Mn\{C_5H_4CHPh\}(CO)_3]^+$	98.3	92.8 {171.0}, 97.6 {188.2} 98.3 {179.4}, 100.3 {169.0}	878
$[Mn(C_5H_4CMeEt)(CO)_3]^+$	97.8	92.2, 99.3	878
$[Mn(C_5H_4CMePh)(CO)_3]^+$	98.3	93.3, 94.6, 98.3, 99.4	878
$[Mn(C_5H_4CPh_2)(CO)_3]^+$	96.8	95.4 {188}, 96.5 {185}	878
$[Mn(C_5H_4CPhC_6H_4OMe\text{-}4)(CO)_3]^+$	94.8	92.8 {188}, 94.8 {182.4}	878
$[\{Mn(C_5H_4)(CO)_3\}_2CH]^+$	90.0	94.3	1702
$[\{Mn(C_5H_4)(CO)_3\}_2CMe]^+$	93.5	92.0, 94.2	1702
$[\{Mn(C_5H_4)(CO)_3\}_2CPh]^+$	91.6	93.6	1702
$[\{Mn(C_5H_4)(CO)_3\}\{Mn(C_5H_4)$-$(CO)_2(PPh_3)\}CH]^+$	94.2 88.6	83.5, 87.6, 89.9, 95.5, 93.0, 102.6	1702

COMPOUND	$\delta(^{13}C)$/p.p.m., J/Hz		REFERENCE
	C^1	$C^{2,3,4,5}$	
$[\{Mn(C_5H_4)(CO)_3\}\{Fe(C_5H_4)$-Cp$\}CH]^+$	89.4	from 77.5, 80.8, 85.5, 88.5, 92.2, 92.8	1702
$[Mn(C_5H_4CHMe)(CO)_2PPh_3]^+$	99.8	93.1, 95.0, 103.9, 104.9	1699
$[Mn(C_5H_4CMe_2)(CO)_2PPh_3]^+$	95.7	90.7, 103.9	1699
$[Mn(C_5H_4CMe_2)(CO)_2PPr^i_3]^+$	95.5	89.6, 99.1	1699
$[Mn(C_5H_4CEt_2)(CO)_2PPh_3]^+$	95.4	90.5, 103.8	1699
$[Mn(C_5H_4CHPh)(CO)_2PPh_3]^+$	96.6	92.3, 96.3, 103.6, 103.9	1699
$[Mn(C_5H_4CMePh)(CO)_2PPh_3]^+$	96.2	91.6, 92.1, 103.4, 104.2	1699
$[Mn(C_5H_4CPh_2)(CO)_2PPh_3]^+$	100.0	93.1, 102.0	1699
$Mn(C_5H_4BF_2)(CO)_3$		86.9, 93.1	885
$Mn(C_5H_4BCl_2)(CO)_3$		87.6, 91.0	885
$Mn(C_5H_4BBr_2)(CO)_3$		87.6, 95.1	885
$Mn(C_5H_4BI_2)(CO)_3$		88.1, 96.7	885
$[Mn(C_5H_4PPh_3)(CO)_3]^+$	75.6 (101)	88.4 (10), 93.3 (10)	1420
$Mn(C_5H_3MeBI_2)(CO)_3$		90.0, 96.6, 95.8, 107.2	885
$Mn\{C_5H_2$-1-Ph-2,3-$(CO_2Me)_2\}$-$(CO)_3$	69.8, 89.4, 92.6, 98.2, 100.0		439, 1116
$Mn(C_5H_2I_3)(CO)_3$	90.9*, 72.1§, 55.1§		1089
$[Mn(C_5Et_5)(CO)_2(PPh_3)H]^+$	107.8		347
$Mn(C_5Et_5)(CO)_2(PPh_3)$	98.8		347
$Mn(C_5Cl_5)(CO)_5$	118		534
$Mn(C_5Cl_5)(CO)_3$	94.6		534
$Mn(C_5Cl_4Br)(CO)_3$	79.8, 95.6, 97.8		534
$Mn(C_5Br_5)(CO)_5$	86.5		1089
Mn(pyrrole)$(CO)_3$	86.5, 106.9		244
	$C^{2,5}$	$C^{3,4}$	
$Mn(C_4H_4NMn(NC_4H_3COMe)(CO)_3\}$-$(CO)_3$	108.5, 109.4	85.5, 86.4	958
Mn(phosphole)$(CO)_3$	96.2 (65.4)	93.8 (7.7)	1412
Mn(3-Me-phosphole)$(CO)_3$	95.5 (62.2) 97.5 (63.8)	113.9 (8.7) 94.4 (7.8)	1412
Mn(3,4-Me_2-phosphole)$(CO)_3$	96.8 (62)	111.8 (7.7)	1412
Mn(3,4-Me_2-phosphole)$(CO)_2PPh_3$	97.4 (61)	107.8 (6.8)	1419
Mn(2-MeCO-3,4-Me_2-phosphole)-$(CO)_3$	103.2 (63.7) 97.1 (62.3)	115.3 (5.5) 113.9 (7.8)	1412
Mn(2-PhCO-3,4-Me_2-phosphole)-$(CO)_3$	111.0 (66.6) 96.6 (64.4)	113.7 (5.0) 111.3 (7.8)	1412
Mn(2,5-Me_2-arsole)$(CO)_3$	128.4, 93.5		1416
Rhenium			
$Re(C_5H_5)(CO)_3$	84.5, 85.9		244, 1712
cis-$[Re(C_5H_5)(CO)_2H(PPh_3)]^+$	89.1		648

*CH, §CI

COMPOUND	δ(^{13}C)/p.p.m., J/Hz	REFERENCE
trans-[Re(C_5H_5)(CO)$_2$H-(PPh$_3$)]$^+$	90.2	648
Re(C_5H_5)(CO)$_2$CHPh	93.8	1423
Re(C_5H_5)(CO)$_2$C=CHPh	90.4 {180.7}	970
Re(C_5H_5)(CO)$_2$CMePh	92.9	832
[Re(C_5H_5)(CO)$_2$CPh]$^+$	94.1	832
Re(C_5H_5)(CO)$_2$C(CN)Ph	98.1	946
[Re(C_5H_5)(CO)$_2$C(PMe$_3$)Ph]$^+$	96.5	1048
[Re(C_5H_5)(CO)$_2$C(PMe$_3$)$_2$Ph]$^+$	85.2	1019
Re(C_5H_5)(CO)$_2$C(SCN)Ph	94.2	946
Re(C_5H_5)(CO)$_2$(CS)	88.4	1712
Re(C_5H_5)(CO)$_2$(CSe)	89.7	1712
[Re(C_5H_4PPh$_3$)(CO)$_3$]$^+$	82.1 (99), 90.7 (10), 95.6 (9)	1420
Re(C_5H_4Cl)(CO)$_3$	104.3, 83.1	1425
Re(C_5H_4Br)(CO)$_3$	84.2, 86.1, 84.5	1425

Iron, [^{57}Fe]

A. Monosubstituted Ferrocenes, Fe(C_5H_5)(C_5H_4X)

X	C_5H_5	C^1	C^{2,3,4,5}		REFERENCE
H	68.2 [4.7]				93, 94, 877, 885, 1754
	69.2				
	67.9				244, 245, 257, 1757
	67.7 [4.9]				496, 497, 1159, 1160
Me	69.3	84.8	67.9	69.9	121
	68.4	83.4	66.8	68.9	497, 1159, 1160
			{172.4, 6.6, 3.2}	{174.0, 6.6}	888
	68.5	83.7	67.0	69.0	499
	68.7	84.0	67.2	69.2	245, 257
	68.7 [4.7]	83.5	69.3	67.4	877
Et	68.2	90.8	66.9 {172.1, 6.6, 3.2}	67.3 {173.9, 6.6}	245, 257
		90.0	67.3	68.3	628, 888
	68.2	90.7	67.2	66.7	497, 1160
	69.1 [4.7]	90.9 [4.8]	68.1 [4.9]	67.3 [4.8]	496
CH_2^+	82.3	110.6	84.6	94.4	1160, 1161
	82.4	110.6	84.8	94.5	492
	83.3	111.8	85.5	95.6	494, 499
	81.4	109.6	83.6	93.5	192

COMPOUND	$\delta(^{13}C)$/p.p.m., J/Hz				REFERENCE
	C_5H_5	C^1	$C^{2,3,4,5}$		
Bu^n	68.7	89.3	67.3	68.3	94, 888
			{172.3, 6.6, 2.8}	{174.0, 6.6}	
CH_2CH_2CN	67.4	85.1	68.3	68.3	245
$CH_2CH_2C_5H_4FeCp$	68.8	-	67.4	68.3	245
$CH_2C_5H_4FeCp$	68.6	88.5	67.3	68.5	245
CH_2Ph			{172.8, 6.6, 3.2}	{174.3, 6.6}	888
CH_2Bu^t	68.2	85.3	66.9	69.9	1160
$CH_2C(OH)Me_2$	69.1	85.4	68.0	70.8	839
$CH_2C(OH)Ph_2$	69.1	83.6	67.8	71.3	839
$CH_2CHMeCO_2H$	68.3	85.2	67.3	69.2	628
CH_2NMe_2	67.8	82.5	67.5	69.4	245
	69.1	84.4	67.8	70.8	121
	68.6	83.9	67.8	70.0	1160
$CH_2NMe_3^+$	69.6	77.9	70.6	72.4	1160
$CH_2PPh_3^+$	69.8	73.7	69.2	70.6	1160
CH_2OH	68.2	88.1	68.2	68.2	192
	68.0	87.8	67.9	68.0	1159, 1160
	68.2	88.0	67.9	68.2	499
CH_2OMe	68.9	84.4	68.7	69.8	121
Pr^i	67.7	96.5	65.8	66.6	245, 257
	68.0	96.6	65.8	66.5	497, 1160
			{172.6, 6.6, 2.8}	{173.8, 6.6}	888
$CHMe^+$	82.3	105.6	81.0	93.9	492
			81.8	94.2	
	83.4	106.9	81.7	95.1	494, 499
			82.8	95.4	
	81.5	104.8	80.0	93.1	192, 1159
			80.9	93.4	
	81.5	104.7	80.9	93.0	496
	[4.3]	[3.3]	[4.5]	[2.0]	
			80.0	93.3	
			[4.1]	[2.1]	
$CHPr^{i+}$	81.3	104.3	80.5	94.6	494, 499
			81.3	95.1	
	79.7	102.4	78.9	92.9	344, 1159
			79.7	93.2	
	82.1	104.7	81.4	95.1	839
			82.1	95.6	
	81.3	104.0	80.6	94.4	187
			81.3	94.8	
$CHCHPh_2^+$	81.7	104.9	79.9	94.5	839
			81.2	94.7	
$CHBu^{t+}$	81.2	101.6	81.1	95.1	494, 499
			82.6	95.3	
	80.2	100.1	80.2	93.6	1159
			81.4	93.8	

COMPOUND	$\delta(^{13}C)$/p.p.m., J/Hz				REFERENCE
	C_5H_5	C^1	$C^{2,3,4,5}$		
$CHC_5H_4Mn(CO)_3^+$		97.6	from 92.2, 92.8, 77.5,		1702
			80.8, 83.9, 85.5, 88.5		
$CHPh^+$	82.6	101.2	79.5	92.8	192
			81.8	93.7	
CHMePh	68.4		65.9	67.3	1723
			66.7	67.6	
$CHMeCH_2CO_2H$	68.3	93.7	66.3	67.1	628
CHMeOH	68.2	76.4	65.6, 66.1	67.8	245
	68.1	94.1	65.4	67.4	1159
			66.0	67.5	
	68.3	94.7	66.0	67.7	192
			66.2		
	68.2	94.4	65.9	67.6	499
			66.3	67.7	
	69.0	95.4	66.8	68.4	496
	[4.6]	[5.3]	[4.8]	[4.8]	
			66.7	68.5	
			[4.7]	[4.8]	
$CH(CH_2CN)(OH)$			65.2	65.5	1755
			66.2	68.0	
$CHPr^iOH$	69.5	76.3	66.9	69.0	187
			68.7	70.2	
	68.2	93.1	64.8	67.7	499
			67.6	68.7	
	69.1	93.7	66.1	67.7	839
			68.3		
	67.9	92.6	67.1	68.3	344, 1159
			67.4	68.5	
$CH(OH)CHPh_2$	69.1	94.5	66.3	67.7	839
			68.6		
$CH(OH)Bu^t$	68.0	91.4	65.2	67.2	1159
			69.4	67.3	
	68.1	91.6	65.3	67.5	499
			67.3	69.7	
$CH(OH)C_5H_4Mn(CO)_3$	-	93.1	66.4, 67.4,	68.4	1702
CH(OH)Ph	68.4		66.0	67.9	245
			67.3	71.9	
	68.2	93.9	65.3	67.7	1160
			67.0	67.8	
	68.5	94.1	66.1	68.0	192
			67.4		
$CH(OH)CH_2CN$	68.3	90.0	65.5	68.1	1160
			66.2	68.2	
CMe_2^+	81.9	100.0	78.7	93.4	1159, 1160
	82.9	101.2	78.7	94.7	494, 499
	82.2	100.1	78.3	93.6	
$CMeEt^+$	82.2	100.3	78.4	93.8	492
			78.9	94.0	
$CMeCH_2CO_2R^+$	82.3	103.3	78.1	94.2	1758
			78.7	94.5	

COMPOUND	$\delta(^{13}C)$/p.p.m., J/Hz				REFERENCE
	C_5H_5	C^1	$C^{2,3,4,5}$		
$CMePr^{i+}$	81.2	100.1	76.9	94.5	494, 499
			77.9	95.9	
$CMePh^+$	81.4	98.5	77.6 {183.8}	92.6 {186.6}	878
				92.9 {186.8}	
$CMeBu^{t+}$	81.7	100.5	79.2,	95.1,	494, 499,
			80.3	95.3	1159, 1160
Bu^t	68.2	101.7	64.8	66.7	497
	68.0	101.8	64.8	66.7	245, 257
CMe_2Et	68.2	100.6	65.6	66.8	245
CMe_2OH	68.0	99.8	65.2	67.3	1159, 1160
	68.2	100.0	65.5	67.6	499
$CMePr^iOH$	68.2	99.3	65.4	67.7	499
			67.4	68.4	
$CMeBu^tOH$	68.3	98.4	66.7	67.3	499
			66.8	69.3	
CMe(OH)Ph	68.1	100.6	66.0	67.2	878
			66.9	67.8	
	68.3	100.3	65.6	67.6	1723
			67.1	68.0	
$CH=CH_2$	68.9	83.2	66.4	68.4	1159, 1160
	69.1	85.3	66.2	68.3	245
CH=CHCN-*cis*	69.6	76.9	69.4	70.9	1160
CH=CHCN-*trans*	69.5	77.7	67.8	70.9	1160
CH=CHCOPh	69.9	79.2	69.2	71.4	355, 423
CH=CMeCOPh	69.9	85.6	67.1	70.6	423
$CH=CHCOPhFe(CO)_4$	69.6	89.0	62.4	70.0	355
			68.0	70.4	
$CH=CHNO_2$-*trans*	70.1		68.6	69.6	245
C≡CH	69.6	63.5	68.3	71.2	245, 1160
meso porphyrin	68.9	76.1	67.6	70.2	1118
		78.8	67.9	70.8	
Ph	69.5	85.7	68.6	66.5	245
	69.5	85.4	68.8	66.4	493
C_6H_4Me-4	69.4	85.7	68.5	66.3	493
C_6H_4Et-4	69.4	85.9	68.7	66.5	493
C_6H_4CHO-4	70.0	82.9	70.2	67.1	493
C_6H_4COMe-4	70.0	83.3	70.0	67.0	493
$C_6H_4CO_2Et$-4	70.1	83.6	70.1	67.0	493
$C_6H_4NMe_2$-4	69.3	86.9	68.1	65.8	493
$C_6H_4NO_2$-4	70.2	81.7	70.7	67.4	245, 493
C_6H_4OH-4	69.6	85.8	68.5	66.2	493
C_6H_4OMe-4	69.3	85.8	68.4	66.0	493
C_6H_4Cl-4	69.7	84.2	69.2	66.5	493
C_6H_4Br-4	69.8	84.2	69.3	66.5	493
C_6H_4Me-3	69.6	85.7	68.8	66.5	493
$C_6H_4CF_3$-3	69.8	83.9	69.6	66.8	493
C_6H_4CHO-3	69.7	83.6	69.5	66.7	493
C_6H_4COMe-3	69.6	84.3	69.3	66.6	493
$C_6H_4NO_2$-3	70.0	82.5	70.0	66.8	493
C_6H_4OMe-3	69.7	85.3	68.8	66.7	493

COMPOUND	δ(^{13}C)/p.p.m., *J*/Hz				REFERENCE
	C_5H_5	C^1	$C^{2,3,4,5}$		
C_5H_4FeCp	69.1	84.6	67.6	66.5	245
$C_5H_4Fe(arene)^+$	69.4 to 70.2	72.8 to 74.7	70.1 to 71.3	67.3 to 67.6	1762
$CW(CO)_4Br$	71.0	89.5	71.6	72.7	751
$CW(CO)_2Cp$	70.1	94.1	70.9	70.3	1043
$C(CO)W(CO)(PMe_3)Cp$	68.7	91.8	70.8	65.7	1037
$CCOW(CO)(PMe_3)Cp$	70.1	86.4	70.8	68.2	1037
$C(OEt)Cr(CO)_5$	71.3	94.9	73.2	75.2	39
$C_5H_4Mn(CO)_3$	69.5		68.9	66.5	245
$CMeN_2CMeC_5H_4FeCp$	69.2	83.8	67.4	69.9	245
CN	70.1	51.4	71.2	70.4	245
	70.2	51.8	70.1	71.4	1159, 1160
CHO	68.7	78.7	68.7	72.3	245
	69.2	79.2	68.0	72.6	1159, 1160
	70.1	80.4	70.1	76.3	645
COMe	69.5	78.8	69.1	71.2	372
	69.9	80.3	69.8	72.0	94
	69.4 [4.6]	79.1 [5.1]	68.1 [4.3]	71.8 [4.7]	496, 1756
	69.7	79.3	69.5	72.1	628
	70.4	80.1	70.1	72.8	645
	69.8	79.5	69.6	72.2	499, 1780
	69.3	78.8	69.7	71.8	245
	69.5	79.3	69.2	71.8	1159, 1160
COEt	69.3	78.9	69.7	71.8	245
$COBu^t$	69.7	76.7	70.9	71.0	499
	69.4	76.6	70.7	70.9	1160
COCH=CHPh	70.2	80.8	69.7	72.9	355, 423
$COCH{=}CHPhFe(CO)_4$	70.5	80.3	69.7 71.2	72.5 72.9	355
$COCH{=}CHPhFe(CO)_3$	70.0	76.0	67.8 71.2	71.5 72.8	355
COPh	70.7	79.0	72.0	73.1	645
CONH	69.9		68.6	70.8	245
CO_2H	70.3	77.8	70.8	72.2	423
CO_2Me	69.5		70.0	71.0	245
$Ti(NMe_2)_3$	60.3	18.9	53.1	59.2	184
$Ti(NEt_2)_3$		18.7	53.5	59.0	184
$Ti(NEt_2)_2C_5H_4FeCp$	60.0		54.0	58.8	184
$AuPPh_3$	67.5	72.4	76.4	69.2	245
HgC_5H_4FeCp	67.6		74.5	70.6	245
BMe_2	68.3		75.5	75.1	885
BMeI	70.6		77.6	77.6	885
BPhI	71.1		78.0	79.4	885
$B(NMe_2)_2$	68.6		76.0	70.1	885
$B(OEt)_2$	68.6		74.7	71.8	885
$B(SMe)_2$	69.4		74.1	72.3	885
BF_2	69.1		74.5	74.7	885
BCl_2	70.9		77.3	76.2	885
BBr_2	71.7		78.4	77.4	885

COMPOUND		δ(^{13}C)/p.p.m., J/Hz					REFERENCE
		C_5H_5	C^1	$C^{2,3,4,5}$			
BI_2		72.5		79.6		79.6	885
$SiMe_3$		68.0		72.7		70.6	245
NH_2		68.7	104.4	59.0		63.2	245
$N_2C_5H_4FeCp$		69.7		64.1		69.4	245
NO_2		71.7		66.5		70.4	244, 245
PHC_5H_4FeCp		68.9		70.6		68.7	245
$P(C_6H_4FeCp)_2$		69.3		72.7		70.3	245
OMe		68.4	125.7	55.0		61.8	245
		68.1	127.3	54.7		61.5	1160
OPh		69.4		60.5		63.4	245
SCN		69.9		72.9		70.5	245
Cl		70.3	83.4	67.9		66.1	245
Br		70.5		70.1		67.0	245
		58.0	51.0	58.2		61.1	184

B. 1,1'-Disubstituted Ferrocenes, $Fe(R^1C_5H_4)(R^2C_5H_4)$*

R^1	R^2	C^1	$C^{2,3,4,5}$				REFERENCE
Me	Me	83.1	69.3		67.3		498
		83.7	71.0		73.5		499
		83.7	69.7		67.7		1757
		83.3	67.7		69.4		257
Et	Et	83.3	69.9		68.0		877
		[4.6]	[4.7]		[4.8]		
		92.1	68.1		68.4		498
		90.3	67.5		67.2		257
		90.7	69.8		67.4		1757
CH_2Ph	CH_2Ph	88.2	69.7		68.7		498
$CH_2-CH_2-CH_2$		85.0	68.9		67.6		1760
CH_2-O-CH_2		83.7	70.2		70.0		1760
$CH_2-CH=CH$			69.5,	70.8,	71.6		1761
$CH_2(CH_2)_3CH_2$		89.2	68.4		66.9		422
		87.5	67.8		66.7		1760
$CH_2CH_2CHPhCH_2CH_2$		88.9	67.8		66.0		1760
			68.9		66.9		
$(CH_2)_7CO(CH_2)_7$		89.8	68.8		67.0		495
Pr^i	Pr^i	96.6	66.3		67.4		1757
		96.5	66.1		67.1		257
$CH(CH_2)-(CH_2)_2CH$		86.9	69.6		68.2		805, 991
			68.4		68.1		
CH_2-CH_2-CO		87.8	70.8		69.0		1759, 1760
		73.9	70.0		72.4		
$CH_2-CHMe-CO$		85.6	66.4,	68.0,	69.6,	70.1,	1759
			72.0,	72.4,	73.2		
		74.6					
$CHMe-CH_2-CO$		92.9					1759
		75.9					

*Shifts for $R^1C_5H_4$ are given above those for $R^2C_5H_4$ when assigned

COMPOUND		$\delta(^{13}C)$/p.p.m., J/Hz			REFERENCE
R^1	R^2	C^1	$C^{2,3,4,5}$		
Bu^t	Bu^t	102.0	65.6	68.0	498
		101.0	64.8	67.0	257
		101.8	65.3	67.5	1757
Ph	Ph	90.3	67.5	67.2	1760
COMe	COMe	80.1	70.4	73.1	1760
		80.9	69.7	67.8	499
		81.7	69.9	73.0	423
		81.7	71.6	74.1	645
$CO-(CH_2)_3-CO$		81.4	69.9	73.4	422, 423, 1760
COCH=CHPh	COCH=CHPh	82.2	71.4	74.2	423
CMe=CHCOPh	CMe=CHCOPh	86.5	68.3	72.1	423
CH=CHCN-*cis*	CH=CHCH-*cis*	79.7	71.7	73.2	423
CH=CHCN-*cis*	CH=CHCN-*trans*	79.7	71.7	73.2	423
		80.5	70.4	73.3	
CH=CHCN-*trans*	CH=CHCN-*trans*	80.4	70.2	73.2	423
CH_2CH_2Ph	CO_2H	89.8	69.1	69.8	628
			70.9	72.3	
$Ti(NEt_2)_3$	$Ti(NEt_2)_3$	109.8	75.0	69.5	1247

C. Other Ferrocenes

COMPOUND	$\delta(^{13}C)$/p.p.m., J/Hz	REFERENCE
$Fe(C_5H_5)\{1,2\text{-}C_5H_3CO(CH_2)_3\}$	Cp 69.9, C^1 75.2, C^2 92.1, $C^{3,4,5}$ 64.9; 70.2	628
$Fe(C_5H_5)\{1,2\text{-}C_5H_3COCMe_2(CH_2)\}$	Cp 69.7, C^1 74.3, C^2 90.0, $C^{3,4,5}$ 65.6; 69.4; 70.4	628
$Fe(C_5H_4Me)\{1,3\text{-}C_5H_3Me(CMe_2OH)\}$	84.2; 70.2; 67.4; 68.8; 70.5; 99.7; 67.4; 83.8; 69.4; 65.3	499
$Fe(indenyl)_2$	$C^{1,2}$ 87.0, $C^{3,5}$ 61.9, C^4 69.8	370, 498,
$Fe(1,3\text{-}C_5H_3Me_2)_2$	$C^{1,3}$ 83.4 {6.1}, $C^{2,4,5}$ 69.9 {171.3}; 71.8 {170.2}	1237
$Fe(1,3\text{-}C_5H_3MePh)_2$	$C^{1,3}$ 86.2; 86.6, $C^{2,4,5}$ 67.9; 69.3; 73.0	498
$Fe(1,3\text{-}C_5H_3Ph_2)_2$	$C^{1,3}$ 85.6, $C^{2,4,5}$ 66.1; 68.5	498
$Fe\{1,2\text{-}C_5H_3CH(CH_2)(CH_2)_2CH\}_2$	$C^{1,2}$ 98.3, $C^{3,4,5}$ 59.5; 68.2	498
$Fe\{1,3\text{-}C_5H_3(C_5H_8)_2C_5H_3\text{-}1,3\}$*	$C^{1,3}$ 83.8; 88.3, $C^{2,4,5}$ 70.5; 72.0; 72.5	805, 991
	$C^{1,3}$ 85.9, $C^{2,4,5}$ 71.4; 71.1	805, 991
$Fe(1,2,4\text{-}C_5H_2Me_3)_2$	$C^{1,2,4}$ 81.2; 81.8, $C^{3,5}$ 71.9 {167.7}	1237

*two isomers

COMPOUND	$\delta(^{13}C)$/p.p.m., J/Hz	REFERENCE
$Fe(C_5HMe_4)_2$	C^{1-4} 80.5; 80.7, C^5 71.2 {163.2}	1237
$Fe(C_5Me_5)_2$	79.0	1237
$Fe(C_5H_5)$(phosphole)	Cp 70.2, $C^{2,5}$ 77.2 (62), $C^{3,4}$ 79.8 (7)	966
$Fe(C_5H_5)(3,4\text{-}Me_2\text{-phosphole})$	Cp 71.4, $C^{2,5}$ 78.2 (61), $C^{3,4}$ 94.7 (7.2)	966, 1467
$Fe(3,4\text{-}Me_2\text{-phosphole})_2$	$C^{2,5}$ 82.1 (61.6) $C^{3,4}$ 97.5 (7.5)	1467
FeCp{3,4-Me_2-phospholeFe$(CO)_4$}	Cp 74.3 $C^{2,5}$ 67.7 (8.1), $C^{3,4}$ 92.5 (3.8)	1474
$Fe(C_5H_5)$(2-Ph-3,4-Me_2-phosphole)	Cp 72.7, C^2 77 (60), $C^{3,4}$ 92 (5.2), 96 (6.1), C^4 100.3 (58)	966
$Fe(C_7SH_4Me)$	61.1, 61.4, 86.1, 95.1, 96.1	976

D. Other $Fe(C_5H_5)$ derivatives containing an unsubstituted C_5H_5 ring

COMPOUND	$\delta(^{13}C)$/p.p.m., J/Hz	REFERENCE
$[Fe(C_5H_5)(C_6H_6)]^+$	76.8, 77.9	244, 260
$[Fe(C_5H_5)(arene)]^+$	75.0 to 79.9	260, 500, 841, 954, 1110, 1464
$[Fe(C_5H_5)(C_8H_8)]^+$	82.6	1003
$Fe(C_5H_5)(C_6H_5R)$	72.5 to 73.4	244, 1763
$[Fe(C_5H_5)(CO)_2C_2H_4]^+$	89.8, 90.1	287, 489, 659
$[Fe(C_5H_5)(CO)_2CH{=}CHPh]^+$	90.4	659
$[Fe(C_5H_5)(CO)_2C_3H_6]^+$	87.6, 91.3	4, 489
$[Fe(C_5H_5)(CO)_2C_4H_8]^+$	87.6	489
$[\{Fe(C_5H_5)(CO)_2\}_2C_4H_4]^{2+}$	92.8 {178}	314, 659
$Fe(C_5H_5)(CO)_2Me$	85.3, 85.7	60, 70, 763, 1394
$Fe(C_5H_5)(CO)_2R$	85.1 to 87.3	34, 60, 70, 78, 80, 93, 244, 280, 356, 763, 856, 980, 1001, 1213, 1232, 1312, 1394
$Fe(C_5H_5)(CO)_2C_3H_5$	110.0	34
$[Fe(C_5H_5)(CO)_3]^+$	90.8, 90.9	287, 295
$[Fe(C_5H_5)(CO)_2CS]^+$	92.1, 92.2	295, 661
$[Fe(C_5H_5)(CO)_2CHPh]^+$	93.2 {185, 7}	1095
$\{Fe(C_5H_5)(CO)_2\}_2SiMeH$	84.3	1455
$Fe(C_5H_5)(CO)_2SiMe_3$	84.2	70
$Fe(C_5H_5)(CO)_2SiPh_3$	84.5	70
$Fe(C_5H_5)(CO)_2GeMe_3$	87.7	70

COMPOUND	$\delta(^{13}C)$/p.p.m., J/Hz	REFERENCE
$Fe(C_5H_5)(CO)_2GePh_3$	84.7	70
$Fe(C_5H_5)(CO)_2SnMe_3$	81.4	70
$Fe(C_5H_5)(CO)_2SnMePhCH_2CMe_2Ph$	82.4	1729
$Fe(C_5H_5)(CO)_2SnPh_3$	83.3	70
$Fe(C_5H_5)(CO)_2CN$	82.3, 85.9	60, 70
$[Fe(C_5H_5)(CO)_2(NCMe)]^+$	87.5	287
$[Fe(C_5H_5)(CO)_2NH_3]^+$	86.7	295
$[Fe(C_5H_5)(CO)_2py]$	88.1	287
$[Fe(C_5H_5)(CO)_2PPh_3]^+$	89.2, 90.8	287, 295
$Fe(C_5H_5)(CO)_2PB_{10}H_{12}$	86.4	485
$[Fe(C_5H_5)(CO)_2(thf)]^+$	85.8	1003
$Fe(C_5H_5)(CO)_2Cl$	85.6, 85.9	60, 70
$Fe(C_5H_5)(CO)_2Br$	85.4, 85.9	60, 70
$Fe(C_5H_5)(CO)_2I$	84.7, 84.8	60, 70
$Fe_2(C_5H_5)_2(CO)_4$	106.7	66
trans-$[Fe_2(C_5H_5)_2(CO)_3CCp]^+$	94.5	1451
cis-$[Fe_2(C_5H_5)_2(CO)_3C(C_5H_5)]^+$	93.0	1451
$Fe_2(C_5H_5)_2(CO)_3CNBu^t$	85.7, 86.5	506
$Fe_2(C_5H_5)_2(CO)_3(\mu\text{-}CNBu^t)$	89.5	506
cis-$Fe_2Cp_2(CO)_3\overline{CNEtCH_2CH_2N}Et$	87.2, 87.4	1224
trans-$Fe_2(C_5H_5)_2(CO)_3\overline{CNEtCH_2CH_2N}Et$	85.9, 88.2	1224
$Fe_2(C_5H_5)_2(CO)_3P(OEt)_3$	86.3, 87.6 89.5	397
$\overline{Fe(C_5H_5)(CO)C(CF_3)=C(CF_3)CH}=CHMe$	83.8	1116
$[Fe(C_5H_5)(CO)(CNCH_2Ph)_2]^+$	85.3	70, 574
$[Fe(C_5H_5)(CO)(CNCy)_2]^+$	85.8	574
$Fe(C_5H_5)(CO)(CNCy)COMe$	84.1	1394
$[Fe(C_5H_5)(CO)(CNR)\{C(NHR^1)_2\}]^+$	83.7 to 84.3	1470
$[Fe(C_5H_5)(CO)(CN)\{C(NHR)_2\}]^+$	82.5 to 82.9	1470
$Fe(C_5H_5)(CO)(CNCy)(7,8\text{-}B_9C_2H_{12})$	85.8	487
$[Fe(C_5H_5)(CO)(CN)_2]^-$	82.6	60
$[Fe(C_5H_5)(CO)(PR_3)\{C(NHR^1)_2\}]^+$	84.1 to 84.9	1470
$Fe(C_5H_5)(CO)H(PPh_3)$	80.5	722
$Fe(C_5H_5)(CO)(PR_3)R$	84.2 to 85.2	482, 776, 856, 980, 1001, 1117, 1276, 1394
$[Fe(C_5H_5)(CO)(PPh_3)CH_2PPh_3]^+$	89.4	1001
$Fe(C_5H_5)(CO)\{P(OPh)_3\}CHPhSiMe_3$	82.7 (1.4), 82.8 (1.5)*	446
$\overline{Fe(C_5H_5)(CO)\{P(OR)_2(OC_6H_4\text{-}2}\})\}$	82.9, 83.0	856
$[Fe(C_5H_5)(CO)(PPh_3)CHPh]^+$	93.3 {181} (7)	1095
$[Fe(C_5H_5)(CO)(PPh_3)CNEt]^+$	85.3	482
$Fe(C_5H_5)(CO)(CN)PR_3$	82.4 to 84.2	482, 490
$[Fe(C_5H_5)(CO)(CS)PR_3]^+$	90.2 to 91.5	661
$Fe(C_5H_5)(CO)\{(PF_2)_2NMe\}Cl$	83.5	1394
$Fe(C_5H_5)(CO)(PF_2NHMe)Cl$	82.7	1390
$Fe(C_5H_5)(CO)S_4Fe(C_5H_5)$	83.4, 86.7	1087

*two isomers

COMPOUND	$\delta(^{13}C)$/p.p.m., J/Hz	REFERENCE
$[Fe(C_5H_5)(CS)(CNPh)_2]^+$	88.7	661
$[Fe(C_5H_5)(CCHMe)(dppe)]^+$	89.5	1446
$[Fe(C_5H_5)\{C(OMe)Ph\}(dppe)]^+$	88.8	980
$[Fe(C_5H_5)(C{\equiv}CMe)(dppe)]^+$	88.0	1446
$Fe(C_5H_5)(C{\equiv}CH)(dppe)$	77.9	1446
$Fe(C_5H_5)(C{\equiv}CMe)(dppe)$	78.8	1446
$Fe(C_5H_5)\{P(OPh)_3\}\{P(OPh)_2(OC_6H_4-2)\}$	80.2	856
$Fe_2(C_5H_5)_2(PF_2NMePF_2)(\mu-PF_2)(NMe{=}PF_2)$	79.2, 81.6	1451
$Fe(C_5H_5)\{(PF_2)_2NMe\}_2Cl$	81.6	1390
E. Other Fe (substituted cyclopentadienyl) derivatives		
$Fe(C_5H_4Me)(CO)S_4Fe(C_5H_4Me)$	86.0, 86.5, 86.8, 87.2	1087
$Fe_2(C_5H_4CHCHCHMe)(CO)_5$	C^1 85.0, $C^{2,5}$ 80.3, 82.3, 87.1, 90.6	697
$[Fe(arene)(C_5H_4)]_2{}^{2+}$	C^1 84.0 to 86.6, $C^{2,5}$ 74.8 to 78.2, $C^{3,4}$ 72.7 to 74.3	1762
$[Fe(C_5H_4C_5H_4FeCp)(arene)]^+$	C^1 96.5 to 98.3, $C^{2,5}$ 75.0 to 76.9, $C^{3,4}$ 71.6 to 72.5	1762
$Me_2Si\{(C_5H_4)Fe(CO)_2C_5H_{11}\}_2$	92.5, 88.8, 85.9	1463
$Fe(C_5H_4I)(CO)_2C_3F_7$	C^1 46.2, $C^{2,5}$ 86.5, $C^{3,4}$ 93.5	1447
$Fe_2(C_5H_4I)_2(CO)_4$	C^1 46.3, $C^{2,5}$ 83.4, $C^{3,4}$ 92.9	1447
$Fe(C_5H_4I)(CO)_2I$	C^1 51.7, $C^{2,5}$ 90.2, $C^{3,4}$ 94.8	1447
$[Fe(indenyl)(CO)_2(C_2H_4)]^+$	$C^{1,3}$ 77.1, C^2 87.9, $C^{4,5}$ 105.3	489
$[Fe(indenyl)(CO)_2(C_3H_6)]^+$	$C^{1,3}$ 74.9, 76.1, C^2 87.3, $C^{4,5}$ 103.5, 107.4	489
$[Fe(indenyl)(CO)_2(C_4H_8)]^+$	$C^{1,3}$ 74.4, C^2 88.4, $C^{4,5}$ 106.7	489
$[Fe(indenyl)(CO)(CNR)\{C(NHR)_2\}]^+$	$C^{1,3}$ 70.8 to 72.6, C^2 91.7 to 92.3, $C^{4,5}$ 102.7 to 104.6	1470
Fe_2(7-hydroindene)$(CO)_5$	C^1 70.6, C^2 78.0, C^3 84.2, $C^{3a,7a}$ 85.8, 87.0	639
$Fe_2(C_9H_6Me_2)(CO)_5$	80.0, 83.7, 84.2, 84.6, 99.4	639
$Fe_2(C_{10}H_8)(CO)_5$	77.6, 81.6, 85.8, 78.3, 84.8	639
$Fe_3(1,3-Me_2-2-ViC_5H_2)(CO)_9CEt$	$C^{1,3}$ 107.8, C^2 102.5, $C^{4,5}$ 89.5	1178
$Fe(C_5Cl_4I)(CO)_2C_3F_7$	C^1 47.8, C^{2-5} 98.5, 106.1	1447

COMPOUND	$\delta(^{13}C)$/p.p.m., J/Hz				REFERENCE
$Fe(C_5Br_4I)(CO)_2C_3F_7$	C^1 57.8, C^{2-5} 91.1, 97.6				1447
Ruthenium					
$Ru(C_5H_5)_2$	70.6, 71.1				244, 1251
$Ru(C_5H_5)(C_5H_4CHO)$	Cp 72.5, C^1 84.9, C^{2-5} 71.4, 74.0				1283
$Ru(C_5H_5)\{C_5H_4CH(OH)Pr^i\}$	Cp 70.5, C^1 98.9, C^{2-5} 68.8; 69.9; 72.3; 73.9				344
$Ru(C_5H_4CHO)_2$	C^1 86.5, C^{2-5} 73.1; 75.6				1283
$Ru\{C_5H_4Ti(NEt_2)_3\}$	C^1 112.6, C^{2-5} 71.5; 76.8				1251
$Ru(C_{10}H_8GeMe_3)(CO)_2GeMe_3$	83.8, 84.0, 85.0, 104.4, 115.6				516
$Ru(C_5Me_4Et)(CO)_2Cl$	100.2, 101.3, 103.3				1499
$Ru(C_5Me_4Et)(CO)(AsPh_3)Cl$	93.4, 95.7, 96.0, 98.3				1499
$[Ru(tetramethylthiophene)(p\text{-}cymene)]^{2+}$	107.9, 113.9				1517
Osmium					
$Os(C_5H_5)_2$	63.9				244
$Os(C_5H_5)\{C_5H_4CH(OH)Pr^i\}$	Cp 64.4, C^1 77.7, C^{2-5} 63.4, 65.5				344
Cobalt					
$[Co(C_5H_5)_2]^+$	85.0				1757
	Cp	C^1	$C^{2,5}$	$C^{3,4}$	
$[Co(C_5H_5)(C_5H_4CPh_2)]^{2+}$	89.0	95.6	93.0	94.8	825
$[Co(C_5H_5)\{C_5H_4C(OH)Ph_2\}]^+$	85.6		84.2	84.2	825
$[Co(C_5H_5)(C_5H_4CHPh_2)]^+$	86.3		85.4	85.0	825
$Co(C_5H_5)(C_5H_4CO_2)$	85.9	96.1	86.4	86.4	825
$[Co(C_5H_5)(C_5H_4NH_3)]^{2+}$	87.4	98.7	83.6	80.8	825
$Co(C_5H_5)(quinone)$	85.4				562
$Co(C_5H_5)(C_{10}H_{12})$	83.9				539
$Co(C_5H_5)(C_4Ph_4)$	83.4				1495
$Co(C_5H_5)\{(EtC_2Et)_2CO\}$	82.2				798
$Co_3(C_5H_5)_3C_2(NEt_2)_2$	82.5				684
$Co(C_5H_5)(CO)OC(OR)=C(CO_2R)CO$	92.6				1020
$Co(C_5H_5)(C_3F_7)(CN\text{-}p\text{-}tol)$	94.8				204
$[\{Co(C_5H_5)\}_2SC_3H_3]^+$	81.2, 83.3				721
$[\{Co(C_5H_5)\}_2SC_3HMeEt]^+$	84.0, 86.0				721
$Co_2(C_5H_5)_2(CHCO_2Et)$	87.2				1494
$Co_2(C_5H_5)_2\{C(CO_2Et)_2\}$	88.3				1494
$[Co(C_5H_4Me)_2]^+$		103.9	85.1	84.1	1757
$[Co(C_5H_4Et)_2]^+$		110.3	83.6	84.0	1757
$[Co(C_5H_4Pr^i)_2]^+$		116.1	82.2	83.7	1757
$[Co(C_5H_4Bu^t)_2]^+$		121.3	81.0	82.3	1757
$Co(C_5H_4CHPh_2)(C_4Ph_4)$		106.2	83.8	83.0	1495
$Co(C_5H_4CPh_2)(C_4Ph_4)$		107.7	105.1	90.8	1495
$[Co_2(C_{10}H_8)_2]^{2+}$		111.2	85.2	83.0	322
$[Co(indenyl)_2]^+$	$C^{1,3}$ 76.7, C^2 83.0, $C^{4,5}$ 98.3				370

COMPOUND	δ(^{13}C)/p.p.m., *J*/Hz			REFERENCE
$[Co(1,3\text{-}Pr^{i}{}_2C_5H_3)H(PMe_3)_2]^+$	$C^{1,3}$ 117.2, 121.4, $C^{2,4,5}$ 71.6 (2.1, 4.4); 83.1 (5.1); 86.2 (2.2)			1016
$Co(1,3\text{-}Pr^{i}{}_2C_5H_3)(PMe_3)_2$	$C^{1,3}$ 107.4, C^2 74.9 (2.2), $C^{4,5}$ 71.3 (1.5)			1016
$Co(1,3\text{-}Bu^{t}{}_2C_5H_3)(PMe_3)_2$	$C^{1,3}$ 110.9 (2.2), C^2 69.2 (3.7), $C^{4,5}$ 69.8 (3.7)			1016
$Co(C_5Me_5)(C_6Me_4O_2)$	95.0 or 96.0			1496
$[Co(C_5Me_5)(C_6Me_4O_2H)]^+$	98.9 or 99.2			1496
$[Co(C_5Me_5)(C_6Me_4O_2H_2)]^{2+}$	104.5 or 106.5			1496
$[Co(C_5Me_5)(C_6Me_4O_2Me)]^+$	99.1			1496
$[Co(C_5Me_5)(C_6Me_4O_2Me_2)]^{2+}$	107.1			1496
$Co(C_5Me_4Et)(CO)_2$	C^1 103.0, C^{2-5} 96.0, 96.8			1499
Rhodium, [^{103}Rh]				
$[Rh_2(C_5H_5)_2H]^+$	87.3, 90.5			299
$Rh(C_5H_5)(C_2H_4)_2$	87.1 [3.6], 87.8			11, 1315
$Rh(C_5H_5)$(nbd)	84.7, 84.8			11, 244, 299
$Rh(C_5H_5)$(cod)	86.5			11
$Rh(C_5H_5)$(cot)	88			298
$[Rh(C_5H_5)(C_8H_{13})]^+$	91.9			1172
$Rh(C_5H_5)(C_6H_4O_2)$	88.2 [9]			562
$Rh(C_5H_5)$(diene)	84.3 to 85.5 [5 to 7]			1519
$Rh(C_5H_5)\{(EtC_2Et)_2CO\}$	84.3 [6]			798
$Rh(C_5H_5)\{(Et_2NC_2NEt_2)_2CO\}$	84.5 [6]			684
$[Rh(C_5H_5)(C_8H_9)]^+$	88.8, 89.8*			298
$Rh(C_5H_5)(CO)_2$	87.6, 87.8 [3.5]			11, 244, 1315
$Rh_2(C_5H_5)_2(C_8H_{10})$	83.1, 87.0			299
$Rh_2(C_5H_5)_2(C_7H_8)$	83.3, 86.3			299
$Rh_2(C_5H_5)_2(C_7H_8)$	87.0			299
$Rh(C_5H_5)(PhC_2Ph)$	86.4 [5.3], 89.8 [5]			556
$\{Rh(C_5H_5)CO\}_2CF_3C{\equiv}CCF_3$	89.6			556
$Rh_2(C_5H_5)_2(CO)_2CH_2$	89.6 [2]			1513
$Rh_3(C_5H_5)_3PhC{\equiv}CPh$	86.5			556
$Rh_3(C_5H_5)_3(CO)PhC{\equiv}CPh$	89.7			206, 556
$Rh_3(C_5H_5)_3(CO)C_6F_5C{\equiv}C_6F_5$	84.9§, 89.0†			206
$Rh_3(C_5H_5)_3C_2(NEt_2)_3$	84.6			684
$[Rh(C_5H_5)(C_3H_4Me)Cl]_2$	88.3, 99.6 [6]			1164
$Rh(C_5H_5)(2\text{-}MeC_3H_4)$py	90.1 [4]			1164
$Rh(C_5H_5)(MeC_3H_4)PPh_3$	93.9, 94.4 [4]			1164
$Rh(C_5H_5)(1\text{-}MeC_3H_4)AsPh_3$	91.9 [4]			1164
$Rh(C_5H_5)(2\text{-}MeC_3H_4)AsPh_3$	90.5 [6]			1164
$Rh(C_5H_4)(S_2CNMe_2)$	96.6 [7.3]			1505
	C^1	$C^{2,5}$	$C^{3,4}$	
$Rh(C_5H_4Me)(C_2H_4)_2$	91.7 [7.7]	83.1 [5.1]	83.1 [5.1]	1315
$Rh(C_5H_4Me)(CO)_2$	106.0 [3.2]	89.5 [3.8]	87.8 [3.5]	1315

*two isomers, §area 1, †area 2

COMPOUND	δ(^{13}C)/p.p.m., *J*/Hz			REFERENCE
	C^1	$C^{2,5}$	$C^{3,4}$	
Rh($C_5H_4Bu^t$)(C_2H_4)$_2$		86.0 [2.9]	84.6 [2.9]	1315
Rh(C_5H_4CN)(C_2H_4)$_2$	72.3	91.1 [3.6]	90.2 [3.6]	1315
Rh(C_5H_4CHO)(C_2H_4)$_2$	103.3 [4.1]	87.8 [3.5]	92.5 [3.8]	1315
Rh(C_5H_4CHO)(2,4-Me_2penta-1,4-diene)	103.8 [4.2]	85.3 [3.7]	92.9 [4.4]	1315
Rh(C_5H_4CHO)(2,3-Me_2-butadiene)	99.4 [5.4]	84.6 [4.9]	88.5 [4.9]	1315
Rh(C_5H_4CHO)(CO)$_2$	107.7 [3.8]	86.7 [3.2]	93.4 [3.2]	1315
Rh($C_5H_4CO_2Me$)(C_2H_4)$_2$	92.4 [4.2]	88.3 [3.5]	90.3 [3.5]	1315
Rh($C_5H_4CO_2Me$)(3-Me-penta-1,4-diene)	91.9 [4.7]	86.1 [4.1]	88.3 [4.4]	1315
	91.9 [4.7]	86.4 [4.1]	88.2 [4.1]	1315
Rh($C_5H_4CO_2Me$)(hexa-1,5-diene)	93.1 [4.2]	88.2 [3.6]	90.3 [3.6]	1315
Rh($C_5H_4CO_2Me$)(cod)	93.2 [4.0]	87.9 [3.6]	90.4 [3.6]	1315
Rh($C_5H_4CO_2Me$)(2,3-Me_2-butadiene)	92.7 [4.9]	85.4 [2.2]	85.6 [2.8]	1315
Rh($C_5H_4CO_2Me$)(CO)$_2$	97.8 [4.1]	87.3 [2.9]	92.2 [3.5]	1315
Rh($C_5H_4COCO_2Et$)(C_2H_4)$_2$	98.1	89.0	94.1	1315
[Rh(indenyl)(C_5Me_5)]$^+$	C_5Me_5 105.5 [9.1], $C^{3a,7a}$ 94.7 [10.7], $C^{1,3}$ 82.6 [6.1], C^2 90.9 [6.1]			742, 1222
[Rh(4,6-Me_2-indenyl)(C_5Me_5)]$^+$	C_5Me_5 105.1 [6.1], $C^{3a,7a}$ 102.6 [4.6]; 105.1 [6.1] $C^{1,3}$ 82.1 [6.1]; 80.4 [6.1], C^2 90.1 [6.1]			1222
[Rh(C_5Me_5)(mesitylene)]$^{2+}$	112.5 [7.6]			1222
[Rh(C_5Me_5)(toluene)]$^{2+}$	114.2 [7.6]			1222
[Rh(C_5Me_5)(indene)]$^{2+}$	112.4 [9.1]			742, 1222
[Rh(C_5Me_5)(indole)]$^{2+}$	109.1 [7.6]			1222
[Rh(C_5Me_5)(C_6H_5O)]$^+$	92.9 [4]			964
[Rh(C_5Me_5)(C_7H_8OMe)]$^+$	104.9 [6.1]			1518
[Rh(C_5Me_5)($C_{10}H_{11}$)]$^+$	104.3 [4.6]			1518
[Rh(C_5Me_5)($C_7H_8CH_2OMe$)]$^+$	104.5 [6.1]			1518
[Rh(C_5Me_5)(nbd)(NCMe)]$^{2+}$	110.9 [7.6]			1518
[Rh(C_5Me_5)(1-MeC_3H_4)Cl]$_2$	98.9 [6]			1164
Rh(C_5Me_5)(1-MeC_3H_4)Cl(py)	100.6 [6]			1164
Rh(C_5Me_5)(1-MeC_3H_4)Cl(PPh_3)	103.9 [4]			1164
Rh(C_5Me_5)(1-MeC_3H_4)Cl($AsPh_3$)	102.9 [6]			1164

COMPOUND	$\delta(^{13}C)$/p.p.m., J/Hz	REFERENCE
$[Rh(C_5Me_5)\{P(OEt)_3\}P(OPh)_2OC_6H_4\text{-}2]^+$	107 [3]	973
$[Rh(C_5Me_5)(py)_2]^{2+}$	100.0 [7.7]	1222
$[Rh\ C_5Me_5)(NCMe)(S_2C_4H_8)]^{2+}$	105.3	1504
$[Rh(C_5Me_5)\{P(OEt)_3\}_3]^{2+}$	111.1	973
$[Rh(C_5Me_5)(SMe_2)_3]^{2+}$	106.5 [7.6]	1504
$[Rh(C_5Me_5)(MeSCH_2CH_2SMe)_2]^{2+}$	102.2 [7.6]	1504
$Rh(C_5Me_5)(S_2CNMe_2)_2$	96.4 [7.6]	1504
$Rh(C_5Me_5)\{S_2C_2(CN)_2\}_2$	101.4 [7.1]	1504
$Rh(C_5Me_5)(S_2C_6H_3Me)_2$	96.5 [4.6], 98.2 [6.1]	1504
$Rh_2(C_5Me_5)_2Cl_4(SMe_2)$	94.0, 96.2	1504
$[Rh(C_5Me_5)Cl(SPh)]_n$	97.9 [7.6]	1504
$[Rh(C_5Me_5)Cl(SCH_2Ph)]_n$	89.4 [6.1]	1504
$[Rh_2(C_5Me_5)_2Cl_2(SMe_2)_2(SHMe)]_n$	97.4 [6.1]	1504
$Rh(C_5Me_5)Cl(S_2CNEt_2)$	96.4 [6.1]	1504
$[Rh(C_5Me_4Et)Cl_2]_2$	97.2 [10.5], 93.9 [9], 94.6 [9]	1499
$Rh(C_5Cl_5)(cod)$	94.1 [4]	534
$[Rh(C_4Me_4S)(C_5Me_5)]^{2+}$	$C^{2,5}$ 118.6 [7.6], $C^{3,4}$ 122.0 [4.5]	1517
$[Rh(C_4Me_4S)(cod)]^+$	$C^{2,5}$ 103.5 [4.6], $C^{3,4}$ 122.4 [3.1]	1517
$[Rh(C_4Me_4S)(nbd)]^+$	$C^{2,5}$ 103.2 [3.1], $C^{3,4}$ 121.2 [4.2]	1517
$[Rh(C_4Me_2H_2S)(cod)]^+$	$C^{2,5}$ 108.2 [1.5]	1517
Iridium		
$Ir(C_5H_5)(EtCH{=}CHCH{=}CHCHO)$	77.3	1519
$Ir(C_5H_5)(EtCH{=}CHCH{=}CHCMeO)$	77.6	1519
$Ir(C_5H_5)(C_6H_4O_2)$	81.8	561
$[Ir(C_5Me_5)H\{P(OMe)_3\}_2]^+$	99.7	973
$[Ir(C_5Me_5)(C_6H_6)]^{2+}$	108.5	1222
$[Ir(C_5Me_5)(phenanthrene)]^{2+}$	104.8	1222
$[Ir(C_5Me_5)(fluorene)]^{2+}$	105.1	1222
$[Ir(C_5Me_5)(indole)]^{2+}$	102.4	1222
$[Ir(C_5Me_5)(indolyl)]^+$	99.7	1222
$[Ir(C_5Me_5)(C_8H_{11})]^+$	99.3	1518
$[Ir(C_5Me_5)(C_3H_5)(C_3H_6)]^+$	101.4	1518
$[Ir(C_5Me_5)(C_7H_8CH_2COMe)]^+$	99.1	1518
$[Ir(C_5Me_5)(C_5H_5MeOH)]^+$	96.8	985
$[Ir(C_5Me_5)(SC_4Me_4)]^{2+}$	104.2	1517
$[Ir(C_5Me_5)(SC_4H_2Me_2)]^{2+}$	99.9	1517
$[Ir(C_5Me_5)\{P(OPh)_3\}P(OPh)_2OC_6H_4\text{-}2]^+$	103.6	973
$[Ir(C_5Me_5)(py)_3]^{2+}$	88.1	1222
$[Ir(C_5Me_5)\{P(OEt)_3\}_3]^{2+}$	106.7 (3)	973
$Ir(C_5Me_5)(S_2CNMe_2)_2$	89.7	1504
$Ir(C_5Me_5)\{S_2C_2(CN)_2\}$	95.4	1504
$[Ir(C_5Me_4Et)Cl_2]_2$	89.0, 86.2, 86.5	1499
$[Ir(C_4Me_4S)(C_5Me_5)]^{2+}$	114.8, 107.3	1517
$[Ir(C_4H_2Me_2S)(C_5Me_5)]^{2+}$	105.3, 112.7	1517

COMPOUND	$\delta(^{13}C)$/p.p.m., J/Hz	REFERENCE
Nickel		
$Ni(C_5H_5)(C_3H_5)$	89.2, 89.3	244, 410
$Ni(C_5H_5)(COCH_2CH_2CH{=}CH_2)$	94.3	284
$Ni_2(C_5H_5)_2(HC{\equiv}CH)$	87.4	363
$Ni_2(C_5H_5)_2(HC{\equiv}CMe)$	87.7	363
$Ni_2(C_5H_5)_2(HC{\equiv}CPh)$	88.1	363
$Ni_2(C_5H_5)_2(MeC{\equiv}CMe)$	87.5	363
$Ni_2(C_5H_5)_2(MeC{\equiv}CPh)$	87.8	363
$Ni_2(C_5H_5)_2(PhC{\equiv}CPh)$	88.3	363
$Ni_2(C_5H_5)_2(CF_3C{\equiv}CCF_3)$	88.6	363
$Ni_2(C_5H_5)_2(MeO_2CC{\equiv}CCO_2Me)$	88.9	363
$Ni(C_5H_4CH_2CH_2CO)(CO)$	C^1 123.5, C^{2-5} 90.5; 92.5	809
$\{Ni(C_5H_4CH_2CH_2)\}_2CO(\mu\text{-}CO)_2$	C^1 114.9, C^{2-5} 92.9; 93.9	809
Palladium		
$Pd(C_5H_5)(C_3H_5)$	94.3, 94.7	118, 119, 193, 244
$Pd(C_5H_5)(2\text{-}MeC_3H_3)$	94.7	118, 119, 193
$Pd(C_5H_5)(C_8H_{11})$	95.2	118, 119, 193
$Pd(C_5H_5)(C_{13}H_8N)$	95.6	564
$Pd(C_5H_5)NEt_2CH_2CH_2CO$	96.1	1125
$Pd_2(C_5H_5)(Br)(PPr^i_3)_2$	88.2	587
$Pd_2(C_5H_5)(Br)\{P(OC_6H_4\text{-}2)_3\}_2$	89.6 {164.7}	1055
$[Pd(C_4Me_4S)(2\text{-}MeC_3H_4)]^+$	128.7, 133.1	1517
$Pd(C_4Me_4S)Cl_2$	119.1, 122.8	1517
Platinum, [^{195}Pt]		
$Pt(C_5H_5)Me(CO)$	94.1 [8.4]	790
$Pt(C_5H_5)Me\{P(OMe)_3\}$	93.2 [3.8]	790
$Pt_3(C_5H_5)\{C_3(C_6H_4OMe\text{-}4)_3\}_3$	90.4 [15]	1535
Silver		
$Ag(C_5H_5)(PPh_3)$	103.2	830
Boron		
$[B(C_5Me_5)I]^+$	115.2	1028
Aluminium		
$Al(C_5H_5)Me_2$	111.6 {167.6, 6.6}	934
$Al(C_5H_5)Et_2$	111.7 {168.3, 6.7}	934
Gallium		
$Ga(C_5H_5)Me_2$	112.7 {165.1, 6.8}	934
$Ga(C_5H_5)Et_2$	112.0 {164.7, 6.8}	934
Indium		
$In(C_5H_5)Et_2$	111.5 {163.3, 7.0}	934

COMPOUND	$\delta(^{13}C)$/p.p.m., J/Hz	REFERENCE
Thallium		
$Tl(C_5H_5)$	107.5	934
$Tl(C_5H_4Me)$	$C^{2,5}$ 107.5, $C^{3,4}$ 105.0	1722
$Tl(C_5H_4CHMePh)$	C^{1} 136.8, $C^{2,5}$ 106.8, $C^{3,4}$ 107.4	1723
Germanium		
$Ge(C_5H_4Me)_2$	C^{1} 126.6, $C^{2,5}$ 113.0, $C^{3,4}$ 110.5	1722
Tin		
$Sn(C_5H_4Me)_2$	C^{1} 123.6, $C^{2,5}$ 110.8, $C^{3,4}$ 108.5	1722
Lead		
$Pb(C_5H_4Me)_2$	C^{1} 114.3, $C^{2,5}$ 104.7, $C^{3,4}$ 103.7	1722

TABLE 2.15

^{13}C N.m.r. Chemical Shift Data for Some Dienyl Groups π-Bonded to Metals

COMPOUND	δ(^{13}C)/p.p.m., *J*/Hz			REFERENCE
Titanium				
Ti(C_7H_9)(C_7H_7)	$C^{1,5}$ 101.6	$C^{2,4}$ 113.3	C^3 102.5	693
Chromium				
Cr(C_5H_5SOMe)$(CO)_3$*	$C^{1,5}$ 60.9	$C^{2,4}$ 106.2	C^3 87.2	1746
	$C^{1,5}$ 45.9	$C^{2,4}$ 108.2	C^3 87.4	
Molybdenum				
Mo(C_5H_5SOMe)$(CO)_3$	$C^{1,5}$ 61.8	$C^{2,4}$ 109.0	C^3 88.3	1746
Tungsten				
W(C_5H_5SOMe)$(CO)_3$	$C^{1,5}$ 58.2	$C^{2,4}$ 103.2	C^3 87.0	1746
Manganese				
Mn(C_6H_7)$(CO)_3$	$C^{1,5}$ 50.1 {168}	$C^{2,4}$ 98.0 {168}	C^3 79.7 {177}	685
Mn(C_7H_9)$(CO)_3$	$C^{1,5}$ 80.2 {164}	$C^{2,4}$ 99.7 {155}	C^3 91.2 {168}	685
Mn(C_7H_7)$(CO)_3$	$C^{1,5}$ 77.1 {154}	$C^{2,4}$ 107.4 {166}	C^3 98.0 {171}	685
Iron				
[Fe($C^1H_2C^2HC^3HC^4HC^5H_2$)-$(CO)_3$]$^+$	$C^{1,5}$ 65.4 {164.4, 165.4}	$C^{2,4}$ 104.6 {170.8}	C^3 98.6 {180.3}	924, 1225, 1432
[Fe(*exo*-Me$C^1HC^2HC^3HC^4$-HC^5H_2)$(CO)_3$]$^+$	C^1 91.0 {170}	C^2 103.2 {170}	C^3 95.3 {170}	675
	C^5 62.0	C^4 103.7		
	C^1 91.3 {166.0}	C^2 104.1 {170.4}	C^3 94.8 {178.5}	924
	C^5 62.2 {165.6, 166.6}	C^4 103.5 {171.1}		
[Fe(*exo*-*exo*-Me$C^1HC^2HC^3$-HC^4HC^5HMe)$(CO)_3$]$^+$	$C^{1,5}$ 87.5 {161}	$C^{2,4}$ 103.2 {174}	C^3 90.9 {165}	675, 1432
[Fe(*exo*-*endo*-C^1HMeC^2-$HC^3HC^4HC^5$HMe)$(CO)_3$]$^+$	C^1 81.9 {170}	C^2 104.7 {157}	C^3 97.4 {174}	675, 1432
	C^5 77.5	C^4 99.0 {165}		

*isomers

COMPOUND	$\delta(^{13}C)$/p.p.m., *J*/Hz			REFERENCE
[Fe(*exo*-*exo*-C^{1}HMeC^{2}HC^3-HC^{4}HC^{5}HPh)(CO)$_3$]$^+$	C^1 87.6	C^2 102.6	C^3 89.7	1432
	C^5 88.1	C^4 94.5		
[Fe(*endo*-*exo*-C^{1}HMeC^2-HC^{3}HC^{4}HC^{5}HPh)(CO)$_3$]$^+$	C^1 77.7	C^2 98.0	C^3 96.7	1432
	C^5 82.6	C^4 97.1		
[Fe(*exo*-*exo*-C^{1}HMeC^{2}HC^3-MeC^{4}HC^{5}HMe)(CO)$_3$]$^+$	C^1 85.1 {165}	C^{2,4} 103.1 {165}	C^{3,5} 109.3	675, 1432
[Fe(*endo*-*exo*-C^{1}HMeC^2-HC^{3}MeC^{4}HC^{5}HPh)(CO)$_3$]$^+$	C^1 74.5 {165}	C^2 96.5 {165}	C^3 117.6	675, 1432
	C^5 80.2 {174}	C^4 104.3 {165}		
[Fe(C_6H_7)(CO)$_3$]$^+$	C^{1,5} 63.7 {175.4}	C^{2,4} 101.4 {175.3}	C^3 89.0 {183.5}	924, 1432
	C^{1,5} 65.4 {180}	C^{2,4} 103.2 {174}	C^3 89.9 {186}	940
Fe(C_6H_7)Cp	C^{1,5} 79.8	C^{2,4} 79.6	C^3 72.1	244
[Fe(1-MeC_6H_6)(CO)$_3$]$^+$	C^1 93.7	C^2 101.3	C^3 86.6	940
	C^5 62.1	C^4 99.6		
[Fe(2-MeC_6H_6)(CO)$_3$]$^+$	C^1 63.8 {174}	C^2 123.4	C^3 90.3 {178}	940
	C^5 66.3 {172}	C^4 101.6 {164}		
[Fe(3-MeC_6H_6)(CO)$_3$]$^+$	C^{1,5} 62.7 {180}	C^{2,4} 103.2 {178}	C^3 109.1	940
[Fe(1-MeO$_2$CC_6H_6)(CO)$_3$]$^+$	C^1 52.0	C^2 101.7 {180}	C^3 88.0 {185}	940
	C^5 68.6 {178}	C^4 99.7 {173}		
[Fe(2-MeO$_2$CC_6H_6)(CO)$_3$]$^+$	C^1 66.3	C^2 105.0	C^3 88.1	940
	C^5 64.2	C^4 99.5		
[Fe(3-MeO$_2$CC_6H_6)(CO)$_3$]$^+$	C^{1,5} 68.6 {174}	C^{2,4} 103.0 {178}	C^3 89.6	940
[Fe(1-MeOC_6H_6)(CO)$_3$]$^+$	C^1 160.6	C^2 65.3 {162}	C^3 83.1 {178}	940
	C^5 53.9 {178}	C^4 95.5 {176}		
[Fe(2-MeOC_6H_6)(CO)$_3$]$^+$	C^1 43.8 {170}	C^2 153.0	C^3 76.9 {178}	940
	C^5 65.3 {172}	C^4 100.7 {176}		
[Fe(3-MeOC_6H_6)(CO)$_3$]$^+$	C^{1,5} 54.3	C^{2,4} 87.7	C^3 141.4	940
[Fe(2-HOC_6H_6)(CO)$_3$]$^+$	C^1 46.1 {172}	C^2 149.8	C^3 77.2 {180}	940
	C^5 64.0 {170}	C^4 100.0 {174}		
[Fe(1-Me-4-MeOC_6H_5)-(CO)$_3$]$^+$	C^1 91.3	C^2 96.2 {174}	C^3 73.0 {184}	1261
	C^5 42.9 {172}	C^4 150.4		
[Fe(C_7H_9)(CO)$_3$]$^+$	C^{1,5} 92.6 {161.8}	C^{2,4} 102.6 {171.4}	C^3 99.4 {179.3}	787, 924
[Fe(C_7H_9)(C_6H_8)(CO)]$^+$	C^{1,2,4,5} 77.9, 88.8, 89.0		C^3 107.6	331

COMPOUND	$\delta(^{13}C)$/p.p.m., J/Hz			REFERENCE
$Fe(C_7H_9)(C_7H_7)$	$C^{1,5}$ 73.0	$C^{2,4}$ 89.3	C^3 91.1	693
	$C^{1,5}$ 64.5, 80.0	$C^{2,4}$ 84.9, 91.9	C^3 91.0	1201
$Fe(C_7H_7)(C_7H_9)$	$C^{1,5}$ 68.7	$C^{2,4}$ 94.3	C^3 95.4	693
	$C^{1,5}$ 63.4 71.8	$C^{5,7}$ 91.3 96.6	C^3 95.1	1201
$[Fe(C_7H_7)(CO)_3]^+$	$C^{1,5}$ 70.9 {159.7}, 73.4 {158.2}	$C^{2,4}$ 94.1 {174.5}, 98.1 {173.0}	C^3 105.7 {184.0}	1225
$[Fe(C_8H_9)(CO)_3]^+$	$C^{1,5}$ 104.0 {164.2}	$C^{2,4}$ 97.9 {173.3}	C^3 99.2 {178.7}	924
Ruthenium				
$Ru(C_7H_9)_2$	$C^{1,5}$ 61.8, 66.2	$C^{2,4}$ 85.6, 90.3	C^3 94.4	606
$[Ru(C_7H_9)(CO)_3]^+$	$C^{1,5}$ 83.3	$C^{2,4}$ 102.0	C^3 88.4	518
$Ru_3(C_7H_9)(C_7H_7)(CO)_6$	$C^{1,5}$ 81.5	$C^{2,4}$ 104.6	C^3 105.3	384
$Ru(6\text{-}Me_3SiC_7H_8)(CO)_2SiMe_3$	83.6, 87.8, 92.8, 104.2, 106.0			517
$Ru(6\text{-}Me_3GeC_7H_8)(CO)_2GeMe_3$	81.0, 88.4, 90.4, 102.8, 105.0			375, 517
Cobalt				
$Co(C_6Me_5O)(C_5Me_5)$	95.0, 96.0			1496
$[Co(C_6Me_5OH)(C_5Me_5)]^+$	98.9, 99.0			1496
Rhodium				
$Rh(C_8H_9)Cp$	$C^{1,5}$ 102.6 [5]	C^{2-4} 88.3 [8], 91.7 [10]		298
$[Rh(6\text{-}MeCOCH_2C_7H_8)\text{-}(C_5Me_5)]^+$	C^1 86.7 [12.2]	C^2 97.6 [6.1]	C^3 101.5 [3.1]	1518
	C^5 94.2 [12.2]	C^4 98.8 [6.1]		
$[Rh(6\text{-}MeOC_7H_8)(C_5Me_5)]^+$	C^1 74.8 [10.7]	C^2 98.7 [4.6]	C^3 101.4	1518
	C^5 87.5 [13.7]	C^4 98.7 [4.6]		
Iridium				
$[Ir\{C^1H_2C^2(OH)C^3HC^4MeC^5\text{-}H_2\}(C_5Me_5)]^+$	$C^{1,5}$ 35.7, 44.7	C^2 97.2	C^3 86.1	985
	C^3 132.5	C^4 132.5		
$[Ir(OC^2MeC^3HC^4MeC^5H_2)\text{-}(C_5Me_5)]^+$		C^2 157.2	C^3 86.2	985
	C^4 111.7	C^5 60.3		
$[Ir(6\text{-}MeCOCH_2C_7H_8)\text{-}(C_5Me_5)]^+$	C^1 67.7	C^2 88.4	C^3 98.2	1518
	C^5 76.5	C^4 88.8		

TABLE 2.16

^{13}C N.m.r. Chemical Shift Data for Some Arenes π-Bonded to Metals

COMPOUND	δ(^{13}C)/p.p.m., J/Hz				REFERENCE
Chromium					
A. Bis Arene Compounds					
$Cr(C_6H_6)_2$	74.2, 74.6, 74.7				886, 1358, 1703
$Cr(C_6H_6)(C_6H_5Et)$	75.2				1703
$Cr(C_6H_6)\{C_6H_4(CH_2)_4\text{-}1,2\}$	76.3				1703
$Cr(C_6H_6)(C_6H_2F_4)$	82.2				1358
$Cr(C_6H_6)(C_6HF_5)$	84.0				1358
$Cr(C_6H_6)\{C_6F_4(CO_2Et)_2\}$	88.6				1358
$Cr(C_6H_6)(C_6F_5CO\text{-furyl})$	87.6				1358
$Cr(C_6H_6)(C_6F_5CO_2Et)$	87.0				1358
$Cr(C_6H_6)(C_6F_6)$	85.8				1358
	C^1	$C^{2,6}$	$C^{3,5}$	C^4	
$Cr(C_6H_5Me)_2$	88.2	77.9	75.0	75.0	886
$Cr(C_6H_5Et)_2$	94.8	75.9	74.9	74.9	886
	95.0	76.6	75.2	75.7	1703
$Cr(C_6H_5Et)(C_6H_6)$	95.1	76.2	74.7	74.5	1703
$Cr(C_6H_5Pr^n)_2$	92.9	77.0	74.9	74.9	886
$Cr(C_6H_5Bu^n)_2$	93.5	77.5	75.5	75.5	886
$Cr\{C_6H_5(CH_2)_4Ph\}_2$	97.1	77.3	75.3	75.3	1357
$Cr(C_6H_5CH_2Ph)_2$	93.0	78.0	75.5	75.5	1357
$Cr(C_6H_5CH_2OCH_2Ph)_2$	88.8	77.6	75.2	75.2	1357
$Cr(C_6H_5Bu^i)_2$	91.8	77.8	74.9	74.9	886
$Cr(C_6H_5Bu^t)_2$	103.0	71.9	73.7	73.7	886
$Cr(C_6H_5CHO)_2$	81.8	79.2	79.2	77.7	1703
$Cr(C_6H_5CO_2H)_2$	77.4	79.3	78.4	75.2	886
$Cr(C_6H_5CF_3)_2$	120.9	73.8	71.5	74.3	886
$Cr(C_6H_5NMe_2)_2$	117.0	64.5	72.2	74.1	886
$Cr(C_6H_5OMe)_2$	131.2	66.1	72.5	74.1	886
$Cr(C_6H_5F)_2$		67.9	72.2	73.8	886
$Cr(C_6H_5Cl)_2$	106.1	75.2	79.3	75.0	886
$Cr(o\text{-}Me_2C_6H_4)_2$	$C^{1,2}$ 88.6, $C^{3,6}$ 79.3, $C^{4,5}$ 75.8				886
$Cr(m\text{-}Me_2C_6H_4)_2$	$C^{1,3}$ 87.9, C^2 80.5, $C^{4,6}$ 77.8, C^5 75.8				886
	$C^{1,3}$ 88.3, C^2 80.6, $C^{4,6}$ 78.0, C^5 76.0				1703
$Cr(p\text{-}Me_2C_6H_4)_2$	$C^{1,4}$ 88.3, $C^{2,3,5,6}$ 75.0				886
$Cr\{1,2\text{-}(CH_2)_4C_6H_4\}(C_6H_6)$	$C^{1,2}$ 91.8, C^{3-6} 75.5, 74.9				1703
$Cr(1,2,3\text{-}Me_3C_6H_3)_2$	$C^{1,3}$ 87.9, C^2 87.6, $C^{4,6}$ 79.8, C^5 76.3				886

COMPOUND	$\delta(^{13}C)$/p.p.m., *J*/Hz				REFERENCE
$Cr(1,2,4\text{-}Me_3C_6H_3)_2$	C^1 88.1, C^2 88.1, C^3 81.7, C^4 87.5, C^5 78.8, C^6 79.8				886
$Cr(1,3,5\text{-}Me_3C_6H_3)_2$	$C^{1,3,5}$ 88.4, $C^{2,4,6}$ 80.4				1703
	$C^{1,3,5}$ 88.0, $C^{2,4,6}$ 80.1				886
<u>B. $Cr(arene)(CO)_3$</u>					
$Cr(C_6H_6)(CO)_3$	93.7				60
	92.4				244
	92.9				595
	93.5				395
	C^1	$C^{2,6}$	$C^{3,5}$	C^4	
$Cr(C_6H_5Me)(CO)_3$	110.4	94.9	93.4	90.2	395
	109.9	93.0	94.5	89.8	595
$Cr\{C_6H_5(CH_2)_4Ph\}(CO)_3$	113.6	92.5	93.7	90.3	290
$Cr(C_6H_5CH_2CH_2Ph)(CO)_3$	112.5	93.7	92.6	90.4	873
$Cr(C_6H_5CH_2Ph)(CO)_3$	112.3	93.4	93.0	90.7	873
$Cr(C_6H_5CH_2C_6H_4Cl\text{-}4)(CO)_3$	111.6	93.3	92.8	90.9	873
$Cr(C_6H_5Pr^i)(CO)_3$	120.2	92.0	93.0	91.8	595
$Cr(C_6H_5Bu^t)(CO)_3$	123.1	92.7	91.0	93.6	595
$Cr(C_6H_5CMe_2OH)(CO)_3$	121.9	92.3	94.6	92.3	927
$Cr(C_6H_5CHPhOH)(CO)_3$	114.9	90.1, 91.7, 91.9, 92.1			920
$Cr(C_6H_5CMePhOH)(CO)_3$	120.4	90.0, 90.6, 92.8, 93.1, 94.1			920
$Cr(C_6H_5CEt_2OH)(CO)_3$	120.5	93.2	90.9	94.4	920
$Cr\{(C_6H_5)_2CHOH\}(CO)_3$	114.6	93.0	94.1	91.8	435
$[Cr(C_6H_5CMe_2)(CO)_3]^+$	101.7	99.5	120.7	96.7	927
$Cr(C_6H_5Ph)(CO)_3$	110.5	92.2	92.6	91.4	873, 1764
$Cr(C_6H_5C_6H_4F\text{-}4)(CO)_3$	109.6	91.8	92.6	91.2	873
$Cr(C_6H_5C_6H_4Cl\text{-}4)(CO)_3$	108.9	91.8	92.4	91.4	873
$Cr(C_6H_5CH{=}CHPh\text{-}cis)(CO)_3$	105.9	93.4	92.0	91.3	873
$Cr(C_6H_5COMe)(CO)_3$	95.7	94.3	89.5	95.3	920
$Cr(C_6H_5COPh)(CO)_3$	96.0	95.6	89.4	94.5	873
	96.3	95.9	90.0	95.1	920
	96.3	95.8	89.6	94.8	1205
$Cr(C_6H_5CO_2Me)(CO)_3$	95.0	95.0	90.2	95.4	873
$Cr(C_6H_5NH_2)(CO)_3$	131.4	77.7	97.0	83.8	395
$Cr(C_6H_5NMe_2)(CO)_3$	135.7	74.8	97.5	83.2	395
$Cr(C_6H_5OMe)(CO)_3$	143.7	78.7 {173}	95.7 {177}	86.0 {177}	395, 787
$Cr(C_6H_5OBu)(CO)_3$	143.4	79.1	95.9	85.7	395
$Cr(C_6H_5F)(CO)_3$	146.2	79.7	93.9	87.0	395
$Cr(C_6H_5Cl)(CO)_3$	113.2	91.7	93.8	88.7	395
$Cr(1,3\text{-}Me_2C_6H_4)(CO)_3$	$C^{1,3}$ 110.6, C^2 94.0, $C^{4,6}$ 90.3, C^5 95.4				595
$Cr\{1\text{-}Me\text{-}4\text{-}(HOCHMe)C_6H_4\}(CO)_3$	C^1 114.6, $C^{2,6}$ 92.1; 92.8, $C^{3,5}$ 92.5, C^4 109.5				1294
$[Cr\{1\text{-}Me\text{-}4\text{-}(MeHC)C_6H_4\}(CO)_3]^+$	C^1 106.3, $C^{2,6}$ 103.2; 106.7, $C^{3,5}$ 101.1; 101.8, C^4 115.4				1294
$Cr\{1\text{-}Me\text{-}4\text{-}(p\text{-}tolCHOH)C_6H_4\}(CO)_3$	C^1 113.3, $C^{2,6}$ 92.3; 94.3, $C^{3,5}$ 93.3, C^4 110.9				1294

COMPOUND	$\delta(^{13}C)$/p.p.m., J/Hz	REFERENCE
$[Cr\{1\text{-}Me\text{-}4(p\text{-}tolCH)C_6H_4\}\text{-}(CO)_3]^+$	C^1 102.1, $C^{2,6}$ 101.3, 106.7, $C^{3,5}$ 100.2, 101.0, C^4 115.0	1294
$Cr\{1\text{-}(HOCMe_2)\text{-}2\text{-}Me\text{-}C_6H_4\}(CO)_3$	C^1 118.0, C^2 109.4, C^3 94.0, C^4 94.4, C^5 88.2, C^6 95.3	920
$Cr\{1\text{-}(HOCEt_2)\text{-}2\text{-}Me\text{-}C_6H_4\}(CO)_3$	C^1 107.6, C^2 140.9, C^3 93.3, C^4 95.1, C^5 83.2, C^6 96.6	920
$\{Cr(CO)_3(1\text{-}Me\text{-}C_6H_4\text{-}4)\}_2CHOH$	C^1 111.2, $C^{2,6}$ 93.1, 92.9, $C^{3,5}$ 94.1, 95.4, C^4 111.2	1361
$[\{Cr(CO)_3(1\text{-}Me\text{-}C_6H_4\text{-}4)\}_2CH]^+$	C^1 117.1, $C^{2,6}$ 96.4, $C^{3,5}$ 97.8, C^4 117.1	1361
$Cr(1\text{-}Me\text{-}4\text{-}p\text{-}tol\text{-}C_6H_4)(CO)_3$	C^1 108.1, $C^{2,6}$ 92.8, $C^{3,5}$ 93.5, C^4 108.1	873
$Cr(1\text{-}Me\text{-}2\text{-}MeCOC_6H_4)(CO)_3$	C^1 98.9, C^2 110.8, C^3 92.5, C^6 96.1, C^5 87.9, C^4 96.7	920
$Cr(1\text{-}Me\text{-}2\text{-}EtCOC_6H_4)(CO)_3$	C^1 99.6, C^2 110.2, C^3 92.1, C^6 95.7, C^5 87.4, C^4 95.7	920
$Cr(1\text{-}Me\text{-}2\text{-}HO_2CC_6H_4)(CO)_3$	C^1 92.2, C^2 113.8, C^3 94.4, C^6 98.7, C^5 90.0, C^4 97.9	920
$Cr(1,2\text{-}\overline{CH_2CH_2CH_2C}_6H_4)(CO)_3$	$C^{1,2}$ 114.4, C^{3-6} 90.4, 91.7	290, 1764
$Cr(1,2\text{-}\overline{CHMeCH_2CHMeC}_6H_4)(CO)_3$	$C^{1,2}$ 122.4, 119.2, $C^{3,6}$ 91.7, 95.3, $C^{4,5}$ 87.0, 88.9	174
Cr(naphthalene)$(CO)_3$	$C^{1,2}$ 105.7, $C^{2,6}$ 92.5, $C^{3,4}$ 90.8,	1764
Cr(phenanthrene)$(CO)_3$	$C^{1,2}$ 96.5, 103.0, C^{3-6} 86.1, 89.7, 92.0	1764
$Cr(1,3\text{-}Bu^t_2C_6H_4)(CO)_3$	$C^{1,3}$ 121.6, C^2 93.5, $C^{4,6}$ 94.3, C^5 89.4	595
$Cr\{1,2\text{-}(HOCHMe)_2C_6H_4\}(CO)_3$*	$C^{1,2}$ 113.1, 120.1, C^{3-6} 89.7, 92.6, 92.5, 94.5	920
	$C^{1,2}$ 118.2, C^{3-6} 90.2, 93.3	920
	$C^{1,2}$ 114.4, C^{3-6} 91.5 to 93.5	920
$Cr(1,2\text{-}Me\overline{BNHNHBMeC}_6H_4)(CO)_3$	$C^{3,6}$ 98.7, $C^{4,5}$ 95.5	1575
$Cr(1,2\text{-}\overline{CHCHCH_2C}_6H_4)(CO)_3$	$C^{1,2}$ 113.2, 115.4, C^{3-6} 88.5, 90.1, 90.6	370
$Cr\{1,2\text{-}O(CHCH)_2C_6H_4\}(CO)_3$	$C^{1,2}$ 104.3, $C^{3,6}$ 94.7, $C^{4,5}$ 92.9	952
$Cr(1,2\text{-}MeO\overline{CCPhCHPhC}_6H_4)(CO)_3$	$C^{1,2}$ 114.5, 116.5, C^{3-6} 82.0, 86.7, 95.7, 97.3	1088
$Cr(1,2\text{-}C_5H_6C_6H_4)(CO)_3$	$C^{1,2}$ 127.4, $C^{3,6}$ 90.8, $C^{4,5}$ 94.6	432
$Cr\{1,2\text{-}(CHO)_2C_6H_4\}(CO)_3$	$C^{1,2}$ 97.3, $C^{3,6}$ 92.8, $C^{4,5}$ 92.7	920
$Cr\{1,2\text{-}(MeO_2C)_2C_6H_4\}(CO)_3$	$C^{1,2}$ 97.3, $C^{3,6}$ 91.4, $C^{4,5}$ 90.8	920
$Cr(1,2\text{-}O\overline{CCH_2CHPr^iCH_2C}_6H_4)\text{-}(CO)_3$*	C^1 92.6, C^2 115.0, C^3 90.1, C^6 91.1, C^5 89.4, C^4 95.0	920

*isomers

COMPOUND	$\delta(^{13}C)$/p.p.m., J/Hz	REFERENCE
	C^1 92.5, C^2 116.8, C^3 90.1, C^6 92.4, C^5 88.7, C^4 96.0	920
$Cr(1,2\text{-}O\overline{COCMe_2\mathit{C}}_6H_4)(CO)_3$	C^1 88.3, C^2 84.5, C^{3-6} 85.4; 85.5; 90.9; 92.8	920
$Cr(1,2\text{-}O\overline{CC_6H_4\mathit{C}}_6H_4)(CO)_3$	C^1 107.6, C^2 92.3, C^3 83.7, C^6 91.6, C^5 87.2, C^4 94.0	1305
$Cr\{1,2\text{-}MeO\overline{CCPhCPhC(OH)\mathit{C}}_6H_4\}(CO)_3$	C^{3-6} 86.2, 92.0, 98.1	1088
$Cr\{1\text{-}(MeCO)\text{-}2\text{-}(MeO)\mathit{C}_6H_4\}(CO)_3$	C^1 89.6, C^2 144.5, C^3 72.9, C^6 95.7, C^5 84.5, C^4 96.0	920
$Cr\{1\text{-}(p\text{-tol})\text{-}4\text{-}F\mathit{C}_6H_4\}(CO)_3$	C^1 104.0, $C^{2,6}$ 92.5, $C^{3,5}$ 78.5, C^4 145.3	873
$Cr(1,3,5\text{-}Me_3\mathit{C}_6H_3)(CO)_3$	$C^{1,3,5}$ 111.5, $C^{2,4,6}$ 92.4	117
	$C^{1,3,5}$ 110.8, $C^{2,4,6}$ 92.2	595
$Cr(1\text{-}Me\text{-}3,5\text{-}Bu^t{}_2\mathit{C}_6H_3)(CO)_3$	$C^{1,3,5}$ 105.5; 122.5, $C^{2,4,6}$ 89.5; 93.4	938
$Cr(1,3,5\text{-}Bu^t{}_3\mathit{C}_6H_3)(CO)_3$	$C^{1,3,5}$ 118.4, $C^{2,4,6}$ 93.2	595
$Cr(1,2,4,5\text{-}Me_4\mathit{C}_6H_2)(CO)_3$	$C^{1,2,4,5}$ 107.6	117
$Cr(1,2\text{-}Me_2\text{-}4,5\text{-}Me\overline{BNHNHBMe\mathit{C}}_6H_2)\text{-}(CO)_3$	$C^{1,2}$ 111.4, $C^{3,6}$ 101.2	1575
$Cr(1,3\text{-}Me_3\text{-}2\text{-}C_6H_2Me_3CO\mathit{C}_6H_2)(CO)_3$	$C^{1,3}$ 107.0, C^2 109.6, $C^{4,6}$ 91.5 {170.9}, C^5 111.0	290
$Cr(\mathit{C}_6Me_6)(CO)_3$	107.5	117
$Cr(1\text{-}HO\text{-}2,3\text{-}Ph_2\text{-}4\text{-}MeO\text{-}5,6\text{-}C_6H_4\mathit{C}_6H_4)(CO)_3$	$C^{1,4}$ 86.9; 93.3; 95.8; 98.5, $C^{5,6}$ 104.1; 107.6	1088
$Cr\{1,1\text{-}(MeO)_2\text{-}2,4,6\text{-}Ph_3\mathit{C}_5H_2P\}\text{-}(CO)_3$	60.0, 95.5, 103.5	1746

C. Other Arene Chromium Compounds

COMPOUND	$\delta(^{13}C)$/p.p.m., J/Hz	REFERENCE
$[Cr(\mathit{C}_6H_6)(CO)_2CPh]^+$	111.1	906, 935
$Cr(\mathit{C}_6H_6)(CO)_2CPhNH_2$	90.1	935
$Cr(\mathit{C}_6H_6)(CO)_2CPhNMe_2$	89.9	935
$[Cr(\mathit{C}_6H_6)(CO)_2CPhPMe_3]^+$	110.6	906, 1048
$Cr(\mathit{C}_6H_6)(CO)_2CPhOMe$	93.0	935
$Cr(\mathit{C}_6H_5Me)(CO)_2CPhNH_2$	C^1 106.7, C^{2-6} 86.2: 87.8; 93.3	935
$Cr(\mathit{C}_6H_5Me)(CO)_2CPhNMe_2$	C^1 104.7, C^{2-6} 88.6; 88.9; 92.5	935
$Cr(\mathit{C}_6H_5Me)(CO)_2CPhOMe$	C^1 109.1, C^{2-6} 90.5; 91.2; 94.6	935
$Cr(\mathit{C}_6H_5Me)(CO)_2PPh_3$	C^1 103.9, $C^{2,6}$ 88.0, $C^{3,5}$ 91.9, C^4 88.7	938

COMPOUND	$\delta(^{13}C)$/p.p.m., J/Hz	REFERENCE
$Cr(C_6H_5Me)(CO)_2AsPh_3$	C^1 103.1, $C^{2,6}$ 87.3, $C^{3,5}$ 90.2, C^4 87.1	938
$Cr(C_6H_5Et)(CO)_2PPh_3$	C^1 109.8, $C^{2,6}$ 88.0, $C^{3,5}$ 91.8, C^4 89.3	938
$Cr(C_6H_5Et)(CO)_2AsPh_3$	C^1 108.4, $C^{2,6}$ 86.1, $C^{3,5}$ 89.9, C^4 86.9	938
$Cr(C_6H_5Bu^t)(CO)_2PPh_3$	C^1 115.9, $C^{2,6}$ 88.0, $C^{3,5}$ 89.5, C^4 92.3	938
$[Cr(C_6H_5OMe)H(CO)_3]^+$	C^1 151.2, $C^{2,6}$ 95.7 {183}, C^4 100.1 {184}, $C^{3,5}$ 110.2 {186}	787
$Cr(1,3\text{-}Me_2C_6H_4)(CO)_2PPh_3$	$C^{1,3}$ 104.6, C^2 93.8, $C^{4,6}$ 89.9, C^5 87.1	938
$[Cr(1,4\text{-}Me_2C_6H_4)(CO)_2CPh]^+$	$C^{1,4}$ 126.1, $C^{2,3,5,6}$ 110.3	935
$Cr(1,4\text{-}Me_2C_6H_4)(CO)_2CPhNH_2$	$C^{1,4}$ 104.0, $C^{2,3,5,6}$ 91.0	935
$Cr(1,4\text{-}Me_2C_6H_4)(CO)_2CPhNMe_2$	$C^{1,4}$ 104.4, $C^{2,3,5,6}$ 91.1	935
$[Cr(1,4\text{-}Me_2C_6H_4)(CO)_2CPhPMe_3]^+$	$C^{1,4}$ 121.6, $C^{2,3,5,6}$ 111.9	1048
$Cr(1,4\text{-}Me_2C_6H_4)(CO)_2CPhOMe$	$C^{1,4}$ 106.2, $C^{2,3,5,6}$ 93.6	935
$Cr(1,2\text{-}C_5H_6C_6H_4)(CO)_2$	76.1, 94.6, 95.6	432
$Cr(1,3\text{-}Bu^t_2C_6H_4)(CO)_2PPh_3$	$C^{1,3}$ 121.3, C^2 78.6, $C^{4,6}$ 86.0, C^5 93.8	938
$Cr(1,3\text{-}Bu^t_2C_6H_4)(CO)_2AsPh_3$	$C^{1,3}$ 118.7, C^2 79.0, $C^{4,6}$ 86.8, C^5 90.1	938
$[Cr(1,3,5\text{-}Me_3C_6H_3)(CO)_2CPh]^+$	$C^{1,3,5}$ 129.4, $C^{2,4,6}$ 107.2	935
$Cr(1,3,5\text{-}Me_3C_6H_3)(CO)_2CPhNH_2$	$C^{1,3,5}$ 107.3, $C^{2,4,6}$ 86.6	935
$Cr(1,3,5\text{-}Me_3C_6H_3)(CO)_2CPhNMe_2$	$C^{1,3,5}$ 106.3, $C^{2,4,6}$ 88.7	935
$[Cr(1,3,5\text{-}Me_3C_6H_3)(CO)_2\text{-}CPhPMe_3]^+$	$C^{1,3,5}$ 123.7, $C^{2,4,6}$ 108.3	1048
$Cr(1,3,5\text{-}Me_3C_6H_3)(CO)_2CPhOMe$	$C^{2,3,5}$ 110.6, $C^{2,4,6}$ 90.4	935
$Cr(1,3,5\text{-}Me_3C_6H_3)(CO)_2PPh_3$	$C^{1,3,5}$ 105.3, $C^{2,4,6}$ 90.1	938
$Cr(1,3\text{-}Me_2\text{-}5\text{-}CH_2{=}CHCH_2CH_2C_6H_4)\text{-}(CO)_2$	$C^{1,3,5}$ 107.1, 109.9 $C^{2,4,6}$ 85.3, 90.9, 99.2	1354
$Cr(1,3\text{-}Me_2\text{-}5\text{-}CH_2{=}CHOCH_2C_6H_4)\text{-}(CO)_2$	$C^{1,3,5}$ 100.9, 105.6 $C^{2,4,6}$ 92.1, 93.1, 97.0	1354
$Cr(1,3,5\text{-}Bu^t_3C_6H_3)(CO)_2PPh_3$	$C^{1,3,5}$ 122.2, $C^{2,4,6}$ 86.5	938
Molybdenum		
$Mo(C_6H_6)Me_2(PMe_3)_2$	88.9	1388
$Mo(C_6H_6)Me_2(PMe_2Ph)_2$	90.0	1388
$Mo(C_6H_6Me)Me_2(PMe_3)_2$	C^1 100.3, $C^{2,6}$ 88.4, $C^{3,5}$ 89.0, C^4 90.8	1388
$Mo(C_6H_5Me)Me_2(PMe_2Ph)_2$	C^1 104.7, $C^{2,6}$ 88.0, $C^{3,5}$ 88.2, C^4 95.2	1388
$Mo(C_6H_5Et)_2$	C^1 97.1, C^{2-6} 75 to 78	1703
$Mo(1,2\text{-}Me_2C_6H_4)Me_2(PMe_3)_2$	$C^{1,2}$ 89.6, $C^{3,6}$ 91.3, $C^{4,5}$ 88.7	1388
$Mo(1,2\text{-}Me_2C_6H_4)Me_2(PMe_2Ph)_2$	$C^{1,2}$ 99.0, $C^{3,6}$ 92.5, $C^{4,5}$ 98.6	1388
$Mo(1,3\text{-}Me_2C_6H_4)(CO)_3$	$C^{1,3}$ 115.6, C^2 96.4, $C^{4,6}$ 92.7, C^5 98.2	117

COMPOUND	$\delta(^{13}C)$/p.p.m., J/Hz	REFERENCE
$Mo(1,4\text{-}Me_2C_6H_4)Me_2(PMe_3)_2$	$C^{1,4}$ 103.7, $C^{2,3,5,6}$ 88.3	1388
$Mo(1,4\text{-}Me_2C_6H_4)Me_2(PMe_2Ph)_2$	$C^{1,4}$ 110.3, $C^{2,3,5,6}$ 86.4	1388
$Mo(1,3,5\text{-}Me_3C_6H_3)(CO)_3$	$C^{1,3,5}$ 111.7, $C^{2,4,6}$ 94.7	117
$Mo(1,2,4,5\text{-}Me_4C_6H_2)(CO)_3$	$C^{1,2,4,5}$ 118.8, $C^{3,6}$ 101.4	117
$Mo(C_6Me_6)(CO)_3$	111.7	117
$Mo(C_5H_5As)(CO)_3$	$C^{2,6}$ 110.8, $C^{3,5}$ 94.2, C^4 87.5	1141
$Mo(C_5H_5Sb)(CO)_3$	$C^{2,6}$ 114.6, $C^{3,5}$ 95.5, C^4 88.0	1141
Tungsten		
$W(C_6H_5Me)(CO)_3$	C^1 109.9, $C^{2,3,5,6}$ 90.7; 92.8, C^4 87.8	1745
$W(1,3,5\text{-}Me_3C_6H_3)(CO)_3$	$C^{1,3,5}$ 111.1, $C^{2,4,6}$ 90.9	117
$W(1,2,4,5\text{-}Me_4C_6H_2)(CO)_3$	$C^{1,2,4,5}$ 107.9, $C^{3,6}$ 97.3	117
$W(C_6Me_6)(CO)_3$	107.9	117
Manganese		
$[Mn(C_6H_6)(CO)_3]^+$	102.5	394
$[Mn(C_6H_5Me)(CO)_3]^+$	C^1 120.6, $C^{2,6}$ 100.9, $C^{3,5}$ 103.5, C^4 98.2	394
$[Mn(1,4\text{-}Me_2C_6H_4)(CO)_3]^+$	$C^{1,4}$ 117.4, $C^{2,3,5,6}$ 101.4	394
$[Mn(1,2\text{-}\overline{CH_2C_6H_4C}_6H_4)(CO)_3]^+$	$C^{1,2}$ 120.7; 121.7, C^{3-6} 92.9; 97.4; 100.8; 101.2	1131
$Mn(1,2\text{-}\overline{CHC_6H_4C}_6H_4)(CO)_3$	$C^{1,2}$ 88.9; 125.8, C^{3-6} 74.1; 87.2; 95.2; 97.9	1131
$[Mn(1,3,5\text{-}Me_3C_6H_3)(CO)_3]^+$	$C^{1,3,5}$ 122.2, $C^{2,4,6}$ 97.5	394
$[Mn(1,2,4,5\text{-}Me_4C_6H_2)(CO)_3]^+$	$C^{1,2,4,6}$ 117.0, $C^{3,6}$ 102.2	394
$[Mn(Me_5C_6H)(CO)_3]^+$	$C^{1,5}$ 113.7, $C^{2,4}$ 107.6, C^3 115.8, C^6 101.0	394
$[Mn(C_6Me_6)(CO)_3]^+$	114.3	394
Iron		
$[Fe(C_6H_6)Cp]^+$	88.4, 88.5	244, 260
$[Fe(C_6H_6)(C_5H_4C_5H_4FeCp)]^+$	88.4	1762
$[Fe(C_6H_6)(C_5H_4)]^{2+}$	89.0	1762
$[Fe(PhC_6H_5)Cp]^+$	87.3, 88.3, 89.3	260
$[CpFeC_6H_5]_2{}^{2+}$	$C^{2,6}$ 88.0, $C^{3,4,5}$ 89.6	260
$[1,4\text{-}(CpFeC_6H_5)_2C_6H_4]^{2+}$	C^1 103.5, $C^{2,6}$ 86.7, $C^{3,4,5}$ 88.6	260
$[Fe(1,2\text{-}Me_2C_6H_4)Cp]^+$	$C^{1,2}$ 103.4, C^{3-6} 86.4; 89.3	260
$[CpFeC_6H_5(CH)_4C_6H_4FeCp]^{2+}$	86.0, 86.7, 87.6, 87.7, 88.5	260
$[Fe\{1,2\text{-}\overline{CH_2(CH_2)_3C}_6H_4\}Cp]^+$	$C^{1,2}$ 102.8, C^{3-6} 86.9; 87.6	500
$[Fe\{1,2\text{-}\overline{CH_2(CH_2)_3C}_6H_4\}\text{-}(C_5H_4C_5H_4FeCp)]^+$	$C^{1,2}$ 98.2, C^{3-6} 85.4; 86.9	1762
$[Fe\{1,2\text{-}\overline{CH_2(CH_2)_3C}_6H_4\}\text{-}(C_5H_4)]_2{}^{2+}$	$C^{1,2}$ 103.1, C^{3-6} 86.9; 87.6	1762
$[Fe(1,2\text{-}\overline{CH_2C_6H_4CH_2C}_6H_4)Cp]^+$	$C^{1,2}$ 102.7, C^{3-6} 85.7; 86.3	500
$[Fe_2(1,2\text{-}\overline{CH_2C_6H_4CH_2C}_6H_4)\text{-}Cp_2]^{2+}$	86.3, 86.5, 99.0	500

COMPOUND	$\delta(^{13}C)$/p.p.m., *J*/Hz	REFERENCE
$[Fe(1,2\text{-}\overline{CHMeC_6H_4CHMe\mathit{C}_6}H_4)\text{-}Cp]^{+*}$	$C^{1,2}$ 105.8, C^{3-6} 86.2; 86.4 $C^{1,2}$ 108.9, C^{3-6} 83.4; 84.4	954
$[Fe_2(1,2\text{-}\overline{CH_2\mathit{C}_6H_4CH_2CH_2\mathit{C}_6}H_4)\text{-}Cp_2]^{2+}$	$C^{1,2}$ 101.5; 102.9; 104.7; 105.9, C^{3-6} 87.4; 87.6; 88.1; 88.3; 89.1; 89.4; 89.6	260
$[Fe(1,2\text{-}\overline{CH_2C_6H_4\mathit{C}_6}H_4)Cp]^+$	$C^{1,2}$ 105.5; 107.5, C^{3-6} 81.0; 86.0; 86.3; 86.7	1110
$Fe(1,2\text{-}\overline{CHC_6H_4\mathit{C}_6}H_4)Cp$	$C^{1,2}$ 67.9; 75.6, C^{3-6} 69.1; 69.9; 76.5; 76.9	1110
$[Fe_2(1,2\text{-}\overline{CH_2\mathit{C}_6H_4\mathit{C}_6}H_4)Cp_2]^{2+}$	$C^{1,2}$ 101.5; 106.7, C^{3-6} 81.8; 86.8; 87.6; 87.9	260
$[Fe(1,3,5\text{-}Me_3\mathit{C}_6H_3)(C_5H_4C_5H_4\text{-}FeCp)]^+$	$C^{1,3,5}$ 101.6, $C^{2,4,6}$ 87.5	1762
$[Fe(1,3,5\text{-}Me_3\mathit{C}_6H_3)(C_5H_4)]_2{}^{2+}$	$C^{1,3,5}$ 102.4, $C^{2,4,6}$ 87.4	1762
$[Fe(naphthalene)Cp]^+$	$C^{4a,8a}$ 96.9, C^{1-4} 86.1; 87.7	260
	$C^{4a,8a}$ 96.6, C^{1-4} 86.0; 87.6	500
$[Fe(naphthalene)(C_5H_4C_5H_4\text{-}FeCp)]^+$	$C^{4a,8a}$ 96.7, C^{1-4} 85.4; 86.9	1762
$[Fe(naphthalene)(C_5H_4)]_2{}^{2+}$	$C^{4a,8a}$ 95.2, C^{1-4} 85.0; 86.5	1762
$[Fe(2\text{-}Ph\text{-}naphthalene)Cp]^+$	$C^{4a,8a}$ 96.5; 102.0, C^{5-8} 89.0; 89.1; 89.4	260
$[Fe(2\text{-}Me\text{-}naphthalene)Cp]^{+*}$	C^{1-4} 85.3; 86.2; 88.1,	841
	C^{5-8} 85.2; 85.7; 86.8; 87.3	841
$[Fe(1\text{-}F\text{-}naphthalene)Cp]^+$	C^{5-7} 85.8; 87.5; 88.5, C^8 79.6 (J_{CF} = 6.1)	841
$[Fe(1\text{-}Cl\text{-}naphthalene)Cp]^+$	C^{5-7} 86.5; 88.2; 88.6, C^8 82.7	841
$[Fe(1\text{-}Br\text{-}naphthalene)Cp]^+$	C^{5-7} 85.8; 87.9; 87.9, C^8 84.6	841
$[Fe(1,2\text{-}C_8H_{10}\mathit{C}_6H_4)Cp]^+$	$C^{1,2}$ 93.1; 93.7, C^{3-6} 80.2; 83.5; 86.5; 86.1	1464
$[Fe(1,2\text{-}C_8H_{10}Me_2\mathit{C}_6H_4)Cp]^+$	$C^{1,2}$ 91.8; 94.7, C^{3-6} 79.5; 81.5; 85.3; 85.9	1464
$[Fe(1,2\text{-}C_8H_{12}\mathit{C}_6H_4)Cp]^+$	$C^{1,2}$ 96.7; 102.6, C^{3-6} 80.0; 86.4; 86.8; 87.5	1464
$[Fe_2(antracene)Cp_2]^{2+}$	$C^{4a,10a}$ 100.4, C^{1-4} 87.8; 88.0	260
$[Fe_2(phenanthrene)Cp_2]^{2+}$	$C^{4a,10a}$ 95.8, C^{1-4} 87.8; 89.0; 90.0,	260
	$C^{4a,10a}$ 92.2; 94.0, C^{1-4} 82.6; 87.2; 87.3; 88.2	1464
$[Fe_2(\mathit{C}_{14}H_{12})Cp_2]^2$	$C^{1,2}$ 91.1; 102.2, C^{3-6} 81.4; 86.2; 87.1	1464
$[Fe_2(\mathit{C}_{14}H_8Me_2)Cp_2]^2$	$C^{4a,10a}$ 94.1; 96.1, C^{1-4} 82.5; 84.0; 87.1; 88.2	1464
$[Fe_2(\mathit{C}_{18}H_{12})Cp_2]^2$	$C^{1a,4a}$ 95.9; 96.3, C^{1-4} 82.4; 87.1; 88.4; 89.1	260
$[Fe_2(3,4\text{-}Me_2\mathit{C}_6H_4)_2Cp_2]^2$	C^1 97.0, $C^{3,4}$ 104.3; 104.7, $C^{2,5,6}$ 84.8; 87.6; 89.8	260

*isomers

COMPOUND	δ(^{13}C)/p.p.m., *J*/Hz	REFERENCE
Ruthenium		
Ru(*C*$_6$H$_6$)(O$_2$CCHMeNH$_2$)Cl*	84.1, 84.8	1006
Ru(*C*$_6$H$_5$BF$_3$)(1-3,5,6-η^5-C$_8$H$_{11}$)	88.6, 89.6, 92.2, 92.9	1180
Ru(*C*$_6$Me$_6$)(cot)	91.3	827
Cobalt		
[Co{1,4-(MeO)$_2$*C*$_6$Me$_4$}-(C$_5$Me$_5$)]$^{2+}$	106.2, 112.5	1496
[Co{1,4-(HO)$_2$*C*$_6$Me$_4$}-(C$_5$Me$_5$)]$^{2+}$	104.5, 106.5	1496
[Co{1-O-4-(MeO)*C*$_6$Me$_4$}-(C$_5$Me$_5$)]$^+$	106.8, 107.1	1496
Rhodium, [^{103}*Rh*]		
[Rh(*C*$_6$H$_6$)(C$_5$Me$_5$)]$^{2+}$	108.7 [0.2] {184}	1222
[Rh(*C*$_6$H$_5$Me)(C$_5$Me$_5$)]$^{2+}$	C^1 125.0, C^{2,3,5,6} 108.1 [4.6]; 108.4 [4.6], C^4 106.5 [3.0]	1222
[Rh(*C*$_6$H$_5$O)(C$_5$Me$_5$)]$^+$	C^{2,6} 104.6 [6], C^{3,5} 106.3 [6], C^4 89.3 [6]	964
[Rh(indene)(C$_5$Me$_5$)]$^{2+}$	C^{1a} 125.7, C^{3a} 129.3 [4.2], C^4 101.1 [3.1], C^7 102.7 [6.1], C^6 105.2 [4.2], C^5 104.1 [4.5]	742, 1222
[Rh(C$_8$H$_6$NH)(C$_5$Me$_5$)]$^{2+}$	C^{1a,3a} 111.9 [9.2]; 119.3 [10.7], C$^{4-7}$ 93.0 [4.6]; 98.5 [6.1]; 99.1 [6.1]; 99.9 [4.6]	1222
[Rh*C*$_6$Ph$_4$(C$_6$H$_4$PPh$_2$)$_2$]$^+$	C^{2,3,5,6} 105	761
Iridium		
[Ir(*C*$_6$H$_6$)(C$_5$Me$_5$)]$^{2+}$	99.7 {188}	1222
[Ir(phenanthrene)(C$_5$Me$_5$)]$^{2+}$	91.1, 96.5, 97.4, 98.0, 104.1, 105.4	
[Ir(fluorene)(C$_5$Me$_5$)]$^{2+}$	90.9, 96.3, 96.7, 97.1, 117.4, 131.0	1222
[Ir(C$_8$H$_6$N)(C$_5$Me$_5$)]$^+$	86.0, 87.0, 87.0, 88.6, 110.0, 123.3	1222
[Ir(C$_8$H$_6$NH)(C$_5$Me$_5$)]$^{2+}$	84.2, 90.8, 91.4, 91.7, 106.5, 115.8	1222
Palladium		
[Pd(*C*$_6$Me$_6$)(2-MeC$_3$H$_4$)]$^+$	125.3, 125.8	1173
Silver		
[Ag(*C*$_6$H$_5$Me)]$^+$	C^1 141.0, C^{2,6} 129.6, C^{3,5} 127.2, C^4 123.4	931
[Ag(*C*$_6$H$_5$CF$_3$)]$^+$	C^{2,6} 125.8, C^{3,5} 128.7, C^4 131.5	931
[Ag(1,2-Me$_2$*C*$_6$H$_4$)]$^+$	C^{1,2} 139.8, C^{3,6} 129.3, C^{4,5} 122.8	931

*isomers

COMPOUND	$\delta(^{13}C)$/p.p.m., *J*/Hz				REFERENCE
$[Ag(1,3\text{-}Me_2C_6H_4)]^+$	$C^{1,3}$ 140.7, C^2 131.5, $C^{4,6}$ 124.6, C^5 124.9				931
$[Ag(1,4\text{-}Me_2C_6H_4)]^+$	$C^{1,4}$ 138.2, $C^{2,3,5,6}$ 126.8				931
$[Ag(1,3,5\text{-}Me_3C_6H_3)]^+$	$C^{1,3,5}$ 140.8, $C^{2,4,6}$ 124.1				931
$[Ag(1,2,4,5\text{-}Me_4C_6H_2)]^+$	$C^{1,2,4,5}$ 136.1, $C^{3,6}$ 126.3				931
$[Ag(C_6Me_6)]^+$	133.5				931
Mercury					
$[Hg(C_6H_6)(O_2CCF_3)]^+$	133.2 {172}				926
	C^1	$C^{2,6}$	$C^{3,5}$	C^4	
$[Hg(C_6H_5Me)(O_2CCF_3)]^+$	168.6	134.8	147.7	90.1	926
$[Hg_2(C_6H_5Me)]^{2+}$	142.8	131.4	130.4	125.3	931
$[Hg_2(C_6H_5CF_3)]^{2+}$		128.0	129.5	132.3	931
$[Hg(C_6H_5OMe)(O_2CCF_3)]^+$	192.6	118.4 {173}	139.0 {177}	105.2 {177}	787, 926
$[Hg(C_6H_5F)(O_2CCF_3)]^+$	176.6	123.5 {176}	152.2 {173}	87.8 {163}	926
$[Hg_2(1,2\text{-}Me_2C_6H_4)]^{2+}$	$C^{1,2}$ 141.6, $C^{3,6}$ 131.9, $C^{4,5}$ 126.0				931
$[Hg_2(1,3\text{-}Me_2C_6H_4)]^{2+}$	$C^{1,3}$ 143.8, C^2 132.9, $C^{4,6}$ 126.1, C^5 129.7				931
$[Hg(1,3\text{-}Me_2C_6H_4)(O_2CCF_3)]^+$	$C^{1,3}$ 169.6, C^2 136.0, $C^{4,6}$ 106.5, C^5 149.7				926
$[Hg_2(1,4\text{-}Me_2C_6H_4)]^{2+}$	$C^{1,4}$ 145.1, $C^{2,3,5,6}$ 131.7				931
$[Hg_2(1,3,5\text{-}Me_3C_6H_3)]^{2+}$	$C^{1,3,5}$ 152.6, $C^{2,4,6}$ 125.9				931
$[Hg(1,3,5\text{-}Me_3C_6H_3)(O_2CCF_3)]^+$	$C^{1,3,5}$ 171.0, $C^{2,4,6}$ 116.1 {165}				926
$[Hg_2(1,2,4,5\text{-}Me_4C_6H_2)]^+$	$C^{1,2,4,5}$ 141.9, $C^{3,6}$ 131.6				931
$[Hg_2(C_6Me_6)]^{2+}$	140.2				931

TABLE 2.17

^{13}C N.m.r. Chemical Shift Data for Some Trienes π-Bonded to Metals

COMPOUND	δ(^{13}C)/p.p.m., J/Hz			REFERENCE
	$C^{1,6}$	$C^{2,5}$	$C^{3,4}$	
Chromium				
$Cr(C_7H_8)(CO)_3$	57.6	101.7	99.5	173, 447
	57.6	101.9	99.0	60, 117
	57.9	102.1	100.0	1765
	58.2	102.3	100.2	927
$Cr(C_7H_8)(CO)_2PMe_3$	56.0	103.0	92.0	1765
$Cr(C_7H_8)(CO)_2P(OMe)_3$	56.0	101.8	94.9	1765
$Cr(C_7H_8)(CO)\{P(OMe)_3\}_2$	52.1	100.1	92.8	1765
$Cr(C_7H_8)(CO)_2AsMe_3$	56.0	103.0	91.3	1765
$Cr(C_7H_8)(PF_3)_3$	49.1 (3.6)	96.1	97.1	693
$Cr(C_6H_6NCO_2Et)(CO)_3$	72.5, 93.8, 94.4, 96.7			1288
$Cr(C_8H_8)(CO)_3$		104.8	102.5	1187
	93.1	106.5	102.7	307
	94.8	106.3	103.8	447
$Cr(C_8H_{10})(CO)_3$	$C^{1,2,5,6}$ 98.3, 99.6, $C^{3,4}$ 100.3			447
$[Cr(C_8H_9PPh_3)(CO)_3]^+$	81 (4.5), 95 (23), 97, 98, 101, 107			663
$Cr(1,3,5,7\text{-}Me_4C_8H_4)(CO)_3$	$C^{1,3,5}$ 112.0, 112.9, 114.2 $C^{2,4,6}$ 98.5, 104.7, 106.7			307
Molybdenum				
$Mo(C_7H_8)(CO)_3$	61.3	103.7	98.1	60, 117
	61.8	103.9	98.7	173, 447
	61.9	104.0	98.8	445
	62.4	104.4	99.7	927
$Mo(C_7H_8)(CO)_2PMe_2Ph$	61.6	105.0	88.2	445
$Mo(C_7H_8)(CO)_2PMePh_2$	62.0	105.4	88.6	445
$Mo(C_7H_8)(CO)_2PPh_3$	65.0	105.4	87.5	445
$Mo(C_7H_8)(CO)_2P(OPh)_3$	61.0	103.0	92.5	445
$Mo(C_6H_6NCO_2Et)(CO)_3$	73.9, 74.3	95.5	95.5	1288
$Mo(C_8H_8)(CO)_3$	88.5	C^{2-5} 101.4, 105.0		307
	88.5	C^{2-5} 101.8, 104.9		318
	88.8	$C^{2,5}$ 105.2	102.0	447
$Mo(C_8H_{10})(CO)_3$	95.4	98.4	100.0	447
Tungsten				
$W(C_7H_8)(CO)_3$	52.3	101.7	94.2	60, 117, 927
	53.5	102.3	95.2	447

COMPOUND	$\delta(^{13}C)$/p.p.m., J/Hz			REFERENCE
	$C^{1,6}$	$C^{2,5}$	$C^{3,4}$	
$W(C_6H_6NCO_2Et)(CO)_3$	65.5	93.3	91.9	1288
	65.9	94.2		
$W(C_8H_8)(CO)_3$	82.1	101.3	98.6	307
	83.1	99.6	99.6	174
Iron				
$[Fe(C_8H_8)Cp]^+$	91.1	103.1	96.7	1003
Ruthenium				
$Ru(C_8H_8)(nbd)$	66.8	92.3, 93.5		650

TABLE 2.18

^{13}C Chemical Shift Data for Some Cycloheptatrienyl Groups π-Bonded to Metals

COMPOUND	$\delta(^{13}C)$/p.p.m., *J*/Hz	REFERENCE
Titanium		
$Ti(C_7H_7)Cp$	86.1 {164}	334, 418
$Ti(C_7H_7)(C_5H_4Me)$	86.8 {166}	334, 418
$Ti(C_7H_7)$(indenyl)	88.7	693
$Ti(C_7H_6Me)Cp$	C^1 96.8, $C^{2,7}$ 87.2 {166}	334, 418
Zirconium		
$Zr(C_7H_7)Cp$	80.2 {166}	334, 418
Chromium		
$Cr(C_7H_7)Cp$	86.9 {159}	334, 418
$[Cr(C_7H_7)(CO)_3]^+$	104.7 {179.3}	927
$Cr_2(C_8H_8)_3$	105.2	693
Molybdenum		
$Mo(C_7H_7)Cp$	80.0 {164}, 80.4	334, 418, 446
$[Mo(C_7H_7)(CO)_3]^+$	100.0 {176.9}	927
Tungsten		
$[W(C_7H_7)(CO)_3]^+$	96.6 {179.9}	927
Manganese		
$MnFe(C_7H_7)(CO)_6$	70.7	716
Rhenium		
$ReFe(C_7H_7)(CO)_6$	64.2	716
Iron		
$FeMn(C_7H_7)(CO)_6$	70.7	716
$FeRe(C_7H_7)(CO)_6$	64.2	716
$FeRh(C_7H_7)(CO)_5$	64.2	716
Ruthenium		
$Ru_3(C_7H_7)(C_7H_9)(CO)_6$	66.3	384
Rhodium		
$RhFe(C_7H_7)(CO)_5$	64.2	716

TABLE 2.19

^{13}C *N.m.r. Data for Some Paramagnetic Organometallic Compounds*

COMPOUND	$\delta(^{13}C)$/p.p.m., J/Hz	REFERENCE
$Eu(C_5H_5)_3(CNC_6H_{11})$	-74.1	1777
$U(C_5H_5)_3Cl$	189	936
$U(C_5H_5)_2(NEt_2)_2$	125 {161}	340
$U(C_5H_5)Cl_2(HBpz_3)$	280	913
$U(C_8H_7OMe)_2$	309.1, 289.7, 288.1, 275.6, 267.8	1124
$U(C_8H_7OPr^i)_2$	330.0, 299.0, 296.1, 283.0, 259.6	1124
$U(C_8H_7OC_3H_5)_2$	341.2, 307.9, 289.2, 275.5, 268.3	1124
$U(C_8H_7CH_2CH_2CH_2NMe_2)_2$	312.1, 301.5, 298.3, 286.3, 281.9	1124
$U(C_8H_7Bu^t)_2$	308.0, 291.8, 288.8, 275.6	1124
$V(C_5H_5)_2$	-508.2, -790	23, 1754
$V(C_5D_5)_2$	-515.9	1754
$V(C_5H_5)_2Cl$	141	1340
$V(C_5H_4Me)_2$	-494, -725	1773
$V(C_5H_4Me)_2Cl$	C^1 59.0, $C^{2,5}$ 166.0, $C^{3,4}$ 189.0	1340
$V(C_5H_4Me)_2Br$	C^1 39.3, $C^{2,5}$ 153.1, $C^{3,4}$ 235.2	1340
$V(C_5H_4Me)_2I$	C^1 39.3, $C^{2,5}$ 134.1, $C^{3,4}$ 278.5	1340
$V(C_5H_4Et)_2Cl$	C^1 -6.9, $C^{2,5}$ 164.2, $C^{3,4}$ 228.7	1340
$V(C_5H_4Bu^n)_2$	C^1 -475, $C^{2,5}$ -725, $C^{3,4}$ -528	755
$V(C_5H_4Bu^t)_2$	C^1 -370, $C^{2,5}$ -484, $C^{3,4}$ -643	1800
$Cr(C_5H_5)_2$	-250.4, -570	23, 1754
$Cr(C_5D_5)_2$	-246.5	1754
$Cr(C_5H_4Me)_2$	-106, -368, -494	1773
$Cr(C_5H_4Bu^n)_2$	C^1 -100, $C^{2,5}$ -475, $C^{3,4}$ -348	755
$[Fe(C_5H_5)_2]^+$	314.5, 322.3	365, 1754
$[Fe(C_5D_5)_2]^+$	323.9	1754
$Fe(C^1{}_5H_5)(2,3\text{-}C^2{}_2B_9H_{11})$	C^1 68.6, C^2 -754	327
$Fe(C^1{}_5H_5)(2,4\text{-}C^2{}_2B_9H_{11})$	C^1 92, C^2 -677	327
$[Fe(C_5H_4Me)_2]^+$	C^1 415.0, C^{2-5} 263.0	365
$[Fe(C_5H_4Et)_2]^+$	C^1 402.5, C^{2-5} 259.8	365
$[Fe(C_5H_4Bu)_2]^+$	C^1 414.0, C^{2-5} 233.6, 250.6	365
$[Fe(2,3\text{-}C_2B_9H_{11})_2]^-$	-630	327
$Co(C_5H_5)_2$	549, 609.5	23, 1754
$Co(C_5D_5)_2$	602.7	1754
$Co(C_5H_4Me)_2$	413, 707	1773
$Co(C_5H_4Bu^n)_2$	C^1 334, $C^{2,5}$ 433, $C^{3,4}$ 706	755
$Ni(C_5H_5)_2$	1430	23
$Ni(C_5H_4Me)_2$	1207	1773

NOTE. For data in this table, extreme caution must be exercised in taking the given values. In many cases the original paper does not state the reference or sign convention used.

REFERENCES

1. M. Akhtar, P.D. Ellis, A.G. MacDiarmid and J.D. Odom, *Inorg. Chem.*, 1972, **11**, 2917.
2. J.P. Albrand and D. Gagnaire, *Chem. Commun.*, 1970, 874.
3. F.A.L. Anet and O.J. Abrams, *Chem. Commun.*, 1970, 1611.
4. K.R. Aris, V. Aris and J.M. Brown, *J. Organomet. Chem.*, 1972, **42**, C67.
5. M. Aritomi, Y. Kawasaki and R. Okawara, *Inorg. Nucl. Chem. Lett.*, 1972, **8**, 69.
6. I.R. Beattie, K.M.S. Livingstone, G.A. Ozin and R. Sabine, *J. Chem. Soc., Dalton Trans.*, 1972, 784.
7. D.C. Beer and L.J. Todd, *J. Organomet. Chem.*, 1972, **36**, 77.
8. R.D. Bertrand, F.B. Ogilvie and J.G. Verkade, *Chem. Commun.*, 1969, 756.
9. R.D. Bertrand, F.B. Ogilvie and J.G. Verkade, *J. Am. Chem. Soc.*, 1970, **92**, 1908.
10. W.E. Billups, L-P. Lin and O.A. Gansow, *Angew. Chem., Int. Ed. Engl.*, 1972, **11**, 637.
11. G.M. Bodner, B.N. Storhoff, D. Doddrell and L.J. Todd, *Chem. Commun.*, 1970, 1530.
12. A.A. Borisenko, N.M. Sergeyev, E.Ye. Nifant'ev and Yu A. Ustynyuk, *J. Chem. Soc., Chem. Commun.*, 1972, 406.
13. R. Bramley, B.N. Figgis and R.S. Nyholm, *Trans. Faraday Soc.*, 1962, **58**, 1893.
14. R. Bramley, B.N. Figgis and R.S. Nyholm, *J. Chem. Soc. (A)*, 1967, 861.
15. P.S. Braterman, D.W. Milne, E.W. Randall and E. Rosenberg, *J. Chem. Soc., Dalton Trans.*, 1973, 1027.
16. J.J. Breen, S.I. Featherman, L.D. Quin and R.C. Stocks, *J. Chem. Soc., Chem. Commun.*, 1972, 657.
17. T.L. Brown and J.C. Puckett, *J. Chem. Phys.*, 1966, **44**, 2238.
18. H.A. Brune, G. Horlbeck and Y.-T. Záhorszky, *Z. Naturforsch., Teil B*, 1971, **26**, 222.
19. H.A. Brune, H. Hütter and H. Hanebeck, *Z. Naturforsch., Teil B*, 1971, **26**, 570.
20. A.D. Buckingham and J.D. Stevens, *J. Chem. Soc.*, 1964, 2747; 4583.

21. T. Bundgaard and H.J. Jakobsen, *Acta Chem. Scand.*, 1972, **26**, 2548.
22. T. Bundgaard and H.J. Jakobsen, *Tetrahedron Lett.*, 1972, 3353.
23. P.K. Burkett, H.P. Fritz, F.H. Köhler and H. Rupp, *J. Organomet. Chem.*, 1970, **24**, C59.
24. C.H. Campbell and M.L.H. Green, *Chem. Commun.*, 1970, 1009.
25. A.J. Canty, B.F.G. Johnson, J. Lewis and J.R. Norton, *J. Chem. Soc., Chem. Commun.*, 1972, 1331.
26. E. Carberry, B.D. Dombek and S.C. Cohen, *J. Organomet. Chem.*, 1972, **36**, 61.
27. F.K. Cartledge and K.H. Riedel, *J. Organomet. Chem.*, 1972, **34**, 11.
28. J. Chatt and L.A. Duncanson, *J. Chem. Soc.*, 1953, 2939.
29. A.J. Cheney, B.E. Mann and B.L. Shaw, *Chem. Commun.*, 1971, 431
30. A.J. Cheney, B.E. Mann and B.L. Shaw, unpublished work.
31. M.H. Chisholm, H.C. Clark, L.E. Manzer and J.B. Stothers, *Chem Commun.*, 1971, 1627.
32. M.H. Chisholm, H.C. Clark, L.E. Manzer and J.B. Stothers, *J. Am. Chem. Soc.*, 1972, **94**, 5087.
33. L.A. Churlyaeva, M.I. Lobach, G.P. Kondratenkov and V.A. Kormer, *J. Organomet. Chem.*, 1972, **39**, C23.
34. D.J. Ciappenelli, F.A. Cotton and L. Kruczynski, *J. Organomet. Chem.*, 1972, **42**, 159.
35. D.J. Ciappenelli, F.A. Cotton and L. Kruczynski, *J. Organomet. Chem.*, 1973, **50**, 171.
36. H.C. Clark, N. Cyr and J.H. Tsai, *Can. J. Chem.*, 1967, **45**, 1073.
37. R.J.H. Clark, A.G. Davies, R.J. Puddephatt and W. McFarlane, *J. Am. Chem. Soc.*, 1969, **91**, 1334.
38. A.O. Clouse, D. Doddrell, S.B. Kahl and L.J. Todd, *Chem. Commun.*, 1969, 729.
39. J.A. Connor, E.M. Jones, E.W. Randall and E. Rosenberg, *J. Chem. Soc., Dalton Trans.*, 1972, 2419.
40. C.D. Cook and K.Y. Wan, *J. Am. Chem. Soc.*, 1970, **92**, 2595.
41. F.A. Cotton, A. Danti, J.S. Waugh and R.W. Fessenden, *J. Chem. Phys.*, 1958, **29**, 1427.
42. F.A. Cotton, L. Kruczynski, B.L. Shapiro and L.F. Johnson, *J. Am. Chem. Soc.*, 1972, **94**, 6191.
43. R. W. Crecely, K.M. Crecely and J.H. Goldstein, *Inorg. Chem.*, 1969, **8**, 252.
44. D.K. Dalling and H.S. Gutowsky, 'U.S. Clearinghouse Federal Scientific Technical Information', 1971, Report 721 717. *Govt. Rep. Announc. (U.S.)*, 1971, **71**,
45. R.R. Dean and W. McFarlane, *Mol. Phys.*, 1967, **12**, 289; 364.
46. R.R. Dean and W. McFarlane, *Mol. Phys.*, 1967, **13**, 343.
47. D. Doddrell, M.L. Bullpitt, C.J. Moore, C.W. Fong, W. Kitching, W. Adcock and B.D. Gupta, *Tetrahedron Lett.*, 1973, 665.
48. R.S. Drago and N.A. Matwiyoff, *J. Organomet. Chem.*, 1965, **3**, 62.

49. H. Dreeskamp, K. Hildenbrand and G. Pfisterer, *Mol. Phys.*, 1969, **17**, 429.
50. H. Dreeskamp and G. Pfisterer, *Mol. Phys.*, 1968, **14**, 295.
51. H. Dreeskamp, C. Schumann and R. Schmutzler, *Chem. Commun.*, 1970, 671.
52. E.A.V. Ebsworth and S.G. Frankiss, *Trans. Faraday Soc.*, 1967, **63**, 1574.
53. P.D. Ellis, G.E. Maciel and J.W. McIver, Jr., *J. Am. Chem. Soc.*, 1972, **94**, 4069.
54. R.V. Emmanuel and E.W. Randall, *J. Chem. Soc. (A)*, 1969, 3002.
55. G.F. Emerson, K. Ehrlich, W.P. Giering and P.C. Lauterbur, *J. Am. Chem. Soc.*, 1966, **88**, 3172.
56. G. Engelhardt, H. Jancke, M. Mägi, T. Pehk and E. Lippmaa, *J. Organomet. Chem.*,1971, **28**, 293.
57. J. Evans, B.F.G. Johnson, J. Lewis and J.R. Norton, *J. Chem. Soc., Chem. Commun.*, 1973, 79.
58. M. Evans, M. Hursthouse, E.W. Randall, E. Rosenberg, L. Milone and M. Valle, *J. Chem. Soc., Chem. Commun.*, 1972, 545.
59. J.W. Faller, A.S. Anderson and C. Chin-Chun, *Chem. Commun.*, 1969, 719.
60. L.F. Farnell, E.W. Randall and E. Rosenberg, *Chem. Commun.*, 1971, 1078.
61. S.I. Featherman and L.D. Quin, *Tetrahedron Lett.*, 1973, 1955.
62. E.G. Finer and R.K. Harris, *Mol. Phys.*, 1967, **13**, 65.
63. N. Flitcroft and H.D. Kaesz, *J. Am. Chem. Soc.*, 1963, **85**, 1377.
64. W.T. Ford and J.B. Grutzner, *J. Org. Chem.*, 1972, **37**, 2561.
65. J.N. Francis, C.J. Jones and M.F. Hawthorne, *J. Am. Chem. Soc.*, 1972, **94**, 4878.
66. O.A. Gansow, A.R. Burke and W.D. Vernon, *J. Am. Chem. Soc.*, 1972, **94**, 2550.
67. O.A. Gansow, A.R. Burke and G.N. LaMar, *J. Chem. Soc., Chem. Commun.*, 1972, 456.
68. O.A. Gansow and B.Y. Kimura, *Chem. Commun.*, 1970, 1621.
69. O.A. Gansow, B.Y. Kimura, G.R. Dobson and R.A. Brown, *J. Am. Chem. Soc.*, 1971, **93**, 5922.
70. O.A. Gansow, D.A. Schexnayder and B.Y. Kimura, *J. Am. Chem. Soc.*, 1972, **94**, 3406.
71. T.A. George and C.D. Turnipseed, *Inorg. Chem.*, 1973, **12**, 394.
72. D.F. Gill, B.E. Mann and B.L. Shaw, *J. Chem. Soc., Dalton Trans.*, 1973, 270; 311.
73. G.A. Gray, *J. Am. Chem. Soc.*, 1971, **93**, 2132.
74. G.A. Gray and S.E. Cremer, *Tetrahedron Lett.*, 1971, 3061.
75. G.A. Gray and S.E. Cremer, *J. Chem. Soc., Chem. Commun.*, 1972, 367.
76. G.A. Gray and S.E. Cremer, *J. Org. Chem.*, 1972, **37**, 3458.
77. G.A. Gray and S.E. Cremer, *J. Org. Chem.*, 1972, **37**, 3470.
78. Yu.K. Grishin, N.M. Sergeyev and Yu.A. Ustynyuk, *J. Organomet. Chem.*, 1970, **22**, 361.

79. Yu.K. Grishin, N.M. Sergeyev and Yu.A. Ustynyuk, *J. Organomet. Chem.*, 1972, **34**, 105.
80. Yu.K. Grishin, N.M. Sergeyev and Yu.A. Ustynyuk, *Org. Magn. Reson.*, 1972, **4**, 377.
81. R.K. Harris, *J. Phys. Chem.*, 1962, **66**, 768.
82. R.K. Harris, *J. Mol. Spectrosc.*, 1963, **10**, 309.
83. B.K. Hunter and L.W. Reeves, *Can. J. Chem.*, 1972, **46**, 1399.
84. H.J. Jakobsen and O. Manscher, *Acta Chem. Scand.*, 1971, **25**, 680.
85. C.J. Jameson and H.S. Gutowsky, *J. Chem. Phys.*, 1969, **51**, 2790.
86. B.F.G. Johnson and J.A. Segal, *J. Chem. Soc., Chem. Commun.*, 1972, 1312.
87. A.J. Jones, D.M. Grant, J.G. Russell and G. Frenkel, *J. Phys. Chem.*, 1969, **73**, 1624.
88. C.H.W. Jones, R.G. Jones, P. Partington and R.M.G. Roberts, *J. Organomet. Chem.*, 1971, **32**, 201.
89. C.G. Kreiter and V. Formáček, *Angew, Chem., Int. Ed. Engl.*, 1972, **11**, 141.
90. P. Krohmer and J. Goubeau, *Z. Anorg. Allg. Chem.*, 1969, **369**, 238.
91. R.L. Lambert, Jr. and D. Seyferth, *J. Am. Chem. Soc.*, 1972, **94**, 9246.
92. P.C. Lauterbur, *J. Chem. Phys.*, 1957, **26**, 217.
93. P.C. Lauterbur and R.B. King, *J. Am. Chem. Soc.*, 1965, **87**, 3266.
94. G.C. Levy, *Tetrahedron Lett.*, 1972, 3709.
95. G.C. Levy and J.D. Cargioli, *Chem. Commun.*, 1970, 1663.
96. G.C. Levy and J.D. Cargioli, *J. Magn. Reson.*, 1972, **6**, 143.
97. G.C. Levy, D.M. White and J.D. Cargioli, *J. Magn. Reson.*, 1972, **8**, 280.
98. E.A.C. Lucken, K. Noack and D.F. Williams, *J. Chem. Soc. (A)*, 1967, 148.
99. W. McFarlane, *Chem. Commun.*, 1967, 58.
100. W. McFarlane, *J. Chem. Soc. (A)*, 1967, 528.
101. W. McFarlane, *J. Chem. Soc. (A)*, 1967, 1148.
102. W. McFarlane, *J. Chem. Soc. (A)*, 1967, 1275.
103. W. McFarlane, *Mol. Phys.*, 1967, **12**, 243.
104. W. McFarlane, *Mol. Phys.*, 1967, **13**, 587.
105. W. McFarlane, *J. Chem. Soc. (A)*, 1968, 1630.
106. W. McFarlane, *Proc. R. Soc., Ser. A*, 1968, **306**, 185.
107. W. McFarlane and J.A. Nash, *Chem. Commun.*, 1969, 127.
108. L.D. McKeever and R. Waack, *Chem. Commun.*, 1969, 750.
109. L.D. McKeever, R. Waack, M.A. Doran and E.B. Baker, *J. Am. Chem. Soc.*, 1968, **90**, 3244.
110. L.D. McKeever, R. Waack, M.A. Doran and E.B. Baker, *J. Am. Chem. Soc.*, 1969, **91**, 1057.
111. K.A. McLauchlan, *Mol. Phys.*, 1966, **11**, 303; 503.
112. K.A. McLauchlan, D.H. Whiffen and L.W. Reeves, *Mol. Phys.*, 1966, **10**, 131.
113. G.E. Maciel, *J. Phys. Chem.*, 1965, **69**, 1947.

114. B.E. Mann, *Chem. Commun.*, 1971, 976.
115. B.E. Mann, *Chem. Commun.*, 1971, 1173.
116. B.E. Mann, *J. Chem. Soc., Perkin Trans. 2*, 1972, 30.
117. B.E. Mann, *J. Chem. Soc., Dalton Trans.*, 1973, 2012.
118. B.E. Mann, R. Pietropaulo and B.L. Shaw, *Chem. Commun.*, 1971, 790.
119. B.E. Mann, R. Pietropaulo, B.L. Shaw and G. Shaw, unpublished work.
120. B.E. Mann, B.L. Shaw and R.E. Stainbank, *J. Chem. Soc., Chem. Commun.*, 1972, 151 and unpublished work.
121. B.E. Mann, B.L. Shaw and M.M. Truelock, unpublished work.
122. B.E. Mann, B.L. Shaw and B. Turtle, unpublished work.
123. G. Mavel and M.J. Green, *Chem. Commun.*, 1968, 742.
124. R.B. Moon and J.H. Richards, *J. Am. Chem. Soc.*, 1972, **94**, 5093.
125. N. Muller, *J. Chem. Phys.*, 1962, **36**, 359.
126. N. Muller and D.E. Pritchard, *J. Chem. Phys.*, 1959, **31**, 768; 1471.
127. Y. Nagai, M.-A. Ohtsuki, T. Nakano and H. Watanbe, *J. Organomet. Chem.*, 1972, **35**, 81.
128. A.N. Nesmeyanov, L.A. Fedorov, R.B. Materikova, E.I. Fedin and N.S. Kochetkova, *Dokl. Akad. Nauk SSSR*, 1968, **183**, 356.
129. G.A. Olah and P.R. Clifford, *J. Am. Chem. Soc.*, 1971, **93**, 2320.
130. R.G. Parker and J.D. Roberts, *J. Am. Chem. Soc.*, 1970, **92**, 743.
131. P.S. Pregosin, unpublished work.
132. H.G. Preston, Jr. and J.C. Davis, Jr., *J. Am. Chem. Soc.*, 1966, **88**, 1585.
133. L.D. Quin, S.G. Borleske and R.C. Stocks, *Org. Magn. Reson.*, 1973, **5**, 161.
134. J.M. Read, Jr., C.T. Mathis and J.H. Goldstein, *Spectrochim. Acta*, 1965, **21**, 85.
135. H.L. Retcofsky, E.N. Frankel and H.S. Gutowsky, *J. Am. Chem. Soc.*, 1966, **88**, 2710.
136. H.L. Retcofsky and C.E. Griffin, *Tetrahedron Lett.*, 1966, 1975.
137. G. Rigatti, G. Boccalon, A. Ceccon and G. Giacometti, *J. Chem. Soc., Chem. Commun.*, 1972, 1165.
138. D. Rosenberg and W. Drenth, *Tetrahedron*, 1971, **27**, 3893.
139. D. Rosenberg, J.W. de Haan and W. Drenth, *Recl. Trav. Chim. Pays-Bas*, 1968, **87**, 1387.
140. H.H. Rupp, quoted by H.H. Keller, *Ber. Bunsenges. Phys. Chem.*, 1972, **76**, 1080.
141. C.D. Schaeffer, Jr. and J.J. Zuckerman, *J. Organomet. Chem.*, 1973, **47**, C1.
142. O.J. Scherer and P. Hornig, *J. Organomet. Chem.*, 1967, **8**, 465.
143. H. Schmidbaur, W. Buchner and D. Scheutzow, *Chem. Ber.*, 1973, **106**, 1251.
144. C. Schumann and H. Dreeskamp, *J. Magn. Reson.*, 1970, **3**, 204.

145. N.M. Sergeyev, Yu.K. Grishin, Yu.N. Luzikov and Yu.A. Ustynyuk, *J. Organomet. Chem.*, 1972, **38**, C1.
146. D. Seyferth, Y.M. Cheng and D.D. Traficante, *J. Organomet. Chem.*, 1972, **46**, 9.
147. M.P. Simonnin, *Bull. Soc. Chim. Fr.*, 1966, 1774.
148. M.-P. Simonnin, R.-M. Lequan and F.W. Wehrli, *J. Chem. Soc., Chem. Commun.*, 1972, 1204.
149. M.-P. Simonnin, R.-M. Lequan and F.W. Wehrli, *Tetrahedron Lett.*, 1972, 1559.
150. G. Singh and G.S. Reddy, *J. Organomet. Chem.*, 1972, **42**, 267.
151. P.S. Skell and P.W. Owen, *J. Am. Chem. Soc.*, 1972, **94**, 1578.
152. G.W. Smith, *J. Chem. Phys.*, 1963, **39**, 2031.
153. S. Sørensen, R.S. Hansen and H.J. Jakobsen, *J. Am. Chem. Soc.*, 1972, **94**, 5900.
154. H. Spiesecke and W.G. Schneider, *J. Chem. Phys.*, 1961, **35**, 722.
155. L.J. Todd, *Pure Appl. Chem.*, 1972, **30**, 587.
156. E.V. Van den Berghe and G.P. Van der Kelen, *J. Organomet. Chem.*, 1971, **26**, 207.
157. D.F. Van de Vondel, *J. Organomet. Chem.*, 1965, **3**, 400.
158. L. Verdonck and G.P. Van der Kelen, *J. Organomet. Chem.*, 1966, **5**, 532.
159. R. Waack, M.A. Doran, E.B. Baker and G.A. Olah, *J. Am. Chem. Soc.*, 1966, **88**, 1272.
160. R. Waack, L.D. McKeever and M.A. Doran, *Chem. Commun.*, 1969, 117.
161. F.J. Weigert and J.D. Roberts, *J. Am. Chem. Soc.*, 1969, **91**, 4940.
162. F.J. Weigert and J.D. Roberts, *Inorg. Chem.*, 1973, **12**, 313; 3021.
163. F.J. Weigert, M. Winokur and J.D. Roberts, *J. Am. Chem. Soc.*, 1968, **90**, 1566.
164. R.B. Wetzel and G.L. Kenyon, *J. Chem. Soc., Chem. Commun.*, 1973, 287.
165. G.M. Whitesides and G. Maglio, *J. Am. Chem. Soc.*, 1969, **91**, 4980.
166. L. Zelta and G. Gatti, *Org. Magn. Reson.*, 1972, **4**, 585.
167. D. Ziessow and M. Carroll, *Ber. Bunsenges Phys. Chem.*, 1972, **76**, 61.
168. N.M. Sergeev, *Russ. Chem. Rev. (Engl. Transl.)*, 1973, **42**, 339.
169. F.A. Cotton, *Bull. Soc. Chim. Fr.*, 1973, 2587.
170. J.R. Llinas, É-J. Vincent and G. Peiffer, *Bull. Soc. Chim. Fr.*, 1973, 3209.
171. J.W. Wilson and E.O. Fischer, *J. Organomet. Chem.*, 1973, **57**, C63.
172. F.R. Kreissl, E.O. Fischer, C.G. Kreiter and K. Weiss, *Angew. Chem., Int. Ed. Engl.*, 1973, **12**, 563.
173. C.G. Kreiter and M. Lang, *J. Organomet. Chem.*, 1973, **55**, C27.
174. E.W. Randall, E. Rosenberg and L. Milone, *J. Chem. Soc., Dalton Trans.*, 1973, 1672.

175. E.O. Fischer, K.R. Schmid, W. Kalbfus and C.G. Kreiter, *Chem. Ber.*, 1973, **106**, 3893.
176. W.O. Fischer, G. Kreis, C.G. Kreiter, J. Müller, G. Huttner and H. Lorenz, *Angew. Chem., Int. Ed. Engl.*, 1973, **12**, 564.
177. A.J. Shortland and G. Wilkinson, *J. Chem. Soc., Dalton Trans.*, 1973, 872.
178. E.O. Fischer, G. Kreis, F.R. Kreissl, C.G. Kreiter and J. Müller, *Chem. Ber.*, 1973, **106**, 3910.
179. E.O. Fischer, G. Kreis and F.R. Kreissl, *J. Organomet. Chem.*, 1973, **56**, C37.
180. F.R. Kreissl, E.O. Fischer and C.G. Kreiter, *J. Organomet. Chem.*, 1973, **57**, C9.
181. P.V. Petrovskii, E.I. Fedin and L.A. Fedorov, *Dokl. Akad. Nauk SSSR*, 1973, **210**, 605.
182. M.J. Webb, R.P. Stewart, Jr. and W.A.G. Graham, *J. Organomet. Chem.*, 1973, **59**, C21.
183. J.J. Dannenberg, M.K. Levenberg and J.H. Richards, *Tetrahedron*, 1973, **29**, 1575.
184. H. Burger and C. Kluess, *J. Organomet. Chem.*, 1973, **56**, 269.
185. M. Anderson, A.D.H. Clague, L.P. Blaauw and P.A. Couperus, *J. Organomet. Chem.*, 1973, **56**, 307.
186. G. Deganello, *J. Organomet. Chem.*, 1973, **59**, 329.
187. V.I. Sokolov, P.V. Petrovskii and O.A. Reutov, *J. Organomet. Chem.*, 1973, **59**, C27.
188. J. Evans, D.V. Howe, B.F.G. Johnson and J. Lewis, *J. Organomet. Chem.*, 1973, **61**, C48.
189. J.P. Collman and S.R. Winter, *J. Am. Chem. Soc.*, 1973, **95**, 4089.
190. N.A. Matwiyoff, P.J. Vergamini, T.E. Needham, C.T. Gregg, J.A. Volpe and W.S. Caughey, *J. Am. Chem. Soc.*, 1973, **95**, 4429.
191. R.D. Adams and F.A. Cotton, *J. Am. Chem. Soc.*, 1973, **95**, 6589.
192. G.H. Williams, D.D. Traficante and D. Seyferth, *J. Organomet. Chem.*, 1973, **60**, C53.
193. D.F. Gill, B.E. Mann and B.L. Shaw, *J. Chem. Soc., Dalton Trans.*, 1973, 311.
194. B.E. Mann, R. Pietropaulo and B.L. Shaw, *J. Chem. Soc., Dalton Trans.*, 1973, 2390.
195. V.I. Sokolov, G.M. Khvostik, I.Ya. Poddubnyi, G.P. Kondratenkov and G.K. Grebenshikov, *J. Organomet. Chem.*, 1973, **54**, 375.
196. R.P. Hughes and J. Powell, *J. Organomet. Chem.*, 1973, **54**, 345.
197. R.M. Moriarty, C.-L. Yeh, K.-N. Chen, E.L. Yeh, K.C. Ramey and C.W. Jefford, *J. Am. Chem. Soc.*, 1973, **95**, 4756.
198. A. Musco, W. Kuran, A. Silvani and M.W. Anker, *J. Chem. Soc., Chem. Commun.*, 1973, 938.
199. R.P. Hughes and J. Powell, *J. Organomet. Chem.*, 1973, **60**, 427.

200. D.G. Cooper, G.K. Hamer, J. Powell and W.F. Reynolds, *J. Chem. Soc., Chem. Commun.*, 1973, 449.
201. C. Masters, *J. Chem. Soc., Chem. Commun.*, 1973, 191.
202. D.J. Cardin, B. Çetinkaya, E. Çetinkaya, M.F. Lappert, E.W. Randall and E. Rosenberg, *J. Chem. Soc., Dalton Trans.*, 1973, 1982.
203. M.H. Chisholm, H.C. Clark, L.E. Manzer, J.B. Stothers and J.E.H. Ward, *J. Am. Chem. Soc.*, 1973, **95**, 8574.
204. H. Brunner and W. Rambold, *J. Organomet. Chem.*, 1973, **60**, 351.
205. T.E. Needham, N.A. Matwiyoff, T.E. Walker and H.P.C. Hogenkamp, *J. Am. Chem. Soc.*, 1973, **95**, 5019.
206. T. Yamamoto, A.R. Garber, G.M. Bodner, L.J. Todd, M.D. Rausch and S.A. Gardner, *J. Organomet. Chem.*, 1973, **56**, C23.
207. J. Evans, B.F.G. Johnson, J. Lewis, J.R. Norton and F.A. Cotton, *J. Chem. Soc., Chem. Commun.*, 1973, 807; 1974, 280.
208. R.J. Foot and B.T. Heaton, *J. Chem. Soc., Chem. Commun.*, 1973, 838.
209. A. J. Leusink, G. Van Koten, J.W. Marsman and J.G. Noltes, *J. Organomet. Chem.*, 1973, **55**, 419.
210. S. Sakaki, *Theor. Chim. Acta*, 1973, **30**, 139.
211. G.A. Olah and P.R. Clifford, *J. Am. Chem. Soc.*, 1973, **95**, 6067.
212. P.A.W. Dean, D.G. Ibbott and J.B. Stothers, *J. Chem. Soc., Chem. Commun.*, 1973, 626.
213. M.P. Brown, A.K. Holliday and G.M. Way, *J. Chem. Soc., Chem. Commun.*, 1973, 532.
214. T.D. Westmoreland, Jr., N.S. Bhacca, J.D. Wander and M.C. Day, *J. Am. Chem. Soc.*, 1973, **95**, 2019.
215. D.E. Axelson, A.J. Oliver and C.E. Holloway, *Org. Magn. Reson.*, 1973, **5**, 255.
216. P.A.J. Gorin and M. Mazurek, *Can. J. Chem.*, 1973, **51**, 3277,
217. S. Kaplan, H.A. Resing and J.S. Waugh, *J. Chem. Phys.*, 1973, **59**, 5681.
218. M.P. Simonnin, G. Guillerm and M. Lequan, *Bull. Soc. Chim. Fr.*, 1973, 1649.
219. J. Pola, J. Schraml and V. Chvalovský, *Collect. Czech. Chem. Commun.*, 1973, **38**, 3158.
220. Vo-Kim-Yen, Z. Papoušková, J. Schraml and V. Chvalovský, *Collect. Czech. Chem. Commun.*, 1973, **38**, 3167.
221. P. Brouant, Y. Limouzin and J.C. Maire, *Helv. Chim. Acta*, 1973, **56**, 2057.
222. K. Kovačević and Z.B. Maksić, *J. Mol. Struct.*, 1973, **17**, 203.
223. G. Engelhardt, R. Radeglia, H. Jancke, E. Lippmaa and M. Mägi, *Org. Magn. Reson.*, 1973, **5**, 561.
224. M. Murray, *J. Magn. Reson.*, 1973, **9**, 326.
225. S.Q.A. Rizvi, B.D. Gupta, W. Adcock, D. Doddrell and W. Kitching, *J. Organomet. Chem.*, 1973, **63**, 67.

226. M.R. Bacon and G.E. Maciel, *J. Am. Chem. Soc.*, 1973, **95**, 2413.
227. T.N. Mitchell, *J. Organomet. Chem.*, 1973, **59**, 189.
228. H.G. Kuivila, J.L. Considine, R.J. Mynott and R.H. Harma, *J. Organomet. Chem.*, 1973, **55**, C11.
229. C.D. Schaeffer, Jr. and J.J. Zuckerman, *J. Organomet. Chem.*, 1973, **55**, 97.
230. T.J. Marks and A.R. Newman, *J. Am. Chem. Soc.*, 1973, **95**, 769.
231. R.M.G. Roberts, *J. Organomet. Chem.*, 1973, **63**, 159.
232. J.D. Kennedy and W. McFarlane, *J. Chem. Soc., Dalton Trans.*, 1973, 2134.
233. G. Märkl and F. Kneidl, *Angew. Chem., Int. Ed. Engl.*, 1973, **12**, 931.
234. G.A. Gray, *J. Am. Chem. Soc.*, 1973, **95**, 7736.
235. D.E. Axelson and C.E. Holloway, *J. Chem. Soc., Chem. Commun.*, 1973, 455.
236. D.G. Cooper and J. Powell, *Can. J. Chem.*, 1973, **51**, 1634.
237. L.D. Quin, S.G. Borleske and R.C. Stocks, *Org. Magn. Reson.*, 1973, **5**, 161.
238. D.H. Lemmon and J.A. Jackson, *Spectrochim. Acta, Part A*, 1973, **29**, 1899.
239. H.R. Allcock, E.C. Bissell and E.T. Shawl, *Inorg. Chem.*, 1973, **12**, 2963.
240. D.F. Gill, B.E. Mann and B.L. Shaw, *J. Chem. Soc., Dalton Trans.*, 1973, 270.
241. G.A. Gray, *J. Am. Chem. Soc.*, 1973, **95**, 5092.
242. R. Radeglia, W. Storek, G. Engelhardt, R. Ritschl, E. Lippmaa, T. Pehk, M. Mägi and D. Martin, *Org. Magn. Reson.*, 1973, **5**, 419.
243. W. McFarlane and D.S. Rycroft, *J. Chem. Soc., Chem. Commun.*, 1973, 10.
244. A.N. Nesmeyanov, E.I. Fedin, L.A. Fedorov and P.V. Petrovskii, *Zh. Strukt. Khim.*, 1972, **13**, 1033.
245. A.N. Nesmeyanov, P.V. Petrovskii, L.A. Fedorov, V.I. Robas and E.I. Fedin, *Zh. Strukt. Khim.*, 1973, **14**, 49.
246. L.A. Fedorov, Z. Stumbreviciute, A.K. Prokof'ev and E.I. Fedin, *Dokl. Akad. Nauk SSSR*, 1973, **209**, 203.
247. G.A. Gray and S.E. Cremer, *J. Magn. Reson.*, 1973, **12**, 5.
248. G.M. Bodner, S.B. Kahl, K. Bork, B.N. Storhoff, J.E. Wuller and L.J. Todd, *Inorg. Chem.*, 1973, **12**, 1071.
249. J. Burdon, J.C. Hotchkiss and W.B. Jennings, *Tetrahedron Lett.*, 1973, 4919.
250. M. Haemers, R. Ottinger, D. Zimmermann and J. Reisse, *Tetrahedron Lett.*, 1973, 2241.
251. M. Haemers, R. Ottinger, D. Zimmermann and J. Reisse, *Tetrahedron*, 1973, **29**, 3539.
252. Y. Yamamoto, D.S. Tarbell, J.R. Fehlner and B.M. Pope, *J. Org. Chem.*, 1973, **38**, 2521.
253. T.N. Ivshina, V.M. Shitkin, S.L. Ioffe, E. Lippmaa and M. Mägi, *Zh. Strukt. Khim.*, 1972, **13**, 933.

254. N.A. Mativiyoff and T.E. Needham, *Biochem. Biophys. Res. Commun.*, 1972, **49**, 1158.
255. H.A. Brune, G. Horlbeck, H. Roettele and U. Tanger, *Z. Naturforsch., Teil B*, 1973, **28**, 68.
256. J. Boersma and J.G. Noltes, *Recl. Trav. Chim. Pays-Bas*, 1973, **92**, 229.
257. A.N. Nesmeyanov, P.V. Petrovskiî, L.A. Fedorov, E.I. Fedin, H.I. Schneiders and N.S. Kochetkova, *Izv. Akad. Nauk SSSR, Ser. Khim.*, 1973, **6**, 1362.
258. A.I. Rezvukhin, V.N. Piottukh-Peletskii, R.N. Berezina and V.G. Shubin, *Izv. Akad. Nauk SSSR, Ser. Khim.*, 1973, **3**, 705.
259. F.A. Cotton and J.M. Troup, *J. Am. Chem. Soc.*, 1974, **96**, 4422.
260. W.H. Morrison, Jr., E.Y. Ho and D.N. Hendrickson, *J. Am. Chem. Soc.*, 1974, **96**, 3603.
261. F.A. Cotton and J.M. Troup, *J. Am. Chem. Soc.*, 1974, **96**, 3438.
262. V.V. Negrebetskii, V.S. Bogdanov, A.V. Kessenikh, P.V. Petrovskii, Yu. N. Bubnov and B.M. Mikhailov, *Zh. Obshch. Khim.*, 1974, **44**, 1882; *J. Gen. Chem. USSR (Engl. Transl.)*, 1974, **44**, 1849.
263. B.M. Mikhailov, V.V. Negrebetskii, V.S. Bogdanov, A.K. Kessenikh, Yu. N. Bubnov, T.K. Baryshnikova and V.N. Smirnov, *Zh. Obshch. Khim.*, 1974, **44**, 1878; *J. Gen. Chem. USSR (Engl. Transl.)*, 1974, **44**, 1849.
264. Yu. K. Grishin, S.V. Ponomarev and S.A. Lebedev, *Zh. Org. Khim.*, 1974, **10**, 404; *J. Org. Chem. USSR (Engl. Transl.)*, 1974, **10**, 402.
265. D. Michel, W. Meiler and D. Hoppach, *Z. Phys. Chem. (Leipzig)*, 1974, **255**, 509.
266. D. Michel, W. Meiler and E. Angelé, *Z. Phys. Chem. (Leipzig)*, 1974, **255**, 389.
267. D. Deininger, D. Geschke and W.-D. Hoffmann, *Z. Phys. Chem. (Leipzig)*, 1974, **255**, 273.
268. H. Schwind, D. Deininger and D. Geschke, *Z. Phys. Chem. (Leipzig)*, 1974, **255**, 149.
269. W. Schäfer, A. Zschunke, H.-J. Kerrinnes and U. Langbein, *Z. Anorg. Allg. Chem.*, 1974, **406**, 105.
270. F.A.L. Anet, J. Krane, W. Kitching, D. Doddrell and D. Praeger, *Tetrahedron Lett.*, 1974, 3255.
271. T. Bundgaard, H.J. Jakobsen, K. Dimroth and H.H. Pohl, *Tetrahedron Lett.*, 1974, 3179.
272. G.-E. Matsubayashi and T. Tanaka, *Spectrochim. Acta, Part A*, 1974, **30**, 869.
273. H.J. Schneider and H. Roswitha, *Justus Liebigs Ann. Chem.*, 1974, 1864.
274. I.D. Gay, *J. Phys. Chem.*, 1974, **78**, 38.
275. H. Mahnke, R.K. Sheline and H.W. Spiess, *J. Chem. Phys.*, 1974, **61**, 55.

276. H. Mahnke, R.J. Clark, R. Rosanske and R.K. Sheline, *J. Chem. Phys.*, 1974, **60**, 2997.
277. S.I. Featherman, S.O. Lee and L.D. Quin, *J. Org. Chem.*, 1974, **39**, 2899.
278. D.J. Hart and W.R. Ford, *J. Org. Chem.*, 1974, **39**, 363.
279. D.A. Redfield, J.H. Nelson and L.W. Cary, *Inorg. Nucl. Chem. Lett.*, 1974, **10**, 727.
280. O.A. Gansow, A.R. Burke, R.B. King and M.S. Saran, *Inorg. Nuclear Chem. Lett.*, 1974, **10**, 291.
281. F.A. Cotton and D.L. Hunter, *Inorg. Chim. Acta*, 1974, **11**, L9.
282. J. Schraml, J. Včelák and V. Chvalovský, *Collect. Czech. Chem. Commun.*, 1974, **39**, 267.
283. F. Heatley, *J. Chem. Soc., Faraday Trans. 2*, 1974, **70**, 148.
284. J.M. Brown, J.A. Conneely and K. Mertis, *J. Chem. Soc., Perkin Trans. 2*, 1974, 905.
285. G. Pirazzini, R. Danieli, A. Ricci and C.A. Boicelli, *J. Chem. Soc., Perkin Trans. 2*, 1974, 853.
286. D.G. Garratt and G.H. Schmid, *Can. J. Chem.*, 1974, **52**, 1027.
287. G.M. Bancroft, K.D. Butler, L.E. Manzer, A. Shaver and J.E.H. Ward, *Can. J. Chem.*, 1974, **52**, 782.
288. S. Sørensen, R.S. Hansen and H.J. Jakobsen, *J. Magn. Reson.*, 1974, **14**, 243.
289. S. Aime and R.K. Harris, *J. Magn. Reson.*, 1974, **13**, 236.
290. D.J. Thoennes, C.L. Wilkins and W.S. Trahanovsky, *J. Magn. Reson.*, 1974, **13**, 18.
291. L. Ernst, *Org. Magn. Reson.*, 1974, **6**, 540.
292. J.D. Odom, L.W. Hall and P.D. Ellis, *Org. Magn. Reson.*, 1974, **6**, 360.
293. D.E. Axelson, A.J. Oliver and C.E. Holloway, *Org. Magn. Reson.*, 1974, **6**, 64.
294. J. Evans and J.R. Norton, *Inorg. Chem.*, 1974, **13**, 3042.
295. G.M. Bodner, *Inorg. Chem.*, 1974, **13**, 2563.
296. A.R. Siedle, G.M. Bodner, A.R. Garber, D.C. Beer and L.J. Todd, *Inorg. Chem.*, 1974, **13**, 2321.
297. J.S. Miller, M.O. Visscher and K.G. Caulton, *Inorg. Chem.*, 1974, **13**, 1632.
298. J. Evans, B.F.G. Johnson, J. Lewis and D.J. Yarrow, *J. Chem. Soc., Dalton Trans.*, 1974, 2375.
299. J. Evans, B.F.G. Johnson, J. Lewis and R. Watt, *J. Chem. Soc., Dalton Trans.*, 1974, 2368.
300. G. Wiger, G. Albels and M.F. Rettig, *J. Chem. Soc., Dalton Trans.*, 1974, 2242.
301. L. Ruiz-Ramirez and T.A. Stephenson, *J. Chem. Soc., Dalton Trans.*, 1974, 1640.
302. W.J. Cherwinski, B.F.G. Johnson and J. Lewis, *J. Chem. Soc., Dalton Trans.*, 1974, 1405.
303. B.F.G. Johnson, J. Lewis, P. McArdle and J.R. Norton, *J. Chem. Soc., Dalton Trans.*, 1974, 1253.
304. T. Onak and E. Wan, *J. Chem. Soc., Dalton Trans.*, 1974, 665.

305. P.E. Cattermole, K.G. Orrell and A.G. Osborne, *J. Chem. Soc., Dalton Trans.*, 1974, 328.
306. V.G. Albano, P. Chini, S. Martinengo, D.J.A. McCaffrey and D. Strumolo, *J. Am. Chem. Soc.*, 1974, **96**, 8106.
307. F.A. Cotton, D.L. Hunter and P. Lahuerta, *J. Am. Chem. Soc.*, 1974, **96**, 7926.
308. C.D. Schaeffer and J.J. Zuckerman, *J. Am. Chem. Soc.*, 1974, **96**, 7160.
309. R.R. Schrock, *J. Am. Chem. Soc.*, 1974, **96**, 6796.
310. H. Schmidbaur, W. Buchner and F.H. Köhler, *J. Am. Chem. Soc.*, 1974, **96**, 6208.
311. M.J. Webb, M.J. Bennett, L.Y.Y. Chan and W.A.G. Graham, *J. Am. Chem. Soc.*, 1974, **96**, 5931.
312. G. Scholes, C.R. Graham and M. Brookhart, *J. Am. Chem. Soc.*, 1974, **96**, 5665.
313. L.J. Guggenberger, P. Meakin and F.N. Tebbe, *J. Am. Chem. Soc.*, 1974, **96**, 5420.
314. A. Sanders and W.P. Giering, *J. Am. Chem. Soc.*, 1974, **96**, 5247.
315. R.B. Wetzel and G.L. Kenyon, *J. Am. Chem. Soc.*, 1974, **96**, 5189.
316. J.E. Bercaw, *J. Am. Chem. Soc.*, 1974, **96**, 5087.
317. F.A. Cotton, B.A. Frenz and D.L. Hunter, *J. Am. Chem. Soc.*, 1974, **96**, 4820.
318. F.A. Cotton, D.L. Hunter and P. Lahuerta, *J. Am. Chem. Soc.*, 1974, **96**, 4723.
319. S.T. Wilson, N.J. Coville, J.R. Shapely and J.A. Osborn, *J. Am. Chem. Soc.*, 1974, **96**, 4038.
320. L. Kruczynski, L.K.K. LiShing-Man and J. Takats, *J. Am. Chem. Soc.*, 1974, **96**, 4006.
321. G.A. Olah, P. Schilling, P.W. Westerman and H.C. Lin, *J. Am. Chem. Soc.*, 1974, **96**, 3581.
322. A. Davison and S.S. Wreford, *J. Am. Chem. Soc.*, 1974, **96**, 3017.
323. C.A. Tolman, P.Z. Meakin, D.L. Lindner and J.P. Jesson, *J. Am. Chem. Soc.*, 1974, **96**, 2762.
324. B.G. McKinnie, N.S. Bhacca, F.K. Cartledge and J. Fayssoux, *J. Am. Chem. Soc.*, 1974, **96**, 2637.
325. D. Doddrell, I. Burfitt, W. Kitching, M. Bullpitt, C.-H. Lee, R.J. Mynott, J.L. Considine, H.G. Kuivila and R.H. Sarma, *J. Am. Chem. Soc.*, 1974, **96**, 1640.
326. L. Kruczynski and J. Takats, *J. Am. Chem. Soc.*, 1974, **96**, 932.
327. R.J. Wiersema and M.F. Hawthorne, *J. Am. Chem. Soc.*, 1974, **96**, 761.
328. J.E. Bercaw, E. Rosenberg and J.D. Roberts, *J. Am. Chem. Soc.*, 1974, **96**, 612.
329. D. Seyferth, G.H. Williams and D.D. Traficante, *J. Am. Chem. Soc.*, 1974, **96**, 604.
330. B.E. Mann, *Adv. Organomet. Chem.*, 1974, **12**, 135.

331. J. Ashley-Smith, D.V. Howe, B.F.G. Johnson, J. Lewis and I.E. Ryder, *J. Organomet. Chem.*, 1974, **82**, 257.
332. M.N. Andrews and P.E. Rakita, *J. Organomet. Chem.*, 1974, **82**, 29.
333. G. Van Koten and J.G. Noltes, *J. Organomet. Chem.*, 1974, **82**, C53.
334. C.J. Groenenboom and F. Jellinek, *J. Organomet. Chem.*, 1974, **80**, 229.
335. E.O. Fischer, H.J. Kalder and F.H. Köhler, *J. Organomet. Chem.*, 1974, **81**, C23.
336. L.J. Todd and J.R. Wilkinson, *J. Organomet. Chem.*, 1974, **80**, C31.
337. L. Kruczynski, J.L. Martin and J. Takats, *J. Organomet. Chem.*, 1974, **80**, C9.
338. J. Daub and J. Kappler, *J. Organomet. Chem.*, 1974, **80**, C5.
339. V.B. Shur, E.G. Berkovitch, L.B. Vasiljeva, R.V. Kudryavtsev and M.E. Vol'pin, *J. Organomet. Chem.*, 1974, **78**, 127.
340. J.D. Jamerson and J. Takats, *J. Organomet. Chem.*, 1974, **78**, C23.
341. R.D. Taylor and J.L. Wardell, *J. Organomet. Chem.*, 1974, **77**, 311.
342. T.J. Todd and J.R. Wilkinson, *J. Organomet. Chem.*, 1974, **77**, 1.
343. F.W.S. Benfield, N.J. Cooper and M.L.H. Green, *J. Organomet. Chem.*, 1974, **76**, 49.
344. V.I. Sokolov, P.V. Petrovskii, A.A. Koridze and O.A. Reutov, *J. Organomet. Chem.*, 1974, **76**, C15.
345. R.P. Roques, C. Segard, S. Combrisson and F. Wehrli, *J. Organomet. Chem.*, 1974, **73**, 327.
346. D.J. Peterson, M.D. Robbins and J.R. Hansen, *J. Organomet. Chem.*, 1974, **73**, 237.
347. A.G. Ginzburg, L.A. Fedorov, P.V. Petrovskii, E.I. Fedin, V.N. Setkina and D.N. Kursanov, *J. Organomet. Chem.*, 1974, **73**, 77.
348. O. Yamamoto, K. Hayamizu and M. Yanagisawa, *J. Organomet. Chem.*, 1974, **73**, 17.
349. B.E. Reichert, *J. Organomet. Chem.*, 1974, **72**, 305.
350. A.J. Hart, D.H. O'Brien and C.R. Russell, *J. Organomet. Chem.*, 1974, **72**, C19.
351. P.W. Hall, R.J. Puddephatt and C.F.H. Tipper, *J. Organomet. Chem.*, 1974, **71**, 145.
352. W. Kitching, D. Praeger, C.J. Moore, D. Doddrell and W. Adcock, *J. Organomet. Chem.*, 1974, **70**, 339.
353. C.T. Lam, C.V. Senoff and J.E.H. Ward, *J. Organomet. Chem.*, 1974, **70**, 273.
354. E. Ban, R.P. Hughes and J. Powell, *J. Organomet. Chem.*, 1974, **69**, 455.
355. A.N. Nesmeyanov, G.B. Shul'pin, L.A. Fedorov, P.V. Petrovskii and M.I. Rybinskaya, *J. Organomet. Chem.*, 1974, **69**, 429.

356. T.Yu. Orlova, P.V. Petrovskii, V.N. Setkina and D.N. Kursanov, *J. Organomet. Chem.*, 1974, **67**, C23.
357. D.J. Peterson and J.F. Ward, *J. Organomet. Chem.*, 1974, **66**, 209.
358. C.G. Kreiter, S. Stuber and L. Wackerle, *J. Organomet. Chem.*, 1974, **66**, C49.
359. W.J. Knebel, R.J. Angelici, O.A. Gansow and D.J. Darensbourg, *J. Organomet. Chem.*, 1974, **66**, C11.
360. Yu.N. Luzikov, N.M. Sergeyev and Yu.A. Ustynyuk, *J. Organomet. Chem.*, 1974, **65**, 303.
361. E.O. Fischer, G. Kreis, F.R. Kreissl, W. Kalbfus and E. Winkler, *J. Organomet. Chem.*, 1974, **65**, C53.
362. W. McFarlane and D.S. Rycroft, *J. Organomet. Chem.*, 1974, **64**, 303.
363. E.W. Randall, E. Rosenberg, L. Milone, R. Rosetti and P.L. Stanghellini, *J. Organomet. Chem.*, 1974, **64**, 271.
364. R.G. Salomon and J.K. Kochi, *J. Organomet. Chem.*, 1974, **64**, 135.
365. F.H. Köhler, *J. Organomet. Chem.*, 1974, **64**, C27.
366. H.H. Karsch, H.-F. Klein, C.G. Kreiter and H. Schmidbaur, *Chem. Ber.*, 1974, **107**, 3692.
367. H.H Karsch and H. Schmidbaur, *Chem. Ber.*, 1974, **107**, 3684.
368. E.O. Fischer, K. Weiss and C.G. Kreiter, *Chem. Ber.*, 1974, **107**, 3554.
369. K. Weiss, E.O. Fischer and J. Müller, *Chem. Ber.*, 1974, **107**, 3548.
370. F.H. Köhler, *Chem. Ber.*, 1974, **107**, 570.
371. F.R. Kreissl and E.O. Fischer, *Chem. Ber.*, 1974, **107**, 183.
372. A. Forster, B.F.G. Johnson, J. Lewis, T.W. Matheson, B.H. Robinson and W.G. Jackson, *J. Chem. Soc., Chem. Commun.*, 1974, 1042.
373. J.D. Kennedy and W. McFarlane, *J. Chem. Soc., Chem. Commun.*, 1974, 983.
374. R.-M. Lequan, M.-J. Pouet and M.-P. Simonnin, *J. Chem. Soc., Chem. Commun.*, 1974, 475; 804.
375. J.A.K. Howard, S.A.R. Knox, V. Riera, B.A. Sosinsky, F.G.A. Stone and P. Woodward, *J. Chem. Soc., Chem. Commun.*, 1974, 673.
376. N.A. Bailey, W.G. Kita, J.A. McCleverty, A.J. Murray, B.E. Mann and N.W.J. Walker, *J. Chem. Soc., Chem. Commun.*, 1974, 592.
377. P.M. Bailey, B.E. Mann, A. Segnitz, K.L. Kaiser and P.M. Maitlis, *J. Chem. Soc., Chem. Commun.*, 1974, 567.
378. G.A. Gray and S.E. Cremer, *J. Chem. Soc., Chem. Commun.*, 1974, 451; 1975, 304.
379. B.F.G. Johnson, J. Lewis and T.W. Matheson, *J. Chem. Soc., Chem. Commun.*, 1974, 441.
380. S. Aime, R.K. Harris, E.M. McVicker and M. Fild, *J. Chem. Soc., Chem. Commun.*, 1974, 426.

381. P.R. Branson, R.A. Cable, M. Green and M.K. Lloyd, *J. Chem. Soc., Chem. Commun.*, 1974, 364.
382. T. Yamamoto, A.R. Garber, J.R. Wilkinson, C.B. Boss, W.E. Streib and L.J. Todd, *J. Chem. Soc., Chem. Commun.*, 1974, 354.
383. P. Chini, S. Martinengo, D.J.A. McCaffrey and B.T. Heaton, *J. Chem. Soc., Chem. Commun.*, 1974, 310; 804.
384. T.H. Whitesides and R.A. Budnik, *J. Chem. Soc., Chem. Commun.*, 1974, 302.
385. T.E. Walker, H.P.C. Hogenkamp, T.E. Needham and N.A. Matwiyoff, *J. Chem. Soc., Chem. Commun.*, 1974, 85.
386. C.R. Bennett and D.C. Bradley, *J. Chem. Soc., Chem. Commun.*, 1974, 29.
387. F.A. Cotton, L. Kruczynski and A.J. White, *Inorg. Chem.*, 1974, **13**, 1402.
388. G.M. Bodner and L.J. Todd, *Inorg. Chem.*, 1974, **13**, 1335.
389. R.D. Adams, M.D. Brice and F.A. Cotton, *Inorg. Chem.*, 1974, **13**, 1080.
390. R.S. Evans, P.J. Hauser and A.F. Schreiner, *Inorg. Chem.*, 1974, **13**, 901.
391. W.J. Knebel and R.J. Angelici, *Inorg. Chem.*, 1974, **13**, 632.
392. W.J. Knebel and R.J. Angelici, *Inorg. Chem.*, 1974, **13**, 627.
393. L. Vancea and W.A.G. Graham, *Inorg. Chem.*, 1974, **13**, 511.
394. T.B. Brill and A.J. Kotlar, *Inorg. Chem.*, 1974, **13**, 470.
395. G.M. Bodner and L.J. Todd, *Inorg. Chem.*, 1974, **13**, 360.
396. K. Henrick and S.B. Wild, *J. Chem. Soc., Dalton Trans.*, 1974, 2500.
397. D.C. Harris, E. Rosenberg and J.D. Roberts, *J. Chem. Soc., Dalton Trans.*, 1974, 2398.
398. F. Fringuelli, S. Gronowitz, A.-B. Hörnfeldt, I. Johnson and A. Taticchi, *Acta Chem. Scand., Ser. B*, 1974, **28**, 175.
399. D. Doddrell, K.G. Lewis, C.E. Mulquiney, W. Adcock, W. Kitching and M. Bullpitt, *Aust. J. Chem.*, 1974, **27**, 417.
400. H.C. Clark, L.E. Manzer and J.E.H. Ward, *Can. J. Chem.*, 1974, **52**, 1973.
401. D.E. Axelson, S.A. Kandil and C.E. Holloway, *Can. J. Chem.*, 1974, **52**, 2968.
402. H.C. Clark and J.E.H. Ward, *Can. J. Chem.*, 1974, **52**, 570.
403. H.C. Clark, L.E. Manzer and J.E.H. Ward, *Can. J. Chem.*, 1974, **52**, 1165.
404. L.D. Quin, M.D. Gordon and S.O. Lee, *Org. Magn. Reson.*, 1974, **6**, 503.
405. W. Adcock, B.D. Gupta, W. Kitching, D. Doddrell and M. Geckle, *J. Am. Chem. Soc.*, 1974, **96**, 7360.
406. J. Kuyper and K. Vrieze, *J. Organomet. Chem.*, 1974, **74**, 289.
407. C.D. Schaeffer, Jr., J.J. Zuckerman and C.H. Yoder, *J. Organomet. Chem.*, 1974, **80**, 29.
408. L. Ernst, *J. Organomet. Chem.*, 1974, **82**, 319.
409. H.C. Clark and J.E.H. Ward, *J. Am. Chem. Soc.*, 1974, **96**, 1741.

410. A.N. Nesmeyanov, E.I. Fedin, L.A. Fedorov, L.S. Isaeva, L.N. Lorens and P.V. Petrovskii, *Dokl. Akad. Nauk SSSR*, 1974, **216**, 816.
411. W. Buchner and W. Wolfsberger, *Z. Naturforsch., Teil B*, 1974, **29**, 328.
412. R.B. Moon and J.H. Richards, *Biochemistry*, 1974, **13**, 3437.
413. E. Antonini, M. Brunori, F. Conti and G. Geraci, *FEBS Lett.*, 1973, **34**, 69.
414. P.J. Vergamini, N.A. Matwiyoff, R.C. Wohl and T. Bradley, *Biochem. Biophys. Res. Commun.*, 1973, **55**, 453.
415. D. Michel, *Surf. Sci.*, 1974, **42**, 453.
416. K. Hildenbrand and H. Dreeskamp, *Z. Naturforsch., Teil B*, 1973, **28**, 226.
417. R. von Ammon and B. Kanellakopulos, *Z. Naturforsch., Teil B*, 1973, **28**, 200.
418. C.J. Groenenboom, H.J. De Liefde Meijer and F. Jellinek, *Proc. Int. Conf. Coord. Chem., 16th, Dublin*, 1974, R2.
419. Yu.N. Luzikov, N.M. Sergeyev and Yu.A. Ustynyuk, *Vestn. Mosk. Univ., Khim.*, 1974, **15**, 109.
420. D. Deininger and D. Michel, *Wiss. Z. Karl-Marx-Univ. Leipzig, Math. - Naturwiss. Reihe*, 1973, **22**, 551.
421. Yu.K. Grishin, Yu.N. Luzikov and Yu.A. Ustynyuk, *Dokl. Akad. Nauk SSSR*, 1974, **216**, 321.
422. A.N. Nesmeyanov, G.B. Shul'pin, P.V. Petrovskii, V.I. Robas and M.I. Rybinskaya, *Dokl. Akad. Nauk SSSR*, 1974, **215**, 865.
423. A.N. Nesmeyanov, G.B. Shul'pin, M.I. Rybinskaya and P.V. Petrovskii, *Dokl. Akad. Nauk SSSR*, 1974, **215**, 599.
424. V.I. Sokolov, V.V. Bashilov, P.V. Petrovskii and O.A. Reutov, *Dokl. Akad. Nauk SSSR*, 1973, **213**, 1103.
425. T. Ibusuki and Y. Saito, *Chem. Lett.*, 1973, 1255.
426. T. Ibusuki and Y. Saito, *Chem. Lett.*, 1974, 311.
427. S.L. Ioffe, V.M. Shitkin, B.N. Khasapov, M.V. Kashutina, V.A. Tartakovskii, M. Magi and E. Lippmaa, *Izv. Akad. Nauk SSSR, Ser. Khim.*, 1973, 2146.
428. S.L. Ioffe, L.M. Leont'eva, A.L. Blyumenfel'd, O.P. Shitov and V.V. Tartakovskii, *Izv. Akad. Nauk SSSR, Ser. Khim.*, 1974, 1659.
429. U. Schubert and E.O. Fischer, *Justus Liebigs Ann. Chem.*, 1975, 393.
430. E.O. Fischer, U. Schubert, W. Kalbfus and C.G. Kreiter, *Z. Anorg. Allg. Chem.*, 1975, **416**, 135.
431. A.R. Siedle, *Inorg. Nucl. Chem. Lett.*, 1975, **11**, 345.
432. B.A. Howell and W.S. Trahanovsky, *J. Magn. Reson.*, 1975, **20**, 141.
433. G. Platbrood and L. Wilputte-Steinert, *J. Organomet. Chem.*, 1975, **85**, 199.
434. D. Seyferth and C.S. Eschbach, *J. Organomet. Chem.*, 1975, **94**, C5.
435. F.R. Kreissl, *J. Organomet. Chem.*, 1975, **99**, 305.

436. K.H. Dötz and C.G. Kreiter, *J. Organomet. Chem.*, 1975, **99**, 309.
437. E.O. Fischer, G. Huttner, W. Kleine and A. Frank, *Angew. Chem., Int. Ed. Engl.*, 1975, **14**, 760.
438. C.V. Senoff and J.E.H. Ward, *Inorg. Chem.*, 1975, **14**, 278.
439. B. Cetinkaya, P.B. Hitchcock, M.F. Lappert and P.L. Pye, *J. Chem. Soc., Chem. Commun.*, 1975, 683.
440. E.O. Fischer and K. Richter, *Angew. Chem., Int. Ed. Engl.*, 1975, **14**, 345.
441. R.B. King and M.S. Saran, *Inorg. Chem.*, 1975, **14**, 1018.
442. G.M. Bodner, *Inorg. Chem.*, 1975, **14**, 2694.
443. G.M. Bodner and M. Gaul, *J. Organomet. Chem.*, 1975, **101**, 63.
444. D.L. Beach and K.W. Barnett, *J. Organomet. Chem.*, 1975, **97**, C27.
445. E.E. Isaacs and W.A.G. Graham, *J. Organomet. Chem.*, 1975, **90**, 319.
446. E.M. Van Dam, W.N. Brent, M.P. Silvon and P.S. Skell, *J. Am. Chem. Soc.*, 1975, **97**, 465.
447. C.G. Kreiter, M. Land and H. Strack, *Chem. Ber.*, 1975, **108**, 1502.
448. W. Majunke, D. Leibfritz, T. Mack and H. Tom Dieck, *Chem. Ber.*, 1975, **108**, 3025.
449. F.H. Köhler, H.J. Kalder and E.O. Fischer, *J. Organomet. Chem.*, 1975, **85**, C19.
450. F.R. Kreissl and W. Held, *J. Organomet. Chem.*, 1975, **86**, C10.
451. T.N. Mitchell, *Org. Magn. Reson.*, 1975, **7**, 610.
452. R. Eujen and H. Bürger, *J. Organomet. Chem.*, 1975, **88**, 165.
453. G.S. Wikholm and L.J. Todd, *J. Organomet. Chem.*, 1974, **71**, 219.
454. J.D. Kennedy and W. McFarlane, *J. Organomet. Chem.*, 1975, **94**, 7; *ibid*; 1975, **99**, C40.
455. C.D. Schaeffer, Jr. and J.J. Zuckerman, *J. Organomet. Chem.*, 1975, **99**, 407.
456. T.N. Mitchell, *Org. Magn. Reson.*, 1975, **7**, 59.
457. M.E. Bishop, C.D. Schaeffer, Jr. and J.J. Zuckerman, *J. Organomet. Chem.*, 1975, **101**, C19.
458. G. Singh, *J. Organomet. Chem.*, 1975, **99**, 251.
459. W. Adcock, B.D. Gupta, W. Kitching and D. Doddrell, *J. Organomet. Chem.*, 1975, **102**, 297.
460. P.E. Rakita and R. Wright, *Inorg. Nucl. Chem. Lett.*, 1975, **11**, 47.
461. D.A. Dawson and W.F. Reynolds, *Can. J. Chem.*, 1975, **53**, 373.
462. R.K. Harris and B.J. Kimber, *J. Magn. Reson.*, 1975, **17**, 174.
463. D. Seyferth and D.C. Annarelli, *J. Am. Chem. Soc.*, 1975, **97**, 2273.
464. E.M. Dexheimer, L. Spialter and L.D. Smithson, *J. Organomet. Chem.*, 1975, **102**, 21.
465. J. Schraml, Nguyen-Duc-Chuy, V. Chvalovský, M. Mägi and E. Lippmaa, *Org. Magn. Reson.*, 1975, **7**, 379.

466. R.K. Harris and B.J. Kimber, *Org. Magn. Reson.*, 1975, **7**, 460.
467. G.A. Taylor and P.E. Rakita, *Org. Magn. Reson.*, 1974, **6**, 644.
468. Yu.A. Ustynyuk, Yu.N. Luzikov, V.I. Mstislavsky, A.A. Azizov and I.M. Pribytkova, *J. Organomet. Chem.*, 1975, **96**, 335.
469. R. Appel and K. Warning, *Chem. Ber.*, 1975, **108**, 1442.
470. A. Ouchi, T. Uehiro and Y. Yoshino, *J. Inorg. Nucl. Chem.*, 1975, **37**, 2347.
471. P. Krommes and J. Lorberth, *J. Organomet. Chem.*, 1975, **97**, 59.
472. Y. Yamamoto and H. Schmidbaur, *J. Chem. Soc., Chem. Commun.*, 1975, 668.
473. H. Schmidbaur and W. Richter, *Angew. Chem., Int. Ed. Engl.*, 1975, **14**, 183.
474. P. Krommes and J. Lorberth, *J. Organomet. Chem.*, 1975, **93**, 339.
475. W.Z.M. Rhee and J.J. Zuckerman, *J. Am. Chem. Soc.*, 1975, **97**, 2291.
476. W.H. Freeman, S.B. Miller and T.B. Brill, *J. Magn. Reson.*, 1975, **20**, 378.
477. H.U. Schwering, J. Weidlein and P. Fischer, *J. Organomet. Chem.*, 1975, **84**, 17.
478. H. Schmidbaur, H.-J. Füller and F.H. Köhler, *J. Organomet. Chem.*, 1975, **99**, 353.
479. W. Siebert, G. Augustin, R. Full, C. Krüger and Y.-H. Tsay, *Angew. Chem., Int. Ed. Engl.*, 1975, **14**, 262.
480. I. Fischler, K. Hildenbrand and E. Koerner von Gustorf, *Angew. Chem., Int. Ed. Engl.*, 1975, **14**, 54.
481. J. Schmetzer, J. Daub and P. Fischer, *Angew. Chem., Int. Ed. Engl.*, 1975, **14**, 487.
482. D.L. Reger, *Inorg. Chem.*, 1975, **14**, 660.
483. G.R. Langford, M. Akhtar, P.D. Ellis, A.G. MacDiarmid and J.D. Odom, *Inorg. Chem.*, 1975, **14**, 2937.
484. A. Cutler, D. Ehntholt, P. Lennon, K. Nicholas, D.F. Marten, M. Madhavarao, S. Raghu, A. Rosan and M. Rosenblum, *J. Am. Chem. Soc.*, 1975, **97**, 3149.
485. T. Yamamoto and L.J. Todd, *J. Organomet. Chem.*, 1974, **67**, 75.
486. R.B. King and C.A. Harmon, *J. Organomet. Chem.*, 1975, **86**, 239.
487. F. Sato, T. Yamamoto, J.R. Wilkinson and L.J. Todd, *J. Organomet. Chem.*, 1975, **86**, 243.
488. J.-Y. Lallemand, P. Laszlo, C. Muzette and A. Stockis, *J. Organomet. Chem.*, 1975, **91**, 71.
489. J.W. Faller and B.V. Johnson, *J. Organomet. Chem.*, 1975, **88**, 101.
490. J.W. Faller and B.V. Johnson, *J. Organomet. Chem.*, 1975, **96**, 99.
491. J.A.S. Howell, J. Lewis, T.W. Matheson and D.R. Russell, *J. Organomet. Chem.*, 1975, **99**, C55.
492. G. Olah and G. Liang, *J. Org. Chem.*, 1975, **40**, 1849.

493. S. Gronowitz, I. Johnson, A. Maholanyiova, S. Toma and E. Solcániová, *Org. Magn. Reson.*, 1975, **7**, 372.
494. S. Braun and W.E. Watts, *J. Organomet. Chem.*, 1975, **84**, C33.
495. A.N. Nesmeyanov, M.I. Rybinskaya, G.B. Shul'pin and A.A. Pogrebnyak, *J. Organomet. Chem.*, 1975, **92**, 341.
496. A.A. Koridze, P.V. Petrovskii, S.P. Gubin and E.I. Fedin, *J. Organomet. Chem.*, 1975, **93**, C26.
497. A.A. Koridze, P.V. Petrovskii, E.I. Fedin and A.I. Mokhov, *J. Organomet. Chem.*, 1975, **96**, C13.
498. F.H. Köhler and G.-E. Matsubayashi, *J. Organomet. Chem.*, 1975, **96**, 391.
499. S. Braun, T.S. Abram and W.E. Watts, *J. Organomet. Chem.*, 1975, **97**, 429.
500. R.G. Sutherland, S.C. Chen, J. Pannekoek and C.C. Lee, *J. Organomet. Chem.*, 1975, **101**, 221.
501. F.A. Cotton, D.L. Hunter and A.J. White, *Inorg. Chem.*, 1975, **14**, 703.
502. G. Deganello, P. Uguagliati, L. Calligaro, P.L. Sandrini and F. Zingales, *Inorg. Chim. Acta*, 1975, **13**, 247.
503. P. Batail, D. Grandjean, D. Astruc and R. Dabard, *J. Organomet. Chem.*, 1975, **102**, 79.
504. H.W. Spiess, R. Grosescu and U. Haeberlen, *Chem. Phys.*, 1974, **6**, 226.
505. L. Milone, S. Aime, E.W. Randall and E. Rosenberg, *J. Chem. Soc., Chem. Commun.*, 1975, 452.
506. J.A.S. Howell, T.W. Matheson and M.J. Mays, *J. Chem. Soc., Chem. Commun.*, 1975, 865.
507. F.A. Cotton, D.L. Hunter and P. Lahuerta, *Inorg. Chem.*, 1975, **14**, 511.
508. R.D. Adams, F.A. Cotton, W.R. Cullen, D.L. Hunter and L. Mihichuk, *Inorg. Chem.*, 1975, **14**, 1395.
509. F.A. Cotton, D.L. Hunter and P. Lahuerta, *J. Am. Chem. Soc.*, 1975, **97**, 1046.
510. F.A. Cotton and D.L. Hunter, *J. Am. Chem. Soc.*, 1975, **97**, 5739.
511. J.P. Hickey, J.R. Wilkinson and L.J. Todd, *J. Organomet. Chem.*, 1975, **99**, 281.
512. S. Sadeh and Y. Gaoni, *J. Organomet. Chem.*, 1975, **93**, C31.
513. B.F.G. Johnson, J. Lewis and J.W. Quail, *J. Chem. Soc., Dalton Trans.*, 1975, 1252.
514. M.J. Webb and W.A.G. Graham, *J. Organomet. Chem.*, 1975, **93**, 119.
515. R.B. King and K.C. Hodges, *J. Am. Chem. Soc.*, 1975, **97**, 2702.
516. S.A.R. Knox, B.A. Sosinsky and F.G.A. Stone, *J. Chem. Soc., Dalton Trans.*, 1975, 1647.
517. A. Brookes, S.A.R. Knox, V. Riera, B.A. Sosinsky and F.G.A. Stone, *J. Chem. Soc., Dalton Trans.*, 1975, 1641.
518. B.A. Sosinsky, S.A.R. Knox and F.G.A. Stone, *J. Chem. Soc., Dalton Trans.*, 1975, 1633.
519. I.D. Gay and J.F. Kriz, *J. Phys. Chem.*, 1975, **79**, 2145.

520. Y. Yamamoto and H. Schmidbaur, *J. Organomet. Chem.*, 1975, **96**, 133.
521. U. Dämmgen and H. Bürger, *J. Organomet. Chem.*, 1975, **101**, 307.
522. A. Zwijnenburg, H.O. Van Oven, C.J. Groenenboom and H.J. de Liefde Meijer, *J. Organomet. Chem.*, 1975, **94**, 23.
523. H. Schmidbaur, W. Richter, W. Wolf and F.H. Köhler, *Chem. Ber.*, 1975, **108**, 2649.
524 S.P. Anderson, H. Goldwhite, D. Ko, A. Letsou and F. Esparza, *J. Chem. Soc., Chem. Commun.*, 1975, 744.
525. H.F. Schröder and J. Müller, *Z. Anorg. Allg. Chem.*, 1975, **418**, 247.
526. K.I. The and R.G. Cavell, *J. Chem. Soc., Chem. Commun.*, 1975, 716.
527. O. Gasser and H. Schmidbaur, *J. Am. Chem. Soc.*, 1975, **97**, 6281.
528. S. Aime, O. Gambino, L. Milone, E. Sappa and E. Rosenberg, *Inorg. Chim. Acta*, 1975, **15**, 53.
529. J. Evans, B.F.G. Johnson, J. Lewis and T.W. Matheson, *J. Organomet. Chem.*, 1975, **97**, C16.
530. J.A. Segal and B.F.G. Johnson, *J. Chem. Soc., Dalton Trans.*, 1975, 677.
531. C.R. Eady, W.G. Jackson, B.F.G. Johnson, J. Lewis and T.W. Matheson, *J. Chem. Soc., Chem. Commun.*, 1975, 958.
532. J.A. Segal and B.F.G. Johnson, *J. Chem. Soc., Dalton Trans.*, 1975, 1990.
533. W.A. Hermann, B. Reiter and H. Biersack, *J. Organomet. Chem.*, 1975, **97**, 245.
534. K.J. Reimer and A. Shaver, *Inorg. Chem.*, 1975, **14**, 2707.
535. J.P. Williams and A. Wojcicki, *Inorg. Chim. Acta*, 1975, **15**, L19.
536. G.M. Bodner, *Inorg. Chem.*, 1975, **14**, 1932.
537. D. Seyferth, C.S. Eschbach and M.O. Nestle, *J. Organomet. Chem.*, 1975, **97**, C11.
538. T.W. Matheson and B.H. Robinson, *J. Organomet. Chem.*, 1975, **88**, 367.
539. D.L. Reger and A. Gabrielli, *J. Am. Chem. Soc.*, 1975, **97**, 4421.
540. M.A. Cohen, D.R. Kidd and T.L. Brown, *J. Am. Chem. Soc.*, 1975, **97**, 4408.
541. D.L. Rabenstein and M.T. Fairhurst, *J. Am. Chem. Soc.*, 1975, **97**, 2086.
542. W. Kitching, D. Praeger, D. Doddrell, F.A.L. Anet and J. Krane, *Tetrahedron Lett.*, 1975, 759.
543. C. Schumann, D. Dreeskamp and K. Hildenbrand, *J. Magn. Reson.*, 1975, **18**, 97.
544. G.A. Olah and S.H. Yu, *J. Org. Chem.*, 1975, **40**, 3638.
545. F.A. Cotton, D.L. Hunter and J.D. Jamerson, *Inorg. Chim. Acta*, 1975, **15**, 245.
546. Yu.S. Shabarov, L.D. Sychkova and S.G. Bandaev, *J. Organomet. Chem.*, 1975, **99**, 213.

547. J.D. Kennedy, W. McFarlane, G.S. Pyne and B. Wrackmeyer, *J. Chem. Soc., Dalton Trans.*, 1975, 386.
548. T. Onak and E. Wan, *J. Magn. Reson.*, 1974, **14**, 66.
549. L.W. Hall, D.W. Lowman, P.D. Ellis and J.D. Odom, *Inorg. Chem.*, 1975, **14**, 580.
550. V.I. Stanks, V.V. Khrapov and T.A. Babushkina, *Russ. Chem. Rev. (Engl. Transl.)*, 1974, **43**, 644.
551. S. Bywater, P. Lachance and D.J. Worsfold, *J. Phys. Chem.*, 1975, **79**, 2148.
552. A. Borg, T. Lindblom and R. Vestin, *Acta Chem. Scand., Ser. A*, 1975, **29**, 475.
553. J.F. Van Baar, K. Vrieze and D.J. Stufkens, *J. Organomet. Chem.*, 1975, **97**, 461.
554. J. Kuyper, P.I. Van Vliet and K. Vrieze, *J. Organomet. Chem.*, 1975, **96**, 289.
555. J.A.S. Howell, T.W. Matheson and M.J. Mays, *J. Organomet. Chem.*, 1975, **88**, 363.
556. L.J. Todd, J.R. Wilkinson, M.D. Rausch, S.A. Gardner and R.S. Dickson, *J. Organomet. Chem.*, 1975, **101**, 133.
557. S.W. Kaiser, R.B. Saillant and P.G. Rasmussen, *J. Am. Chem. Soc.*, 1975, **97**, 425.
558. J. Evans, B.F.G. Johnson, J. Lewis and T.W. Matheson, *J. Chem. Soc., Chem. Commun.*, 1975, 576.
559. B.T. Heaton, A.D.C. Towl, P. Chini, A. Fumagalli, D.J.A. McCaffrey and S. Martinengo, *J. Chem. Soc., Chem. Commun.*, 1975, 523.
560. N.W. Alcock, J.M. Brown, J.A. Conneely and J.J. Stofko, Jr., *J. Chem. Soc., Chem. Commun.*, 1975, 234.
561. G.M. Bodner and T.R. Engelmann, *J. Organomet. Chem.*, 1975, **88**, 391.
562. N.W. Alcock, J.M. Brown, J.A. Conneely and D.H. Williamson, *J. Chem. Soc., Chem. Commun.*, 1975, 792.
563. S.A. Dias, A.W. Downs and W.R. McWhinnie, *J. Chem. Soc., Dalton Trans.*, 1975, 162.
564. A.R. Garber, P.E. Garrou, G.E. Hartwell, M.J. Smas, J.R. Wilkinson and L.J. Todd, *J. Organomet. Chem.*, 1975, **86**, 219.
565. Nguyen-Duc-Chuy, V. Chvalovský, J. Schraml, M. Mägi and E. Lippmaa, *Collect. Czech. Chem. Commun.*, 1975, **40**, 875.
566. W. McFarlane, B. Wrackmeyer and H. Nöth, *Chem. Ber.*, 1975, **108**, 3831.
567. Y. Yamamoto and I. Moritani, *J. Org. Chem.*, 1975, **40**, 3434.
568. A.N. Nesmeyanov, L.A. Fedorov, N.P. Avakyan, P.V. Petrovskii, E.I. Fedin, E.V. Arshavskaya and I.I. Kritskaya, *J. Organomet. Chem.*, 1975, **101**, 121.
569. M.H. Chisholm, H.C. Clark, J.E.H. Ward and K. Yasufuku, *Inorg. Chem.*, 1975, **14**, 893.
570. W. Kuran and A. Musco, *Inorg. Chim. Acta*, 1975, **12**, 187.
571. J. Casanova, H.R. Rogers and K.L. Servis, *Org. Magn. Reson.*, 1975, **7**, 57.

572. C. Bied-Charreton, B. Septe and A. Gaudemer, *Org. Magn. Reson.*, 1975, **7**, 116.
573. M.H. Chisholm, H.C. Clark, L.E. Manzer, J.B. Stothers and J.E.H. Ward, *J. Am. Chem. Soc.*, 1975, **97**, 721.
574. D.L. Cronin, J.R. Wilkinson and L.J. Todd, *J. Magn. Reson.*, 1975, **17**, 353.
575. O. Yamamoto, *J. Chem. Phys.*, 1975, **63**, 2988.
576. T. Iwayanagi and Y. Saito, *Inorg. Nucl. Chem. Lett.*, 1975, **11**, 459.
577. M.A.M. Meester, D.J. Stufkens and K. Vrieze, *Inorg. Chim. Acta*, 1975, **15**, 137.
578. M. Green, J.A.K. Howard, J.L. Spencer and F.G.A. Stone, *J. Chem. Soc., Chem. Commun.*, 1975, 449.
579. W.J. Cherwinski, B.F.G. Johnson, J. Lewis and J.R. Norton, *J. Chem. Soc., Dalton Trans.*, 1975, 1156.
580. H.C. Clark, J.E.H. Ward and K. Yasufuku, *Can. J. Chem.*, 1975, **53**, 186.
581. R.A. Sheldon and J.A. van Doorn, *J. Organomet. Chem.*, 1975, **94**, 115.
582. H.D. Empsall, B.L. Shaw and A.J. Stringer, *J. Organomet. Chem.*, 1975, **94**, 131.
583. A. Sonoda, B.E. Mann and P.M. Maitlis, *J. Chem. Soc., Chem. Commun.*, 1975, 108.
584. A.D.H. Clague and C. Masters, *J. Chem. Soc., Dalton Trans.*, 1975, 858.
585. A.W. Verstuyft, L.W. Cary and J.H. Nelson, *Inorg. Chem.*, 1975, **14**, 1495.
586. D.A. Redfield, L.W. Cary and J.H. Nelson, *Inorg. Chem.*, 1975, **14**, 50.
587. A. Ducruix, H. Felkin, C. Pascard and G.K. Turner, *J. Chem. Soc., Chem. Commun.*, 1975, 615.
589. H. Minematsu, S. Takahashi and N. Hagihara, *J. Organomet. Chem.*, 1975, **91**, 389.
590. A. Sonoda, B.E. Mann and P.M. Maitlis, *J. Organomet. Chem.*, 1975, **96**, C16.
591. R.R. Schrock, *J. Am. Chem. Soc.*. 1975, **97**, 6577.
592. F.H. Köhler, *J. Organomet. Chem.*, 1975, **91**, 57.
593. F.A. Cotton, D.L. Hunter and P. Lahuerta, *J. Organomet. Chem.*, 1975, **87**, C42.
594. J. Evans, B.F.G. Johnson, J. Lewis and T.W. Matheson, *J. Am. Chem. Soc.*, 1975, **97**, 1245.
595. W.R. Jackson, C.F. Pincombe, I.D. Rae and S. Thapebin Karn, *Aust. J. Chem.*, 1975, **28**, 1535.
596. J.R. Shapley, S.I. Richter, M. Tachikawa and J.B. Keister, *J. Organomet. Chem.*, 1975, **94**, C43.
597. R.B. King and C.A. Harmon, *J. Organomet. Chem.*, 1975, **88**, 93.
598. H.G. Kuivila, J.E. Dixon, P.L. Maxfield, N.M Scarpa, T.M. Topka, K.-H. Tsai and K.R. Wursthorn, *J. Organomet. Chem.*, 1975, **86**, 89.

599. C. Couret, J. Escudie, J. Satge, N.T. Anh and G. Soussan, *J. Organomet. Chem.*, 1975, **91**, 11.
600. H. Dreeskamp and K. Hildenbrand, *Justus Liebigs Ann. Chem.*, 1975, 712.
601. R.T. Gray and D.N. Reinhoudt, *Tetrahedron Lett.*, 1975, 2109.
602. R.M. Atkins, R. Mackenzie, P.L. Timms and T.W. Turney, *J. Chem. Soc., Chem. Commun.*, 1975, 764.
603. C.A. Tolman, A.D. English and L.E. Manzer, *Inorg. Chem.*, 1975, **14**, 2353.
604. J.F. Van Baar, K. Vrieze and D.J. Stufkens, *J. Organomet. Chem.*, 1975, **85**, 249.
605. J. Müller, H.-O. Stühler and W. Goll, *Chem. Ber.*, 1975, **108**, 1074.
606. J. Müller, C.G. Kreiter, B. Mertschenk and S. Schmitt, *Chem. Ber.*, 1975, **108**, 273.
607. J.D. Edwards, R. Goddard, S.A.R. Knox, R.J. McKinney, F.G.A. Stone and P. Woodward, *J. Chem. Soc., Chem. Commun.*, 1975, 828.
608. R. Goddard, A.P. Humphries, S.A.R. Knox and P. Woodward, *J. Chem. Soc., Chem. Commun.*, 1975, 507.
609. K. Stanley and M.C. Baird, *J. Am. Chem. Soc.*, 1975, **97**, 6598.
610. H.J. Reich and J.E. Trend, *Can. J. Chem.*, 1975, **53**, 1922.
611. H.J. Kroth, H. Schumann, H.G. Kuivala, C.D. Schaeffer, Jr. and J.J. Zuckerman, *J. Am. Chem. Soc.*, 1975, **97**, 1754.
612. M.H. Chisholm and S. Godleski, *Prog. Inorg. Chem.*, 1975, **20**, 299.
613. F.A. Cotton, *J. Organomet. Chem.*, 1975, **100**, 29.
614. W. Kitching, C.J. Moore, D. Doddrell and W. Adcock, *J. Organomet. Chem.*, 1975, **94**, 469.
615. R.K. Sheline and H. Mahnke, *Angew. Chem., Int. Ed. Engl.*, 1975, **14**, 314.
616. P.S. Pregosin and R. Kunz, *Helv. Chim. Acta*, 1975, **58**, 423.
617. G. Balimann and P.S. Pregosin, *Helv. Chim. Acta*, 1975, **58**, 1913.
618. J. Kagan, W.-L. Lin, S.M. Cohen and R.N. Schwartz, *J. Organomet. Chem.*, 1975, **90**, 67.
619. M. Garreau, G.J. Martin, M.L. Martin, J. Morel and C. Paulmier, *Org. Magn. Reson.*, 1974, **6**, 648.
620. H.G. de Graaf and F. Bickelhaupt, *Tetrahedron*, 1975, **31**, 1097
621. W. Winter, *Angew. Chem., Int. Ed. Engl.*, 1975, **14**, 170.
622. R.K. Harris, E.M. McVicker and M. Fild, *J. Chem. Soc., Chem. Commun.*, 1975, 886.
623. A.H. Cowley, M. Cushner, M. Fild and J.A. Gibson, *Inorg. Chem.*, 1975, **14**, 1851.
624. J. Martin and J.B. Robert, *Org. Magn. Reson.*, 1975, **7**, 76.
625. P.S. Pregosin and L.M. Venanzi, *Helv. Chim. Acta*, 1975, **58**, 1548.
626. A. Salzer and H. Werner, *J. Organomet. Chem.*, 1975, **87**, 101.
627. R.B. King and J.C. Cloyd, Jr., *J. Chem. Soc., Perkin Trans. 2*, 1975, 938.

628. M.L. Martin, J. Tirouflet and B. Gautheron, *J. Organomet. Chem.*, 1975, **97**, 261.
629. Y. Yamamoto and I. Moritani, *Chem. Lett.*, 1975, 439.
630. Y. Yamamoto and I. Moritani, *Chem. Lett.*, 1975, 57.
631. J. Puskar, T. Saluvere, E. Lippmaa, A.B. Permin and V.S. Petrosyan, *Dokl. Akad. Nauk SSSR*, 1975, **220**, 112.
632. M.G. Edelev, T.M. Filippova, V.N. Robos, I.K. Shmyrev, A.S. Guseva, S.G. Verenikina and A.M. Yurkevich, *Zh. Obshch. Khim.*, 1974, **44**, 2321.
633. O. Yamamoto, *Chem. Lett.*, 1975, 511.
634. H. Schmidbaur and W. Wolf, *Chem. Ber.*, 1975, **108**, 2842.
635. L.J. Todd, J.P. Hickey, J.R. Wilkinson, J.C. Huffman and K. Folting, *J. Organomet. Chem.*, 1976, **112**, 167.
636. J.F. van Baar, J.M. Klerks, P. Overbosch, D.J. Stufkens and K. Vrieze, *J. Organomet. Chem.*, 1976, **112**, 95.
637. E.O. Fischer, T.L. Lindner and F.R. Kreissl, *J. Organomet. Chem.*, 1976, **112**, C27.
638. A.R.L. Bursico, M. Murray and F.G.A. Stone, *J. Organomet. Chem.*, 1976, **111**, 31.
639. D.G. Leppard, H.-J. Hansen, K. Bachmann and W.v. Philipsborn, *J. Organomet. Chem.*, 1976, **110**, 359.
640. R. Meij, J. Kuyper, D.J. Stufkens and K. Vrieze, *J. Organomet. Chem.*, 1976, **110**, 219.
641. P. Krommes and J. Lorberth, *J. Organomet. Chem.*, 1976, **110**, 195.
642. P.W. Clark, *J. Organomet. Chem.*, 1976, **110**, C13.
643. J.K. Stille and D.E. James, *J. Organomet. Chem.*, 1976, **108**, 401.
644. E.G. Bryan, W.G. Jackson, B.F.G. Johnson, J.W. Kelland, J. Lewis and K.T. Schorpp, *J. Organomet. Chem.*, 1976, **108**, 385.
645. J. Sandström and J. Seita, *J. Organomet. Chem.*, 1976, **108**, 371.
646. J. Kuyper, L.G. Hubert-Pfalzgraf, P.C. Keijzer and K. Vrieze, *J. Organomet. Chem.*, 1976, **108**, 271.
647. H. Lumbroso and D.M. Bertin, *J. Organomet. Chem.*, 1976, **108**, 111.
648. N.I. Pyshnograyeva, V.N. Setkina, G.A. Panosyan, P.V. Petrovskii, Yu.V. Makarov, N.E. Kolobova and D.N. Kursanov, *J. Organomet. Chem.*, 1976, **108**, 85.
649. E.M. Dexheimer and K. Spialter, *J. Organomet. Chem.*, 1976, **107**, 229.
650. F.A. Cotton and J.R. Kolb, *J. Organomet. Chem.*, 1976, **107**, 113.
651. A. Roubineau and J.C. Pommier, *J. Organomet. Chem.*, 1976, **107**, 63.
652. E.O. Fischer, W. Kleine and F.R. Kreissl, *J. Organomet. Chem.*, 1976, **107**. C23.
653. W. Kitching, D. Doddrell and J.B. Grutzner, *J. Organomet. Chem.*, 1976, **107**, C5.

654. E. Pfeiffer, J. Kuyper and K. Vrieze, *J. Organomet. Chem.*, 1976, **105**, 371.
655. U. Schroer and W.P. Neumann, *J. Organomet. Chem.*, 1976, **105**, 183.
656. J.H. Eekhof, H. Hogeveen, R.M. Kellogg and E.P. Schudde, *J. Organomet. Chem.*, 1976, **105**, C35.
657. W. Petz, *J. Organomet. Chem.*, 1976, **105**, C19.
658. A. Forster, B.F.G. Johnson, J. Lewis and T.W. Matheson, *J. Organomet. Chem.*, 1976, **104**, 225.
659. A. Sanders and W.P. Giering, *J. Organomet. Chem.*, 1976, **104**, 49.
660. M.A.M. Meester, H. van Dam, D.J. Stufkens and A. Oskam, *Inorg. Chim. Acta*, 1976, **20**, 155.
661. L. Busetto and A. Palazzi, *Inorg. Chim. Acta*, 1976, **19**, 233.
662. M.A.M. Meester, R.C.J. Vriends, D.J. Stufkens and K. Vrieze, *Inorg. Chim. Acta*, 1976, **19**, 95.
663. A. Salzer, *Inorg. Chim. Acta*, 1976, **18**, L31.
664. S. Aime, L. Milone and M. Valle, *Inorg. Chim. Acta*, 1976, **18**, 9.
665. S. Aime, L. Milòne and E. Sappa, *Inorg. Chim. Acta*, 1976, **16**, L7.
666. M.A.M. Meester, D.J. Stufkens and K. Vrieze, *Inorg. Chim. Acta*, 1976, **16**, 191.
667. L.E. Manzer and P.Z. Meakin, *Inorg. Chem.*, 1976, **15**, 3117.
668. R.E. Wasylishen, G.S. Birdi and A.F. Janzen, *Inorg. Chem.*, 1976, **15**, 3054.
669. B.A. Amero and E.P. Schram, *Inorg. Chem.*, 1976, **15**, 2842.
670. M.A. Sens, J.D. Odom and M.H. Goodrow, *Inorg. Chem.*, 1976, **15**, 2825.
671. F.A. Cotton and B.E. Hanson, *Inorg. Chem.*, 1976, **15**, 2806.
672. K.I. The and R.G. Cavell, *Inorg. Chem.*, 1976, **15**, 2518.
673. B.D. Dombek and R.J. Angelici, *Inorg. Chem.*, 1976, **15**, 2403.
674. B.D. Dombek and R.J. Angelici, *Inorg. Chem.*, 1976, **15**, 2397.
675. P.A. Dobosh, D.G. Gresham, C.P. Lillya and E.S. Magyar, *Inorg. Chem.*, 1976, **15**, 2311.
676. M.H. Chisholm, F.A. Cotton, M. Extine, M. Millan and B.R. Stults, *Inorg. Chem.*, 1976, **15**, 2244.
677. F.A. Cotton, P. Lahuerta and B.R. Stults, *Inorg. Chem.*, 1976, **15**, 1866.
678. A.D. English, J.P. Jesson and C.A. Tolman, *Inorg. Chem.*, 1976, **15**, 1730.
679. M. Brookhart, T.H. Whitesides and J.M. Crockett, *Inorg. Chem.*, 1976, **15**, 1550.
680. M.A. Cohen and T.L. Brown, *Inorg. Chem.*, 1976, **15**, 1417.
681. A.W. Verstuyft, D.A. Redfield, L.W. Cary and J.H. Nelson, *Inorg. Chem.*, 1976, **15**, 1128.
682. B.D. Dombek and R.J. Angelici, *Inorg. Chem.*, 1976, **15**, 1089.
683. Y. Souma, J. Iyoda and H. Sano, *Inorg. Chem.*, 1976, **15**, 968.
684. R.B. King and C.A. Harmon, *Inorg. Chem.*, 1976, **15**, 879.
685. T.H. Whitesides and R.A. Budnik, *Inorg. Chem.*, 1976, **15**, 874.

686. F.A. Cotton, D.L. Hunter, P. Lahuerta and A.J. White, *Inorg. Chem.*, 1976, **15**, 557.
687. A.J. Canty and A. Marker, *Inorg. Chem.*, 1976, **15**, 425.
688. T.A. George and C.D. Sterner, *Inorg. Chem.*, 1976, **15**, 165.
689. M.H. Quick and R.J. Angelici, *Inorg. Chem.*, 1976, **15**, 160.
690. S.S. Eaton and G.R. Eaton, *Inorg. Chem.*, 1976, **15**, 134.
691. F.A. Cotton, D.L. Hunter and J.M. Troup, *Inorg. Chem.*, 1976, **15**, 63.
692. D.J. Mabbott and P.M. Maitlis, *J. Chem. Soc., Dalton Trans.*, 1976, 2156.
693. P.L. Timms and T.W. Turney, *J. Chem. Soc., Dalton Trans.*, 1976, 2021.
694. R. Goddard, M. Green, R.P. Hughes and P. Woodward, *J. Chem. Soc., Dalton Trans.*, 1976, 1890.
695. S.C. Nyburg, K. Simpson and W. Wong-Ng, *J. Chem. Soc., Dalton Trans.*, 1976, 1865.
696. E. Sappa, L. Milone and A. Tiripicchio, *J. Chem. Soc., Dalton Trans.*, 1976, 1843.
697. J.D. Edwards, S.A.R. Knox and F.G.A. Stone, *J. Chem. Soc., Dalton Trans.*, 1976, 1813.
698. H.D. Empsall, B.L. Shaw and B.L. Turtle, *J. Chem. Soc., Dalton Trans.*, 1976, 1500.
699. J.A. Gibson and G.-V. Roschenthaler, *J. Chem. Soc., Dalton Trans.*, 1976, 1440.
700. B.F.G. Johnson, J. Lewis, B.E. Reichert and K.T. Schorpp, *J. Chem. Soc., Dalton Trans.*, 1976, 1403.
701. J.A. McCleverty, D. Seddon, N.A. Bailey and N.W.J. Walker, *J. Chem. Soc., Dalton Trans.*, 1976, 898.
702. M. Mickiewicz, K.P. Wainwright and S.B. Wild, *J. Chem. Soc., Dalton Trans.*, 1976, 262.
703. P.R. Branson, R.A. Cable, M. Green and M.K. Lloyd, *J. Chem. Soc., Dalton Trans.*, 1976, 12.
704. K.A. Ostoja Starzewski and H. Bock, *J. Am. Chem. Soc.*, 1976, **98**, 8486.
705. M.S. Brookhart, G.W. Koszalka, G.O. Nelson, G. Scholes and R.A. Watson, *J. Am. Chem. Soc.*, 1976, **98**, 8155.
706. A.J. Ashe, III and T.W. Smith, *J. Am. Chem. Soc.*, 1976, **98**, 7861.
707. N.L. Holy, N.C. Baenziger, R.M. Flynn and D.C. Swenson, *J. Am. Chem. Soc.*, 1976, **98**, 7823.
708. H. Sakurai, Y. Kamiyama and Y. Nakadaira, *J. Am. Chem. Soc.*, 1976, **98**, 7453.
709. R.J. Lawson and J.R. Shapley, *J. Am. Chem. Soc.*, 1976, **98**, 7433.
710. T.P. Fehlmer, J. Ragaini, M. Mangion and S.G. Shore, *J. Am. Chem. Soc.*, 1976, **98**, 7085.
711. D. Seyferth, D.C. Annarelli and S.C. Vick, *J. Am. Chem. Soc.*, 1976, **98**, 6382.
712. O.A. Gansow, A.R. Burke and W.D. Vernon, *J. Am. Chem. Soc.*, 1976, **98**, 5817.

713. B.F. Spielvogel, L. Wojnowich, M.K. Das, A.T. McPhail and K.D. Hargrave, *J. Am. Chem. Soc.*, 1976, **98**, 5702.
714. A.J. Ashe, III, R.R. Sharp and J.W. Tolan, *J. Am. Chem. Soc.* 1976, **98**, 5451.
715. V.G. Albano, P. Chini, G. Ciani, M. Sansoni, D. Strumolo, B.T. Heaton and S. Martinengo, *J. Am. Chem. Soc.*, 1976, **98**, 5027.
716. M.J. Bennett, J.L. Pratt, K.A. Simpson, L.K.K. Li Shing Man and J. Takats, *J. Am. Chem. Soc.*, 1976, **98**, 4810.
717. J.R. Anglin and W.A.G. Graham, *J. Am. Chem. Soc.*, 1976, **98**, 4678.
718. B.D. Dombek and R.J. Angelici, *J. Am. Chem. Soc.*, 1976, **98**, 4110.
719. R.T. Conlin and P.P. Gaspar, *J. Am. Chem. Soc.*, 1976, **98**, 3715.
720. G.K. Barker, M. Green, J.A.K. Howard, J.L. Spencer and F.G.A. Stone, *J. Am. Chem. Soc.*, 1976, **98**, 3373.
721. D.C. Dittmer, K. Takahashi, M. Iwanami, A.I. Tsai, P.L. Chang, B.B. Blidner and I.K. Stamos, *J. Am. Chem. Soc.*, 1976, **98**, 2795.
722. D.L. Reger and E.C. Culbertson, *J. Am. Chem. Soc.*, 1976, **98**, 2789.
723. A.J. Pribula and R.S. Drago, *J. Am. Chem. Soc.*, 1976, **98**, 2784.
724. W.T. Ford, *J. Am. Chem. Soc.*, 1976, **98**, 2727.
725. R.B. King and C.A. Harmon, *J. Am. Chem. Soc.*, 1976, **98**, 2409.
726. F.A. Cotton and D.L. Hunter, *J. Am. Chem. Soc.*, 1976, **98**, 1413.
727. L. Vancea, R.K. Pomeroy and W.A.G. Graham, *J. Am. Chem. Soc.*, 1976, **98**, 1407.
728. B.E. Mann and P.M. Maitlis, *J. Chem. Soc., Chem. Commun.*, 1976, 1058.
729. C.R. Eady, B.F.G. Johnson, J. Lewis, M.C. Malatesta, P. Machin and M. McPartlin, *J. Chem. Soc., Chem. Commun.*, 1976, 945.
730. M. McPartlin, C.R. Eady, B.F.G. Johnson and J. Lewis, *J. Chem. Soc., Chem. Commun.*, 1976, 883.
731. C.R. Eady, J.J. Guy, B.F.G. Johnson, J. Lewis, M.C. Malatesta and G. M. Sheldrick, *J. Chem. Soc., Chem. Commun.*, 1976, 807.
732. J.A. Segal, M.L.H. Green, J.-C. Daran and K. Prout, *J. Chem. Soc., Chem. Commun.*, 1976, 766.
733. M. Green, D.M. Grove, J.A.K. Howard, J.L. Spencer and F.G.A. Stone, *J. Chem. Soc., Chem. Commun.*, 1976, 759.
734. H.G. Raubenheimer, S. Lotz and J. Coetzer, *J. Chem. Soc., Chem. Commun.*, 1976, 732.
735. J.H. Eekhof, H. Hogeveen and R.M. Kellogg, *J. Chem. Soc., Chem. Commun.*, 1976, 657.
736. P.B. Hitchcock, M.F. Lappert and P.L. Pye, *J. Chem. Soc., Chem. Commun.*, 1976, 644.

737. R.J. Puddephat, M.A. Quyser and C.F.H. Tipper, *J. Chem. Soc., Chem. Commun.*, 1976, 626.
738. A.J. Birch and A.J. Pearson, *J. Chem. Soc., Chem. Commun.*, 1976, 601.
739. R. Goddard, S.A.R. Knox, F.G.A. Stone, M.J. Winter and P. Woodward, *J. Chem. Soc., Chem. Commun.*, 1976, 559.
740. P. McArdle and H. Sherlock, *J. Chem. Soc., Chem. Commun.*, 1976, 537.
741. T. Mitsudo, H. Nakanishi, T. Inubushi, I. Morishima, Y. Watanabe and Y. Takegami, *J. Chem. Soc., Chem. Commun.*, 1976, 416.
742. C. White, S.J. Thompson and P.M. Maitlis, *J. Chem. Soc., Chem. Commun.*, 1976, 409.
743. H.J. Reich and J.E. Trend, *J. Chem. Soc., Chem. Commun.*, 1976, 310.
744. C.R. Eady, B.F.G. Johnson and J. Lewis, *J. Chem. Soc., Chem. Commun.*, 1976, 302.
745. R.A. Cable, M. Green, R.E. Mackenzie, P.L. Timms and T.W. Turney, *J. Chem. Soc., Chem. Commun.*, 1976, 270.
746. E.G. Bryan, B.F.G. Johnson, J.W. Kelland, J. Lewis and M. McPartlin, *J. Chem. Soc., Chem. Commun.*, 1976, 254.
747. J.R. Blackborow, K. Hildenbrand, E.K. von Gustorf, A. Scrivanti, C.R. Eady and D. Ehntolt, *J. Chem. Soc., Chem. Commun.*, 1976, 16.
748. K. Jonas, R. Mynott, C. Krüger, J.C. Sekutowski and Y.-H. Tsay, *Angew. Chem., Int. Ed. Engl.*, 1976, **15**, 767.
749. H.G. Alt, *Angew. Chem., Int. Ed. Engl.*, 1976, **15**, 759.
750. R. Appel and M. Halstenburg, *Angew. Chem., Int. Ed. Engl.*, 1976, **15**, 696.
751. E.O. Fischer, M. Schulge and J.O. Besenhard, *Angew. Chem., Int. Ed. Engl.*, 1976, **15**, 683.
752. H. Berke, *Angew. Chem., Int. Ed. Engl.*, 1976, **15**, 624.
753. L. Rösch and H. Müller, *Angew. Chem., Int. Ed. Engl.*, 1976, **15**, 620.
754. A.D. Redhouse and W.A. Hermann, *Angew. Chem., Int. Ed. Engl.*, 1976, **15**, 615.
755. K. Eberl, F.H. Köhler and L. Mayring, *Angew. Chem., Int. Ed. Engl.*, 1976, **15**, 554.
756. E.O. Fischer, R.L. Clough, G. Besl and F.R. Kreissl, *Angew. Chem., Int. Ed. Engl.*, 1976, **15**, 543.
757. H. Schmidbaur and O. Gasser, *Angew. Chem., Int. Ed. Engl.*, 1976, **15**, 502.
758. H. Schmidbaur and H.-J. Füller, *Angew. Chem., Int. Ed. Engl.*, 1976, **15**, 501.
759. H. Schmidbaur and M. Heimann, *Angew. Chem., Int. Ed. Engl.*, 1976, **15**, 367.
760. R. Appel, F. Knoll and H. Veltmann, *Angew. Chem., Int. Ed. Engl.*, 1976, **15**, 315.
761. W. Winter, *Angew. Chem., Int. Ed. Engl.*, 1976, **15**, 241.

762. E.O. Fischer, G. Huttner, T.L. Lindner, A. Frank and F.R. Kreissl, *Angew. Chem., Int. Ed. Engl.*, 1976, **15**, 231.
763. L. Pope, P. Sommerville, M. Laing, K.J. Hindson and J.R. Moss, *J. Organomet. Chem.*, 1976, **112**, 309.
764. J. Kärger, D. Michel, A. Petzold, J. Caro, H. Pfeifer and R. Schöllner, *Z. Phys. Chem. (Leipzig)*, 1976, **257**, 1009.
765. W.-D. Hoffmann, *Z. Phys. Chem. (Leipzig)*, 1976, **257**, 817.
766. D.F. Kusharev, G.A. Kalabin, T.G. Mannafov, V.A. Mullin, M.F. Larin and V.A. Pestunovich, *Zh. Org. Khim.*, 1976, **12**, 1482; *J. Org. Chem. USSR (Engl. Transl.)*, 1976, **12**, 1465.
767. J. Schraml, J. Včelák, G. Engelhardt and V. Chvalovský, *Collect. Czech. Chem. Commun.*, 1976, **41**, 3758.
768. S. Heřmánek, V. Gregor, B. Štíbr, J. Plešek, Z. Janoušek and V.A. Antonovich, *Collect. Czech. Chem. Commun.*, 1976, **41**, 1492.
769. J. Pola and V. Chvalovský, *Collect. Czech. Chem. Commun.*, 1976, **41**, 581.
770. J. Schraml, J. Pola, H. Jancke, G. Engelhardt, M. Cerný and V. Chvalovský, *Collect. Czech. Chem. Commun.*, 1976, **41**, 360.
771. J. Pola, Z. Papoušková and V. Chvalovský, *Collect. Czech. Chem. Comm.*, 1976, **41**, 239.
772. J. Souček, G. Engelhardt, K. Stránský and J. Schraml, *Collect. Czech. Chem. Commun.*, 1976, **41**, 234.
773. D. Michel, W. Meiler and H. Pfeifer, *J. Mol. Catal.*, 1976, **1**, 85.
774. B.R. Gragg and K. Niedenzu, *Synth. React. Inorg. Metal-Org. Chem.*, 1976, **6**, 275.
775. L.I. Zakharkin, V.N. Kalinin and N.I. Kobel'kova, *Synth. React. Inorg. Metal-Org. Chem.*, 1976, **6**, 65.
776. D.L. Reger and E.C. Culbertson, *Synth. React. Inorg. Metal-Org. Chem.*, 1976, **6**, 1.
777. J. Kuyper and K. Vrieze, *Transition Met. Chem.*, 1976, **1**, 208.
778. S. Aime, G. Gervasio, L. Milone and E. Rosenberg, *Transition Met. Chem.*, 1976, **1**, 177.
779. M. Herberhold and H. Alt, *Justus Liebigs Ann. Chem.*, 1976, 292.
780. J. Martin, J.B. Robert and C. Taieb, *J. Phys. Chem.*, 1976, **80**, 2417.
781. M. Evers, R. Weber, Ph. Thibaut, L. Christiaens, M. Renson, A. Croisy and P. Jacquignon, *J. Chem. Soc., Perkin Trans. 1*, 1976, 2452.
782. P. Frøyen and D.G. Morris, *Acata Chem. Scand., Ser. B*, 1976, **30**, 790.
783. F. Fringuelli, S. Gronowitz, A.-B. Hörnfeldt, I. Johnson and A. Taticchi, *Acta Chem. Scand., Ser. B*, 1976, **30**, 605.
784. P. Frøyen and D.G. Morris, *Acta Chem. Scand., Ser. B*, 1976, **30**, 435.
785. G. Olah and G. Liang, *J. Org. Chem.*, 1976, **41**, 2659.
786. R.B. King, J.C. Cloyd, Jr. and R.H. Reimann, *J. Org. Chem.*, 1976, **41**, 972.

787. G.A. Olah and S.H. Yu, *J. Org. Chem.*, 1976, **41**, 717.
788. A.E. Lemire and J.C. Thompson, *Can. J. Chem.*, 1975, **53**, 3732.
789. J.E. Drake, C. Riddle and L. Coatsworth, *Can. J. Chem.*, 1975, **53**, 3602.
790. H.C. Clark and A. Shaver, *Can. J. Chem.*, 1976, **54**, 2068.
791. T. Yamamoto, Y. Nakamura and A. Yamamoto, *Bull. Chem. Soc. Jpn.*, 1976, **49**, 191.
792. H. Alper, *Org. Magn. Reson.*, 1976, **8**, 587.
793. J.Y. Lallemand and M. Duteil, *Org. Magn. Reson.*, 1976, **8**, 328.
794. P.E. Rakita, L.S. Worsham and J.P. Srebro, *Org. Magn. Reson.*, 1976, **8**, 310.
795. D. Knol, N.J. Koole and M.J.A. de Bie, *Org. Magn. Reson.*, 1976, **8**, 213.
796. B.L. Shapiro and T.W. Proulx, *Org. Magn. Reson.*, 1976, **8**, 40.
797. M.T.W. Hearn, *Aust. J. Chem.*, 1976, **29**, 2315.
798. R.S. Dickson and S.H. Johnson, *Aust. J. Chem.*, 1976, **29**, 2189.
799. J.E. Fergusson and C.T. Page, *Aust. J. Chem.*, 1976, **29**, 2159.
800. A.T.T. Hsieh, C.A. Rogers and B.O. West, *Aust. J. Chem.*, 1976, **29**, 49.
801. Y. Kashman and A. Rudi, *Tetrahedron Lett.*, 1976, 2819.
802. J. Elzinga and H. Hogeveen, *Tetrahedron Lett.*, 1976, 2383.
803. T.N. Mitchell and B. Kleine, *Tetrahedron Lett.*, 1976, 2173.
804. C. Symmes, Jr. and L.D. Quin, *Tetrahedron Lett.*, 1976, 1853.
805. D. Astruc, R. Dabard, M. Martin, P. Batail and D. Grandjean, *Tetrahedron Lett.*, 1976, 829.
806. G. Buono and J.R. Llinas, *Tetrahedron Lett.*, 1976, 749.
807. F.B. Iteke, L. Christiaens and M. Renson, *Tetrahedron*, 1976, **32**, 689.
808. H. Schmidbaur and P. Holl, *Chem. Ber.*, 1976, **109**, 3151.
809. P. Eilbracht, *Chem. Ber.*, 1976, **109**, 3136.
810. K.H. Dötz and C.G. Kreiter, *Chem. Ber.*, 1976, **109**, 2026.
811. K. Weiss and E.O. Fischer, *Chem. Ber.*, 1976, **109**, 1868.
812. H. tom Dieck and M. Svoboda, *Chem. Ber.*, 1976, **109**, 1657.
813. H. Meier, M. Layer, W. Combrink and S. Schniepp, *Chem. Ber.*, 1976, **109**, 1650.
814. J. Müller, W. Holzinger and F.H. Köhler, *Chem. Ber.*, 1976, **109**, 1222.
815. J. Müller, H.O. Stühler, G. Huttner and K. Scherzer, *Chem. Ber.*, 1976, **109**, 1211.
816. R. Aumann and J. Knecht, *Chem. Ber.*, 1976, **109**, 174.
817. J.C. Dewan and J. Silver, *Inorg. Nucl. Chem. Lett.*, 1976, **12**, 647.
818. D.C. van Beeleb, D. de Vos, G.J.M. Bots, L.J. van Doom and J. Wolters, *Inorg. Nucl. Chem. Lett.*, 1976, **12**, 581.
819. T.A. Gerken and W.M. Ritchey, *J. Magn. Reson.*, 1976, **24**, 155.
820. R.K. Harris and B. Lemarié, *J. Magn. Reson.*, 1976, **23**, 371.
821. M.T.W. Hearn, *J. Magn. Reson.*, 1976, **22**, 521.

822. D. Seyferth, R.L. Lambert, Jr. and D.C. Annarelli, *J. Organomet. Chem.*, 1976, **122**, 311.
823. E.G. Bryan, B.F.G. Johnson and J. Lewis, *J. Organomet. Chem.*, 1976, **122**, 249.
824. P.W. Clark and A.J. Jones, *J. Organomet. Chem.*, 1976, **122**, C41.
825. J.E. Sheats, E.J. Sabol, Jr., D.Z. Denney and N. El. Murr, *J. Organomet. Chem.*, 1976, **121**, 73.
826. W.I. Bailey, F.A. Cotton, J.D. Jamerson and J.R. Kolb, *J. Organomet. Chem.*, 1976, **121**, C23.
827. M.A. Bennett, T.W. Matheson, G.B. Robertson, A.K. Smith and P.A. Tucker, *J. Organomet. Chem.*, 1976, **121**, C18.
828. B.F.G. Johnson, J. Lewis and S.R. Postle, *J. Organomet. Chem.*, 1976, **121**, C7.
829. A.R. Siedle, *J. Organomet. Chem.*, 1976, **120**, 369.
830. H.K. Hofstee, J. Boersma and G.J.M. Van der Kerk, *J. Organomet. Chem.*, 1976, **120**, 313.
831. M.J. Buchanan, R.H. Cragg and A. Steltner, *J. Organomet. Chem.*, 1976, **120**, 189.
832. E.O. Fischer, R.L. Clough and P. Stückler, *J. Organomet. Chem.*, 1976, **120**, C6.
833. A.G. Davies, B. Muggleton, B.P. Roberts, M.-W. Tse and J.N. Winter, *J. Organomet. Chem.*, 1976, **118**, 289.
834. J.R. Wilkinson and L.J. Todd, *J. Organomet. Chem.*, 1976, **118**, 199.
835. M. Ishikawa, T. Fuchikami and M. Kumada, *J. Organomet. Chem.*, 1976, **118**, 139.
836. P.J. Burke, R.W. Matthews and D.G. Gillies, *J. Organomet. Chem.*, 1976, **118**, 129.
837. G.L. Kuykendall and J.L. Mills, *J. Organomet. Chem.*, 1976, **118**, 123.
838. T. Ikariya, Y. Nakamura and A. Yamamoto, *J. Organomet. Chem.*, 1976, **118**, 101.
839. C.C. Lee, S.C. Chen, W.J. Pannekoek and R.G. Sutherland, *J. Organomet. Chem.*, 1976, **118**, C47.
840. E.O. Fischer, K. Scherzer and F.R. Kreissl, *J. Organomet. Chem.*, 1976, **118**, C33.
841. R.G. Sutherland, S.C. Chen, W.J. Pannekoek and C.C. Lee, *J. Organomet. Chem.*, 1976, **117**, 61.
842. L.K.K. Li Shing Man and J. Takats, *J. Organomet. Chem.*, 1976, **117**, C104.
843. P. Laszlo and A. Stockis, *J. Organomet. Chem.*, 1976, **117**, C41.
844. W. McFarlane, *J. Organomet. Chem.*, 1976, **116**, 315.
845. E.S. Magyar and C.P. Lillya, *J. Organomet. Chem.*, 1976, **116**, 99.
846. P. McArdle and H. Sherlock, *J. Organomet. Chem.*, 1976, **116**, C23.
847. F.R. Young, tert, D.H. O'Brien, R.C. Pettersen, R.A. Levenson and D.L. von Minden, *J. Organomet. Chem.*, 1976, **114**, 157.

848. H. Schaper and H. Behrens, *J. Organomet. Chem.*, 1976, **113**, 361.
849. C. Potvin, J.M. Manoli, G. Pannetier, R. Chevalier and N. Platzer, *J. Organomet. Chem.*, 1976, **113**, 273.
850. D.L. Reger and M.D. Dukes, *J. Organomet. Chem.*, 1976, **113**, 173.
851. F.R. Kreissl, W. Uedelhoven and A. Ruhs, *J. Organomet. Chem.*, 1976, **113**, C55.
852. H. Goldwhite, J. Grey and R. Teller, *J. Organomet. Chem.*, 1976, **113**, C1.
853. T. Bottin-Strzalko, J. Seyden-Penne and M.-P. Simonnin, *J. Chem. Soc., Chem. Commun.*, 1976, 905.
854. J.B. Lambert, D.A. Netzel, H.-N. Sun and K.K. Lilianstrom, *J. Am. Chem. Soc.*, 1976, **98**, 3778.
855. J.B. Lambert and D.A. Netzel, *J. Am. Chem. Soc.*, 1976, **98**, 3783.
856. R.P. Stewart, Jr., L.R. Isbrandt, J.J. Benedict and J.G. Palmer, *J. Am. Chem. Soc.*, 1976, **98**, 3215.
857. D.R. Coulson, *J. Am. Chem. Soc.*, 1976, **98**, 3111.
858. G.A. Gray, S.E. Cremer and K.L. Marsi, *J. Am. Chem. Soc.*, 1976, **98**, 2109.
859. S. Aime, L. Milone and E. Sappa, *J. Chem. Soc., Dalton Trans.*, 1976, 838.
860. T.A. Albright, M.D. Gordon, W.J. Freeman and E.E. Schweizer, *J. Am. Chem. Soc.*, 1976, **98**, 6249.
861. A.J. Brown, O.W. Haworth and P. Moore, *J. Chem. Soc., Dalton Trans.*, 1976, 1589.
862. M. Green and R.P. Hughes, *J. Chem. Soc., Dalton Trans.*, 1976, 1880.
863. S. Aime, R.K. Harris, E.M. McVicker and M. Fild, *J. Chem. Soc., Dalton Trans.*, 1976, 2144.
864. A.W. Verstuyft, L.W. Cary and J.H. Nelson, *Inorg. Chem.*, 1976, **15**, 3161.
865. L. Kruczynski and J. Takats, *Inorg. Chem.*, 1976, **15**, 3140.
866. L.E. Manzer, *Inorg. Chem.*, 1976, **15**, 2354.
867. D.G. Cooper and J. Powell, *Inorg. Chem.*, 1976, **15**, 1959.
868. S. Aime, L. Milone, D. Osella, M. Valle and E.W. Randall, *Inorg. Chim. Acta*, 1976, **20**, 217.
869. C. Eschbach, D. Seyferth and P.C. Reeves, *J. Organomet. Chem.*, **104**, 363.
870. D. de Vos, *J. Organomet. Chem.*, 1976, **104**, 193.
871. R. Bonnaire and N. Platzer, *J. Organomet. Chem.*, 1976, **104**, 107.
872. P.E. Rakita, J.P. Srebro and L.S. Worsham, *J. Organomet. Chem.*, 1976, **104**, 27.
873. D.A. Brown, N.J. Fitzpatrick, I.J. King and N.J. Mathews, *J. Organomet. Chem.*, 1976, **104**, C9.
874. D. Leibfritz and H. tom Dieck, *J. Organomet. Chem.*, 1976, **105**, 255.
875. H.G. Kuivila, J.L. Considine, R.H. Sarma and R.J. Mynott, *J. Organomet. Chem.*, 1976, **111**, 179.
876. F.H. Köhler, *J. Organomet. Chem.*, 1976, **110**, 235.
877. P.S. Nielsen, R.S. Hansen and H.J. Jakobsen, *J. Organomet. Chem.*, 1976, **114**, 145.

878. N.M. Loim, P.V. Petrovskii, V.I. Robas, Z.N. Parnes and D.N. Kursanov, *J. Organomet. Chem.*, 1976, **117**, 265.
879. M. Bullpitt, W. Kitching, W. Adcock and D. Doddrell, *J. Organomet. Chem.*, 1976, **116**, 187.
880. M. Bullpitt, W. Kitching, W. Adcock and D. Doddrell, *J. Organomet. Chem.*, 1976, **116**, 161.
881. W. Beck, H. Brix and F.H. Köhler, *J. Organomet. Chem.*, 1976, **121**, 211.
882. T.N. Mitchell and G. Walter, *J. Organomet. Chem.*, 1976, **121**, 177
883. F.H. Köhler, *J. Organomet. Chem.*, 1976, **121**, C61.
884. J. Schraml, V. Chvalovsky, M. Mägi, E. Lippmaa, R. Calas, J. Dunogues and P. Bourgeois, *J. Organomet. Chem.*, 1976, **120**, 41.
885. T. Renk, W. Ruf and W. Siebert, *J. Organomet. Chem.*, 1976, **120**, 1.
886. V. Graves and J.J. Lagowski, *Inorg. Chem.*, 1976, **15**, 577.
887. R. Rottler, C.G. Kreier and G. Fink, *Z. Naturforsch., Teil B*, 1976, **31**, 730.
888. F.H. Köhler, *Z. Naturforsch., Teil B*, 1976, **31**, 1151.
889. F.H. Kohler and G. Matsubayashi, *Z. Naturforsch., Teil B*, 1976, **31**, 1153.
890. E.G. Hoffmann, P.W. Jolly, A. Kuesters, R. Mynott and G. Wilke, *Z. Naturforsch., Teil B*, 1976, **31**, 1712.
891. E.O. Fischer and K. Richter, *Chem. Ber.*, 1976, **109**, 3079.
892. E.O. Fischer and K. Richter, *Chem. Ber.*, 1976, **109**, 2547.
893. E.O. Fischer and G. Kreis, *Chem. Ber.*, 1976, **109**, 1673.
894. E.O. Fischer and K. Richter, *Chem. Ber.*, 1976, **109**, 1140.
895. A.L. Galyer and G. Wilkinson, *J. Chem. Soc., Dalton Trans.*, 1976, 2235.
896. E.O. Fischer, W. Kleine and F.R. Kreissl, *Angew. Chem., Int. Ed. Engl.*, 1976, **15**, 616.
897. R.A. Andersen, A.L. Galyer and G. Wilkinson, *Angew. Chem., Int. Ed. Engl.*, 1976, **15**, 609.
898. V.I. Zakharov, Yu.V. Belov, Yu.L. Kleiman, N.V. Morkovin and B.I. Ionin, *Zh. Obshch. Khim.*, 1976, **46**, 1415; *J. Gen. Chem. USSR (Engl. Transl.)*, 1976, **46**, 1391.
900. T. Pehk, E. Lippmaa, E. Lukevics and L.I. Simchenko, *Zh. Obshch. Khim.*, 1976, **46**, 602; *J. Gen. Chem. USSR (Engl. Transl.)*, 1976, **46**, 600.
901. P.S. Braterman, 'Metal Carbonyl Spectra', in 'Organometallic Chemistry: A Series of Monographs', Ed. P.M. Maitlis, F.G.A. Stone and R. West, Academic Press, New York, 1974.
902. H.O. House, A.V. Prabhu and W.V. Phillips, *J. Org. Chem.*, 1976, **41**, 1209.
903. J.M. Manriquez, D.R. McAlister, R.D. Sanner and J.E. Bercaw, *J. Am. Chem. Soc.*, 1976, **98**, 6733.
904. K. Richter, E.O. Fischer and C.G. Kreiter, *J. Organomet. Chem.*, 1976, **122**, 187.
905. R. Lett, G. Chassaing and A. Marquet, *J. Organomet. Chem.*, 1976, **111**, C17.

906. F.R. Kreissl and P. Stückler, *J. Organomet. Chem.*, 1976, **110**, C9.
907. E.O. Fischer, H. Hollfelder, F.R. Kreissl and W. Uedelhoven, *J. Organomet. Chem.*, 1976, **113**, C31.
908. F.R. Kreissl, E.W. Meineke, E.O. Fischer and J.D. Roberts, *J. Organomet. Chem.*, 1976, **108**, C29.
909. J.F. Sebastian, B. Hsu and J.R. Grunwell, *J. Organomet. Chem.* 1976, **105**, 1.
910. D.G. Cooper, S.C. Nyburg, J. Powell and K.A. Simpson, *Proc. Int. Conf. Coord. Chem., 16th, Dublin*, 1974, R81.
911. J. Holton, M.F. Lappert, G.R. Scollary, D.G.H. Ballard, R. Pearce, J.L. Atwood and W.E. Hunter, *J. Chem. Soc., Chem. Commun.*, 1976, 425.
912. I.D. Gay and S. Liang, *J. Catal.*, 1976, **44**, 306.
913. K.W. Bagnall, J. Edwards and F. Heatley, *Transplutonium 1975, Proc. Int. Transplutonium Elem. Symp., 4th*, Ed. W. Mueller and R. Lindner, North-Holland, Amsterdam, 1976, p.119.
914. R.K. Harris and B.J. Kimber, *Adv. Mol. Relaxation Processes*, 1976, **8**, 23.
915. G. Barbieri, *Atti Soc. Nat. Mat. Modena*, 1974, **105**, 127.
916. Y. Kawasaki, M. Aritomi and J. Iyoda, *Bull. Chem. Soc. Jpn.*, 1976, **49**, 3478.
917. K.D. Summerhays and D.A. Deprez, *J. Organomet. Chem.*, 1976, **118**, 19.
918. K. Bachmann and W. von Philipsborn, *Org. Magn. Reson.*, 1976, **8**, 648.
919. T.A. Albright, *Org. Magn. Reson.*, 1976, **8**, 489.
920. J. Tirouflet, J. Besancon, F. Mabon and M.L. Martin, *Org. Magn. Reson.*, 1976, **8**, 444.
921. T.N. Mitchell, *Org. Magn. Reson.*, 1976, **8**, 34.
922. W. Kitching, M. Marriott, W. Adcock and D. Doddrell, *J. Org. Chem.*, 1976, **41**, 1671.
923. T.A. Albright, W.J. Freeman and E.E. Schweizer, *J. Org. Chem.*, 1976, **41**, 2716.
924. G.A. Olah, S.H. Yu and G. Liang, *J. Org. Chem.*, 1976, **41**, 2383.
925. G.A. Olah, G. Liang and S.H. Yu, *J. Org. Chem.*, 1976, **41**, 2227.
926. G.A. Olah, S.H. Yu and D.G. Parker, *J. Org. Chem.*, 1976, **41**, 1983.
927. G.A. Olah and S.H. Yu, *J. Org. Chem.*, 1976, **41**, 1695.
928. M.D. Gordon and L.D. Quin, *J. Org. Chem.*, 1976, **41**, 1690.
929. P.A.W. Dean and D.G. Ibbott, *Can. J. Chem.*, 1976, **54**, 177.
930. G.W. Buchanan and C. Benezra, *Can. J. Chem.*, 1976, **54**, 231.
931. P.A.W. Dean, D.G. Ibbott and J.B. Stothers, *Can. J. Chem.*, 1976, **54**, 166.
932. A.N. Nesmeyanov, E.I. Fedin, P.V. Petrovskii, V.A. Dubovitskii, O.V. Nogina and N.S. Kochetkova, *Zh. Strukt. Khim.*, 1975, **16**, 759; *J. Struct. Chem. (Engl. Transl.)*, 1975, **16**, 705.

933. F.H. Köhler, H.J. Kalder and E.O Fischer, *J. Organomet. Chem.*, 1976, **113**, 11.
934. P. Fischer, J. Stadelhofer and J. Weidlein, *J. Organomet. Chem.*, 1976, **116**, 65.
935. E.O. Fischer, P. Stücker, H.-J. Beck and F.R. Kreissl, *Chem. Ber.*, 1976, **109**, 3089.
936. E. Fukushima and S.D. Larsen, *Chem. Phys. Lett.*, 1976, **44**, 285.
937. Y. Senda, H. Suda, J. Ishiyama and S. Imaizumi, *Tetrahedron Lett.*, 1976, 1983.
938. W.R. Jackson, C.F. Pincombe, I.D. Rae, D. Rash and B. Wilkinson, *Aust. J. Chem.*, 1976, **29**, 2431.
939. A.J. Pearson, *Aust. J. Chem.*, 1976, **29**, 1679.
940. A.J. Birch, P.W. Westerman and A.J. Pearson, *Aust. J. Chem.*, 1976, **29**, 1671.
941. E.O. Fischer and K. Weiss, *Chem. Ber.*, 1976, **109**, 1128.
942. G. Balimann and P.S. Pregosin, *J. Magn. Reson.*, 1976, **22**, 235.
943. N.K. Wilson, R.D. Zehr and P.D. Ellis, *J. Magn. Reson.*, 1976, **21**, 437.
944. L.J. Todd, A.R. Siedle, G.M. Bodner, S.B. Kahl and J.P. Hickey, *J. Magn. Reson.*, 1976, **23**, 301.
945. R. Victor, V. Usieli and S. Sarel, *J. Organomet. Chem.*, 1977, **129**, 387.
946. E.O. Fischer, P. Stückler and F.R. Kreissl, *J. Organomet. Chem.*, 1977, **129**, 197.
947. P. Powell, *J. Organomet. Chem.*, 1977, **129**, 175.
948. H.G. Kuivila and P.P. Patnode, *J. Organomet. Chem.*, 1977, **129**, 145.
949. P. Hübener and E. Weiss, *J. Organomet. Chem.*, 1977, **129**, 105.
950. W.C. Kaska, R.F. Reichelderfer and L. Prizant, *J. Organomet. Chem.*, 1977, **129**, 97.
951. R. Grüning and J. Lorberth, *J. Organomet. Chem.*, 1977, **129**, 55.
952. C.G. Kreiter and S. Özkar, *J. Organomet. Chem.*, 1977, **129**, C14.
953. R. Meyer, L. Gorrichon and P. Maroni, *J. Organomet. Chem.*, 1977, **129**, C7.
954. R.G. Sutherland, W.J. Pannekoek and C.C. Lee, *J. Organomet. Chem.*, 1977, **129**, C1.
955. C. Blandy, R. Guerreiro and D. Gervais, *J. Organomet. Chem.*, 1977, **128**, 415.
956. K. Hoffmann and E. Weiss, *J. Organomet. Chem.*, 1977, **128**, 399.
957. K. Hoffmann and E. Weiss, *J. Organomet. Chem.*, 1977, **128**, 389.
958. N.I. Pyshnograeva, V.N. Setkina, V.G. Andrianov, Yu.T. Struchkov and D.N. Kursanov, *J. Organomet. Chem.*, 1977, **128**, 381.
959. K. Hoffmann and E. Weiss, *J. Organomet. Chem.*, 1977, **128**, 237.
960. K. Hoffmann and E. Weiss, *J. Organomet. Chem.*, 1977, **128**, 225.
961. R. Meij, T.A.M. Kaandorp, D.J. Stufkens and K. Vrieze, *J. Organomet. Chem.*, 1977, **128**, 203.

962. E.O. Fischer, W. Kleine, F.R. Kreissl, H. Fischer, P. Friedrich and G. Huttner, *J. Organomet. Chem.*, 1977, **128**, C49.
963. M. Tachikawa, S.I. Richter and J.R. Shapley, *J. Organomet. Chem.*, 1977, **128**, C9.
964. C. White, S.J. Thompson and P.M. Maitlis, *J. Organomet. Chem.*, 1977, **127**, 415.
965. P.W.J. de Graaf, J. Boersma and G.J.M. van der Kerk, *J. Organomet. Chem.*, 1977, **127**, 391.
966. F. Mathey, *J. Organomet. Chem.*, 1977, **139**, 77.
967. P.M. Treichel and D.B. Shaw, *J. Organomet. Chem.*, 1977, **139**, 21.
968. P.W. Clark, *J. Organomet. Chem.*, 1977, **137**, 235.
969. N.E. Kolobova, A.B. Antonova, O.M. Khitrova, M.Yu. Antipin and Yu.T. Struchkov, *J. Organomet. Chem.*, 1977, **137**, 69.
970. A.B. Antonova, N.E. Kolobova, P.V. Petrovsky, B.V. Lokshin and N.S. Obezyuk, *J. Organomet. Chem.*, 1977, **137**, 55.
971. D.J. Darensbourg, L.J. Todd and J.P. Hickey, *J. Organomet. Chem.*, 1977, **137**, C1.
972. M.A. de Paoli, H.-W. Fruhauf, F.-W. Grevels, E.A. Koerner von Gustorf, W. Riemer and C. Krüger, *J. Organomet. Chem.*, 1977, **136**, 219.
973. S.J. Thompson, C. White and P.M. Maitlis, *J. Organomet. Chem.*, 1977, **136**, 87.
974. B.P. Roques, *J. Organomet. Chem.*, 1977, **136**, 33.
975. C. Glidewell, *J. Organomet. Chem.*, 1977, **136**, 7.
976. H. Volz and H. Kowarsch, *J. Organomet. Chem.*, 1977, **136**, C27.
977. P.J. Burke, L.A. Gray, P.J.C. Hayward, R.W. Matthews, M. McPartlin and D.G. Gillies, *J. Organomet. Chem.*, 1977, **136**, C7.
978. P. Binger, M.J. Doyle, J. McMeeking, C. Kruger and Y.-H. Tsay, *J. Organomet. Chem.*, 1977, **135**, 405.
979. R.H. Crabtree and G.E. Morris, *J. Organomet. Chem.*, 1977, **135**, 395.
980. H. Felkin, B. Meunier, C. Pascard and T. Prange, *J. Organomet. Chem.*, 1977, **135**, 361.
981. A. Forchioni, V. Galasso, K.J. Irgolic and G.C. Pappalardo, *J. Organomet. Chem.*, 1977, **135**, 387.
982. W.K. Dean, *J. Organomet. Chem.*, 1977, **135**, 195.
983. S. Bywater, D.J. Patmore and D.J. Worsfold, *J. Organomet. Chem.*, 1977, **135**, 145.
984. W.I. Bailey, Jr., D.M. Collins and F.A. Cotton, *J. Organomet. Chem.*, 1977, **135**, C53.
985. C. White, S.J. Thompson and P.M. Maitlis, *J. Organomet. Chem.*, 1977, **134**, 319.
986. L. Vancea and W.A.G. Graham, *J. Organomet. Chem.*, 1977, **134**, 219.
987. D. Neibecker and B. Castro, *J. Organomet. Chem.*, 1977, **134**, 105.

988. S. Hietkamp, D.J. Stufkens and K. Vrieze, *J. Organomet. Chem.*, 1977, **134**, 95.
989. H. Jancke, G. Engelhardt, S. Wagner, W. Dirnens, G. Herzog, E. Thieme and K. Rühlmann, *J. Organomet. Chem.*, 1977, **134**, 21.
990. M. Wieber, N. Baumann, H. Wunderlich and H. Rippstein, *J. Organomet. Chem.*, 1977, **133**, 183.
991. D. Astruc, P. Batail and M.L. Martin, *J. Organomet. Chem.*, 1977, **133**, 77.
992. H. Müller and L. Rösch, *J. Organomet. Chem.*, 1977, **133**, 1.
993. T. Boschi, P. Vogel and R. Roulet, *J. Organomet. Chem.*, 1977, **133**, C36.
994. B.A. Grigor and A.J. Nielson, *J. Organomet. Chem.*, 1977, **132**, 439.
995. R.G.W. Gingerich and R.J. Angelici, *J. Organomet. Chem.*, 1977, **132**, 377.
996. E. Glozbach and J. Lorberth, *J. Organomet. Chem.*, 1977, **132**, 359.
997. G. Deganello, L.K.K. Li Shing Man and J. Takats, *J. Organomet. Chem.*, 1977, **132**, 265.
998. F. Hohmann, H. tom Dieck, T. Mack and D. Leibfritz, *J. Organomet. Chem.*, 1977, **132**, 255.
999. L. Killian and B. Wrackmeyer, *J. Organomet. Chem.*, 1977, **132**, 213.
1000. R. Gassend, J.C. Maire and J.C. Pommier, *J. Organomet. Chem.*, 1977, **132**, 69.
1001. D.L. Reger and E.C. Culbertson, *J. Organomet. Chem.*, 1977, **131**, 297.
1002. E.J. Stampf and J.D. Odom, *J. Organomet. Chem.*, 1977, **131**, 171.
1003. D.L. Reger and C. Coleman, *J. Organomet. Chem.*, 1977, **131**, 153.
1004. P. Krommes and J. Lorberth, *J. Organomet. Chem.*, 1977, **127**, 19.
1005. J. Dubac, P. Mazerolles, M. Joly and F. Piau, *J. Organomet. Chem.*, 1977, **127**, C69.
1006. D.F. Dersnah and M.C. Baird, *J. Organomet. Chem.*, 1977, **127**, C55.
1007. E.O. Fischer, V.N. Postnov and F.R. Kreissl, *J. Organomet. Chem.*, 1977, **127**, C19.
1008. M.S. Brookhart, C.P. Lewis and A. Eisenstadt, *J. Organomet. Chem.*, 1977, **127**, C14.
1009. C.W. Allen, *J. Organomet. Chem.*, 1977, **125**, 215.
1010. D. Seyferth and S.C. Vick, *J. Organomet. Chem.*, 1977, **125**, C11.
1011. J. Boersma, F.J.A. Des Tombe, F. Weyers and G.J.M. Van Der Kerk, *J. Organomet. Chem.*, 1977, **124**, 229.
1012. M.L. Martin, M. Mabon, B. Gautheron and P. Dodey, *J. Organomet. Chem.*, 1977, **124**, 175.
1013. A. Segnitz, E. Kelly, S.H. Taylor and P.M. Maitlis, *J. Organomet. Chem.*, 1977, **124**, 113.

1014. C.S. Hoad, R.W. Matthews, M.M. Thakur and D.G. Gillies, *J. Organomet. Chem.*, 1977, **124**, C31.
1015. E. Niecke, R. Kröher and S. Pohl, *Angew. Chem., Int. Ed. Engl.*, 1977, **16**, 864.
1016. W. Hofmann, W. Buchner and H. Werner, *Angew. Chem., Int. Ed. Engl.*, 1977, **16**, 795.
1017. W.E. Carroll, M. Green, J.A.K. Howard, M. Pfeffer and F.G.A. Stone, *Angew. Chem., Int. Ed. Engl.*, 1977, **16**, 793.
1018. H. Schmidbaur, P. Holl and F.H. Köhler, *Angew. Chem., Int. Ed. Engl.*, 1977, **16**, 722.
1019. F.R. Kreissl, K. Eberl and P. Stückler, *Angew. Chem., Int. Ed. Engl.*, 1977, **16**, 654.
1020. M.L. Ziegler, K. Weidenhammer and W.A. Herrmann, *Angew. Chem., Int. Ed. Engl.*, 1977, **16**, 555.
1021. F.R. Kreissl and P. Friedrich, *Angew. Chem., Int. Ed. Engl.*, 1977, **16**, 543.
1022. W. Walter and H.-W. Lüke, *Angew. Chem., Int. Ed. Engl.*, 1977, **16**, 535.
1023. E.O. Fischer, A. Ruhs, P. Friedrich and G. Huttner, *Angew. Chem., Int. Ed. Engl.*, 1977, **16**, 465.
1024. G. Becker and G. Gutekunst, *Angew. Chem., Int. Ed. Engl.*, 1977, **16**, 463.
1025. R. Appel, F. Knoll and H.-D. Wihler, *Angew. Chem., Int. Ed. Engl.*, 1977, **16**, 402.
1026. E.O. Fischer, H. Hollfelder, P. Friedrich, F.R. Kreissl and G. Huttner, *Angew. Chem., Int. Ed. Engl.*, 1977, **16**, 401.
1027. R.L. Hillard, III and K.P.C. Vollhardt, *Angew. Chem., Int. Ed. Engl.*, 1977, **16**, 399.
1028. P. Jutzi and A. Seufert, *Angew. Chem., Int. Ed. Engl.*, 1977, **16**, 330.
1029. F.R. Kreissl, P. Friedrich, T.L. Lindner and G. Huttner, *Angew. Chem., Int. Ed. Engl.*, 1977, **16**, 314.
1030. I. Ruppert, *Angew. Chem., Int. Ed. Engl.*, 1977, **16**, 311.
1031. G. Schmid and J. Schulze, *Angew. Chem., Int. Ed. Engl.*, 1977, **16**, 249.
1032. M. Herberhold, M. Süss-Fink and C.G. Kreiter, *Angew. Chem., Int. Ed. Engl.*, 1977, **16**, 193.
1033. R. Appel and W. Morbach, *Angew. Chem., Int. Ed. Engl.*, 1977, **16**, 180.
1034. F.R. Kreissl, P. Friedrich and G. Huttner, *Angew. Chem., Int. Ed. Engl.*, 1977, **16**, 102.
1035. D. Hänssgen and D. Hajduga, *Chem. Ber.*, 1977, **110**, 3961.
1036. E.O. Fischer, W. Held and F.R. Kreissl, *Chem. Ber.*, 1977, **110**, 3842.
1037. F.R. Kreissl, K. Eberl and W. Uedelhoven, *Chem. Ber.*, 1977, **110**, 3782.
1038. P. Paetzold and H.-P. Biermann, *Chem. Ber.*, 1977, **110**, 3678
1039. H. Schmidbaur, H.-J. Füller, V. Bejenke, A. Franck and G. Huttner, *Chem. Ber.*, 1977, **110**, 3536.
1040. H. Schmidbaur and H.-J. Füller, *Chem. Ber.*, 1977, **110**, 3528

1041. H. Schmidbaur, O. Gasser, C. Krüger and J.C. Sekutowski, *Chem. Ber.*, 1977, **110**, 3517.
1042. E.O. Fischer, H. Hollfelder, P. Friedrich, F.R. Kreissl and G. Huttner, *Chem. Ber.*, 1977, **110**, 3467.
1043. E.O. Fischer, T.L. Lindner, G. Huttner, P. Friedrich, F.R. Kreissl and J.O. Besenhard, *Chem. Ber.*, 1977, **110**, 3397
1044. R. Appel and J.-R. Lundehn, *Chem. Ber.*, 1977, **110**, 3205.
1045. R. Appel and R. Milker, *Chem. Ber.*, 1977, **110**, 3201.
1046. H. Fussstetter, H. Noth, B. Wrackmeyer and W. McFarlane, *Chem. Ber.*, 1977, **110**, 3172.
1047. E.O. Fischer, T.L. Lindner, F.R. Kreissl and P. Braunstein, *Chem. Ber.*, 1977, **110**, 3139.
1048. F.R. Kreissl, P. Stückler and E.W. Meineke, *Chem. Ber.*, 1977, **110**, 3040.
1049. E.O. Fischer, T. Selmayr and F.R. Kreissl, *Chem. Ber.*, 1977, **110**, 2947.
1050. H.G. Alt, *Chem. Ber.*, 1977, **110**, 2862.
1051. G. Schmid and J. Schulze, *Chem. Ber.*, 1977, **110**, 2744.
1052. E.O. Fischer, T. Selmayr, F.R. Kreissl and U. Schubert, *Chem. Ber.*, 1977, **110**, 2574.
1053. R. Appel, R. Milker and I. Ruppert, *Chem. Ber.*, 1977, **110**, 2385.
1054. R. Appel and J. Halstenberg, *Chem. Ber.*, 1977, **110**, 2374.
1055. H. Werner, A. Kühn, D.J. Tune, C. Krüger, D.J. Brauer, J.C. Sekutowski and Y.-H. Tsay, *Chem. Ber.*, 1977, **110**, 1763.
1056. E.O. Fischer, S. Walz, G. Kreis and F.R. Kreissl, *Chem. Ber.*, 1977, **110**, 1651.
1057. H. Schmidbaur and H.P. Scherm, *Chem. Ber.*, 1977, **110**, 1576.
1058. E.O. Fischer, E.W. Meineke and F.R. Kreissl, *Chem. Ber.*, 1977, **110**, 1140.
1059. B. Wrackmeyer and H. Nöth, *Chem. Ber.*, 1977, **110**, 1086.
1060. W.A. Herrmann and H. Biersack, *Chem. Ber.*, 1977, **110**, 896.
1061. E.O. Fischer, A. Ruhs and F.R. Kreissl, *Chem. Ber.*, 1977, **110**, 805.
1062. F.R. Kreissl and W. Held, *Chem. Ber.*, 1977, **110**, 799.
1063. H. Schmidbaur, J. Eberlein and W. Richter, *Chem. Ber.*, 1977, **110**, 677.
1064. H.C. Clark and K.V. Werner, *Chem. Ber.*, 1977, **110**, 667.
1065. E.O. Fischer, W. Held, F.R. Kreissl, A. Frank and G. Huttner, *Chem. Ber.*, 1977, **110**, 656.
1066. H. Kanter, W. Mach and K. Dimroth, *Chem. Ber.*, 1977, **110**, 395.
1067. E.O. Fischer, A. Schwanzer, H. Fischer, D. Neugebauer and G. Huttner, *Chem. Ber.*, 1977, **110**, 53.
1068. J. Paasivirta, R. Versterinen, L. Virkki and P. Pyykkö, *Org. Magn. Reson.*, 1977, **10**, 265.
1069. K. Issleib, P. Thorausch and H. Meyer, *Org. Magn. Reson.*, 1977, **10**, 172.
1070. P.Y. Appriou, A.M. Samat, J.Y. Le Gall and R.J. Guglielmetti, *Org. Magn. Reson.*, 1977, **10**, 39.

1071. J. Martin and J.B. Roberts, *Org. Magn. Reson.*, 1977, **9**, 637.
1072. U. Edlund, *Org. Magn. Reson.*, 1977, **9**, 593.
1073. V.S. Petrosyan and J.D. Roberts, *Org. Magn. Reson.*, 1977, **9**, 555.
1074. M. Mägi, E. Lippmaa, E. Lukevics and N.P. Erĉak, *Org. Magn. Reson.*, 1977, **9**, 297.
1075. J. Schraml, V. Chvalovský, H. Jancke and G. Engelhardt, *Org. Magn. Reson.*, 1977, **9**, 237.
1076. M. Mazurek, T.M. Mallard and P.A.J. Gorin, *Org. Magn. Reson.*, 1977, **9**, 193.
1077. C. Brown, B.T. Heaton and J. Sabounchei, *J. Organomet. Chem.*, 1977, **142**, 413.
1078. S. Sørensen and H.J. Jakobsen, *Org. Magn. Reson.*, 1977, **9**, 101.
1079. W.H. de Roode, M.L. Beekes, A. Oskam and K. Vrieze, *J. Organomet. Chem.*, 1977, **142**, 337.
1080. H. Nöth and T. Taeger, *J. Organomet. Chem.*, 1977, **142**, 281.
1081. P.W.J. de Graaf, A.J. de Koning, J. Boersma and G.J.M. Van der Kerk, *J. Organomet. Chem.*, 1977, **141**, 345.
1082. B.F.G. Johnson, J. Lewis and D.G. Parker, *J. Organomet. Chem.*, 1977, **141**, 319.
1083. B.G. Ramsey and S.J. O'Neill, *J. Organomet. Chem.*, 1977, **141**, 257.
1084. C. Krüger, J.C. Sekutowski, H. Hoberg and R. Krause-Göing, *J. Organomet. Chem.*, 1977, **141**, 141.
1085. B.E. Mann, *J. Organomet. Chem.*, 1977, **141**, C33.
1086. J. Kiji, Y. Miura and J. Furukawa, *J. Organomet. Chem.*, 1977, **140**, 317.
1087. C. Giannotti, A.M. Ducourant, H. Chanaud, A. Chiaroni and C. Riche, *J. Organomet. Chem.*, 1977, **140**, 289.
1088. K.H. Dötz, *J. Organomet. Chem.*, 1977, **140**, 177.
1089. W.A. Herrmann and M. Huber, *J. Organomet. Chem.*, 1977, **140**, 55.
1090. W. Walter, H.-W. Luke and G. Adiwidjaja, *J. Organomet. Chem.*, 1977, **140**, 11.
1091. K. Jurkschat, C. Mügge, A. Tzschach, A. Zschunke, M.F. Larin, V.A. Pestunovich and M.G. Voronkov, *J. Organomet. Chem.*, 1977, **139**, 279.
1092. J. Dubac, P. Mazerolles and P. Fagoaga, *J. Organomet. Chem.*, 1977, **139**, 271.
1093. W.G. Jackson, B.F.G. Johnson and J. Lewis, *J. Organomet. Chem.*, 1977, **139**, 125.
1094. F.A. Cotton, B.E. Hanson and J.D. Jamerson, *J. Am. Chem. Soc.*, 1977, **99**, 6588.
1095. M. Brookhart and G.O. Nelson, *J. Am. Chem. Soc.*, 1977, **99**, 6099.
1096. M. Bottrill and M. Green, *J. Am. Chem. Soc.*, 1977, **99**, 5795.
1097. R.B. Calvert and J.R. Shapley, *J. Am. Chem. Soc.*, 1977, **99**, 5225.

1098. L.M. Jackman and N.M. Szeverenyi, *J. Am. Chem. Soc.*, 1977, **99**, 4954.
1099. T.C. Flood, E. Rosenberg and A. Sarhangi, *J. Am. Chem. Soc.*, 1977, **99**, 4334.
1100. R.L. Hillard, III and K.P.C. Vollhardt, *J. Am. Chem. Soc.*, 1977, **99**, 4058.
1101. F.A. Cotton, B.E. Hanson, J.R. Kolb, P. Lahuerta, G.G. Stanley, B.R. Stults and A.J. White, *J. Am. Chem. Soc.*, 1977, **99**, 3673.
1102. S.J. McLain, C.D. Wood and R.R. Schrock, *J. Am. Chem. Soc.*, 1977, **99**, 3519.
1103. F.A. Cotton, B.E. Hanson, J.D. Jamerson and B.R. Stults, *J. Am. Chem. Soc.*, 1977, **99**, 3293.
1104. D.G. Gorenstein, *J. Am. Chem. Soc.*, 1977, **99**, 2254.
1105. G.W. Rice and R.S. Tobias, *J. Am. Chem. Soc.*, 1977, **99**, 2141.
1106. T. Yoshida and S. Otsuka, *J. Am. Chem. Soc.*, 1977, **99**, 2134.
1107. C.P. Casey, T.J. Burkhardt, C.A. Bunnell and J.C. Calabrese, *J. Am. Chem. Soc.*, 1977, **99**, 2127.
1108. S. Brownstein, J. Dunogues, D. Lindsay and K.U. Ingold, *J. Am. Chem. Soc.*, 1977, **99**, 2073.
1109. K. Stanley and M.C. Baird, *J. Am. Chem. Soc.*, 1977, **99**, 1808
1110. J.W. Johnson and P.M. Treichel, *J. Am. Chem. Soc.*, 1977, **99**, 1427.
1111. P.A. Milton and T.L. Brown, *J. Am. Chem. Soc.*, 1977, **99**, 1390.
1112. M.H. Chisholm, L.A. Rankel, W.I. Bailey, Jr., F.A. Cotton and C.A. Murillo, *J. Am. Chem. Soc.*, 1977, **99**, 1261.
1113. D.M. Collins, F.A. Cotton, S. Koch, M. Millar and C.A. Murillo, *J. Am. Chem. Soc.*, 1977, **99**, 1259.
1114. A.P. Sattelberger and J.P. Fackler, *J. Am. Chem. Soc.*, 1977, **99**, 1258.
1115. D.J. Darensbourg, H.H. Nelson, III and M.A. Murphy, *J. Am. Chem. Soc.*, 1977, **99**, 896.
1116. J.P. Williams and A. Wojcicki, *Inorg. Chem.*, 1977, **16**, 3116.
1117. D.L. Reger and E.C. Culbertson, *Inorg. Chem.*, 1977, **16**, 3104
1118. R.G. Wollmann and D.N. Hendrickson, *Inorg. Chem.*, 1977, **16**, 3079.
1119. I. The and R.G. Cavell, *Inorg. Chem.*, 1977, **16**, 2887.
1120. F.A. Cotton and B.E. Hanson, *Inorg. Chem.*, 1977, **16**, 2820.
1121. R.B. King and K.N. Chen, *Inorg. Chem.*, 1977, **16**, 2648.
1122. J.C. Wilburn and R.H. Neilson, *Inorg. Chem.*, 1977, **16**, 2519.
1123. J.R. Anglin, H.P. Calhoun and W.A.G. Graham, *Inorg. Chem.*, 1977, **16**, 2281.
1124. C.A. Harmon, D.P. Bauer, S.R. Berryhill, K. Hagiwara and A. Streitwieser, Jr., *Inorg. Chem.*, 1977, **16**, 2143.
1125. L.S. Hegedus, O.P. Anderson, K. Zetterberg, G. Allen, K. Siirala-Hansen, D.J. Olsen and A.B. Packard, *Inorg. Chem.*, 1977, **16**, 1887.
1126. F.A. Cotton and B.E. Hanson, *Inorg. Chem.*, 1977, **16**, 1861.

1127. M.E. Bishop and J.J. Zuckerman, *Inorg. Chem.*, 1977, **16**, 1749
1128. R.K. Pomeroy, L. Vancea, H.P. Calhoun and W.A.G. Graham, *Inorg. Chem.*, 1977, **16**, 1508.
1129. R.B. King and K.N. Chen, *Inorg. Chem.*, 1977, **16**, 1164.
1130. L. Vancea, M.J. Bennett, C.E. Jones, R.A. Smith and W.A.G. Graham, *Inorg. Chem.*, 1977, **16**, 897.
1131. P.M. Treichel and J.W. Johnson, *Inorg. Chem.*, 1977, **16**, 749.
1132. R.C. Stewart and L.G. Marzilli, *Inorg. Chem.*, 1977, **16**, 424.
1133. A.B. Burg, *Inorg. Chem.*, 1977, **16**, 379.
1134. F.A. Cotton, B.E. Hanson, J.R. Kolb and P. Lahuerta, *Inorg. Chem.*, 1977, **16**, 89.
1135. J.H. Smith and T.B. Brill, *Inorg. Chem.*, 1977, **16**, 20.
1136. Y. Kobayashi, S. Fujino, H. Hamana, I. Kumadaki and Y. Hanzawa, *J. Am. Chem. Soc.*, 1977, **99**, 8511.
1137. J.E. Figard, J.V. Paukstelis, E.F. Byrne and J.D. Petersen, *J. Am. Chem. Soc.*, 1977, **99**, 8417.
1138. L.D. Quin and E.D. Middlemas, *J. Am. Chem. Soc.*, 1977, **99**, 8370.
1139. D.L. Lichtenberger and T.L. Brown, *J. Am. Chem. Soc.*, 1977, **99**, 8187.
1140. G.R. Dobson, K.J. Asali, J.L. Marshall and C.R. McDaniel, Jr., *J. Am. Chem. Soc.*, 1977, **99**, 8100.
1141. A.J. Ashe, III and J.C. Colburn, *J. Am. Chem. Soc.*, 1977, **99**, 8099.
1142. J.R. Shapley, G.A. Pearson, M. Tachikawa, G.E. Schmidt, M.R. Churchill and F.J. Hollander, *J. Am. Chem. Soc.*, 1977, **99**, 8064.
1143. H. Goff, *J. Am. Chem. Soc.*, 1977, **99**, 7723.
1144. J.P. Collman, R.K. Rothrock, R.G. Finke and F. Rose-Munch, *J. Am. Chem. Soc.*, 1977, **99**, 7381.
1145. S.W. Kirtley, M.A. Andrews, R. Bau, G.W. Grynkewich, T.J. Marks, D.L. Tipton and B.R. Whittlesey, *J. Am. Chem. Soc.*, 1977, **99**, 7154.
1146. W. Ando, A. Sekiguchi, A.J. Rothschild, R.R. Gallucci, M. Jones, Jr., T.J. Barton and J.A. Kilgour, *J. Am. Chem. Soc.*, 1977, **99**, 6995.
1147. M.P. Li, R.S. Drago and A.J. Pribula, *J. Am. Chem. Soc.*, 1977, **99**, 6900.
1148. G.L. Geoffrey and W.L. Gladfelter, *J. Am. Chem. Soc.*, 1977, **99**, 6775.
1149. M. Green, S.M. Heathcock, T.W. Turney and D.M.P. Mingos, *J. Chem. Soc., Dalton Trans.*, 1977, 204.
1150. R.G. Brown, R.V. Chaudhari and J.M. Davidson, *J. Chem. Soc., Dalton Trans.*, 1977, 176.
1151. E.G. Bryan, B.F.G. Johnson and J. Lewis, *J. Chem. Soc., Dalton Trans.*, 1977, 144.
1152. R.B. King and K.-N. Chen, *Inorg. Chem.*, 1977, **16**, 3372.
1153. F.A. Cotton and B.E. Hanson, *Inorg. Chem.*, 1977, **16**, 3369.
1154. K. Henrick, R.W. Matthews and P.A. Tasker, *Inorg. Chem.*, 1977, **16**, 3293.

1155. M.J. Albright, J.N. St. Denis and J.P. Oliver, *J. Organomet. Chem.*, 1977, **125**, 1.
1156. S. Ruh and W. von Philipsborn, *J. Organomet. Chem.*, 1977, **127**, C59.
1157. P.E. Rakita and L.S. Worsham, *J. Organomet. Chem.*, 1977, **137**, 145.
1158. M.L. Filleux-Blanchard, N.-D. An and G. Manuel, *J. Organomet. Chem.*, 1977, **137**, 11.
1159. A.A. Koridze, P.V. Petrovskii, S.P. Gubin, V.I. Sokolov and A.A. Mokhov, *J. Organomet. Chem.*, 1977, **136**, 65.
1160. A.A. Koridze, P.V. Petrovskii, A.I. Mokhov and A.I. Lutsenko, *J. Organomet. Chem.*, 1977, **136**, 57.
1161. W. Adcock, D.P. Cox and W. Kitching, *J. Organomet. Chem.*, 1977, **133**, 393.
1162. P.F. Barron, D. Doddrell and W. Kitching, *J. Organomet. Chem.*, 1977, **132**, 351.
1163. B.R. Gragg, W.J. Layton and K. Niedenzu, *J. Organomet. Chem.*, 1977, **132**, 29.
1164. P. Powell and L.J. Russell, *J. Organomet. Chem.*, 1977, **129**, 415.
1165. V.N. Setkina, A.Zh. Zhakaeva, G.A. Panosyan, V.I. Zdanovitch, P.V. Petrovskii and D.N. Kursanov, *J. Organomet. Chem.*, 1977, **129**, 361.
1166. R.J. Puddephatt and P.J. Thompson, *J. Chem. Soc., Dalton Trans.*, 1977, 1219.
1167. Y.-S. Wong, A.J. Carty and P.C. Chieh, *J. Chem. Soc., Dalton Trans.*, 1977, 1157.
1168. E.W. Ainscough, A.M. Brodie and S.T. Wong, *J. Chem. Soc., Dalton Trans.*, 1977, 915.
1169. R.A. Andersen and G. Wilkinson, *J. Chem. Soc., Dalton Trans.*, 1977, 809.
1170. B.F.G. Johnson, J. Lewis, D.G. Parker and S.R. Postle, *J. Chem. Soc., Dalton Trans.*, 1977, 794.
1171. C.J. Cardin, D.J. Cardin and M.F. Lappert, *J. Chem. Soc., Dalton Trans.*, 1977, 767.
1172. J. Evans, B.F.G. Johnson and J. Lewis, *J. Chem. Soc., Dalton Trans.*, 1977, 510.
1173. D.J. Mabbott, B.E. Mann and P.M. Maitlis, *J. Chem. Soc., Dalton Trans.*, 1977, 294.
1174. M. Green, J.A.K. Howard, A. Laguna, L.E. Smart, J.L. Spencer and F.G.A. Stone, *J. Chem. Soc., Dalton Trans.*, 1977, 278.
1175. M. Green, J.A.K. Howard, J.L. Spencer and F.G.A. Stone, *J. Chem. Soc., Dalton Trans.*, 1977, 271.
1176. E.R. Hamner, R.D.W. Kemmitt and M.A.R. Smith, *J. Chem. Soc., Dalton Trans.*, 1977, 261.
1177. D.J. Mabbott and P.M. Maitlis, *J. Chem. Soc., Dalton Trans.*, 1977, 254.
1178. S. Aime, L. Milone, E. Sappa and A. Tiripicchio, *J. Chem. Soc., Dalton Trans.*, 1977, 227.

1179. J.A. Gibson, G.-V. Röschenthaler and V. Wray, *J. Chem. Soc., Dalton Trans.*, 1977, 1492.
1180. T.V. Ashworth, M.J. Notte, R.H. Reimann and E. Singleton, *J. Chem. Soc., Chem. Commun.*, 1977, 937.
1181. N.M. Boag, M. Green, J.A.K. Howard, J.L. Spencer, R.F.D. Stansfield, F.G.A. Stone, M.D.O. Thomas, J. Vicente and P. Woodward, *J. Chem. Soc., Chem. Commun.*, 1977, 930.
1182. J.-M. Bassett, M. Green, J.A.K. Howard and F.G.A. Stone, *J. Chem. Soc., Chem. Commun.*, 1977, 853.
1183. D.H.R. Barton and H. Patin, *J. Chem. Soc., Chem. Commun.*, 1977, 799.
1184. J. Elzinga and H. Hogeveen, *J. Chem. Soc., Chem. Commun.*, 1977, 705.
1185. J.W. Byrne, J.R.M. Kress, J.A. Osborn, L. Ricard and R.E. Weiss, *J. Chem. Soc., Chem. Commun.*, 1977, 662.
1186. D. Mansuy, M. Lange, J.-C. Chottard, P. Guerin, P. Morliere, D. Brault and M. Rougee, *J. Chem. Soc., Chem. Commun.*, 1977, 648.
1187. B.E. Mann, *J. Chem. Soc., Chem. Commun.*, 1977, 626.
1188. S. Bhaduri, B.F.G. Johnson, J. Lewis, D.J. Watson and C. Zuccaro, *J. Chem. Soc., Chem. Commun.*, 1977, 477.
1189. I.B. Benson, S.A.R. Knox, R.F.D. Stansfield and P. Woodward, *J. Chem. Soc., Chem. Commun.*, 1977, 404.
1190. A. Agapiou, S.E. Pedersen, L.A. Zyzyck and J.R. Norton, *J. Chem. Soc., Chem. Commun.*, 1977, 393.
1191. C.R. Eady, B.F.G. Johnson, J. Lewis, R. Mason, P.B. Hitchcock and K.M. Thomas, *J. Chem. Soc., Chem. Commun.*, 1977, 385.
1192. E.G. Bryan, B.F.G. Johnson and J. Lewis, *J. Chem. Soc., Chem. Commun.*, 1977, 329.
1193. J.C. Wilburn and R.H. Neilson, *J. Chem. Soc., Chem. Commun.*, 1977, 308.
1194. E.A. Kelly, P.M. Bailey and P.M. Maitlis, *J. Chem. Soc., Chem. Commun.*, 1977, 289.
1195. G.K. Barker, A.M.R. Galas, M. Green, J.A.K. Howard, F.G.A. Stone, T.W. Turney, A.J. Welch and P. Woodward, *J. Chem. Soc., Chem. Commun.*, 1977, 256.
1196. P.B. Hitchcock, M.F. Lappert and P.L. Pye, *J. Chem. Soc., Chem. Commun.*, 1977, 196.
1197. S. Martinengo, B.T. Heaton, R.J. Goodfellow and P. Chini, *J. Chem. Soc., Chem. Commun.*, 1977, 39.
1198. D. Cunningham, P. McArdle, H. Sherlock, B.F.G. Johnson and J. Lewis, *J. Chem. Soc., Dalton Trans.*, 1977, 2340.
1199. H.D. Empsall, E.M. Hyde, E. Mentzer and B.L. Shaw, *J. Chem. Soc., Dalton Trans.*, 1977, 2285.
1200. M. Green, D.M. Grove, J.L. Spencer and F.G.A. Stone, *J. Chem. Soc., Dalton Trans.*, 1977, 2228.
1201. J.R. Blackborow, R.H. Grubbs, K. Hildenbrand, E.A. Koerner von Gustorf, A. Miyashita and A. Scrivanti, *J. Chem. Soc., Dalton Trans.*, 1977, 2205.

1202. J. Browning, P.L. Goggin, R.J. Goodfellow, M.G. Norton, A.J.M. Rattray, B.F. Taylor and J. Mink, *J. Chem. Soc., Dalton Trans.*, 1977, 2061.
1203. M.J.S. Gynane, M.F. Lappert, S.J. Miles, A.J. Carty and N.J. Taylor, *J. Chem. Soc., Dalton Trans.*, 1977, 2009.
1204. M.J.S. Gynane, D.H. Harris, M.F. Lappert, P.P. Power, P. Rivière and M. Rivière-Baudet, *J. Chem. Soc., Dalton Trans.*, 1977, 2004.
1205. Y.S. Wong, A.J. Carty and C. Chieh, *J. Chem. Soc., Dalton Trans.*, 1977, 1801.
1206. Md. Giasuddin, P.W. Hickmott and M. Cais, *J. Chem. Soc., Dalton Trans.*, 1977, 1557.
1207. M. Green, J.L. Spencer, F.G.A. Stone and C.A. Tsipis, *J. Chem. Soc., Dalton Trans.*, 1977, 1519.
1208. M. Green, J.A.K. Howard, M. Murray, J.L. Spencer and F.G.A. Stone, *J. Chem. Soc., Dalton Trans.*, 1977, 1509.
1209. B.F.G. Johnson, J. Lewis, B.E. Reichert, K.T. Schorpp and G.M. Sheldrick, *J. Chem. Soc., Dalton Trans.*, 1977, 1417.
1210. M.F. Lappert and P.L. Pye, *J. Chem. Soc., Dalton Trans.*, 1977, 1283.
1211. M. Bottrill, R. Davies, R. Goddard, M. Green, R.P. Hughes, B. Lewis and P. Woodward, *J. Chem. Soc., Dalton Trans.*, 1977, 1252.
1212. F. Glockling and V.B. Mahale, *Inorg. Chim. Acta*, 1977, **25**, L117.
1213. L.S. Chen, D.W. Lichtenberg, P.W. Robinson, Y. Yamamoto and A. Wojcicki, *Inorg. Chim. Acta*, 1977, **25**, 165.
1214. S. Aime, L. Milone, R. Rossetti and P.L. Stanghellini, *Inorg. Chim. Acta*, 1977, **25**, 103.
1215. G. Deganello, J. Lewis, D.G. Parker and P.L. Sandrini, *Inorg. Chim. Acta*, 1977, **24**, 165.
1216. T. Nalesnik, L. Warfield, N. Haly, J. Layton and S. Smith, *Inorg. Nucl. Chem. Lett.*, 1977, **13**, 523.
1217. B. Mathiasch, *Inorg. Nucl. Chem. Lett.*, 1977, **13**, 271.
1218. D. Dykes, T.N. Huckerby and C. Oldham, *Inorg. Nucl. Chem. Lett.*, 1977, **13**, 63.
1219. T. Chivers and P.L. Timms, *Can. J. Chem.*, 1977, **55**, 3509.
1220. W.F. Reynolds and R.A. McClelland, *Can. J. Chem.*, 1977, **55**, 536.
1221. G. Schramm and J.I. Zink, *J. Magn. Reson.*, 1977, **26**, 513.
1222. C. White, S.J. Thompson and P.M. Maitlis, *J. Chem. Soc., Dalton Trans.*, 1977, 1654.
1223. P.B. Hitchcock, M.F. Lappert and P.L. Pye, *J. Chem. Soc., Dalton Trans.*, 1977, 2160.
1224. M.F. Lappert and P.L. Pye, *J. Chem. Soc., Dalton Trans.*, 1977, 2172.
1225. G.A. Olah, G. Liang and S. Yu, *J. Org. Chem.*, 1977, **42**, 4262.
1226. J.R. Bartels-Keith, M.T. Burgess and J.M. Stevenson, *J. Org. Chem.*, 1977, **42**, 3725.

1227. H.K. Spencer and M.P. Cava, *J. Org. Chem.*, 1977, **42**, 2937.
1228. G.W. Kramer and H.C. Brown, *J. Org. Chem.*, 1977, **42**, 2832.
1229. L. Brener and H.C. Brown, *J. Org. Chem.*, 1977, **42**, 2702.
1230. M.A.M. Meester, D.J. Stufkens and K. Vrieze, *Inorg. Chim. Acta*, 1977, **21**, 251.
1231. R.G. Cavell, J.A. Gibson and K.I. The, *J. Am. Chem. Soc.*, 1977, **99**, 7841.
1232. P. Lennon, A.M. Rosan and M. Rosenblum, *J. Am. Chem. Soc.*, 1977, **99**, 8426.
1233. R.B. King and W.F. Masler, *J. Am. Chem. Soc.*, 1977, **99**, 4001.
1234. G.F. Stuntz and J.R. Shapley, *J. Am. Chem. Soc.*, 1977, **99**, 607.
1235. T. Iwayanagi, T. Ibusuki and Y. Saito, *J. Organomet. Chem.*, 1977, **128**, 145.
1236. L.F. Wuyts, D.F. Van de Vondel and G.P. Van der Kelen, *J. Organomet. Chem.*, 1977, **129**, 163.
1237. R.B. Materikova, V.N. Babin, I.R. Lyatifov, T.Kh. Kurbanov, E.I. Fedin, P.V. Petrovskii and A.I. Lutsenko, *J. Organomet Chem.*, 1977, **142**, 81.
1238. T.N. Mitchell, *J. Organomet. Chem.*, 1977, **141**, 289.
1239. H. Müller, L. Rösch, W. Erb and R. Zeisberg, *J. Organomet. Chem.*, 1977, **140**, C17.
1240. P.F. Barron, D. Doddrell and W. Kitching, *J. Organomet. Chem.*, 1977, **139**, 361.
1241. P.E. Rakita and L.S. Worsham, *J. Organomet. Chem.*. 1977, **139**, 135.
1242. J.A. McCleverty and J. Williams, *Transition Met. Chem.*, 1976, **1**, 288.
1243. B.R. Steele and K. Vrieze, *Transition Met. Chem.*, 1977, **2**, 169.
1244. B.R. Steele and K. Vrieze, *Transition Met. Chem.*, 1977, **2**, 140.
1245. E. Fluck, E. Schmid and W. Haubold, *Z. Anorg. Allg. Chem.*, 1977, **434**, 95.
1246. G. Fritz and U. Finke, *Z. Anorg. Allg. Chem.*, 1977, **430**, 121.
1247. G. Becker, *Z. Anorg. Allg. Chem.*, 1977, **430**, 66.
1248. B. Sarry and P. Velling, *Z. Anorg. Allg. Chem.*, 1976, **426**, 107.
1249. G. Fritz and E. Matern, *Z. Anorg. Allg. Chem.*, 1976, **426**, 28.
1250. G. Becker, *Z. Anorg. Allg. Chem.*, 1976, **423**, 242.
1251. H. Burger and C. Kluess, *Z. Anorg. Allg. Chem.*, 1976, **423**, 112.
1252. D.J. Collins, W.R. Jackson and R.N. Timms, *Aust. J. Chem.*, 1977, **30**, 2167.
1253. E. Schaumann and F.-F. Grabley, *Tetrahedron Lett.*, 1977, 4307.
1254. D. Neibecker and B. Castro, *Tetrahedron Lett.*, 1977, 2351.

1255. I. Ruppert, *Tetrahedron Lett.*, 1977, 1987.
1256. T.H. Chan. W. Mychajlowskij and R. Amouroux, *Tetrahedron Lett.*, 1977, 1605.
1257. J.P. Dutasta, A.C. Guimaraes and J.B. Robert, *Tetrahedron Lett.*, 1977, 801.
1258. A.J. Ashe, III and T.W. Smith, *Tetrahdron Lett.*, 1977, 407.
1259. K.M. Rapp and J. Daub, *Tetrahedron Lett.*, 1977, 227.
1260. J.A. Miller and D. Stewart, *J. Chem. Soc., Perkin Trans. 1*, 1977, 2416.
1261. A.J. Pearson, *J. Chem. Soc., Perkin Trans. 1*, 1977, 2069.
1262. J.I.G. Cadogan, A.G. Rowley, R.J. Scott and N.H. Wilson, *J. Chem. Soc., Perkin Trans. 1*, 1977, 1044.
1263. A.J. Bloodworth and M.E. Loveitt, *J. Chem. Soc., Perkin Trans. 1*, 1977, 1031.
1264. C. Jogsma, J.J. de Kok, R.J.M. Weustink, M. Van der Ley, J. Bulthuis and F. Bickelhaupt, *Tetrahedron*, 1977, **33**, 205.
1265. P.J. Hammond and C.D. Hall, *Phosphorus Sulfur*, 1977, **3**, 351.
1266. J.P. Albrand, A. Cogne and J.B. Robert, *Chem. Phys. Lett.*, 1977, **48**, 524.
1267. V.S. Petrosyan, *Prog. Nucl. Magn. Reson. Spectrosc.*, 1977, **11**, 115.
1268. S. Aime and L. Milone, *Prog. Nucl. Magn. Reson. Spectrosc.*, 1977, **11**, 183.
1269. D. Florentin, B.P. Roques and M.C. Fournie-Zaluski, *Bull. Soc. Chim. Fr.*, 1976, 1999.
1270. F. Glockling, N.S. Hosmane, V.B. Mahale, J.J. Swindall, L. Magos and T.J. King, *J. Chem. Res. (S)*, 1977, 116.
1271. R. Köster and G. Seidel, *Justus Liebigs Ann. Chem.*, 1977, 1837.
1272. H. Mahnke, R.J. Clark and R.K. Sheline, *J. Chem. Phys.*, 1977, **66**, 4822.
1273. E.O. Fischer, T. Selmayr and F.R. Kreissl, *Monatsch. Chem.*, 1977, **108**, 759.
1274. P. Pyykko, *Chem. Phys.*, 1977, **22**, 289.
1275. K. Moedritzer and R.E. Miller, *Synth. React. Inorg. Metal-Org. Chem.*, 1977, **7**, 311.
1276. D.L. Reger, D.J. Fauth and M.D. Dukes, *Synth. React. Inorg. Metal-Org. Chem.*, 1977, **7**, 151.
1277. B. Costisella and H. Gross, *J. Prakt. Chem.*, 1977, **319**, 8.
1278. B. Młotkowska, H. Gross, B. Costisella, M. Mikołajczyk, S. Grezejszczak and A. Zatorski, *J. Prakt. Chem.*, 1977, **319**, 17.
1279. R.G. Pearson and M. Laurent, *Isr. J. Chem.*, 1977, **15**, 243.
1280. F.A. Cotton and B.E. Hanson, *Isr. J. Chem.*, 1977, **15**, 165.
1281. D.W. Asknes, *Acta Chem. Scand., Ser. A*, 1977, **31**, 845.
1282. P. Frøgen and D.G. Morris, *Acta Chem. Scand., Ser. B*, 1977, **31**, 256.
1283. J. Sandström and J. Seita, *Acta Chem. Scand., Ser. B*, 1977, **31**, 86.

1284. G.-e. Matsubayashi and J. Iyoda, *Bull. Chem. Soc. Jpn.*, 1977, **50**, 3055.
1285. M. Uetsuki and Y. Fujiwara, *Bull. Chem. Soc. Jpn.*, 1977, **50**, 673.
1286. A. Ricci, D. Pietropaolo, G. Distefano, D. Macciantelli and F.P. Colonna, *J. Chem. Soc., Perkin Trans. 2*, 1977, 689.
1287. W. Buchner and W. Wolfsberger, *Z. Naturforsch., Teil B*, 1977, **32**, 967.
1288. C.G. Kreiter and S. Oezkar, *Z. Naturforsch, Teil B*, 1977, **32**, 408.
1289. H. Schmidbaur, G. Blaschke and F.H. Köhler, *Z. Naturforsch., Teil B*, 1977, **32**, 757.
1290. H. Schmidbaur, B. Zimmer-Gasser, F.H. Köhler and W. Buchner, *Z. Naturforsch., Teil B*, 1977, **32**, 481.
1291. W. McFarlane, D.S. Rycroft and C.J. Turner, *Bull. Soc. Chim. Belg.*, 1977, **86**, 457.
1292. Y. Senda, H. Suda, J. Ishiyama, S. Imaizumi, A. Kasahara, T. Izumi and T. Kato, *Bull. Chem. Soc. Jpn.*, 1977, **50**, 1608.
1293. W.F. Reynolds, G.K. Hamer and A.R. Bassindale, *J. Chem. Soc., Perkin Trans. 2*, 1977, 971.
1294. M. Acampora, A. Ceccon, M. Dal Farra, G. Giacometti and G. Rigatti, *J. Chem. Soc., Perkin Trans. 2*, 1977, 483.
1295. K. Bachmann, W. von Philipsborn, C. Amith and D. Ginsburg, *Helv. Chim. Acta*, 1977, **60**, 400.
1296. P.S. Pregosin, S.N. Sze, P. Salvadori and R. Lazzaroni, *Helv. Chim. Acta*, 1977, **60**, 2514.
1297. G.W. Buchanan and J.H. Bowen, *Can. J. Chem.*, 1977, **55**, 604.
1298. G.W. Buchanan and F.G. Morin, *Can. J. Chem.*, 1977, **55**, 2885.
1299. T.A. Modro, *Can. J. Chem.*, 1977, **55**, 3681.
1300. D. Cozak and I.S. Butler, *Can. J. Chem.*, 1977, **55**, 4056.
1301. C.R. Lassigne and E.J. Wells, *J. Magn. Reson.*, 1977, **26**, 55.
1302. A.A.V. Gibson, T.A. Scott and E. Fukushima, *J. Magn. Reson.*, 1977, **27**, 29.
1303. A.J. Brown, O.W. Howarth, P. Moore and A.D. Bain, *J. Magn. Reson.*, 1977, **28**, 317.
1304. A.J. Pearson, *Aust. J. Chem.*, 1977, **30**, 407.
1305. M. Coletta, G. Granozzi and G. Rigatti, *Inorg. Chim. Acta*, 1977, **24**, 195.
1306. S. Aime, L. Milone, R. Rossetti and P.L. Stanghellini, *Inorg. Chim. Acta*, 1977, **22**, 135.
1307. A.B. Murg, *Inorg. Nucl. Chem. Lett.*, 1977, **13**, 199.
1308. P.E. Rakita and L.S. Worsham, *Inorg. Nucl. Chem. Lett.*, 1977, **13**, 547.
1309. A. Rudi and Y. Kashman, *Org. Magn. Reson.*, 1977, **10**, 245.
1310. L. Ernst, *Org. Magn. Reson.*, 1977, **9**, 35.
1311. D.A. Brown, J.P. Chester, N.J. Fitzpatrick and I.J King, *Inorg. Chem.*, 1977, **16**, 2497.
1312. J.P. Williams and A. Wojcicki, *Inorg. Chem.*, 1978, **16**, 2506.
1313. D.G. Cooper and J. Powell, *Inorg. Chem.*, 1977, **16**, 142.

1314. R.P. Hughes, N. Krishnamachari, C.J.L. Lock, J. Powell and G. Turner, *Inorg. Chem.*, 1977, **16**, 314.
1315. M. Arthurs, S.M. Nelson and M.G.B. Drew, *J. Chem. Soc., Dalton Trans.*, 1977, 779.
1316. A. Rahm, M. Pereyre, M. Petraud and B. Barbe, *J. Organomet. Chem.*, 1977, **139**, 49.
1317. C.L. Khetrapal, A.C. Kunwar and A. Saupe, *Mol. Cryst. Liq. Cryst.*, 1976, **35**, 215.
1318. P. Loftus, W.H. Bearden and J.D. Roberts, *Nouv. J. Chim.*, 1977, **1**, 283.
1319. Y. Souma, J. Iyoda and H. Sano, *Osaka Kogyo Gijutsu Shikensho Kiho*, 1976, **27**, 277.
1320. Y. Saito and T. Iwayanagi, *Seisan Kenkyu*, 1976, **28**, 440.
1321. T. Saluvere, J. Puskar, J. Past, E. Lippmaa and G. Engelhardt, *Magn. Reson. Relat. Phenom., Proc. Congr. Ampère, 19th*, Ed. H. Brunner, K.H. Hausser and D. Schweitzer, Group Ampere, Heidelberg, 1976, p.333 (Eng.).
1322. J. Schraml, V. Chvalovsky, M. Mägi and E. Lippmaa, *Collect. Czech. Chem. Commun.*, 1977, **42**, 306.
1323. V.G. Gibb and L.D. Hall, *Carbohydr. Res.*, 1977, **55**, 239.
1324. E. Lippmaa, M. Mägi, V. Chvalovsky and J. Schraml, *Collect. Czech. Chem. Commun.*, 1977, **42**, 318.
1325. T. Hitomi and S. Kozima, *J. Organomet. Chem.*, 1977, **127**, 273
1326. L.F. Kuyper and E.L. Eliel, *J. Organomet. Chem.*, 1978, **156**, 245.
1327. L. Lochmann, R.L. De and J. Trekoval, *J. Organomet. Chem.*, 1978, **156**, 307.
1328. F.T. Oakes and J.F. Sebastian, *J. Organomet. Chem.*, 1978, **159**, 363.
1329. F.W. Wehrli, *Org. Magn. Reson.*, 1978, **11**, 106.
1330. G. Chassaing, R. Lett and A. Marquet, *Tetrahedron Lett.*, 1978, 471.
1331. R. Rauscher, T. Clark, D. Poppinger and P.V.R. Schleyer, *Angew. Chem., Int. Ed. Engl.*, 1978, **17**, 276.
1332. K. Takahashi, Y. Kondo and R.Asami, *J. Chem. Soc., Perkin Trans. 2*, 1978, 577.
1333. E.A. Sadurski, W.I. Ilsley, R.D. Thomas, M.D. Glick and J.P. Oliver, *J. Am. Chem. Soc.*, 1978, **100**, 7761.
1334. J.L. Atwood, W.E. Hunter, R.D. Rogers, J. Holton, J. McMeeking, R. Pearce and M.F. Lappert, *J. Chem. Soc., Chem. Commun.*, 1978, 140.
1335. D.G.H. Ballard, A. Courtis, J. Holton, J. McMeeking and R. Pearce, *J. Chem. Soc., Chem. Commun.*, 1978, 994.
1336. E.W. Abel, M. Booth and K.G. Orrell, *J. Organomet. Chem.*, 1978, **160**, 75.
1337. W. Scharf, D. Neugebauer, U. Schubert and H. Schmidbaur, *Angew. Chem., Int. Ed. Engl.*, 1978, **17**, 601.
1338. J.N. Armor, *Inorg. Chem.*, 1978, **17**, 203.
1339. G. Erker and F. Rosenfeldt, *Angew. Chem., Int. Ed. Engl.*, 1978, **17**, 605.

1340. F.H. Kohler and W. Prossdorf, *Chem. Ber.*, 1978, **111**, 3464.
1341. J.D. Fellmann, G.A. Rupprecht, C.D. Wood and R.R. Schrock, *J. Am. Chem. Soc.*, 1978, **100**, 5964.
1342. J. Amaudrut, J.E. Guerchais and J. Sala-Pala, *J. Organomet. Chem.*, 1978, **157**, C10.
1343. S.J. McLain and R.R. Schrock, *J. Am. Chem. Soc.*, 1978, **100**, 1315.
1344. R.R. Schrock and P.R. Scharp, *J. Am. Chem. Soc.*, 1978, **100**, 2389.
1345. R.R. Schrock and J.D. Fellmann, *J. Am. Chem. Soc.*, 1978, **100**, 3359.
1346. R.R. Schrock, L.W. Messerle, C.D. Wood and L.J. Guggenberger, *J. Am. Chem. Soc.*, 1978, **100**, 3793.
1347. S.J. McLain, C.D. Wood, L.W. Messerle, R.R. Schrock, F.J. Hollander, W.J. Youngs and M.R. Churchill, *J. Am. Chem. Soc.*, 1978, **100**, 5962.
1348. E.O. Fischer, W. Kleine, U. Schubert and D. Neugebauer, *J. Organomet. Chem.*, 1978, **149**, C40.
1349. J.S. Bradley, *J. Organomet. Chem.*, 1978, **150**, C1.
1350. W. Geibel, G. Wilke, R. Goddard, C. Krüger and R. Mynott, *J. Organomet. Chem.*, 1978, **160**, 139.
1351. S. Özkar, H. Kurz, D. Neugebauer and C.G. Kreiter, *J. Organomet. Chem.*, 1978, **160**, 115.
1352. J.H. Eekhof, H. Hogeveen, R.M. Kellogg and E. Klei, *J. Organomet. Chem.*, 1978, **161**, 183.
1353. J.H. Eekhof, H. Hogeveen and R.M. Kellogg, *J. Organomet. Chem.*, 1978, **161**, 361.
1354. A.N. Nesmeyanov, V.V. Krivykh, P.V. Petrovskii, V.S. Kaganovich and M.I. Rybinskaya, *J. Organomet. Chem.*, 1978, **162**, 323.
1355. P. Diehl and A.C. Kunwar, *Org. Magn. Reson.*, 1978, **11**, 47.
1356. H. Daamen and A. Oskam, *Inorg. Chim. Acta*, 1978, **26**, 81.
1357. J.A. Gladysz, J.G. Fulcher and A.B. Bocarsly, *Tetrahedron Lett.*, 1978, 1725.
1358. N. Hao and M.J. McGlinchey, *J. Organomet. Chem.*, 1978, **161**, 381.
1359. W. Malisch, H. Blau and S. Voran, *Angew. Chem., Int. Ed. Engl.*, 1978, **17**, 780.
1360. B.W.S. Kolthammer and P. Legzdins, *J. Am. Chem. Soc.*, 1978, **100**, 2247.
1361. D. Seyferth, J.S. Merola and C.S. Eschbach, *J. Am. Chem. Soc.*, 1978, **100**, 4124.
1362. M.M. Maricq, J.S. Waugh, J.L. Fletcher and M.J. McGlinchey, *J. Am. Chem. Soc.*, 1978, **100**, 6902.
1363. F.R. Kreissl, W. Uedelhoven and G. Kreis, *Chem. Ber.*, 1978, **111**, 3283.
1364. U. Koelle, W.-D.H. Beiersdorf and G.E. Herberich, *J. Organomet. Chem.*, 1978, **152**, 7.
1365. H.G. Raubenheimer, S. Lotz, H.W. Viljoen and A.A. Chalmers, *J. Organomet. Chem.*, 1978, **152**, 73.

1366. W.H. de Roode, D.G. Prins, A. Oskam and K. Vrieze, *J. Organomet. Chem.*, 1978, **154**, 273.
1367. L.H. Staal, D.J. Stufkens and A. Oskam, *Inorg. Chim. Acta*, 1978, **26**, 255.
1368. C.G. Kreiter and R. Aumann, *Chem. Ber.*, 1978, **111**, 1223.
1369. E.O. Fischer, M. Schluge, J.O. Bensenhard, P. Friedrich, G. Huttner and F.R. Kreissl, *Chem. Ber.*, 1978, **111**, 3530.
1370. C.C. Frazier, III and H. Kisch, *Inorg. Chem.*, 1978, **17**, 2736
1371. R.B. King and W.M. Rhee, *Inorg. Chem.*, 1978, **17**, 2961.
1372. A.J. Hartshorn, M.F. Lappert and K. Turner, *J. Chem. Soc., Dalton Trans.*, 1978, 348.
1373. R.A. Andersen, R.A. Jones and G. Wilkinson, *J. Chem. Soc., Dalton Trans.*, 1978, 446.
1374. C.J. Hawkins, R.M. Peachey and C.L. Szoredi, *Aust. J. Chem.*, 1978, **31**, 973.
1375. K. Fukushima, T. Miyamoto and Y. Sasaki, *Bull. Chem. Soc. Jpn.*, 1978, **51**, 499.
1376. E.O. Fischer, W. Kleine, G. Kreis and F.R. Kreissl, *Chem. Ber.*, 1978, **111**, 3542.
1377. H. Brunner and J. Wachter, *J. Organomet. Chem.*, 1978, **155**, C29.
1378. D.L. Reger and M.D. Dukes, *J. Organomet. Chem.*, 1978, **153**, 67.
1379. M.H. Chisholm, F.A. Cotton, M.W. Extine and L.A. Rankel, *J. Am. Chem. Soc.*, 1978, **100**, 807.
1380. M.H. Chisholm, F.A. Cotton, M.W. Extine and W.W. Reichert, *J. Am. Chem. Soc.*, 1978, **100**, 1727.
1381. R.W. Light and R.T. Paine, *J. Am. Chem. Soc.*, 1978, **100**, 2230.
1382. M.H. Chisholm, W.W. Reichert and P. Thornton, *J. Am. Chem. Soc.*, 1978, **100**, 2744.
1383. W.I. Bailey, Jr., M.H. Chisholm, F.A. Cotton and L.A. Rankel, *J. Am. Chem. Soc.*, 1978, **100**, 5764.
1384. S. Datta, B. Dezube, J.K. Kouba and S.S. Wreford, *J. Am. Chem. Soc.*, 1978, **100**, 4404.
1385. R.D. Adams and D.F. Chodosh, *Inorg. Chem.*, 1978, **17**, 41.
1386. B.J. Brisdon and A.A. Woolf, *J. Chem. Soc., Dalton Trans.*, 1978, 291.
1387. M.M. Hunt, W.G. Kita, B.E. Mann and J.A. McCleverty, *J. Chem. Soc., Dalton Trans.*, 1978, 467.
1388. E. Carmona-Guzman and G. Wilkinson, *J. Chem. Soc., Dalton Trans.*, 1978, 1139.
1389. J.L. Thomas, *Inorg. Chem.*, 1978, **17**, 1507.
1390. R.B. King and J. Gimeno, *Inorg. Chem.*, 1978, **17**, 2396.
1391. R.W. Balk, D.J. Stufkens and A. Oskam, *Inorg. Chim. Acta*, 1978, **28**, 133.
1392. W.I. Bailey, Jr., M.H. Chisholm, F.A. Cotton, C.A. Murillo and L.A. Rankel, *J. Am. Chem. Soc.*, 1978, **100**, 802.
1393. W. Malisch and R. Janta, *Angew. Chem., Int. Ed. Engl.*, 1978, **17**, 211.

1394. L.J. Todd, J.R. Wilkinson, J.P. Hickey, D.L. Beach and K.W. Barnett, *J. Organomet. Chem.*, 1978, **154**, 151.
1395. J.W. Faller and D.A. Haitko, *J. Organomet. Chem.*, 1978, **149**, C19.
1396. S. Braun, P. Dahler and P. Eilbracht, *J. Organomet. Chem.*, 1978, **146**, 135.
1397. W.C. Kaska and C.S. Creaser, *Transition Met. Chem.*, 1978, **3**, 360.
1398. S.S. Woodward, R.J. Angelici and B.D. Dombek, *Inorg. Chem.*, 1978, **17**, 1634.
1399. J. Chatt, R.A. Head, G.J. Leigh and C.J. Pickett, *J. Chem. Soc., Dalton Trans.*, 1978, 1638.
1400. E.O. Fischer, S. Walz, A. Ruhs and F.R. Kreissl, *Chem. Ber.*, 1978, **111**, 2765.
1401. C.P. Casey and S.W. Polichnowski, *J. Am. Chem. Soc.*, 1978, **100**, 7565.
1402. D.N. Clark and R.R. Schrock, *J. Am. Chem. Soc.*, 1978, **100**, 6774.
1403. F.R. Kreissl, K. Eberl and W. Uedelhoven, *Angew. Chem., Int. Ed. Engl.*, 1978, **17**, 860.
1404. F.R. Kreissl, U. Uedelhoven and K. Eberl, *Angew. Chem., Int. Ed. Engl.*, 1978, **17**, 859.
1405. P.C. Bevan, J. Chatt, M. Hidai and G.J. Leigh, *J. Organomet. Chem.*, 1978, **160**, 165.
1406. J.A. Connor and G.A. Hudson, *J. Organomet. Chem.*, 1978, **160**, 159.
1407. N.G. Connelly and C. Gardner, *J. Organomet. Chem.*, 1978, **159**, 179.
1408. U. Koemm, C.G. Kreiter and H. Strack, *J. Organomet. Chem.*, 1978, **148**, 179.
1409. J. Levisalles, H. Rudler and D. Villemin, *J. Organomet. Chem.*, 1978, **146**, 259.
1410. W.H. de Roode and K. Vrieze, *J. Organomet. Chem.*, 1978, **145**, 207.
1411. J. Levisalles, H. Rudler, D. Villemin, J. Daran, Y. Jeannin and L. Martin, *J. Organomet. Chem.*, 1978, **155**, C1.
1412. F. Mathey, A. Mitschler and R. Weiss, *J. Am. Chem. Soc.*, 1978, **100**, 5748.
1413. D.W. Kuty and J.J. Alexander, *Inorg. Chem.*, 1978, **17**, 1489.
1414. A. Oudeman and T.S. Sorensen, *J. Organomet. Chem.*, 1978, **156**, 259.
1415. J.C. Smart and J.L. Robbins, *J. Am. Chem. Soc.*, 1978, **100**, 3936.
1416. G. Thiollet, R. Poilblanc, D. Voigt and F. Mathey, *Inorg. Chim. Acta*, 1978, **30**, L294.
1417. E.O. Fischer and G. Besl, *J. Organomet. Chem.*, 1978, **157**, C33.
1418. S. Fontana, U. Schubert and E.O. Fischer, *J. Organomet. Chem.*, 1978, **146**, 39.

1419. A. Breque and F. Mathey, *J. Organomet. Chem.*, 1978, **144**, C9.
1420. V.I. Zdanovich, N.E. Kolobova, N.I. Vasyukova, Yu.S. Nekrasov, G.A. Panosyan, P.V. Petrovskii and A.Zh. Zhakaeva, *J. Organomet. Chem.*, 1978, **148**, 63.
1421. R.A. Jones and G. Wilkinson, *J. Chem. Soc., Dalton Trans.*, 1978, 1063.
1422. A.M. Bond, R. Colton and M.E. McDonald, *Inorg. Chem.*, 1978, **17**, 2842.
1423. E.O. Fischer and A. Frank, *Chem. Ber.*, 1978, **111**, 3740.
1424. M. Herberhold, G. Süss, J. Ellermann and H. Gäbelein, *Chem. Ber.*, 1978, **111**, 2931.
1425. W.A. Herrman, *Chem. Ber.*, 1978, **111**, 2458.
1426. J.A. Gladysz and W. Tann, *J. Am. Chem. Soc.*, 1978, **100**, 2545.
1427. C.P. Casey and S.M. Neumann, *J. Am. Chem. Soc.*, 1978, **100**, 2544.
1428. M. Von Bueren, P. Bischofberger and H.J. Hansen, *Helv. Chim. Acta*, 1978, **61**, 1695.
1429. D.J. Cane, W.A.G. Graham and L. Vancea, *Can. J. Chem.*, 1978, **56**, 1538.
1430. M.B. Stringer and D. Wege, *Aust. J. Chem.*, 1978, **31**, 1607.
1431. Y. Nakamura, K. Bachmann, H. Heimgartner, H. Schmid and J.J. Daly, *Helv. Chim. Acta*, 1978, **61**, 589.
1432. P.A. Dobosh, D.C. Gresham, D.J. Kowalski, C.P. Lillya and E.S. Magyar, *Inorg. Chem.*, 1978, **17**, 1775.
1433. H. Le Bozec, A. Gorgues and P. Dixneuf, *J. Chem. Soc., Chem. Commun.*, 1978, 573.
1434. B.E. Mann, *J. Chem. Soc., Dalton Trans.*, 1978, 1761.
1435. T. Mitsudo, Y. Watanabe, H. Nakanishi, I. Morishima, T. Inubushi and Y. Takegami, *J. Chem. Soc., Dalton Trans.*, 1978, 1298.
1436. C.S. Cundy and M.F. Lappert, *J. Chem. Soc., Dalton Trans.*, 1978, 665.
1437. S. Aime, G. Gervasio, L. Milone, R. Rossetti and P.L. Stanghellini, *J. Chem. Soc., Dalton Trans.*, 1978, 534.
1438. A. Bonny and K.M. MacKay, *J. Chem. Soc., Dalton Trans.*, 1978, 506.
1439. B.F.G. Johnson, J. Lewis, G.R. Stephenson and E.J.S. Vicki, *J. Chem. Soc., Dalton Trans.*, 1978, 369.
1440. R. Aumann and J. Knecht, *Chem. Ber.*, 1978, **111**, 3429.
1441. H.A. Hodali, D.F. Shriver and C.A. Ammlung, *J. Am. Chem. Soc.*, 1978, **100**, 5239.
1442. D.L. Gallup and J.G. Morse, *Inorg. Chem.*, 1978, **17**, 3438.
1443. R.B. King and J. Gimeno, *Inorg. Chem.*, 1978, **17**, 2390.
1444. F.A. Cotton, R.J. Haines, B.E. Hanson and J.C. Sekutowski, *Inorg. Chem.*, 1978, **17**, 2010.
1445. T.C. Flood, F.J. DiSanti and K.D. Campbell, *Inorg. Chem.*, 1978, **17**, 1643.
1446. A. Davison and J.P. Selegue, *J. Am. Chem. Soc.*, 1978, **100**, 7763.
1447. W.A. Hermann and M. Huber, *Chem. Ber.*, 1978, **111**, 3124.

1448. J.P. Collman, R.G. Finke, J.N. Cawse and J.I. Brauman, *J. Am. Chem. Soc.*, 1978, **100**, 4766.
1449. D. Mansuy, J.P. Battioni and J.C. Chottard, *J. Am. Chem. Soc.*, 1978, **100**, 4311.
1450. H. Le Bozec, A. Gorgues and P.H. Dixneuf, *J. Am. Chem. Soc.*, 1978, **100**, 3946.
1451. M. Nitay, W. Priester and M. Rosenblum, *J. Am. Chem. Soc.*, 1978, **100**, 3620.
1452. D. Mansuy, M. Lange and J.C. Chottard, *J. Am. Chem. Soc.*, 1978, **100**, 3213.
1453. M.G. Newton, R.B. King, M. Chang and J. Gimeno, *J. Am. Chem. Soc.*, 1978, **100**, 1632.
1454. J.P. Collman, R.G. Finke, P.L. Matlock, R. Wahren, R.G. Komoto and J.I. Brauman, *J. Am. Chem. Soc.*, 1978, **100**, 1119.
1455. W. Malisch and W. Ries, *Angew. Chem., Int. Ed. Engl.*, 1978, **17**, 120.
1456. A. Salzer and W. von Philipsborn, *J. Organomet. Chem.*, 1978, **161**, 39.
1457. J.P. Hickey, J.C. Huffman and L.J. Todd, *Inorg. Chim. Acta*, 1978, **28**, 77.
1458. S. Aime, L. Milone, D. Osella and A. Poli, *Inorg. Chim. Acta*, 1978, **30**, 45.
1459. J. Elzinga and H. Hogeveen, *J. Org. Chem.*, 1978, **43**, 745.
1460. Z. Goldschmidt and S. Antebi, *Tetrahedron Lett.*, 1978, 271.
1461. A.A. Pinkerton, P.A. Carrupt, P. Vogel, T. Boschi, N.H. Thuy and R. Roulet, *Inorg. Chim. Acta*, 1978, **28**, 123.
1462. R. Victor and I. Ringel, *Org. Magn. Reson.*, 1978, **11**, 31.
1463. P.A. Wegner and G.P. Sterling, *J. Organomet. Chem.*, 1978, **162**, C31.
1464. C.C. Lee, K.J. Demchuk, W.J. Pannekoek and R.G. Sutherland, *J. Organomet. Chem.*, 1978, **162**, 253.
1465. H. Behrens, G. Thiele, A. Pürzer, P. Würstl and M. Moll, *J. Organomet. Chem.*, 1978, **160**, 255.
1466. B.F.G. Johnson, K.D. Karlin, J. Lewis and D.G. Parker, *J. Organomet. Chem.*, 1978, **157**, C67.
1467. G. De Lauzon, F. Mathey and M. Simalty, *J. Organomet. Chem.*, 1978, **156**, C33.
1468. U. Oehmichen, T.G. Southern, H. Le Bozec and P. Dixneuf, *J. Organomet. Chem.*, 1978, **156**, C29.
1469. D.H. Gibson and T.-S. Ong, *J. Organomet. Chem.*, 1978, **155**, 221.
1470. B.V. Johnson, D.P. Sturtzel and J.E. Shade, *J. Organomet. Chem.*, 1978, **154**, 89.
1471. D. Dong, D.A. Slack and M.C. Baird, *J. Organomet. Chem.*, 1978, **153**, 219.
1472. W. Petz, *J. Organomet. Chem.*, 1978, **146**, C23.
1473. B.F.G. Johnson, K.D. Karlin and J. Lewis, *J. Organomet. Chem.*, 1978, **145**, C23.
1474. F. Mathey, *J. Organomet. Chem.*, 1978, **154**, C13.

1475. P.D. Gavens and M.J. Mays, *J. Organomet. Chem.*, 1978, **162**, 389.
1476. A.P. Humphries and S.A.R. Knox, *J. Chem. Soc., Dalton Trans.* 1978, 1514.
1477. P.B. Hitchcock, M.F. Lappert and P.L. Pye, *J. Chem. Soc., Dalton Trans.*, 1978, 826.
1478. P.J. Harris, J.A.K. Howard, S.A.R. Knox, R.J. McKinney, R.P. Phillips, F.G.A. Stone and P. Woodward, *J. Chem. Soc., Dalton Trans.*, 1978, 403.
1479. K.E. Inkrott and S.G. Shore, *J. Am. Chem. Soc.*, 1978, **100**, 3954.
1480. L.D. Brown, C.F.J. Barnard, J.A. Daniels, R.J. Mawby and J.A. Ibers, *Inorg. Chem.*, 1978, **17**, 2932.
1481. S. Aime, L. Milone, D. Osella and E. Sappa, *Inorg. Chim. Acta*, 1978, **29**, L211.
1482. S. Aime, G. Gervasio, L. Milone, E. Sappa and M. Franchini-Angela, *Inorg. Chim. Acta*, 1978, **26**, 223.
1483. M.I. Bruce and R.C. Wallis, *J. Organomet. Chem.*, 1978, **161**, C1.
1484. M.F. Lappert and P.L. Pye, *J. Chem. Soc., Dalton Trans.*, 1978, 837.
1485. K.A. Azam, A.J. Deeming, I.P. Rothwell, M.B. Hursthouse and L. New, *J. Chem. Soc., Chem. Commun.*, 1978, 1086.
1486. E.G. Bryan, A. Forster, B.F.G. Johnson, J. Lewis and T.W. Matheson, *J. Chem. Soc., Dalton Trans.*, 1978, 196.
1487. C.R. Eady, B.F.G. Johnson, J. Lewis and M.C. Malatesta, *J. Chem. Soc., Dalton Trans.*, 1978, 1358.
1488. A.J. Deeming, *J. Organomet. Chem.*, 1978, **150**, 123.
1489. M.P. Atkins, B.T. Golding and P.J. Sellars, *J. Chem. Soc., Chem. Commun.*, 1978, 954.
1490. S. Bhaduri, B.F.G. Johnson, J. Lewis, P.R. Raithby and D.J. Watson, *J. Chem. Soc., Chem. Commun.*, 1978, 343.
1491. C.J. Cooksey, D. Dodd, M.D. Johnson and B.L. Lockman, *J. Chem. Soc., Dalton Trans.*, 1978, 1814.
1492. R.B. King, J. Gimeno and T.J. Lotz, *Inorg. Chem.*, 1978, **17**, 2401.
1493. D.C. Miller and T.B. Brill, *Inorg. Chem.*, 1978, **17**, 240.
1494. W.A. Herrmann, *Chem. Ber.*, 1978, **111**, 1077.
1495. D. Seyferth and J.S. Merola, *J. Organomet. Chem.*, 1978, **160**, 275.
1496. G. Fairhurst and C. White, *J. Organomet. Chem.*, 1978, **160**, C17.
1497. D. Seyferth and R.S. Henderson, *J. Organomet. Chem.*, 1978, **162**, C35.
1498. F.H. Köhler, *J. Organomet. Chem.*, 1978, **160**, 299.
1499. T. Dooley, G. Fairhurst, C.D. Chalk, K. Tabatabaian and C. White, *Transition Met. Chem.*, 1978, **3**, 299.
1500. B.F.G. Johnson and R.E. Benfield, *J. Chem. Soc., Dalton Trans.*, 1978, 1554.

1501. J. Evans, B.F.G. Johnson, J. Lewis, T.W. Matheson and J.R. Norton, *J. Chem. Soc., Dalton Trans.*, 1978, 626.
1502. R. Bishop, *Aust. J. Chem.*, 1978, **31**, 1485.
1503. C. Bied-Charreton, A. Gaudemer, C.A. Chapman, D. Dodd, M.D. Johnson, B.L. Lockman and B. Septe, *J. Chem. Soc., Dalton Trans.*, 1978, 1807.
1504. M.J.H. Russell, C. White, A. Yates and P.M. Maitlis, *J. Chem. Soc., Dalton Trans.*, 1978, 849.
1505. D.R. Robertson and T.A. Stephenson, *J. Chem. Soc., Dalton Trans.*, 1978, 486.
1506. J.L. Vidal, R.A. Fiato, L.A. Cosby and R.L. Pruett, *Inorg. Chem.*, 1978, **17**, 2574.
1507. R. Grigg, B. Kongkathip and T.J. King, *J. Chem. Soc., Dalton Trans.*, 1978, 333.
1508. R.J. Lawson and J.R. Shapley, *Inorg. Chem.*, 1978, **17**, 772.
1509. P.E. Eaton and D.R. Patterson, *J. Am. Chem. Soc.*, 1978, **100**, 2573.
1510. A.C. Jesse, H.P. Gijben, D.J. Stufkens and K. Vrieze, *Inorg. Chim. Acta*, 1978, **31**, 203.
1511. A. Maisonnat and R. Poilblanc, *Inorg. Chim. Acta*, 1978, **29**, 203.
1512. H. Schmidbauer, G. Blaschke, H.J. Füller and H.P. Scherm, *J. Organomet. Chem.*, 1978, **160**, 41.
1513. R.J. Lawson and J.R. Shapley, *Inorg. Chem.*, 1978, **17**, 2963.
1514. S. Hietkamp, D.J. Stufkens and K. Vrieze, *J. Organomet. Chem.*, 1978, **152**, 347.
1515. A.C. Jesse, M.A.M. Meester, D.J. Stufkens and K. Vrieze, *Inorg. Chim. Acta*, 1978, **26**, 129.
1516. N. Platzer, N. Goasdoue and R. Bonnaire, *J. Organomet. Chem.*, 1978, **160**, 455.
1517. M.J.H. Russell, C. White, A. Yates and P.M. Maitlis, *J. Chem. Soc., Dalton Trans.*, 1978, 857.
1518. C. White, S.J. Thompson and P.M. Maitlis, *J. Chem. Soc., Dalton Trans.*, 1978, 1305.
1519. P. Powell and L.J. Russell, *J. Chem. Res. (S)*, 1978, 283.
1520. J.A. van Doorn, C. Masters and C. van der Woude, *J. Chem. Soc., Dalton Trans.*, 1978, 1213.
1521. L. Dahlenburg, V. Sinnwell and D. Thoennes, *Chem. Ber.*, 1978, **111**, 3367.
1522. G.F. Stuntz, J.R. Shapley and C.G. Pierpont, *Inorg. Chem.*, 1978, **17**, 2596.
1523. U. Griebsch and H. Hoberg, *Angew. Chem., Int. Ed. Engl.*, 1978, **17**, 950.
1524. H. Hoberg, R. Krause-Göing and R. Mynott, *Angew. Chem., Int. Ed. Engl.*, 1978, **17**, 123.
1525. P.W. Jolly, S. Stobbe, G. Wilke, R. Goddard, C. Krüger, J.C. Sekutowski and Y.-H. Tsay, *Angew. Chem., Int. Ed. Engl.*, 1978, **17**, 124.
1526. R.H. Grubbs and A. Miyashita, *J. Organomet. Chem.*, 1978, **161**, 371.

1527. K. Hiraki, M. Onishi, K. Sewaki and K. Sugino, *Bull. Chem. Soc. Jpn.*, 1978, **51**, 2548.
1528. W.R. Jackson and J.U. Strauss, *Aust. J. Chem.*, 1978, **31**, 1073.
1529. S.H. Taylor and P.M. Maitlis, *J. Am. Chem. Soc.*, 1978, **100**, 4700.
1530. H.C. Clark, C.R.C. Milne and N.C. Payne, *J. Am. Chem. Soc.*, 1978, **100**, 1164.
1531. H. Schmidbaur and H.P. Scherm, *Chem. Ber.*, 1978, **111**, 797.
1532. Y. Becker and J.K. Stille, *J. Am. Chem. Soc.*, 1978, **100**, 838.
1533. P.S. Manchand, H.S. Wong and J.F. Blount, *J. Org. Chem.*, 1978, **43**, 4769.
1534. S.H. Taylor, B.E. Mann and P.M. Maitlis, *J. Organomet. Chem.*, 1978, **145**, 255.
1535. A. Keasey and P.M. Maitlis, *J. Chem. Soc., Dalton Trans.*, 1978, 1830.
1536. I.M. Al-Najjar and M. Green, *Transition Met. Chem.*, 1978, **3**, 142.
1537. C. Crocker, P.L. Goggin and R.J. Goodfellow, *J. Chem. Soc., Chem. Commun.*, 1978, 1056.
1538. A. Shaver, *Can. J. Chem.*, 1978, **56**, 2281.
1539. N.C. Harrison, M. Murray, J.L. Spencer and F.G.A. Stone, *J. Chem. Soc., Dalton Trans.*, 1978, 1337.
1540. H.D. Empsall, P.N. Heys, W.S. McDonald, M.C. Norton and B.L. Shaw, *J. Chem. Soc., Dalton Trans.*, 1978, 1119.
1541. C. Eaborn, K.J. Odell and A. Pidcock, *J. Chem. Soc., Dalton Trans.*, 1978, 357.
1542. T.G. Appleton, M.A. Bennett, A. Singh and T. Yoshida, *J. Organomet. Chem.*, 1978, **154**, 369.
1543. G.K. Barker, M. Green, J.A.K. Howard, J.L Spencer and F.G.A. Stone, *J. Chem. Soc., Dalton Trans.*, 1978, 1839.
1544. A. Keasey, P.M. Bailey and P.M. Maitlis, *J. Chem. Soc., Chem. Commun.*, 1978, 142.
1545. R.L. Phillips and R.J. Puddephatt, *J. Chem. Soc., Dalton Trans*, 1978, 1736.
1546. W.E. Carroll, M. Green, J.A.K. Howard, M. Pfeffer and F.G.A. Stone, *J. Chem. Soc., Dalton Trans.*, 1978, 1472.
1547. I.J.B. Lin, H.W. Chen and J.P. Fackler, Jr., *Inorg. Chem.*, 1978, **17**, 394.
1548. R.J. Al-Essa, R.J. Puddephatt, C.F.H. Tipper and P.J. Thompson, *J. Organomet. Chem.*, 1978, **157**, C40.
1549. T. Ito and A. Yamamoto, *J. Organomet. Chem.*, 1978, **161**, 61.
1550. J.B. Heyns and F.G.A. Stone, *J. Organomet. Chem.*, 1978, **160**, 337.
1551. R. Meij, D.J. Stufkens and K. Vrieze, *J. Organomet. Chem.*, 1978, **144**, 239.
1552. H.K. Hofstee, J. Boersma and G.J.M. van der Kerk, *J. Organomet. Chem.*, 1978, **144**, 255.
1553. G.S. Lewandos, *Tetrahedron Lett.*, 1978, 2279.

1554. H. Schmidbaur, A.A.M. Aly and U. Schubert, *Angew. Chem., Int. Ed. Engl.*, 1978, **17**, 846.
1555. D.M. Heinekey and S.R. Stobart, *Inorg. Chem.*, 1978, **17**, 1463.
1556. H. Schmidbaur, G. Müller, U. Schubert and O. Orama, *Angew. Chem., Int. Ed. Engl.*, 1978, **17**, 126.
1557. H.K. Hofstee, J. Boersma, J.D. van der Meulen and G.J.M. van der Kerk, *J. Organomet. Chem.*, 1978, **153**, 245.
1558. J. Jokisaari, K. Räisänen, L. Lajunen, A. Passoja and P. Pyykkö, *J. Magn. Reson.*, 1978, **31**, 121.
1559. A.J. Bloodworth and M.E. Loveitt, *J. Chem. Soc., Perkin Trans. 1*, 1978, 522.
1560. L.C. Damude and P.A.W. Dean, *J. Chem. Soc., Chem. Commun.*, 1978, 1083.
1561. J. Jokisaari and K. Räisänen, *Mol. Phys.*, 1978, **36**, 113.
1562. J. Browning, P.L. Goggin, R.J. Goodfellow, N.W. Hurst, L.G. Mallinson and M. Murray, *J. Chem. Soc., Dalton Trans.*, 1978, 872.
1563. N.W. Alcock, P.A. Lampe and P. Moore, *J. Chem. Soc., Dalton Trans.*, 1978, 1324.
1564. I. Morishima, T. Inubushi, S. Uemura and H. Miyoshi, *J. Am. Chem. Soc.*, 1978, **100**, 354.
1565. T.N. Mitchell and H.C. Marsmann, *J. Organomet. Chem.*, 1978, **150**, 171.
1566. J.A. Soderquist and K.L. Thompson, *J. Organomet. Chem.*, 1978, **159**, 237.
1567. K. Niedenzu, K.-D. Müller, W.J. Layton and L. Komorowski, *Z. Anorg. Allg. Chem.*, 1978, **439**, 112.
1568. K. Niedenzu and B.K. Christmas, *Z. Anorg. Allg. Chem.*, 1978, **439**, 103.
1569. K.-D. Müller, L. Komorowski and K. Niedenzu, *Synth. React. Inorg. Metal-Org. Chem.*, 1978, **8**, 149.
1570. E.-I. Negishi, M.J. Idacavage, K.-W. Chiu, T. Yoshida, A. Abramovitch, M.E. Goettel, A. Silveira, Jr. and H.D. Bretherick, *J. Chem. Soc., Perkin Trans. 2*, 1978, 1225.
1571. D.P. Emerick, L. Komorowski and K. Niedenzu, *J. Organomet. Chem.*, 1978, **154**, 147.
1572. H. Kessler, G. Zimmermann, H. Fietze and H. Moehrle, *Chem. Ber.*, 1978, **111**, 2605.
1573. D.S. Matteson, M.S. Biernbaum, R.A. Bechtold, J.D. Campbell and R.J. Wilcsek, *J. Org. Chem.*, 1978, **43**, 950.
1574. L. Killian and B. Wrackmeyer, *J. Organomet. Chem.*, 1978, **153**, 153.
1575. H. Schmidt and W. Siebert, *J. Organomet. Chem.*, 1978, **155**, 157.
1576. G.E. Herberich, H.J. Becker and C. Engelke, *J. Organomet. Chem.*, 1978, **153**, 265.
1577. D.P. Emerick, B.R. Gragg and K. Niedenzu, *J. Organomet. Chem.*, 1978, **153**, 9.
1578. B.R. Gragg and K. Niedenzu, *J. Organomet. Chem.*, 1978, **149**, 271.

1579. L. Komorowski and K. Niedenzu, *J. Organomet. Chem.*, 1978, **149**, 141.
1580. C. Brown, R.H. Cragg, T. Miller, D.O'N. Smith and A. Steltner, *J. Organomet. Chem.*, 1978, **149**, C34.
1581. H.-O. Berger, H. Nöth and B. Wrackmeyer, *J. Organomet. Chem.*, 1978, **145**, 17.
1582. D. Freude, U. Lohse, D. Michel, H. Pfeifer and H.J. Zaehr, *Z. Phys. Chem. (Leipzig)*, 1978, **259**, 225.
1583. O. Yamamoto and M. Yanagisawa, *Bull. Chem. Soc. Jpn.*, 1978, **51**, 1231.
1584. H. Hoberg and S. Krause, *Angew. Chem., Int. Ed. Engl.*, 1978, **12**, 949.
1585. P.B.J. Driessen and H. Hogeveen, *J. Am. Chem. Soc.*, 1978, **100**, 1193.
1586. W. Fries, W. Schwarz, H.-D. Hausen and J. Weidlein, *J. Organomet. Chem.*, 1978, **159**, 373.
1587. P.B.J. Driessen and H. Hogeveen, *J. Organomet. Chem.*, 1978, **156**, 265.
1588. H. Hoberg, V. Gotor and R. Mynott, *J. Organomet. Chem.*, 1978, **149**, 1.
1589. G. Fritz and K.-H. Schmid, *Z. Anorg. Allg. Chem.*, 1978, **441**, 125.
1590. U. Klingbiel, L. Skoda and A. Meller, *Z. Anorg. Allg. Chem.*, 1978, **441**, 113.
1591. J.E. Drake and H.E. Henderson, *J. Inorg. Nucl. Chem.*, 1978, **40**, 137.
1592. F. Glockling and V.B. Mahale, *J. Chem. Res. (S)*, 1978, 170.
1593. A. Ekouya, J. Dunoguès and R. Calas, *J. Chem. Res. (S)*, 1978, 296.
1594. M. Laguerra, J. Dunoguès and R. Calas, *J. Chem. Res. (S)*, 1978, 295.
1595. W. Kantlehner, P. Fischer, W. Kugel, E. Möhring and H. Bredereck, *Justus Liebigs Ann. Chem.*, 1978, 512.
1596. J. Schraml, N.-D.-Chuy, P. Novák, V. Chvalovský, M. Mägi and E. Lippmaa, *Collect. Czech. Chem. Commun.*, 1978, **43**, 3202.
1597. J. Schraml, J. Včelák, V. Chvalovský, G. Engelhardt, H. Jancke, L. Vodička and J. Hlavatý, *Collect. Czech. Chem. Commun.*, 1978, **43**, 3179.
1598. J.E. Drake, B.M. Glavinčevski and H.E. Henderson, *Synth. React. Inorg. Metal-Org. Chem.*, 1978, **8**, 7.
1599. M.D. Beer and R. Grinter, *J. Magn. Reson.*, 1978, **31**, 187.
1600. J. Kroner, W. Schneid, N. Wiberg, B. Wrackmeyer and G. Ziegleder, *J. Chem. Soc., Faraday Trans. 2*, 1978, **74**, 1909.
1601. J. Dubac, P. Mazerolles, M. Joly, F.K. Cartledge and J.M. Wolcott, *J. Organomet. Chem.*, 1978, **154**, 203.
1602. A. Bonny, S.R. Stobart and P.C. Angus, *J. Chem. Soc., Dalton Trans.*, 1978, 938.

1603. R.H. Neilson, R.D. Jacobs, R.W. Scheirman and J.C. Wilburn, *Inorg. Chem.*, 1978, **17**, 1880.
1604. I. Ruppert and R. Appel, *Chem. Ber.*, 1978, **111**, 751.
1605. H. Schmidbaur and M. Heimann, *Chem. Ber.*, 1978, **111**, 2696.
1606. W.D. Hounshell, L.D. Iroff, R.J. Wroczynski and K. Mislow, *J. Am. Chem. Soc.*, 1978, **100**, 5212.
1607. E. Block and L.K. Revelle, *J. Am. Chem. Soc.*, 1978, **100**, 1630.
1608. G. Büchi and H. Wüest, *J. Am. Chem. Soc.*, 1978, **100**, 294.
1609. T.J. Barton and C.R. Tully, *J. Org. Chem.*, 1978, **43**, 3649.
1610. S.-M.L. Chen, R.E. Schaub and C.V. Grudzinskas, *J. Org. Chem.*, 1978, **43**, 3450.
1611. R.P. Woodbury and M.W. Rathke, *J. Org. Chem.*, 1978, **43**, 1947.
1612. E. Niecke and D.A. Wildbredt, *Angew. Chem., Int. Ed. Engl.*, 1978, **17**, 199.
1613. G.-V. Röschenthaler, K. Sauerbrey and R. Schmutzler, *Chem. Ber.*, 1978, **111**, 3105.
1614. H. Schmidbaur and G. Hasslberger, *Chem. Ber.*, 1978, **111**, 2702.
1615. G. Boche and F. Heidenkain, *Angew. Chem., Int. Ed. Engl.*, 1978, **17**, 283.
1616. M. Reiffen and R.W. Hoffmann, *Tetrahedron Lett.*, 1978, 1107.
1617. A. Wissner, *Tetrahedron Lett.*, 1978, 2749.
1618. A. Adcock, G.L. Aldous and W. Kitching, *Tetrahedron Lett.*, 1978, 3397.
1619. R.W. Hillyard, Jr., C.M. Ryan and C.H. Yoder, *J. Organomet. Chem.*, 1978, **153**, 369.
1620. D. Seyferth, D.P. Duncan, H. Schimdbaur and P. Holl, *J. Organomet. Chem.*, 1978, **159**, 137.
1621. R.T. Oakley, D.A. Stanislawski and R. West, *J. Organomet. Chem.*, 1978, **157**, 389.
1622. T.J. Barton and D.S. Banasiak, *J. Organomet. Chem.*, 1978, **157**, 255.
1623. F. Duboudin and O. Laporte, *J. Organomet. Chem.*, 1978, **156**, C25.
1624. J. Dubac, P. Mazerolles and M. Joly, *J. Organomet. Chem.*, 1978, **153**, 289.
1625. L. Killian and B. Wrackmeyer, *J. Organomet. Chem.*, 1978, **148**, 137.
1626. F.G. Yusupova, G.A. Gailyunas, I.I. Furley, A.A. Panasenko, V.D. Sheludyakov, G.A. Tolstikov and V.P. Yurjev, *J. Organomet. Chem.*, 1978, **155**, 15.
1627. F.K. Cartledge, J.M. Wolcott, J. Dubac, P. Mazerolles and M. Joly, *J. Organomet. Chem.*, 1978, **154**, 187.
1628. W. Kitching, H. Olszowy, J. Waugh and D. Doddrell, *J. Org. Chem.*, 1978, **43**, 898.
1629. J. Dubac, P. Mazerolles and J. Cavezzan, *Tetrahedron Lett.*, 1978, 3255.
1630. C.R. Lassigne and E.J. Wells, *J. Magn. Reson.*, 1978, **31**, 195.

1631. B. Wrackmeyer and R. Zentgraf, *J. Chem. Soc., Chem. Commun.*, 1978, 402.
1632. U. Ansorge, E. Lindner and J. Strähle, *Chem. Ber.*, 1978, **111**, 3048.
1633. K. Kamienska-Trela, *J. Organomet. Chem.*, 1978, **159**, 15.
1634. T.N. Mitchell and C. Kummetat, *J. Organomet. Chem.*, 1978, **157**, 275.
1635. B. Mathiasch and M. Dräger, *Angew. Chem. Int. Ed. Engl.*, 1978, **17**, 767.
1636. H.G. Kuivila and G.H. Lein, Jr., *J. Org. Chem.*, 1978, **43**, 750.
1637. A. Bonny and S.R. Stobart, *Inorg. Chim. Acta*, 1978, **31**, L437.
1638. B.E. Mann, B.F. Taylor, N.A. Taylor and R. Wood, *J. Organomet. Chem.*, 1978, **162**, 137.
1639. G. Domazetis, R.J. Magee and B.D. James, *J. Organomet. Chem.*, 1978, **162**, 239.
1640. M. Lehnig, W.P. Neumann and P. Seifert, *J. Organomet. Chem.*, 1978, **162**, 145.
1641. R.W.M. Ten Hoedt, G. Van Koten and J.G. Noltes, *J. Organomet. Chem.*, 1978, **161**, C13.
1642. S. Onaka, Y. Yoshikawa and H. Yamatera, *J. Organomet. Chem.*, 1978, **157**, 187.
1643. B. Wrackmeyer, *J. Organomet. Chem.*, 1978, **145**, 183.
1644. I. Beletskaya, A.N. Kashin and O.A. Reutov, *J. Organomet. Chem.*, 1978, **155**, 31.
1645. T.N. Mitchell, J. Gmehling and F. Huber, *J. Chem. Soc., Dalton Trans.*, 1978, 960.
1646. D.C. Van Beelen, J. Wolters and A. Van der Gen, *J. Organomet. Chem.*, 1978, **145**, 359.
1447. H.J. Padberg, J. Lindber and G. Bergerhoff, *J. Chem. Res. (S)*, 1978, 445.
1648. G.D. MacDonell, K.D. Berlin, S.E. Ealich and D. Van der Helm, *Phosphorus Sulfur*, 1978, **4**, 187.
1649. S. Samaan, *Chem. Ber.*, 1978, **111**, 579.
1650. L.D. Quin and L.B. Littlefield, *J. Org. Chem.*, 1978, **43**, 3508.
1651. L.D. Quin and S.O. Lee, *J. Org. Chem.*, 1978, **43**, 1424.
1652. M.A. Calcagno and E.E. Schweizer, *J. Org. Chem.*, 1978, **43**, 4207.
1653. T. Bottin-Strzalko, J. Seyden-Penne, M.-J. Pouet and M.-P. Simonnin, *J. Org. Chem.*, 1978, **43**, 4346.
1654. G.A. Krudy and R.S. Macomber, *J. Org. Chem.*, 1978, **43**, 4656.
1655. H. Fritz and C.D. Weis, *J. Org. Chem.*, 1978, **43**, 4900.
1656. R.K. Hertz, M.L. Denniston and S.G. Shore, *Inorg. Chem.*, 1978, **17**, 2673.
1657. R.K. Harris, E.M. McVicker and G. Hägele, *J. Chem. Soc., Dalton Trans.*, 1978, 9.
1658. J.P. Dutasta and J.B. Robert, *J. Am. Chem. Soc.*, 1978, **100**, 1925.

1659. R.G. Cavell, J.A. Gibson and K.I. The, *Inorg. Chem.*, 1978, **17**, 2880.
1660. W.J. Parr, *J. Chem. Soc., Faraday Trans. 2*, 1978, **74**, 933.
1661. I.J. Colquhoun and W. McFarlane, *J. Magn. Reson.*, 1978, **31**, 63.
1662. S. Yanai, D. Vofsi and M. Halmann, *J. Chem. Soc., Perkin Trans. 2*, 1978, 511.
1663. T. Hayashi and H. Nakanishi, *Bull. Chem. Soc. Jpn.*, 1978, **51**, 1133.
1664. I.J. Colquhoun and W. McFarlane, *J. Chem. Res. (S)*, 1978, 368.
1665. G. Fronza, P. Bravo and C. Ticozzi, *J. Organomet. Chem.*, 1978, **157**, 299.
1666. J.E. Ferguson and P.F. Heveldt, *Inorg. Chim. Acta*, 1978, **31**, 145.
1667. R. Appel and H.-F. Schöler, *Chem. Ber.*, 1978, **111**, 2056.
1668. F.R. Kreissl, K. Eberl and W. Kleine, *Chem. Ber.*, 1978, **111**, 2451.
1669. Th.C. Klebach, R. Lourens and F. Bickelhaupt, *J. Am. Chem. Soc.*, 1978, **100**, 4886.
1670. W. Winter, A. Scheller, T. Butlers and W. Holweger, *Angew. Chem., Int. Ed. Engl.*, 1978, **17**, 196.
1671. J.B. Robert and H. Weichmann, *J. Org. Chem.*, 1978, **43**, 3031.
1672. D.A. Evans, K.M. Hurst and J.M. Takacs, *J. Am. Chem. Soc.*, 1978, **100**, 3467.
1673. G.D. Macdonell, K.D. Berlin, J.R. Baker, S.E. Ealick, D. Van der Helm and K.L. Marsi, *J. Am. Chem. Soc.*, 1978, **100**, 4535.
1674. G.D. Macdonell, A. Radhakrishna, K.D. Berlin, J. Barycki, R. Tyka and P. Mastalerz, *Tetrahedron Lett.*, 1978, 857.
1675. D. Redmore, *J. Org. Chem.*, 1978, **43**, 996.
1676. M. Rotem and Y. Kashman, *Tetrahedron Lett.*, 1978, 63.
1677. F. Mathey, *Tetrahedron Lett.*, 1978, 133.
1678. D. Redmore, *J. Org. Chem.*, 1978, **43**, 992.
1679. R.S. Jain, H.F. Lawson and L.D. Quin, *J. Org. Chem.*, 1978, **43**, 108.
1680. A.C. Guimaraes, J.B. Robert, C. Taieb and J. Tebony, *Org. Magn. Reson.*, 1978, **11**, 411.
1681. A. Rudi and Y. Kashman, *Tetrahedron Lett.*, 1978, 2209.
1682. R. Ketari and A. Foucard, *Tetrahedron Lett.*, 1978, 2563.
1683. J. Thiem and B. Meyer, *Org. Magn. Reson.*, 1978, **11**, 50.
1684. R. Paasonen, J. Enqvist, M. Karhu, E. Rahkamaa, M. Sundberg and R. Uggla, *Org. Magn. Reson.*, 1978, **11**, 42.
1685. K. Issleib, H. Schmidt and H. Meyer, *J. Organomet. Chem.*, 1978, **160**, 47.
1686. H. Schmidbaur, H.-P. Scherm and U. Schubert, *Chem. Ber.*, 1978, **111**, 764.
1687. G.M.L. Cragg, B. Davidowitz, G.V. Fazakerley, L.R. Nassimbeni and R.J. Haines, *J. Chem. Soc., Chem. Commun.*, 1978, 510.

1688. D. Hellwinkel and W. Krapp, *Chem. Ber.*, 1978, **111**, 13; 1231.
1689. D. Denney, V.M. Mastikhin, S. Namba and J. Turkevich, *J. Phys. Chem.*, 1978, **82**, 1752.
1690. K. Salzer, *Z. Phys. Chem. (Leipzig)*, 1978, **259**, 795.
1691. J. Reffy, T. Veszpremi, P. Hencsei and J. Nagy, *Acta Chim. Acad. Sci. Hung.*, 1978, **96**, 95.
1692. A.D.H. Clague, I.E. Maxwell, J.P.C.M. Van Dongen and J. Binsma, *Appl. Surf. Sci.*, 1978, **1**, 288.
1693. T.A. Holak, B. Kawalek and J. Mirek, *Bull. Acad. Pol. Sci., Ser. Sci. Chim.*, 1978, **26**, 449.
1694. P.W. Hickmott, M. Cais and A. Modiano, *Ann. Rep. Nucl. Magn. Reson. Spectrosc.*, 1977, **6C**,
1695. E. Sappa and M.L.N. Marchino, *Atti Acad. Sci. Torino, Cl. Sci. Fis., Mat. Nat.*, 1977, **111**, 53.
1696. D. Michel, W. Meiler, H. Pfeifer and H.J. Rauscher, *Acta Phys. Chem.*, 1978, **24**, 221.
1697. D. Liebfritz, B.O. Wagner and J.D. Roberts, *Justus Liebigs Ann. Chem.*, 1972, **763**, 173.
1698. G. Fraenkel, D.G. Adams and J. Williams, *Tetrahedron Lett.*, 1963, 767.
1699. A.G. Ginzburg, V.N. Setkina, P.V. Petrovskii, Sh.G. Kasumov, G.A. Panosyan and D.N. Kursanov, *J. Organomet. Chem.*, 1976, **121**, 381.
1700. S.L. Ioffe, A.S. Shaskov, A.L. Blyumenfel'd, L.M. Leont'eva, L.M. Makarenkova, O.B. Belkina and V.A. Tartakovskii, *Izv. Akad. Nauk SSSR, Ser. Khim.*, 1976, **11**, 2547.
1701. L.I. Zakharkin, V.N. Kalinin, V.A. Antonovich and E.G. Rys, *Izv. Akad. Nauk SSSR, Ser. Khim.*, 1976, **5**, 1036.
1702. A.G. Ginzburg, V.N. Setkina, Sh.G. Kasumov, G.A. Panosyan, P.V. Petrovskii and D.N. Kursanov, *Dokl. Akad. Nauk SSSR*, 1976, **228**, 1368.
1703. A.N. Nesmeyanov, L.P. Yur'eva, B.I. Kozyrkin, N.N. Zaitseva, V.A. Svoren, V.I. Robas, E.B. Zavelovich, and P.V. Petrovskii, *Izv. Akad. Nauk SSSR, Ser. Khim.*, 1975, **9**, 2006.
1704. V.I. Stanko, Yu.V. Gol'tyapin, V.K. Pavlovich and G.L. Gal'chenko, *Dokl. Akad. Nauk SSSR*, 1975, **224**, 1360.
1705. S.I. Pombrik, E.D. Korniets, D.N. Kravtsov, L.A. Fedorov and E.I. Fedin, *Izv. Akad. Nauk SSSR, Ser. Khim.*, 1976, **7**, 1667.
1706. B. Boddenberg and J.A. Moreno, *ACS Symp. Ser. 34 Magn. Reson. Colloid Interface Sci., Symp.*, 1976, 198 (*Chem. Abs.*, 1976, **85**, 198760).
1707. Y. Yamamoto, H. Toi, A. Sonoda and S. Murahashi, *Chem. Lett.*, 1975, **11**, 1199.
1708. T. Iwayanagi and Y. Saito, *Chem. Lett.*, 1976, **11**, 1193.
1709. R. Gruning, P. Krommes and J. Lorberth, *J. Organomet. Chem.*, 1977, **128**, 167.

1710. V.M. Bzhezovskii, G.A. Kalabin, G.A. Chmutova and B.A. Trofimov, *Izv. Akad. Nauk SSSR, Ser. Khim.*, 1977, 586.
1711. E.V. Arshavskaya, N.A. Vasneva and A.M. Sladkov, *Dokl. Akad. Nauk SSSR*, 1977, **234**, 833.
1712. D. Cozak and I.S. Butler, *Spectrosc. Lett.*, 1976, **9**, 673.
1713. J.R. Bartels-Keith and M.T. Burgess, U.S. 4003910 260-308D; C07D257/04 180177 ZZ 180875 (*Chem. Abs.*, 1977, **86**, 171459).
1714. E.A. Williams, J.D. Cargioli and S.Y. Hobbs, *Macromolecules*, 1977, **10**, 782.
1715. Y. Koma, K. Iimura, S. Kondo and M. Takeda, *J. Polym. Sci., Polym. Chem. Ed.*, 1977, **15**, 1697.
1716. R. Aumann, H. Averbeck and C. Krüger, *J. Organomet. Chem.*, 1978, **160**, 241.
1717. G.A. Olah, G.K.S. Prakash, G. Liange, K.L. Henold and G.B. Haigh, *Proc. Natl. Acad. Sci. USA*, 1977, **74**, 5217.
1718. C.E. May, K. Niedenzu and S. Trofimenko, *Z. Naturforsch., Teil B*, 1978, **33**, 220.
1719. L.A. Fedorov, Z. Stumbreviciute and E.I. Fedin, *Zh. Strukt. Khim.*, 1974, **15**, 1063.
1720. S. Bywater and D.J. Worsfold, *J. Organomet. Chem.*, 1978, **159**, 229.
1721. R. Wicke and K.F. Elgert, *Makromol. Chem.*, 1977, **178**, 3063.
1722. A. Bonny, A.D. McMaster and S.R. Stobart, *Inorg. Chem.*, 1978, **17**, 935.
1723. A.A. Koridze, N.A. Ogorodnikova and P.V. Petrovsky, *J. Organomet. Chem.*, 1978, **157**, 145.
1724. A. Fadini, E. Glozbach, P. Krommes and J. Lorberth, *J. Organomet. Chem.*, 1978, **149**, 297.
1725. A.G. Proidakov, G.A. Kalabin, R.G. Mirskov, N.P. Ivanova and A.L. Kuznetsov, *Izv. Akad. Nauk SSSR, Ser. Khim.*, 1978, 94.
1726. A.N. Nesmeyanov, V.A. Blinova, E.I. Fedin, I.I. Kritskaya and L.A. Fedorov, *Dokl. Akad. Nauk SSSR*, 1975, **220**, 1336.
1727. G. Domazetis, R.J. Magee and B.D. James, *J. Organomet. Chem.*, 1978, **148**, 339.
1728. G.A. Panosyan, P.V. Petrovskii, A.Zh. Zhakaeva, V.N. Setkina, V.I. Zdanovich and D.N. Kursanov, *J. Organomet. Chem.*, 1978, **146**, 253.
1729. M. Gielen, C. Hoogzand and I. Van den Eynde, *Bull. Soc. Chim. Belg.*, 1975, **84**, 939.
1730. A. Schmidt and B. Wrackmeyer, *Z. Naturforsch., Teil B*, 1978, **33**, 855.
1731. G. Becker and O. Mundt, *Z. Anorg. Allg. Chem.*, 1978, **443**, 53.
1732. R. Appel, M. Halstenberg and F. Knolle, *Z. Naturforsch., Teil B*, 1977, **32**, 1030.
1733. V.A. Pestunovich, S.N. Tandura, M.G. Voronkov, V.P. Baryshok, G.I. Zelchan, V.I. Glukhikh, G. Engelhardt and M. Witanowski, *Spectrosc. Lett.*, 1978, **11**, 339.

1734. R.K. Harris, J. Jones and S. Ng, *J. Magn. Reson.*, 1978, **30**, 521.
1735. A.A. Panasenko, L.M. Khalilov, I.M. Salimgareeva and V.P. Yur'ev, *Izv. Akad. Nauk, SSSR, Ser. Khim.*, 1978, 938.
1736. S.N. Tandura, V.A. Pestunovich, M.G. Voronkov, G. Zelcans, I.I. Solomennikova and E. Lukevics, *Khim. Geterotsikl. Soedin.*, 1977, 1063.
1737. H. Ahlbrecht, B. König and H. Simon, *Tetrahedron Lett.* 1978, 1191.
1738. M. Căpka, J. Schraml and H. Jancke, *Collect. Czech. Chem. Commun.*, 1978, **43**, 3347.
1739. J. Schraml, V. Chvalovský, M. Mägi and E. Lippmaa, *Collect. Czech. Chem. Commun.*, 1978, **43**, 3365.
1740. H. Konishi, S. Matsumoto, Y. Kamitori, H. Ogoshi and Z. Yoshida, *Chem. Lett.*, 1978, 241.
1741. E.I. Fedin, V.A. Antonovich, E.G. Rys, V.N. Kalinin and L.I. Zakharkin, *Izv. Akad. Nauk SSSR, Ser. Khim.*, 1975, 801.
1742. V.I. Glukhikh, S.N. Tandura, G.A. Kuznetsova, V.V. Keiko, V.M. D'yakov and M.G. Voronkov, *Dokl. Akad. Nauk SSSR*, 1978, **239**, 1129.
1743. D. De Vos and J. Wolters, *Recl. Trav. Chim. Pays-Bas*, 1978, **97**, 219.
1744. K.P. Darst and C.M. Lukehart, *J. Organomet. Chem.*, 1978, **161**, 1.
1745. J.P. Hickey, I.M. Baibich, I.S. Butler and L.J. Todd, *Spectrosc. Lett.*, 1978, **11**, 671.
1746. L. Weber, C. Kruger and Y.-H. Tsay, *Chem. Ber.*, 1978, **111**, 1709.
1747. E. Linder and W. Nagel, *Z. Naturforsch., Teil B*, 1977, **32**, 1116.
1748. W.H. de Roode, J. Berke, A. Oskam and K. Vrieze, *J. Organomet. Chem.*, 1978, **155**, 307.
1749. W.H. de Roode and K. Vrieze, *J. Organomet. Chem.*, 1978, **153**, 345.
1750. A. Keasey, B.E. Mann, A. Yates and P.M. Maitlis, *J. Organomet. Chem.*, 1978, **152**, 117.
1751. C.G. Kreiter and S. Öskar, *J. Organomet. Chem.*, 1978, **152**, C13.
1752. U. Franke and E. Weiss, *J. Organomet. Chem.*, 1978, **153**, 39.
1753. P. McArdle and H. Sherlock, *J. Chem. Soc., Dalton Trans.*, 1978, 1678.
1754. F.H. Köhler and W. Prössdorf, *J. Am. Chem. Soc.*, 1978, **100**, 5970.
1755. A.A. Koridze, A.I. Mokhov, P.V. Petrovskii and E.I. Fedin, *Izv. Akad. Nauk SSSR, Ser. Khim.*, 1974, **9**, 2156.
1756. A.A. Koridze, P.V. Petrovskii, S.P. Gubin, E.I. Fedin, A.I. Lutsenko, I.P. Amitin and P.O. Okalevich, *Izv. Akad. Nauk SSSR, Ser. Khim.*, 1975, **7**, 1675.
1757. A.N. Nesmeyanov, E.V. Leonova, E.I. Fedin, N.S. Kochetkova, P.V. Petrovskii and A.I. Lutsenko, *Dokl. Akad. Nauk SSSR*, 1977, **237**. 143.

1758. A.N. Nesmeyanov, B.A. Surkov, I.F. Leshcheva and V.A. Sazonova, *Dokl. Akad. Nauk SSSR*, 1975, **222**, 848.
1759. T. Onak and E. Wan, *U.S. Nat. Tech. Inform. Serv., Ad. Rep.*, 1973, No. 755464.
1760. A.N. Nesmeyanov, G.B. Shul'pin and K.I. Rybinskaya, *Dokl. Akad. Nauk SSSR*, 1974, **218**, 1107.
1761. S. Allenmark and A. Grunstrom, *Chem. Scr.*, 1977, **11**, 135.
1762. C.C. Lee, K.J. Demchuk and R.G. Sutherland, *Synth. React. Inorg. Metal-Org. Chem.*, 1978, **8**, 361.
1763. A.N. Nesemyanov, E.I. Fedin, P.V. Petrovskii, B.V. Lokshin, L.S. Kotova and N.A. Vol'kenau, *Koord. Khim.*, 1975, **1**, 550.
1764. F. Coletta, A. Gambaro, G. Rigatti and A. Venzo, *Spectrosc. Lett.*, 1977, **10**, 971.
1765. J.A. Pople, *Faraday Discuss. Chem. Soc.*, 1962, **34**, 7.
1766. J.A. Pople, *J. Chem. Phys.*, 1962, **37**, 53.
1767. D. Seyferth and J.S. Merola, *J. Am. Chem. Soc.*, 1978, **100**, 6783.
1768. H. Bürger and W. Kilian, *J. Organomet. Chem.*, 1969, **18**, 299.
1769. T.A. Albright, W.J. Freeman and E.E. Schweizer, *J. Am. Chem. Soc.*, 1975, **97**, 940.
1770. T.A. Albright, W.J. Freeman and E.E. Schweizer, *J. Org. Chem.*, 1975, **40**, 3437.
1771. J.J. Breen, S.O. Lee and L.D. Quin, *J. Org. Chem.*, 1975, **40**, 2245.
1772. A.A. Borisenko and N.M. Sergeyev, *Zh. Obshch. Khim.*, 1974, **44**, 2781.
1773. S.E. Anderson, Jr. and N.A. Matwiyoff, *Chem. Phys. Letts.*, 1972, **13**, 150.
1774. B. Crociani and R.L. Richards, *J. Organomet. Chem.*, 1978, **144**, 85.
1775. H. Schmidbaur and P. Holl, *Z. Naturforsch., Teil B*, 1978, **33**, 855.
1776. K.H. Jogan, J.J. Stezowski, E. Fluck and H.J. Weissgraeber, *Z. Naturforsch., Teil B*, 1978, **33**, 1257.
1777. R.V. Ammon, unpublished results.
1778. H. Quast and N. Heuschmann, *Angew. Chem., Int. Ed. Engl.*, 1978, **17**, 867.
1779. R. Appel and G. Erbelding, *Tetrahedron Lett.*, 1978, 2689.
1780. A. Zschunke, H. Meyer, E. Leissring, H. Oehme and K. Issleib, *Phosphorus Sulfur*, 1978, **5**, 81.
1781. K. Moedritzer and R.E. Miller, *Synth. React. Inorg. Metal-Org. Chem.*, 1978, **8**, 167.
1782. I. Ruppert and V. Bastian, *Angew. Chem. Int. Ed. Engl.*, 1978, **17**, 214.
1783. D.B. Denney, D.Z. Denney and J.-H. Tsai, *J. Chem. Res. (S)*, 1978, 458.
1784. J. Heinicke and A. Tzschach, *J. Organomet. Chem.*, 1978, **154**, 1.
1785. S.I. Radchenko, A.S. Khachaturov and B.I. Ionin, *Zh. Org. Khim.*, 1978, **14**, 680.

1786. G. Llabres, M. Baiwir, L. Christiaens, J. Denoel, L. Laitem and J.L. Piette, *Can. J. Chem.*, 1978, **56**, 2008.
1787. G.H. Schmid and D.G. Garratt, *Tetrahedron*, 1978, **34**, 2869.
1788. G.A. Kalabin, D.F. Kushnarev, T.G. Mannafov and A.A. Retinskii, *Izv. Akad. Nauk SSSR, Ser. Khim.*, 1978, 2410.
1789. H. Petersen and H. Meier, *Chem. Ber.*, 1978, **111**, 3423.
1790. V.V. Bairov, G.A. Kalabon, M.L. Al'pert, V.M. Bzhezovskii, I.D. Sadekov, B.A. Trofimov and V.I. Minkin, *Zh. Org. Khim.*, 1978, **14**, 671.
1791. J.C. Dewan, W.B. Jennings, J. Silver and M.S. Tolley, *Org. Magn. Reson.*, 1978, **11**, 449.
1792. K. Moedritzer and R.E. Miller, *Synth. React. Inorg. Metal-Org. Chem.*, 1978, **8**, 167.
1793. E.M. Engler, V.V. Patel, J.R. Andersen, R.R. Schumaker and A.A. Fukushima, *J. Am. Chem. Soc.*, 1978, **100**, 3769.
1794. T. Laitalainen, T. Simonen and R. Kivekäs, *Tetrahedron Lett.*, 1978, 3079.
1795. W. Lohner and K. Praefcke, *Chem. Ber.*, 1978, **111**, 3745.
1796. S. Gronowitz, I. Johnson and A.B. Hornfeldt, *Chem. Scr.*, 1975, **7**, 111.
1797. R.M. Lequan, M.-J. Pouet and M.-P. Simonnin, *Org. Magn. Reson.*, 1975, **7**, 392.
1798. D.A. Bowman, D.B. Denney and D.Z. Denney, *Phosphorus Sulfur*, 1978, **4**, 229.
1799. N.S. Dance and C.H.W. Jones, *J. Organomet. Chem.*, 1978, **152**, 175.